BASIC CIRCUIT THEORY

PRENTICE-HALL COMPUTER APPLICATIONS IN ELECTRICAL ENGINEERING SERIES

Franklin F. Kuo, editor

BASIC CIRCUIT THEORY

second edition

LAWRENCE P. HUELSMAN

Department of Electrical and Computer Engineering
University of Arizona

PRENTICE-HALL, INC., Englewood Cliffs, N. J. 07632

Library of Congress Cataloging in Publication Data

Huelsman, Lawrence P.
    Basic circuit theory.

    (Prentice-Hall computer applications in electrical
engineering series)
    Rev. ed. of: Basic circuit theory, with digital
computations. 1972.
    Bibliography: p.
    Includes index.
    1. Electric circuit analysis.   I. Title.
II. Series.
TK454.H84   1984   621.319′2   83-8646
ISBN 0-13-057711-1

Editorial/production supervision
  and interior design: *Mary Carnis*
Manufacturing buyer: *Anthony Caruso*

Printed in the United States of America

10   9   8   7   6   5   4   3   2

ISBN 0-13-057711-1

Prentice-Hall International, Inc., *London*
Prentice-Hall of Australia Pty. Limited, *Sydney*
Editora Prentice-Hall do Brasil, Ltda., *Rio de Janeiro*
Prentice-Hall Canada Inc., *Toronto*
Prentice-Hall of India Private Limited, *New Delhi*
Prentice-Hall of Japan, Inc., *Tokyo*
Prentice-Hall of Southeast Asia Pte. Ltd., *Singapore*
Whitehall Books Limited, *Wellington, New Zealand*

*To **Jo** and **David***

Contents

*Sections marked with an asterisk may be omitted without loss of continuity.

Contents

Preface

This book was written to provide an introduction to the basic concepts of modern circuit theory. The level at which the developments are made and the quantity of material which is included make this book suitable for a one-year course beginning either in the sophomore or junior year. Such a course is usually referred to as a "traditional" circuit theory course.

It is almost trite to say that there are many texts available which cover the same general area of subject matter that this one docs. It seems probable that if all the introductory circuit theory books that have been written in the last decade were gathered on a single shelf, the shelf would break. This being the case, before undertaking to write still another circuit theory book, the author should have strong personal convictions that his efforts will result in a work that, in some way, is an improvement over the existing ones. The same strong premise obviously should hold for the preparation of a second edition such as this. A dominant motivation which guided me in preparing this revision of the original book was my firm conviction that any presentation of modern circuit theory should have as its goal the development of a background for other courses. Thus, I feel that a circuit theory course should not only introduce techniques, but should illustrate their application to as many subsequent electrical engineering courses as possible. To this end, one of the major additions made during the revision was that of application examples taken from many electrical engineering subject areas. These examples include the use of resistance networks in an R-$2R$ D/A converter, the use of controlled sources to model transistors and operational amplifiers, the use of second-order time-domain techniques to analyze an automobile ignition system, and the application of sinusoidal steady-state methods to passive and active filters, to mention just a few.

In making the many decisions involved in selecting and eliminating material to be included in the text it was assumed that the student had taken the usual lower division courses in physics and mathematics. Specifically, these courses were assumed to include:

1. A basic physics course including a treatment of the electrical variables of voltage and current, the circuit elements of resistance, capacitance, and inductance, and the use of Kirchhoff's laws to describe simple networks
2. The usual introductory engineering mathematics courses including differential and integral calculus

In the writing of this text, several departures were made from the procedure followed in most of the existing basic circuit theory texts. One such departure is the extended use which is made in the text of matrix methods for writing and solving network equations. The decision to include this type of material was based on two factors: first, the convenience of such methods and the resulting simplicity of the mathematical operations, and, second, the very real need to develop in the student, early in his undergraduate career, a facility for the use of such methods. This is most important, since such techniques become increasingly useful in later studies. In general, it has been found that most students entering courses at this level have already encountered enough applications of matrices to be familiar with the matrix operations of addition and multiplication and the mechanics of writing a set of simultaneous equations in matrix form. Thus, this material has been assumed to be part of the student's background. For the individuals who do not have such a background, an introductory treatment of matrix algebra, suitable for either study or review, is given in Appendix A. The operations connected with the solution of a matrix equation, i.e., the process of matrix inversion, are covered in the text at the point the problem arises.

The subject of complex numbers is treated in a manner similar in philosophy to that chosen for the matrix algebra material. Thus, it is assumed that most of the students at this level will have a working knowledge of the use of complex numbers. For this reason it was not felt desirable to break the flow of text material by inserting a treatment of this subject in the body of the book. An appendix describing this topic (Appendix B), however, is included, and may be assigned for study if required by the students.

The topics covered in this book have been ordered so as to emphasize a logical or algorithmic approach to the procedures of network analysis. Thus, the first three chapters emphasize the mechanics of formulating network equations. The resistance network is used as a vehicle to illustrate the techniques. In this group of chapters, Chapter 1 serves as an introduction to the book and also provides a review of some of the basic concepts such as Kirchhoff's law. Also in this chapter, the all-important concept of reference directions is heavily emphasized, since experience has shown this to be a stumbling block for many students. In Chapter 2, sources and resistors are introduced and simple interconnections of these elements are discussed. In the third chapter, mesh and node equations are presented. Topics pertaining to controlled sources, and the multi-terminal devices such as operational amplifiers and transistors that they can be used to model, are emphasized.

The next group of chapters extends the formulation methods developed in the first three chapters to the time-domain treatment of circuits which include capacitors and inductors. Chapter 4 presents the characteristics of these elements and discusses the constraints they impose on various time-domain waveforms. Chapter 5 treats the solution of first-order circuits, and Chapter 6 covers the properties of higher-ordered ones. In these chapters, Secs. 5-7 (on convolution), 6-6 (on higher-order circuits), 6-7 (on state-variable formulation), and 6-8 (on dual and analog circuits), may be omitted without loss of continuity. Chapter 7 covers the sinusoidal steady-state case. In this chapter, Secs. 7-9 (on filters) and 7-11 (on three-phase circuits) may be omitted without loss of continuity. Chapter 8 introduces the Laplace transformation and its applications. Chapter 9 discusses two-port parameters, and Chapter 10 presents a treatment of the Fourier series and the Fourier transformation.

The majority of the material contained in this book may be covered in a two-semester course of three one-hour periods per week. In the original text it has been used in this manner for the past ten years at the University of Arizona in a course on basic circuit theory presented in the junior year. The pace may be slowed if desired by eliminating some or all of the sections referred to above which treat more advanced topics. For the instructor's convenience, these sections are shown with an asterisk in the Contents. Other variations, such as eliminating the chapter on two-port parameters or some of the material on the Fourier transformation, are also possible. Some rearrangement of the order of the chapters to meet individual preferences or satisfy particular curriculum requirements may also be made. To make such a rearrangement easier, an effort has been made to keep the separate treatments of time-domain and frequency-domain topics as self-contained as possible, although complete independence of these topics is obviously not practical or desirable.

One of the features which dominated *Basic Circuit Theory with Digital Computations*, the 1972 predecessor of this book, was the inclusion in considerable depth of digital computer implementations of the various analysis techniques developed in the text. Over the preceding decade, however, changes relating to these digital computational methods have taken place in the engineering curriculum. In the main, it has been found desirable to introduce numerical methods, similar to those given in *Basic Circuit Theory with Digital Computations*, at an earlier point. Thus, the majority of students entering their basic circuit theory studies have already been exposed to many of these topics. For this reason, the digital computer material has been removed from the main flow of topics in this second edition. Descriptions and listings for some of the most important subroutines, however, have been included in Appendix C to accommodate the instructor who desires to include this material.

The author would like to express his thanks to the many persons who assisted him in the preparation of this second edition, especially Dr. Roy H. Mattson, Head of the Department of Electrical Engineering of the University of Arizona, whose enthusiasm and continuing support made the project possible; and Dr. D. Hamilton, who provided many of the applications examples. Finally, the author would like to acknowledge his debt of gratitude to the many students who contributed to the original text and to this revision. The final form of the book owes much to their collective efforts.

L. P. Huelsman

1

Introduction

An electrical network may be considered as a particular example of a broad class of entities called systems. There are many different kinds of systems, such as hydraulic systems, mechanical systems, air traffic control systems, nuclear reactors, computers, etc. In order to be able to discuss the behavior of a system, it is necessary to specify a set of variables which can be used to describe the conditions which prevail in that particular system. Thus, for a mechanical system we might be interested in variables such as the position and velocity of a mass. In an air traffic control system, the density of airplanes would be an important variable. In a nuclear reactor, the fission rate would be a most critical variable, etc. In this chapter we shall consider some of the variables which are most useful in characterizing those electrical systems referred to as circuits or networks.[1] In addition, we shall see that certain constraining relations called Kirchhoff's laws exist relating the variables of a given network.

1-1 NETWORK VARIABLES

In this section we present a brief description of the basic electrical variables that we shall use throughout the text. There are several systems of units which may be used to describe electrical network variables. The one which is most commonly used by electrical engineers is the MKS system, which is based on the use of the meter, the

---

[1] In modern practice the words "network" and "circuit" are used interchangeably. We will follow such practice throughout this text.

kilogram, and the second as basic units of length, mass, and time, respectively.[2] In actual physical situations, the ranges of these variables may vary from very small numbers to quite large numbers. In such cases, it is customary to avoid the use of large numbers of zeros by using an appropriate multiplier of ten raised to the proper positive or negative power. Thus, 0.0000035 second (s) is more rapidly visualized as 3.5×10^{-6} s. In addition, it is sometimes convenient to replace the multiplying factor of ten by redefining the units. For example, 3.5×10^{-6} s may also be expressed as 3.5 microseconds (μs). A table of the commonly accepted multiplying factors, the associated prefix for the units, and the symbol used are given in Table 1-1.1. The

TABLE 1-1.1 Common Multiplying Factors for Network Theory Variables

Multiplying Factor	Name of Prefix	Symbol
10^{12}	tera	T
10^9	giga	G
10^6	mega	M
10^3	kilo	k
10^{-3}	milli	m
10^{-6}	micro	μ
10^{-9}	nano	n
10^{-12}	pico	p
10^{-15}	femto	f
10^{-18}	atto	a

names in this table apply to all the network variables that we shall deal with in this text.

Now let us consider in some detail the variables that are used in network studies.

Charge

This electrical quantity is a property of the atomic particles of which all matter consists. We know that the atom may be pictured as composed of a nucleus with a positive charge surrounded by negatively charged particles called electrons. When there is a balance between positive and negative charges in a given quantity of matter, we say that it is uncharged or neutral. If electrons are removed from a neutral quantity of matter, the result is to leave it positively charged. Similarly, if electrons are added to a quantity of matter, the result is to leave it negatively charged. The MKS unit of charge is the coulomb (abbreviated C). It is the charge contained in 6.24×10^{18} electrons. When referring to the charge that a given quantity of matter possesses, we

---

[2]Actually, there are two MKS systems, the "rationalized" and the "unrationalized" MKS system. The units used for electrical networks are the same no matter which of these systems is used.

shall use the literal symbol $q(t)$. It is a basic physical postulate that charge cannot be created or destroyed, only transferred. This is referred to as the *conservation of charge*.

Current

When charged particles are transferred through a given two-dimensional surface (such as the cross section of a conductor), the net time-rate of transference of charge is referred to as the flow of current. Frequently, especially in solid-state phenomena, charge may be transferred by the movement of both positively charged particles and negatively charged ones. Thus it is important to keep in mind the fact that current refers to the *net* transference of charge. The unit of current in the MKS system is the ampere (abbreviated A). It is the transference of one coulomb of charge per second. We will use the literal symbol $i(t)$ for current. From the above definition, we see that

$$i(t) = \frac{dq}{dt} \quad \text{and} \quad q(t) = \int_{-\infty}^{t} i(\tau)\, d\tau \qquad (1\text{-}1)$$

Energy

It is a basic physical postulate that energy cannot be created, only transformed. This is referred to as the *conservation of energy*. Electrical energy may be produced from many other types of energy such as chemical energy (as in a battery), mechanical energy (as in a hydroelectric generator), atomic energy (as in a nuclear reactor), etc. The basic MKS unit of energy (or work) is the joule (abbreviated J). The literal symbol that we shall use for energy is $w(t)$.

Voltage

If energy is expended (as work) on a quantity of charge, the ratio of work to charge is given the name voltage. For example, a battery uses chemical processes to do work on charged particles, thus a voltage appears across its terminals. The unit of voltage in the MKS system is the volt (abbreviated V). It is equal to an energy of one joule given to a charge of one coulomb.[3] We shall use the literal symbol $v(t)$ for voltage. Thus

$$v(t) = \frac{dw}{dq} \qquad (1\text{-}2)$$

Power

The time rate of performing work is defined as power. The MKS unit of power is the watt (abbreviated W). It is equal to one joule of work per second. We shall use the literal symbol $p(t)$ for power. Thus, in terms of energy, we see that

$$p(t) = \frac{dw}{dt} \quad \text{and} \quad w(t) = \int_{-\infty}^{t} p(\tau)\, d\tau \qquad (1\text{-}3)$$

[3] In other words, if a total energy of one joule is required to move a group of charged particles with a total charge of one coulomb from one point to another in a given circuit, then a potential difference of one volt is produced between the two points.

Since the specification of conditions at $t = -\infty$ (as required by the lower limit of the integral in the above relation) is not always convenient, an alternative integral relation between power and energy that is frequently seen is

$$w(t) = w(t_0) + \int_{t_0}^{t} p(\tau)\,d\tau \qquad (1\text{-}4)$$

Similar alternative expressions may be written for any of the other integral relations used in this section. Since current is the time-rate of transference of charge (dq/dt), and voltage is a quantity giving the ratio of work to charge (dw/dq), we see that the product of voltage and current is dw/dt, i.e., power. Thus we may write

$$p(t) = v(t)i(t) \qquad (1\text{-}5)$$

This is a relation for which we shall have frequent use in this text.

Flux Linkages

A current flowing in a conductor will produce magnetic lines of flux which link the conductor as shown in Fig. 1-1.1. The MKS unit for such flux linkages is the weber-turn. This is the product of the flux in webers and the number of turns of the circuit

Flux

Current

Figure 1-1.1 Lines of flux produced by a current flowing in a conductor.

which are linked by such a flux. Thus, a single loop of wire with a current $i(t)$ flowing in it as shown in Fig. 1-1.2(a) will produce the same number of flux linkages as the two-turn loop with a current $i(t)/2$ flowing in it as shown in Fig. 1-1.2(b). When the

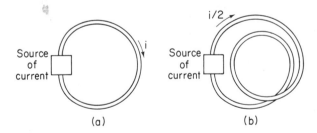

Source of current

i

i/2

Source of current

(a)

(b)

Figure 1-1.2 Flux linkages produced by a current flowing in a coil.

flux linkages of an electric circuit are changed, either by moving the conductors of the circuit so that fewer lines of flux-linkages are enclosed, or by changing the flux, then a voltage is induced in the circuit. Using the literal symbol $\phi(t)$ for flux, the relations between flux linkages and voltage are

$$v(t) = \frac{d\phi}{dt} \qquad \text{and} \qquad \phi(t) = \int_{-\infty}^{t} v(\tau)\,d\tau \qquad (1\text{-}6)$$

The above expressions are frequently referred to as Faraday's law.

The basic electrical variables with which we shall be concerned in this text are summarized in Table 1-1.2.

TABLE 1-1.2 The Basic Electrical Variables

Variable	Unit	Abbreviation	Symbol	Relation
Charge	coulomb	C	$q(t)$	$q(t) = \int i(t)\,dt$
Current	ampere	A	$i(t)$	$i(t) = \dfrac{dq}{dt}$
Energy	joule	J	$w(t)$	$w(t) = \int p(t)\,dt$
Voltage	volt	V	$v(t)$	$v(t) = \dfrac{d\phi}{dt} = \dfrac{dw}{dq}$
Power	watt	W	$p(t)$	$p(t) = v(t)i(t) = \dfrac{dw}{dt}$
Flux linkages	weber-turn	Wb-turn	$\phi(t)$	$\phi(t) = \int v(t)\,dt$

1-2 REFERENCE DIRECTIONS

In the last section we defined the basic electrical variables which we shall use in our study of networks. We now need to develop some conventions by means of which functions of time such as $q(t)$, $i(t)$, $v(t)$, etc., may be related to the physical quantities which these functions represent. This requires the use of reference directions (or reference polarities). In this section we shall indicate more fully what we mean by these "signposts," and define them for the various electrical variables.

Reference Directions for Charge

Let us consider a pair of conducting plates separated by an air space as shown in Fig. 1-2.1(a). We may define a variable $q(t)$ with the positive and negative reference markings shown in the figure. The reference polarities have nothing to do with the actual presence (or absence) of charge on the plates. They simply provide a convention

Figure 1-2.1 (a) Reference directions for charge $q(t)$ on a pair of conducting plates; (b) $q(t) > 0$; (c) $q(t) < 0$.

by means of which we can correlate the parity (or sign) of the *function* $q(t)$ at a given instant of time with the *actual* charge polarity. Thus, as shown in Fig. 1-2.1(b), if the actual charge distribution at a given instant of time is such that the upper plate is positively charged while the bottom plate is negatively charged, since the actual charge distribution agrees with the reference markings, we say that at the given instant of time $q(t)$ is positive. Similarly, if at some other given instant of time the top plate is charged negatively while the bottom plate is charged positively, as shown in Fig. 1-2.1(c), then, since the charge distribution is the opposite of that indicated by the reference polarity markings, we say that at the given instant of time $q(t)$ is negative. The opposite viewpoint may also be taken, namely, if for a certain value of t, the function $q(t)$ as given by some expression is positive, then we say that the actual charge distribution must be as shown in Fig. 1-2.1(b), etc.

Reference Directions for Current

Let us consider a section of some conductor. We may define a variable $i(t)$ and place a reference direction arrow adjacent to the conductor, as shown in Fig. 1-2.2(a). If, at a given instant of time, the actual net current flow in the conductor is from left to right (as indicated by the movement of *negatively* charged particles from *right to left*), then, as shown in Fig. 1-2.2(b), we say that the variable $i(t)$ is positive at that instant of time. Similarly, if at some other given instant of time the actual net current flow is from right to left, as shown in Fig. 1-2.2(c), we say that $i(t)$ is negative at that instant of time. We may also take the opposite viewpoint, and say that if the function $i(t)$ is positive at a given instant of time, then the net current flow must be in the direction indicated by the reference arrow, as shown in Fig. 1-2.2(b), etc.

Figure 1-2.2 (a) Reference directions for current $i(t)$ flowing in a conductor; (b) $i(t) > 0$; (c) $i(t) < 0$.

Reference Polarities for Voltage

Let us consider a pair of terminals connected to some network as shown in Fig. 1-2.3(a). We may define a variable $v(t)$ by marking the terminals with positive and negative reference markings as shown in the figure. When the actual potential difference between the two terminals at a given instant of time is such that the upper terminal is at a higher potential than the lower terminals, as shown in Fig. 1-2.3(b), then we

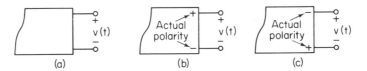

Figure 1-2.3 (a) Reference directions for the voltage $v(t)$ at a pair of terminals; (b) $v(t) > 0$; (c) $v(t) < 0$.

say that the variable $v(t)$ is positive. Similarly, if, at a given instant of time, the potential difference between the two terminals is such that the lower terminal is at the higher potential, as shown in Fig. 1-2.3(c), we then say that the voltage $v(t)$ is negative. The same logic holds true if we consider the function $v(t)$ itself. At any instant of time when this function is positive, then the relative polarities of the two terminals must be as indicated in Fig. 1-2.3(b). If the function is negative, then the relative polarites must be as shown in Fig. 1-2.3(c).

Reference Conventions for Two-Terminal Elements

The first type of network element that we shall study in this text is the two-terminal network element. Some examples of two-terminal network elements are resistors, capacitors, inductors, diodes, batteries, sources, etc. In the chapters that follow, we shall develop the properties of such various two-terminal elements in detail. Here we shall only point out that many of these elements have the property of relating two or more of the network variables defined in the last section. For example, the resistor provides a relation between voltage and current specified by Ohm's law. Thus, we may write

$$v(t) = Ri(t) \qquad (1\text{-}7)$$

Any such element must have reference polarities specified for each of the variables that is involved in its defining relation. Furthermore, the *relative* directions of these reference polarities with respect to each other must be chosen in such a way that the relation between the variables is satisfied. For two-terminal elements, such as a resistor, in which a relation is specified between the variables of voltage and current, the relative reference directions of voltage and current are taken to be as shown in Fig. 1-2.4. As we see from this figure, the reference arrow for the current variable enters the terminal which is identified by the *plus* sign of the voltage reference polarity markings. Two-terminal network elements such as resistors, inductors, and capacitors must always have the *relative* reference directions shown in this figure. They are usually referred to as *associated* reference directions. We shall see, however, that this convention need not be followed if there is no defining expression relating the two variables and identifying the two-terminal element. We shall show that such a situation occurs for the case of ideal sources.

Figure 1-2.4 Associated reference directions for voltage and current at a pair of terminals.

Reference Polarities for Power

In (1-5) we defined power as the product $v(t)i(t)$. Since power is a function of two variables, namely, voltage and current, the relative reference directions of both variables must be taken into account in determining a reference direction for power. Let us consider the situation where two networks are connected together at a pair of terminals with the reference polarities for $v(t)$, $i(t)$, and $p(t)$ as shown in Fig. 1-2.5(a). If, at a given instant of time, $v(t)$ and $i(t)$ have polarities such that $p(t)$ is positive, then there will be a flow of energy from the network on the left to the network on the right as shown in Fig. 1-2.5(b). Similarly, if $p(t)$ is negative, the actual flow of energy is from right to left as indicated in Fig. 1-2.5(c). Note also that if the network shown on the right in Fig. 1-2.5(a) is a simple two-terminal element, the relative reference directions for $v(t)$ and $i(t)$ used to define the reference direction for $p(t)$ are the same as those specified for a two-terminal element as shown in Fig. 1-2.4. Thus we see that, for a two-terminal element, a positive value of $p(t)$ at a given instant of time indicates that energy is flowing into the two-terminal element. There are two things that can happen to this energy, depending on the type of two-terminal element. If the element is a resistor, the energy that flows into it is dissipated as heat. Thus, the power associated with a resistor is always positive.[4] On the other hand, two-terminal elements such as an inductor, a capacitor, a battery, or a source, are able to store this energy and deliver it back to the network at some later time. Thus, the power associated with such elements may be positive or negative. We shall explore such considerations in more detail when we discuss the properties of these elements.

Figure 1-2.5 (a) Reference directions for power $p(t)$; (b) $p(t) > 0$; (c) $p(t) < 0$.

1-3 CLASSIFICATION OF NETWORK ELEMENTS

Before we develop tools to use in the analysis of networks, we have to specify the types of networks which are to be analyzed. In this section, therefore, we shall define some basic properties which may be used to divide networks into classes. It should be noted that the classifications we shall present here are directly applicable to the *individual elements* of a network, as well as to the network considered as a whole.

Linear vs. Non-Linear

One of the most frequently encountered classifications used to categorize network elements is based on whether or not the network elements exhibit *linear* behavior. Actually, in the physical or "real-life" world, there is no such thing as a linear element or a linear system. Apply enough voltage to any electrical component and it will be ruined. Place too much stress on a beam and it will buckle. Allow too many airplanes

---

[4]We shall see that it is possible to define a negative-valued resistor for which the power is always negative.

to be under the control of an air traffic control system, and a collision will result. There are many other examples. Such examples might be termed "destructive" non-linearities. If we avoid these extreme cases, then the elements of most systems fall into two classes: (1) those which over a useful range of their defining variables exhibit a behavior which is so close to linear that it can be treated (within this range) as linear, and (2) those which over even the smallest useful range of their variables display characteristics which can only be described as non-linear. Thus, we see that in a practical sense linearity implies a consideration of how an element is used as well as a consideration of the characteristics of the element itself. In treating the situation specified in (1) above, we say that we may represent an actual physical element, i.e., one which is non-linear, by an *idealized model* which is *defined* as being linear for all ranges of its variables. This modeling concept enables us to use linear analysis techniques for a large portion of our system studies.

The property of linearity may be defined in terms of the following test. Consider a network element as shown symbolically in Fig. 1-3.1(a), where an input $i_1(t)$ is applied, and an output $o_1(t)$ is produced as a result. Similarly, assume that a second (different) input $i_2(t)$ produces an output $o_2(t)$ as shown in Fig. 1-3.1(b). Now let us apply an input $i_1(t) + i_2(t)$ to the element. If it is a linear element, then the output must have the form $o_1(t) + o_2(t)$.[5] This must hold true for any arbitrary inputs $i_1(t)$ and $i_2(t)$. In describing this test, we may simply say that the property of *superposition* holds true for linear elements.[6] An element which does not satisfy the above test is said to be *non-linear*. The assumption of linearity will apply to all of the networks to be studied in this text.

Figure 1-3.1 Two tests of a network element.

Time-Invariant vs. Time-Varying

A second characteristic which we may use to classify network elements is the characteristic of time-invariance. If an element has parameters whose values do not change with time, then it may be said to be a time-invariant element. Actually, in the physical or "real-life" world, there is no such thing as a time-invariant element, in the same sense (as we pointed out in the paragraphs above) that there is no such thing as a linear element. Over a sufficiently long span of time, the parameters of all physical elements must change. For purposes of analysis, however, such elements may be modeled by time-invariant elements by assuming that their parameters remain substantially constant over the time range of interest. Thus, we consider such elements as *time-invariant*. On the other hand, elements in which the change in their parameters is so rapid as to affect our analysis during the time span of interest are called *time-varying*. This text will consider only time-invariant network elements.

[5]We assume that there are no initial conditions present in the element.

[6]If such a test is made for the case where $i_2(t)$ is some multiple of $i_1(t)$, i.e., where $i_2(t) = ki_1(t)$, and if the output $o_2(t) = ko_1(t)$, the system is sometimes said to satisfy the property of *homogeneity*.

Lumped vs. Distributed

The propagation of signal information, whether it be an electrical signal in a conductor, a shock wave in a metal beam, or a flow of fluid down a pipe, is never accomplished instantaneously. Thus, in the physical or real-life world the dimensions of the elements of a system must be taken into consideration. For example, a resistor with a total resistance of 1000 ohms (Ω) and a length of 1 in. would be expected to have quite different properties than a wire with the same resistance but with a length of 1 mi. When the physical dimensions of an element of a system become significant with respect to the propagation of signal information in the element, then we refer to such an element as a *distributed* element. When the physical dimensions are of little significance, and may be safely ignored without invalidating the analysis of the system, then such a component is referred to as a *lumped* element. The mathematics that must be used to treat distributed elements is, in general, considerably more complicated than the mathematics which is necessary to describe lumped elements. In brief, *partial* differential equations must be used to describe distributed elements, while *ordinary* differential equations are used to treat lumped ones. In this text we shall assume that the network elements may be represented by lumped models. Such a representation is a valid one over the portion of the frequency spectrum which is of major interest in most network analyses.

Active vs. Passive

The final classification that may be applied to the network elements we shall study in this text is the classification of active vs. passive. The defining criterion is most easily stated in terms of energy. Such a definition, of course, must be based on the reference conventions for power and energy which were developed in Sec. 1-2. If the total energy supplied to a given network element is always non-negative regardless of the type of circuit to which the element is connected, i.e., if

$$w(t) = \int_{-\infty}^{t} p(\tau)\, d\tau \geq 0 \qquad (1\text{-}8)$$

then we shall say that such an element is *passive*. If it is possible for $w(t)$ to be negative, then the element is *active*. In this text we will be concerned with both active and passive network elements.

1-4 TOPOLOGY—KIRCHHOFF'S LAWS

The network variables which were introduced in Sec. 1-1 may have many interrelationships among themselves. Some of these relationships are due to the nature of the variables. For example, there is a relation between current and charge, namely $i(t) = dq/dt$, which is the result of the definition of these variables. A different class of relationship occurs because of the restriction that some specific type of network element places on the variables. For example, for a two-terminal resistor the variables of voltage and current are related by Ohm's law, $v(t) = Ri(t)$. Still another class of relationship is one between several variables of the same type which occurs as the result of the network configuration, i.e., the manner in which the various elements of the network are interconnected. Such a relation is said to be based on the *topology* of the network. In this section we shall introduce some of the most basic concepts of network topology.

The first topological relationship among network variables that we shall introduce is Kirchhoff's current law, which we shall refer to as KCL. Before presenting this relationship, let us introduce the concept of a node and a branch. A *node* is a point in a circuit where two or more network elements are connected together. A *branch* is any two-terminal network element.[7]

As an example of the concepts of branches and nodes, consider Fig. 1-4.1. Node *a* is readily seen to be the point where branches 1 and 3 are connected. Similarly, node *b* is the point where branches 1, 2, and 4 are connected, etc. Now let us see how KCL applies to branches and nodes. If we define a set of reference currents in each of the branches connected to a node, then KCL tells us that there is a relation among these currents which may be summarized as follows:

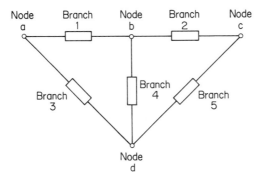

Figure 1-4.1 Nodes and branches.

SUMMARY 1-4.1

Kirchhoff's Current Law (KCL). The algebraic sum of the branch currents at a node is zero at every instant of time.

Kirchhoff's Current Law (alternate form). The sum of the branch currents entering a node at a given instant of time is equal to the sum of the currents leaving the node at that instant of time. ■

Thus, we see that a linear constraint exists among the branch currents at a given node.

As an example of an application of KCL, consider the three branches connected to the node shown in Fig. 1-4.2(a) (we are not interested in the remainder of the network so this is merely shown as a box). If, at a given instant of time, the

---

[7]Another term which will be used interchangeably in this text for node is *terminal*. Similarly, we will frequently refer to a branch as an *element*. In the literature, the terms *vertex* and *junction* are also used to refer to a node, and the term *edge* is used to refer to a branch.

Figure 1-4.2 Examples of KCL.

currents in branches 1 and 2 have the values and directions indicated on the figure, then we may find the current i by applying KCL. We obtain

$$1 - 2 - i = 0 \qquad (1\text{-}9)$$

From this we see that the current i at the specified instant of time equals -1 A, i.e., a current of 1 A flows in a direction opposite to that of the reference arrow for i in branch 3. The preceding example considered constant currents, however, as indicated in Summary 1-4.1, KCL also applies to the case where the currents are functions of time. As an example of this, consider the set of branches and the node shown in Fig. 1-4.2(b). At every instant of time, KCL says that the three currents shown in the figure must obey the relationship

$$i_1(t) - i_2(t) + i_3(t) = 0 \qquad \text{or} \qquad i_1(t) + i_3(t) = i_2(t) \qquad (1\text{-}10)$$

where the second equation illustrates the alternate form. For example, if $i_1(t)$ and $i_2(t)$ were specified by the relations

$$i_1(t) = 1 - 3 \sin t \text{ A} \qquad \text{and} \qquad i_2(t) = 4 \sin 3t \text{ A} \qquad (1\text{-}11)$$

then from (1-10) we see that

$$i_3(t) = 4 \sin 3t - 1 + 3 \sin t \text{ A} \qquad (1\text{-}12)$$

In addition to having constant values, or being specified by mathematical expressions, the currents in a network may also be specified by graphical plots. For example, for the network branches shown in Fig. 1-4.2(b), if the currents $i_2(t)$ and $i_3(t)$ are specified by the plots shown in Fig. 1-4.3(a) and (b), then the plot for $i_1(t)$ must be as shown in Fig. 1-4.3(c).

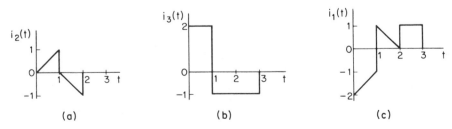

Figure 1-4.3 Set of currents for the network shown in Fig. 1-4.2(b).

Introduction   Chap. 1

An interesting corollary which results from KCL may be obtained by integrating both sides of the second equation of (1-10). Thus, we obtain

$$\int i_1(t)\, dt + \int i_3(t)\, dt = q_1(t) + q_3(t) = \int i_2(t)\, dt = q_2(t) \qquad (1\text{-}13)$$

From (1-13) we see that KCL basically implies the conservation of charge at a node, i.e., the charge delivered to the node equals the charge taken from the node.

Because KCL is based completely on the topology of the network, and not on the properties of the network elements, it applies to all networks with lumped elements, independent of whether the elements are linear or non-linear, time-varying or time-invariant. By implication, KCL tells us that the current into one terminal of a two-terminal element equals the current out of the other terminal at every instant of time. Such a conclusion, however, does not apply to distributed elements. For example, consider a microwave antenna placed on the roof of an automobile for use in mobile communication. When the antenna is being used for transmission, the current at the base of the antenna must certainly be non-zero. However, at the other end of the antenna, the current must always be zero. If we consider the two ends of the antenna as comprising a two-terminal element, we see that because of the distributed nature of the antenna (at the frequency at which it is used), it will not satisfy KCL. We conclude that networks containing distributed elements may not satisfy KCL.

A generalization of KCL is frequently useful in many network situations. This requires the definition of the topological term called a cutset. To see what such a term implies, let us construct the *graph* of a given network by replacing every two-terminal element with a simple line as shown in Fig. 1-4.4(a) and (b). If we now assign a reference direction to each of the branches in the graph, then we may refer to the result as an *oriented graph*. A convenient way of doing this is by drawing an arrowhead on each branch. Such arrowheads can be thought of as establishing the reference directions for the currents in the branches. Such directions may be established arbitrarily; however, once they have been chosen, then the associated voltage reference polarities for the branches are fixed, as was pointed out in Sec. 1-2. An oriented graph for the network shown in Fig. 1-4.4(a) is drawn in Fig. 1-4.4(c). Now let us look for collections (a mathematical term for a collection is a *set*) of branches such that if they are removed from the network, the branches and nodes that remain from two separate parts, i.e., the network is *cut* into two parts. Such a collection of branches shall be called a *cutset*. It must have the property that if any branch is excluded from the cutset, i.e., if any branch of the set is replaced in the network, then the network is no longer separated. Thus we may define a cutset by the following statement:

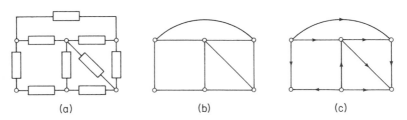

(a)   (b)   (c)

Figure 1-4.4   Network and its graph.

Definition of a Cutset. A cutset is a set of n branches with the property that if all n branches are removed from the network graph, it is separated into two parts, but if any $n - 1$ branches of the set are removed, the graph remains connected.

∎

In the network graph shown in Fig. 1-4.5, for example, we have indicated the various cutsets by drawing dashed lines through the graph interesecting the branches of the different cutsets. Some examples of these various cutsets are: cutset A, branches 1-2; cutset B, branches 2-3-4; cutset D, branches 4-5; cutset F, branches 1-3-5; etc. From a study of these cutsets, two facts should be apparent concerning the cutsets that exist for a given network graph. The first of these is that the number of branches in the various cutsets will not be the same. Cutset A for example contains two branches, while cutset B contains three branches. The second fact which should be noticed concerning cutsets is that a single node is considered as a part of the network graph; thus, a cutset, such as cutset A, which separates this node from the rest of the network is simply the set of branches which are directly connected to that node.

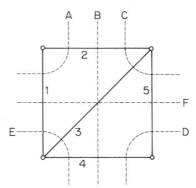

Figure 1-4.5 Examples of cutsets.

One of the reasons for learning about cutsets is that KCL may be applied to cutsets as well as to simple nodes. The reasoning for this may be developed intuitively. A cutset separates the network into two parts; therefore the *net* current flow between these two parts must be zero, since a movement of charge in one direction "through" the cutset along a branch must be matched by a movement of charge in an opposite direction through the cutset along some other branch or branches. Thus we may state the following version of KCL:

Kirchhoff's Current Law (generalized to cutsets). The algebraic sum of the branch currents of a cutset is zero at every instant of time. ∎

When the cutset is such as to separate a single node from the rest of the network, then KCL as applied to cutsets simply becomes the standard version of KCL as given in the previous definition.

As an example of the application of KCL to cutsets, in Table 1-4.1 we have listed the cutsets for the network graph shown in Fig. 1-4.5 with the reference current direction for the branches shown in Fig. 1-4.6. Application of KCL to these cutsets yields the equations listed in the right column of the table.

TABLE 1-4.1 KCL Equations for the Cutsets of Fig. 1-4.5

Cutset	Branches	KCL Applied to Cutset
A	1-2	$i_1(t) + i_2(t) = 0$
B	2-3-4	$i_2(t) + i_3(t) + i_4(t) = 0$
C	2-3-5	$i_2(t) + i_3(t) - i_5(t) = 0$
D	4-5	$i_4(t) + i_5(t) = 0$
E	1-3-4	$i_1(t) - i_3(t) - i_4(t) = 0$
F	1-3-5	$i_1(t) - i_3(t) + i_5(t) = 0$

The second of the two topological relationships among network variables which we will present in this section is Kirchhoff's voltage law. We shall refer to this as KVL. Before introducing this relationship, let us introduce the concept of a loop. A *loop* is a set of branches which forms a connected path in the network, and which has the property that exactly two branches of the set are connected to every node which is encountered in the path. This latter restriction prevents the possibility of multiple paths being included in the definition of a loop. For example, for the network graph shown in Fig. 1-4.5 there are three possible loops, namely a loop formed by branches 1-2-3, a loop formed by branches 3-4-5, and a loop formed by branches 1-2-4-5. We may now state KVL as follows:

SUMMARY 1-4.4

Kirchhoff's Voltage Law (KVL). The algebraic sum of the branch voltages around a loop is zero at every instant of time.

Kirchhoff's Voltage Law (alternate form). The sum of the voltage rises around a loop is equal to the sum of the voltage drops around the loop at every instant of time.

∎

As an example, consider the loop consisting of the branches labeled 1-2-3 of the network shown in Fig. 1-4.5. The reference polarities and the branch voltages for these elements are indicated as $v_1(t)$, $v_2(t)$, $v_3(t)$ as shown in Fig. 1-4.6. From KVL we may say that

$$v_1(t) - v_2(t) + v_3(t) = 0 \quad \text{or} \quad v_1(t) + v_3(t) = v_2(t) \qquad (1\text{-}14)$$

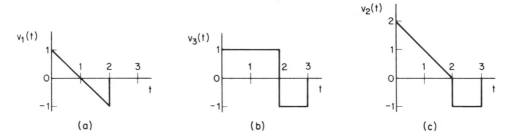

Figure 1-4.6 Branch current and voltage reference directions for the network graph shown in Fig. 1-4.5.

for all values of t. For example, if $v_1(t) = t^2 + 1$ V, and $v_2(t) = \sin t + 3$ V, then $v_3(t) = \sin t - t^2 + 2$ V. Similarly, for this loop, if $v_1(t)$ and $v_3(t)$ are defined by the graphical plots indicated in Fig. 1-4.7(a) and (b), then $v_2(t)$ will have the waveshape indicated in Fig. 1-4.7(c).

Figure 1-4.7 Set of voltages for the network shown in Fig. 1-4.6.

It should be noted that KVL, like KCL, does not apply to circuits containing distributed elements. It does, however, apply irrespective of whether the network is linear or non-linear, time-varying or time-invariant, active or passive.

Another classification of network graphs which is frequently of interest is the distinction between planar and non-planar graphs. A *planar graph* is one which has the property (or can be redrawn so that it has the property) that no lines cross. Correspondingly, a *non-planar graph* is one that cannot be redrawn so as to eliminate the crossing of lines. For example, the graph shown in Fig. 1-4.8(a) is a planar graph, since it can be redrawn as shown in Fig. 1-4.8(b) without any crossing lines. On the other hand, the graph shown in Fig. 1-4.9 is a non-planar graph, since there is no way in which it can be redrawn so that no lines cross. We shall see in a later chapter that, in some cases, the analysis techniques which we use for networks with planar graphs must be modified when networks with non-planar graphs are encountered.

Subgraphs of a given graph may be defined by removing one or more branches from the original graph. One of the most important subgraphs is called a *tree*. For a connected graph of n nodes, a tree has the following properties: (1) It connects all the nodes of the original graph; (2) it contains $n - 1$ branches; and (3) it has no closed paths. A graph and some of its trees (the heavy lines) is shown in Fig. 1-4.10. The branches removed from the graph in forming the tree are called *chords* or *links*. If the network has b branches and n nodes, it will have $b - n + 1$ links.

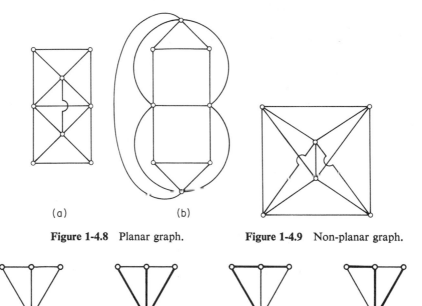

(a)	(b)

Figure 1-4.8 Planar graph.   **Figure 1-4.9** Non-planar graph.

Figure 1-4.10 Graph and some of its trees.

1-5 CONCLUSION

The major result of this chapter has been the presentation of Kirchhoff's laws. The importance of these laws cannot be overemphasized. They are based only on the topological form of the network, i.e., on the manner in which the various branches of the network are interconnected. Thus, they are completely independent of the nature of the network elements, requiring only that they be lumped. In the following chapters, we shall see that Kirchhoff's laws form the basis of all our network analysis techniques.

PROBLEMS

Problem 1-1 (Sec. 1-1)

The charge $q(t)$ in coulombs present in a two-terminal network element is defined by the waveshape shown in Fig. P1-1.

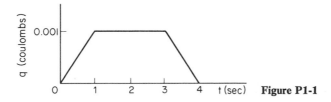

Figure P1-1

(a) Find a waveshape of current (in mA) which must be applied to the element to produce the waveshape of $q(t)$.

(b) Find another different waveshape of charge which could be produced by the current waveshape which was found in part (a).

Problem 1-2 (Sec. 1-1)

Find an expression for the total charge $q(t)$ in coulombs which has passed a point in a branch of a circuit, if the current in the branch is equal to $i(t) = 4e^{-4t}$ for $t \geq 0$.

Problem 1-3 (Sec. 1-1)

The voltage and current variables at a pair of network terminals are defined with the relative reference polarities shown in Fig. P1-3. It is assumed that the network has no initial energy stored in it. If for the period $0 \leq t \leq 3$ ms, the voltage $v(t) = 10 - 200t$ V and the current $i(t) = 10t^2$ mA find:

Figure P1-3

(a) The power $p(t)$ transferred from the source to the network.

(b) The energy $w(t)$ supplied to the network.

(c) The total energy stored in the network at $t = 3$ ms.

Problem 1-4 (Sec. 1-1)

Waveshapes of voltage and current defined as shown in Fig. P1-4 are measured at the terminals of the network shown in Fig. P1-3. Draw waveshapes for the power $p(t)$ and the energy $w(t)$. Identify the units used on the ordinates of these plots.

Problem 1-5 (Sec. 1-1)

Repeat Problem 1-4 using the waveshape of current shown in Fig. P1-5.

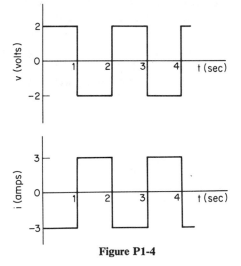

Figure P1-4   **Figure P1-5**

Problem 1-6 (Sec. 1-2)

A voltage $v(t) = \sin 10t$ V and a current $i(t) = 5 \sin 10t$ A are the variables (with associated reference directions) of a two-terminal element.

(a) Find an expression for the power dissipated in the two-terminal element.

(b) Find the total energy supplied to the element over the time range $0 \le t \le 0.2$ s.

(c) Repeat parts (a) and (b) for the case where the current $i(t) = 5 \cos 10t$ A.

Problem 1-7 (Sec. 1-2)

Repeat Problem 1-3 for the case where the reference direction for the current $i(t)$ shown in Fig. P1-3 is reversed.

Problem 1-8 (Sec. 1-4)

Write the equations determined by applying KCL at nodes $a, b, c,$ and d of the network shown in Fig. P1-8.

Figure P1-8

Problem 1-9 (Sec. 1-4)

Add the four equations found in Problem 1-8. What conclusion can you make from the result?

Problem 1-10 (Sec. 1-4)

In the circuit shown in Fig. P1-8, $i_2(t) = 3 + 2t$, and $i_3(t) = -3 + \sin t$ A. Find the currents $i_1(t), i_4(t),$ and $i_5(t)$.

Problem 1-11 (Sec. 1-4)

In the circuit shown in Fig. P1-11 on page 20, the waveshapes for $i_1(t), i_2(t),$ and $i_3(t)$ are shown. Make a sketch of the current $i_4(t)$.

Problem 1-12 (Sec. 1-4)

Apply KVL to obtain three different equations for the circuit shown in Fig. P1-12 on page 20.

Problem 1-13 (Sec. 1-4)

Add the three KVL equations determined in Problem 1-12. What conclusion can you make from the result?

Problem 1-14 (Sec. 1-4)

In the circuit shown in Fig. P1-12, $v_1(t) = 2 + 7t$, $v_4(t) = -4 + \sin t$, and $v_5(t) = 3 + e^{-t}$ V. Find $v_2(t)$ and $v_3(t)$.

Problem 1-15 (Sec. 1-4)

Draw an oriented graph for the circuit shown in Fig. P1-8.

Problem 1-16 (Sec. 1-4)

Draw an oriented graph for the circuit shown in Fig. P1-11.

Problem 1-17 (Sec. 1-4)

Determine all possible cutsets for the circuit shown in Fig. P1-8 and write the KCL equations for each cutset.

Problem 1-18 (Sec. 1-4)

Determine all possible cutsets for the circuit shown in Fig. P1-11 and write the KCL equations for each cutset.

Figure P1-11

Figure P1-12

Problem 1-19 (Sec. 1-4)

In the circuit shown in Fig. P1-11, $i_1(t) = 2 + 3t$, $i_6(t) = -1 + 2\sin t$, and $i_7(t) = -3t + t^2$ A. Find $i_2(t)$.

Problem 1-20 (Sec. 1-4)

In the network graph shown in Fig. P1-20, four of the branch currents have the indicated values. Find the value of the current i.

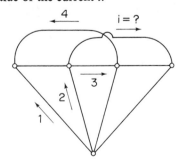

Figure P1-20

Problem 1-21 (Sec. 1-4)

Find all the cutsets for the network graph shown in Fig. P1-21 and write the KCL equations for each.

Introduction Chap. 1

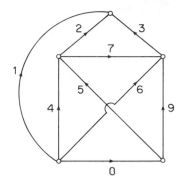

Figure P1-21

Problem 1-22 (Sec. 1-4)

Identify all possible cutsets in the network graph shown in Fig. P1-22 by listing the numbers of the branches in each such cutset.

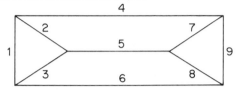

Figure P1-22

Problem 1-23 (Sec. 1-4)

In the network graph shown in Fig. P1-23, the currents $i_1(t)$ and $i_2(t)$ have the waveshapes illustrated. Sketch the waveshape for $i_3(t)$. Give all the necessary values on the ordinate of the sketch.

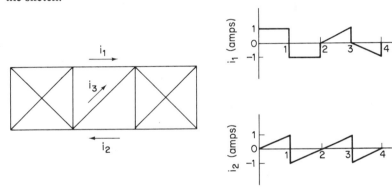

Figure P1-23

Problem 1-24 (Sec. 1-4)

For the network shown in Fig. P1-24, find:
(a) The number of nodes.
(b) The number of branches.
(c) The number of cutsets.

Figure P1-24

Problem 1-25 (Sec. 1-4)

    Classify each of the networks shown in Fig. P1-25 as either planar or non-planar.

(a)

(b)

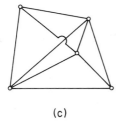

(c)

(d)   **Figure P1-25**

Problem 1-26 (Sec. 1-4)

    For each of the networks shown in Fig. P1-25:

(a) Find the number of links.

(b) Find the number of tree branches.

(c) Draw all possible trees.

2

Resistors, Sources,

and Simple Circuits

In order to determine the properties of any specific type of system, we must start by determining the answers to two general questions. The first of these is: What are the basic elements or building blocks of which the system is constituted? The second is: How are these basic elements interconnected? Once the answers to these questions have been found, then we may turn our attention to determining the properties of the system, i.e., finding how it operates under various conditions of input. In this and the following chapter we shall apply this method of approach to determine the properties of resistance networks. We will begin by examining the basic elements of such networks, namely resistors and sources.

2-1 THE RESISTOR

A *resistor* is a two-terminal element which has the property that its branch voltage and current variables are related by the expressions

$$v(t) = Ri(t) \quad \text{and} \quad i(t) = Gv(t) \tag{2-1}$$

These expressions are frequently referred to as *Ohm's law*. The unit used for specifying the value of the *resistance R* is the *ohm* (abbreviated Ω), and the unit used for expressing the *conductance G* ($= 1/R$) is the *mho* (abbreviated $\mho$). The circuit symbol and the reference polarities for a resistor are shown in Fig. 2-1.1. Note that the reference polarities are associated, i.e., they follow the convention for two-terminal elements which was set up in Sec. 1-2.

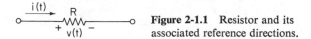

Figure 2-1.1 Resistor and its associated reference directions.

We may illustrate the properties of a resistor more fully by plotting the locus of Eq. (2-1) on an *i-v* plane as shown in Fig. 2-1.2. A *linear* resistor is characterized by a straight line passing through the origin. Actually, in the physical or real-life world, there is no such thing as a linear resistor. For example, at high current levels, the resistance of a resistor may increase markedly over its zero current value (a light bulb is a good example of this). Such behavior is definitely *nonlinear*, i.e., it is represented by a characteristic on the *i-v* plane which is not a straight line, and/or which does not pass through the origin. An example of a device with a non-linear characteristic is the *solid-state diode*. Its circuit symbol and (associated) reference polarities are shown in Fig. 2-1.3. It has a static relation between its terminal variables of voltage *v* and current *i* that may be closely approximated by the expressions[1]

$$i = I_s(e^{v/V_o} - 1) \quad \text{and} \quad v = V_o \ln\left(\frac{i}{I_s} + 1\right) \tag{2-2}$$

where I_s and V_o are constants whose values depend on the given diode. Typical values are $I_s = 0.0001$ A (0.1 mA), and $V_o = \frac{1}{40}$ V. A plot of the relations of Eq. (2-2) is given on the *v-i* plane shown in Fig. 2-1.4. Frequently, this characteristic is idealized as shown in Fig. 2-1.5. In this case the device is referred to as an *ideal diode*, i.e., an *idealized model* of a physical or real-life diode.

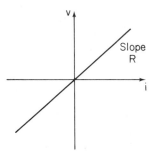

Figure 2-1.2 Linear resistor volt-ampere characteristic.

Figure 2-1.3 Circuit symbol and reference directions for a diode.

In addition to the possibility of being nonlinear, physical resistors may also be *time-varying*. A *linear*, time-varying resistor is represented by a characteristic which is exactly the same as that shown in Fig. 2-1.2, except that the slope of the straight-line characteristic varies with time. In this case, *R* of Eq. (2-1) may be written as *R(t)*. For many applications, the terminal behavior of physical resistors so closely approximates that of a linear or ideal resistor that little is gained by trying to include the non-linear and/or time-varying behavior. Indeed, much may be lost, since the mathe-

---

[1] By "static" here we refer to the case where the voltage and current variables have constant values, i.e., they are not functions of time.

Resistors, Sources, and Simple Circuits Chap. 2

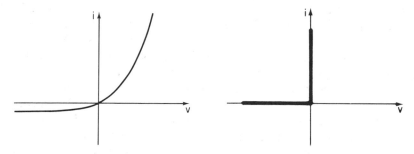

Figure 2-1.4 Volt-ampere characteristic for a solid-state diode.

Figure 2-1.5 Idealization of the diode characteristic shown in Fig. 2-1.4.

matical treatment required to characterize the physical element may become so complicated that it completely conceals the overall properties of the circuit in which the element appears. To avoid this, we will assume that physical resistors are "modeled" by ideal elements obeying the relations given in Eq. (2-1). Thus, from this point on, when we use the word resistor, we imply a *lumped, linear, time-invariant* model of some physical resistor, unless otherwise specified.

The instantaneous power which is supplied to a resistor is easily determined from the relation $p(t) = v(t)i(t)$. Substituting from Eq. (2-1) we obtain

$$p(t) = i^2(t)R \qquad \text{and} \qquad p(t) = v^2(t)G \qquad (2\text{-}3)$$

Since the expressions for power given above are always non-negative for positive values of R (and G), the energy supplied to a resistor must also always be non-negative. Thus, from the discussion given in Sec. 1-3, we conclude that a resistor must be a passive element. In future developments of this text we shall see that it is possible to realize two-terminal network elements which have the property that they may be modeled as negative-valued resistors (over certain ranges of their variables). Such an element is characterized by the relations

$$v(t) = -|R|i(t) \qquad \text{and} \qquad p(t) = -|R|i^2(t) \qquad (2\text{-}4)$$

where, for emphasis, we have indicated that the value of the resistor is negative by using magnitude symbols and a minus sign. Thus we see that the power being supplied to such a negative-valued resistor is negative. From the conventions established for the reference polarity of power in Sec. 1-2, we may say that such a resistor *supplies* energy to the circuit to which it is connected. Thus, it may be considered to be an active element. We shall discuss the means by which negative-valued resistors may be realized in later portions of this text.

The passive resistor is sometimes referred to as a *dissipative* element in that the energy $w(t)$ (where $w = \int p \, dt$) which is absorbed by it is turned into heat, and cannot be reclaimed. We shall see that two other types of basic network elements, namely, the inductor and the capacitor, have a quite different property in that they can store energy, which can then be used to excite a network at some future time.

Since real-world resistors are called upon to dissipate energy, they must be rated not only with respect to their resistance, i.e., their nominal value in ohms, but also with respect to the maximum amount of power that they can dissipate without reaching a temperature which is so high that their characteristics may drastically change, or that they may even "burn out." For radio, television, and communication applications, wattage ratings are usually in fractions of a watt, such as $\frac{1}{4}$ W, $\frac{1}{2}$ W, etc. For industrial control and power applications, the rating of a resistor may be many watts. Obviously, the power rating greatly influences the size and physical construction of a given resistor.

The values of resistors used in most practical circuit applications range from hundreds of ohms to several megohms. In many of our examples, however, we shall use resistors whose values are restricted to being only a few ohms. Such a restriction still gives us complete generality in the analysis of circuits, however, since we shall see in Sec. 7-9 that a process called "normalization" may be applied to transform such circuits to ones having resistors of any desired range of values. Thus, the use of small numbers gives us the advantage of numerical simplicity without sacrificing practicality.

2-2 SOURCES

In this section we shall introduce a basic network element called a *source*. The first class of source which we shall consider here is referred to as an *independent* source, since its output characteristics are *not dependent* on any other network variable such as a current or a voltage. Its characteristics, however, may be time-varying. There are two types of independent sources. The first of these is called a *voltage* source. It is a two-terminal element which has the property that a specified voltage appears across its terminals, and that the value of this voltage at any instant of time is *independent of the value or direction of the current that flows through it*. Since there is no relation between the voltage and the current, it is not necessary for us to observe any relative reference directions between these variables. A circuit symbol for a voltage source is shown in Fig. 2-2.1(a). It should be noted that the terminal voltage $v_s(t)$ may be a constant, or it may be some specified function of time. Thus, we may refer to a *constant* (dc) source or a *time-varying* source. The constant-valued (dc) voltage source is sometimes referred to as a model for an ideal battery, which is indicated by the circuit symbol shown in Fig. 2-2.1(b). We may illustrate the properties of the voltage source more fully by drawing its characteristic on the i-v plane. As shown in Fig. 2-2.1(c), this is simply a horizontal line. If the source is time-varying, the position of

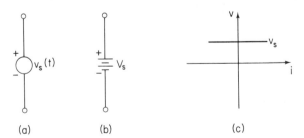

Figure 2-2.1 Circuit symbols and volt-ampere characteristic for a voltage source.

the horizontal-line characteristic will vary for different values of time. When the terminal voltage is zero, the characteristic coincides with the abscissa. In this case it is exactly the same as the characteristic of a resistor which has zero resistance, i.e., a *short circuit*. Thus, we may say that the *internal resistance* of a voltage source, i.e., the resistance measured between its terminals when its output voltage is zero, is zero. This, of course, is implied in the statement that the terminal voltage of such a source is independent of the current flowing through it.

A second type of independent source is the *current* source. A current source is a two-terminal element which has the property that a specified current flows through it, and that the value and direction of this current at any instant of time is *independent of the value or direction of the voltage that appears across the terminals of the source.* Just as in the case of the voltage source, there is no required relation between the reference polarities of voltage and current for such an element. A circuit symbol for a current source is shown in Fig. 2-2.2(a). The output current $i_s(t)$ may be constant, or it may be a function of time, thus we may refer to a *constant* (dc) or a *time-varying* current source. On the *i-v* plane, the characteristics of the current source are indicated by a vertical line as shown in Fig. 2-2.2(b). For the case of a time-varying source, the position of this vertical line will vary with time. When the source current is zero, then the characteristic coincides with the ordinate. In this case, the source has properties that are identical with those of a resistor of infinite resistance, i.e., an *open circuit*. Thus we may say that the internal resistance of a current source, i.e., the resistance measured between its terminals when its output current is zero, is infinite. This, of course, is implied in the statement that the current flowing through such a source is independent of the value of the voltage which is impressed across it.

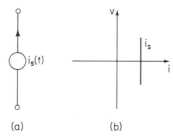

(a)   (b)

Figure 2-2.2 Circuit symbol and volt-ampere characteristic for a current source.

Sources are generally considered to be active elements. For example, consider the connection of a constant voltage source and a resistor as shown in Fig. 2-2.3. For the indicated variables, application of KVL around the loop tells us that $v_s = iR$. Thus, for a 10-V dc source, 0.1 A flows through the 100-Ω resistor. The power dissipated in the resistor is found from the expression $p = i^2R$. It is equal to 1 W. In calculating the power for the two-terminal voltage source, however, we see that, following the associated reference polarity conventions for voltage and current specified in the definition of power in Sec. 1-2, the power for the source is given by the expression $p = -iv_s$, and thus the source power is -1 W. From our definition, power *into* a two-terminal element is positive; therefore we see that 1 W of power flows *out of* the source in this case. Note that it is possible for a source to receive power from the circuit. For example, for the network shown in Fig. 2-2.4, application of KVL shows that power associated with source 1 is -1.2 W, and that the power associated with source 2 is 0.2 W. Thus, source 2 is, in effect, being "charged," i.e., it is *receiving* energy from source 1. In such a case, unlike a resistor, the source does not dissipate this energy. Instead it converts it into some other form of energy. This might be chem-

Figure 2-2.3 Simple circuit with a voltage source.

Figure 2-2.4 Circuit with two voltage sources.

ical energy (as in the case of a battery), or mechanical energy (as in the case of a motor-driven generator), etc.

Similar conclusions hold true for the case of a current source. For example, let us analyze the network shown in Fig. 2-2.5 using KCL. We see that $i_s = vG$. For the 0.1-A source, a voltage of 10 V appears across the 0.01-mho resistor. The power dissipated in the resistor is found from the expression $p = v^2G$, and is equal to 1 W. It is readily shown that the power for the current source is -1 W, thus the current source supplies power to the circuit.

Figure 2-2.5 Simple circuit with a current source.

In general we will refer to voltage and current sources as *active* elements, since they are capable of supplying energy to a network. We should remember, however, that in a given situation, such sources may actually receive energy from the remainder of the circuit. Since sources are not defined by a relation between two variables, but simply by a specification of a single variable, the question of linear vs. non-linear behavior does not arise. Specifically, a test for superposition as outlined in Sec. 1-3 is not defined, since there is only one variable associated with the source.

As we have seen in the preceding paragraphs, an independent source is characterized as one whose output is either a constant or a specified function of time. The second class of sources to be considered in this section is the *dependent* (or *controlled*) source. Its output is proportional to some controlling voltage or current variable. The constant of proportionality is called the *gain*. Both voltage and current types of sources may be dependent, and either may be controlled by a voltage or a current. Thus, as shown in Fig. 2-2.6, there are four types of dependent sources. These are the voltage-controlled voltage source (VCVS), current-controlled voltage source (ICVS), voltage-controlled current source (VCIS), and current-controlled current source

Figure 2-2.6 Types of dependent sources.

Resistors, Sources, and Simple Circuits   Chap. 2

(ICIS). In the figure, dashed lines are used to indicate that although the controlling variables may be located anywhere in the network, frequently they are connected to the source so as to form a three-terminal element. From the figure we see that the gain constants have the dimensions of resistance (ICVS), conductance (VCIS), or are dimensionless (VCVS and ICIS). As an example of the use of a controlled source, consider a transistor used in a common (grounded) emitter configuration. A circuit symbol for it is shown in Fig. 2-2.7(a). For simplicity, the dc biasing circuit needed for operation of the transistor has been omitted (see Sec. 2-6). An idealized model using an ICIS is shown in Fig. 2-2.7(b). The value of β is of the order of 100. From the model we see that the input resistance measured between the base and ground is zero, i.e., a short circuit. Similarly, the output resistance measured between the collector and ground is infinite, i.e., the resistance of the current source. An actual transistor has input and output resistance values which are similar. As another example of a dependent source, consider an inverting operational amplifier. A circuit symbol for it is shown in Fig. 2-2.8(a). An ideal model is shown in Fig. 2-2.8(b). In practice, the gain K shown in the model has a very large value (10^4 to 10^5) and is usually treated as being infinite. From the model we see that the input resistance measured between the input terminal and ground is infinite, i.e., the resistance of an open circuit. The output resistance is zero, i.e., that of a voltage source. These values are similar to those found in an actual operational amplifier. We shall cover more details on transistors and operational amplifiers in Sec. 3-9.

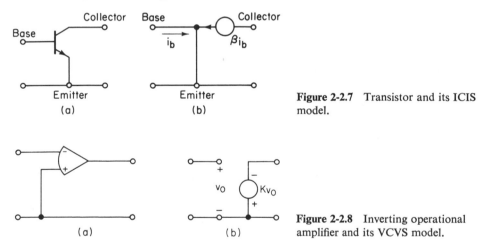

Figure 2-2.7 Transistor and its ICIS model.

Figure 2-2.8 Inverting operational amplifier and its VCVS model.

2-3 NON-IDEAL SOURCES

The voltage and current sources which were discussed in the preceding section should actually be considered as *ideal* sources in the sense that they have no dissipation associated with them. We will now define *non-ideal* voltage and current sources as ones which include some internal dissipation. The non-ideal voltage source may be shown schematically by a resistor in series with an ideal voltage source, as shown in Fig. 2-3.1(a), and the non-ideal current source by a resistor in parallel (or shunt) with

Figure 2-3.1 Non-ideal voltage and current sources.

(a)   (b)

an ideal current source, as shown schematically in Fig. 2-3.1(b).[2] Non-ideal sources come considerably closer to modeling the properties of physical energy-supplying devices than do the ideal sources described in the last section. For example, an automobile battery may be modeled (approximately) as an ideal voltage source of about 12 V, in series with a resistor of about 0.05 Ω. Similarly, a sinusoidal signal generator as used in testing electronic circuits might be modeled as an ideal voltage source which has a terminal voltage $v(t) = V_0 \sin(\omega_0 t + \phi)$ where V_0 is the magnitude of the sinusoidal output (this may usually be varied by an "amplitude" or "gain" control), and a series resistance of 600 Ω. In both examples, the resistance is referred to as the *internal resistance* of the non-ideal source.

Actually, the two non-ideal source models shown in Fig. 2-3.1 may be used interchangeably in network studies. To see this, consider the effect of connecting each of them to a pair of terminals of some arbitrary network. Let the network variables be $v_0(t)$ and $i_0(t)$ as shown in Figs. 2-3.2 and 2-3.3. For the non-ideal voltage

Figure 2-3.2 Non-ideal voltage source connected to a network.

Figure 2-3.3 Non-ideal current source connected to a network.

source, we may apply KVL to obtain the relation

$$v_0(t) = v_s(t) - i_0(t)R_s \tag{2-5}$$

For the non-ideal current source, we may apply KCL to obtain

$$v_0(t) = \frac{i_s(t) - i_0(t)}{G_s} \tag{2-6}$$

[2]It should be noted that a non-ideal source consisting of an ideal voltage source with a resistor in *parallel* with it is indistinguishable from an ideal voltage source alone, since the specified voltage will always be present at the terminals and it will be independent of the current (the voltage source, however, will have to supply the extra current required by the resistor). Similarly, a non-ideal source consisting of an ideal current source in *series* with a resistor is indistinguishable from an ideal current source by itself, since it will still supply the specified current to any externally connected circuit independent of any voltage that appears across it (there will, however, be a voltage drop across the resistor).

If we compare the two relations given above, we see that they will be the same for all values of $v_0(t)$ and $i_0(t)$, providing that

$$v_s(t) = \frac{i_s(t)}{G_s} \quad \text{and} \quad G_s = \frac{1}{R_s} \tag{2-7}$$

Thus, *at a given pair of terminals*, either of the non-ideal sources shown in Fig. 2-3.1 may be replaced by the other without affecting the behavior of the rest of the network. As an example of such an equivalence, the automobile battery referred to above may also be represented by an ideal current source of $i_s = v_s G_s = (12)(20) = 240$ A in parallel with a conductance of $G_s = 1/R_s = 1/0.05 = 20$ mhos. The two equivalent models are shown in Fig. 2-3.4. To emphasize this equivalence note the following:

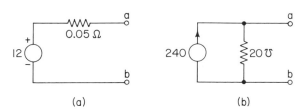

(a)   (b)

Figure 2-3.4 Examples of non-ideal sources.

1. The open-circuit voltage that appears at the terminals *a-b* for the two sources is the same (12 V).

2. The short-circuit current that results when the terminals *a-b* of either source are shorted together is the same (240 A).

3. If an arbitrary resistor is connected across the output terminals of either source, the same power will be dissipated in it.

4. The sources are equivalent only as concerns their behavior at the external terminals.

As an illustration of this last point, note that under open-circuit conditions, the model shown in Fig. 2-3.4(a) does not dissipate any internal power, while the one shown in Fig. 2-3.4(b) dissipates 2880 W! Under short-circuit conditions the situation is reversed. Thus, even though the two models have identical properties in terms of their behavior at their external terminals, their internal behavior is considerably different.

Let us consider that a source has a constant open-circuit voltage of $v_s(t) = V_s$ V and an internal resistance of R_s Ω. If we connect such a source to a pair of terminals of some arbitrary network, and define the variables at the pair of terminals as $v_0(t)$ and $i_0(t)$ as shown in Fig. 2-3.2, then, applying KVL to the loop formed by the non-ideal source and the pair of terminals, we see that

$$v_0(t) = i_0(t)(-R_s) + V_s \tag{2-8}$$

This equation defines a straight line on an i_0-v_0 plane as shown in Fig. 2-3.5. This plot may be used to determine the voltage appearing at the terminals of the non-ideal source for any given value of output current $i_0(t)$. It is frequently used as a "load line" in solving biasing problems for non-linear devices. An example is given in Sec. 2-6.

Figure 2-3.5 $I_0 - v_0$ characteristic for the circuit shown in Fig. 2-3.2.

2-4 SERIES AND PARALLEL CONNECTIONS OF RESISTORS

In this section we shall discuss some simple circuits formed by the interconnection of resistors. There are two basic connections to be considered, namely, series connections and parallel connections. In Fig. 2-4.1 we consider the series connection of n resistors

Figure 2-4.1 Series connection of n resistors.

R_i $(i = 1, 2, \ldots, n)$. The series connection forms an overall two-terminal branch with variables $v(t)$ and $i(t)$ as shown. If we apply KCL at node 1, we see that the branch current variables $i(t)$ and $i_1(t)$ are equal. Similarly, at node 2, we see that $i_1(t)$ equals $i_2(t)$, etc. Thus, we may write

$$i(t) = i_1(t) = i_2(t) = \cdots = i_n(t) \tag{2-9}$$

Now let us apply KVL to the "loop" formed by the voltages $v(t)$, $v_1(t)$, $v_2(t)$, etc. We obtain

$$v(t) = \sum_{i=1}^{n} v_i(t) \tag{2-10}$$

The defining relations for the voltage and current variables for each of the n resistive branches are given by Ohm's law. They have the form

$$v_i(t) = R_i i_i(t) \qquad i = 1, 2, \ldots, n \tag{2-11}$$

From the relations (2-9), (2-10), and (2-11), we see that we may write

$$v(t) = \sum_{i=1}^{n} R_i i_i(t) = i(t) \sum_{i=1}^{n} R_i \tag{2-12}$$

We conclude that if we define an equivalent resistor R_{eq}, where

$$R_{eq} = \sum_{i=1}^{n} R_i \tag{2-13}$$

then the variables $v(t)$ and $i(t)$ are related by the equation

$$v(t) = R_{eq} i(t) \tag{2-14}$$

Thus, we may say that a series connection of resistors R_i may be replaced by a single equivalent resistor R_{eq} with no change in the terminal variables of the overall two-terminal network.

A similar analysis holds for the case where n resistors are connected in parallel as shown in Fig. 2-4.2. Such a connection may be treated as a single overall two-

Figure 2-4.2 Parallel connection of n resistors.

terminal element. If we apply KVL around the "loop" consisting of the voltage $v(t)$ on the one side and the branch voltage $v_1(t)$ on the other side, we see that $v(t)$ equals $v_1(t)$. Similarly, we see that $v_1(t)$ equals $v_2(t)$, etc. Thus, we may write

$$v(t) = v_1(t) = v_2(t) = \cdots = v_n(t) \tag{2-15}$$

Now let us apply KCL to the upper node. We obtain

$$i(t) = \sum_{i=1}^{n} i_i(t) \tag{2-16}$$

The defining relations for the n resistive branches are conveniently written using units of conductance rather than resistance; thus, we obtain

$$i_i(t) = G_i v_i(t) \qquad i = 1, 2, \ldots, n \tag{2-17}$$

From the above three sets of relations we see that we may write

$$i(t) = \sum_{i=1}^{n} G_i v_i(t) = v(t) \sum_{i=1}^{n} G_i \tag{2-18}$$

We conclude that if we define a constant G_{eq}, where

$$G_{eq} = \sum_{i=1}^{n} G_i \tag{2-19}$$

then the variables $i(t)$ and $v(t)$ are related by the equation

$$i(t) = G_{eq} v(t) \tag{2-20}$$

Thus we may say that a parallel connection of a set of resistors G_i may be replaced by a single equivalent resistor G_{eq} with no change in the overall network variables.

Sometimes it is desirable to deal with resistance values in ohms while computing the equivalent resistance of a set of paralleled resistors. In such a case, Eq. (2-19) may be written as

$$\frac{1}{R_{eq}} = \sum_{i=1}^{n} \frac{1}{R_i} \tag{2-21}$$

For the frequently encountered case of two resistors R_1 and R_2 in parallel, Eq. (2-21) becomes

$$R_{eq} = \frac{R_1 R_2}{R_1 + R_2} \tag{2-22}$$

In the above paragraphs we have concluded that if resistors are connected in series, their resistances add, while if resistors are connected in parallel, their conductances add. Thus, in some analyses it is convenient to treat resistance values in ohms, while in other analyses the computations may be simplified by utilizing values in mhos. This is only the first of many illustrations that we shall make showing how calculations may be simplified by having available reciprocal units to describe a given network element. We shall see more examples of the convenience of such different units in the later chapters of this text. Here, however, we insert a word of caution, in that to take advantage of this flexibility, network elements must be labeled not only with their numerical values, but also with an indication such as the word "ohms" or the word "mhos" to indicate which type of units are being used.

Networks which consist of various configurations of resistors in series and parallel may sometimes be reduced to a single equivalent resistance with respect to a specific pair of nodes by successive application of the series and parallel rules given above. For example, consider the network shown in Fig. 2-4.3(a). Working from

(a)   (b)

Figure 2-4.3   Ladder network of resistors.

right to left we see that the series resistance of R_5 and R_6 is 10 Ω. Similarly, the equivalent resistance of this 10 Ω in parallel with R_4 is 5 Ω. Continuing, the equivalent resistance of R_3 in series with the 5 Ω is 10 Ω, and the parallel combination of the 10 Ω with R_2 gives an equivalent 5 Ω. Finally, the series combination of R_1 and the 5 Ω gives 10 Ω. We conclude that an equivalent single resistance of 10 Ω may be placed between nodes a-b as shown in Fig. 2-4.3(b), and that this single resistance will have the same effect as the original resistance network in relating the variables $v(t)$ and $i(t)$ defined for this node-pair.

The type of network illustrated in Fig. 2-4.3(a) is called a *ladder* network. The process of finding the equivalent resistance of a ladder network described above is easily generalized. Consider the network shown in Fig. 2-4.4. Starting with the resistor R_1, and proceeding through the network from left to right, we see that we may write an expression for the *input resistance* R_{in} in the form

Figure 2-4.4   Ladder network of resistors.

Resistors, Sources, and Simple Circuits   Chap. 2

$$R_{\text{in}} = R_1 + \cfrac{1}{G_2 + \cfrac{1}{R_3 + \cfrac{1}{G_4}}} \qquad (2\text{-}23)$$

Such an expression is called a *continued fraction*.

It is also possible to calculate a *voltage transfer function* for a ladder network. For the variables v_2 and v_1 defined in Fig. 2-4.4, such a function specifies the ratio v_2/v_1. Because the network is linear, this ratio can be determined for a specific numerical value of v_2, rather than for a literal one. Letting $v_2 = 1$ greatly simplifies the computations. With this choice, starting at the right end of Fig. 2-4.4, we obtain:

1. $i_{G4} = G_4 v_2 = G_4$
2. $i_{R3} = i_{G4} = G_4$
3. $v_{R3} = R_3 i_{R3} = R_3 G_4$
4. $v_{G2} = v_{R3} + v_2 = R_3 G_4 + 1$
5. $i_{G2} = G_2 v_{G2} = G_2 R_3 G_4 + G_2$
6. $i_{R1} = i_{G2} + i_{R3} = G_2 R_3 G_4 + G_2 + G_4$
7. $v_{R1} = R_1 i_{R1} = R_1 G_2 R_3 G_4 + R_1 G_2 + R_1 G_4$
8. $v_1 = v_{R1} + v_{G2} = R_1 G_2 R_3 G_4 + R_1 G_2 + R_1 G_4 + R_3 G_4 + 1$

Thus, the voltage transfer function is

$$\frac{v_2}{v_1} = \frac{1}{R_1 G_2 R_3 G_4 + R_1 G_2 + R_1 G_4 + R_3 G_4 + 1} \qquad (2\text{-}24)$$

This method is called "working your way through" the ladder network. We shall find frequent application for it in this text.

2-5 SERIES AND PARALLEL CONNECTIONS OF SOURCES

In the last section we considered interconnections of resistors. In this section we will consider the effects of interconnecting various kinds of sources. Our conclusions will apply to both independent and dependent sources. In Fig. 2-5.1 we show a series

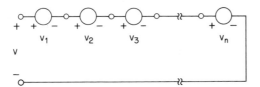

Figure 2-5.1 Series connection of n ideal voltage sources.

connection of n voltage sources. The overall voltage $v(t)$ may be found in terms of the voltages of the individual sources from KVL. Thus, we obtain

$$v(t) = \sum_{i=1}^{n} v_i(t) \qquad (2\text{-}25)$$

We conclude that for such a series connection the overall voltage is the sum of the voltages from the individual sources. The voltage resulting from a parallel connection of voltage sources is not defined, except for the case where all the sources have exactly the same voltage for all values of time. To see why this is so, consider what would happen if a 1-V source and a 2-V source were connected in parallel. Applying KVL around the loop formed by the two sources, there would be a set voltage of 1 V. However, the total resistance of the loop, i.e., the sum of the internal resistances of the two sources, is zero. Thus, the loop current that would flow is $i = v/R = 1/0 = \infty$, which is clearly not possible.

Now let us consider the current source. A parallel connection of n current sources is shown in Fig. 2-5.2. The overall current may be found in terms of the

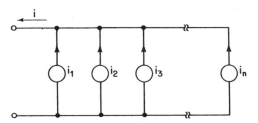

Figure 2-5.2 Parallel connection of n ideal current sources.

currents of the individual sources from KCL. Thus, we obtain

$$i(t) = \sum_{i=1}^{n} i_i(t) \qquad (2\text{-}26)$$

We conclude that the overall current is the sum of the currents from the individual sources. The current resulting from a series connection of current sources is not defined except for the case where all the sources have exactly the same output current for all values of time, since otherwise such a connection would violate KCL at any node connecting two sources.

In the above paragraphs we have discussed the replacement of several sources by a single source. Frequently in our studies of network analysis it will be convenient to follow the opposite procedure, namely to replace a single source with an equivalent connection of more than one source. For the case of a voltage source, such an equivalence is shown in Fig. 2-5.3. From these figures we see that we may replace a single voltage source with output voltage $v(t)$ as shown in Fig. 2-5.3(a) with a parallel connection of an arbitrary number of voltage sources which have the same output voltage $v(t)$ as shown in Fig. 2-5.3(b). If some branches of a network are connected to the single source in the manner shown in Fig. 2-5.4(a), then an equivalent connection is that shown in Fig. 2-5.4(b). This is readily verified by application of Kirchhoff's

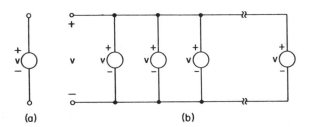

(a)   (b)

Figure 2-5.3 Parallel connection of ideal voltage sources.

     Resistors, Sources, and Simple Circuits   Chap. 2

laws, since the potential difference between nodes *a* and *b* in Fig. 2-5.4(b) must be zero; therefore, if a short circuit is placed between these two points, no current will flow through it. Thus, it is immaterial to the operation of the circuit whether such a conductor is present or absent. The same conclusion holds true for nodes *b* and *c* and nodes *a* and *c*.

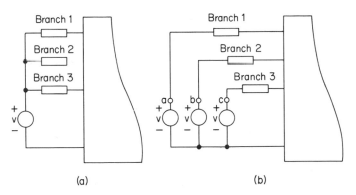

(a)   (b)

Figure 2-5.4   Replacing a single ideal voltage source with several ideal voltage sources.

A similar equivalence holds true for the case of current sources. The single current source shown in Fig. 2-5.5(a) is equivalent to the series connection of current sources which have the same output current as shown in Fig. 2-5.5(b). At each internal

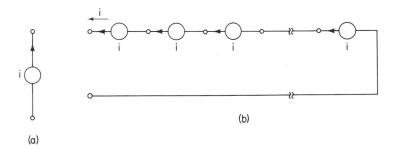

(b)

(a)

Figure 2-5.5   Series connection of ideal current sources.

node of the equivalent source, KCL is completely satisfied by the source currents. Since the voltage between the sources is not related to the currents, any arbitrary network nodes which are not originally connected to the current sources, such as nodes *a* and *b* shown in Fig. 2-5.6(a), may be connected between two of the component sources of the equivalent source as shown in Fig. 2-5.6(b) without producing any change in any of the network variables. Note especially that no current flows between nodes *a* and *c* or between nodes *b* and *d*. It is thus immaterial to the operation of the circuit whether these connections are present or absent. We shall find the use of equivalent voltage and current sources of the type described above to be of help in our analysis of circuits.

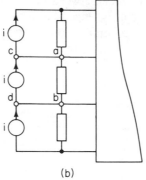

(a)   (b)

Figure 2-5.6 Replacing a single ideal current source with several ideal current sources.

As a final topic to be considered in this section, let us investigate what happens when a voltage source and a current source are connected in parallel as shown in Fig. 2-5.7. In such a configuration, since the internal resistance of the voltage source is zero, the output current from the current source flows through the voltage source without any effect on its output voltage. Thus the combination has exactly the same properties as a voltage source alone. Similarly, we may examine the behavior of a series-connected voltage source and current source as shown in Fig. 2-5.8. In this case, since the voltage across the current source is undefined, the addition of the output voltage of the voltage source still leaves a quantity which is undefined, and the combination has the same properties as a current source alone.

Figure 2-5.7 Voltage source and a current source in parallel.

Figure 2-5.8 Voltage source and a current source in series.

2-6 SOME APPLICATIONS OF RESISTANCE NETWORKS

In this section we present some simple applications of resistance networks which may be analyzed using the techniques presented in this chapter. The first example is the *voltage divider*. Such a network is shown in Fig. 2-6.1. The current $i(t)$ that flows from the source $v_1(t)$ is determined by the relation

$$i(t) = \frac{v_1(t)}{R_1 + R_2} \tag{2-27}$$

The voltage $v_2(t)$ (sometimes referred to as the *output* of the voltage divider) is determined by the expression

$$v_2(t) = i(t)R_2 \tag{2-28}$$

Figure 2-6.1   Voltage divider.

We may combine the two expressions given above to determine the relationship between the input voltage $v_1(t)$ and the output voltage $v_2(t)$, i.e., the voltage transfer function. We obtain

$$\frac{v_2(t)}{v_1(t)} = \frac{R_2}{R_1 + R_2} \qquad (2\text{-}29)$$

It should be noted that this expression is only valid if the current $i(t)$ flows through both of the resistors, i.e., *if no current flows at the output terminals at which $v_2(t)$ is measured.*

Another frequently encountered basic circuit is the *current divider*. Such a circuit is shown in Fig. 2-6.2. The voltage $v(t)$ that appears across the current source

Figure 2-6.2   Current divider.

$i_0(t)$ is determined by the relation

$$v(t) = \frac{i_0(t)}{G_1 + G_2} \qquad (2\text{-}30)$$

The current $i_1(t)$ is determined by the relation

$$i_1(t) = G_1 v(t) \qquad (2\text{-}31)$$

Combining the above two expressions, we obtain a relation for the ratio between the input current $i_0(t)$ and the current $i_1(t)$, namely, the current transfer function

$$\frac{i_1(t)}{i_0(t)} = \frac{G_1}{G_1 + G_2} \qquad (2\text{-}32)$$

Similarly, the expression for the ratio between the input current $i_0(t)$ and the current $i_2(t)$ is

$$\frac{i_2(t)}{i_0(t)} = \frac{G_2}{G_1 + G_2} \qquad (2\text{-}33)$$

As another example of the application of resistance networks to modern electrical engineering problems, consider the problem of converting a digital signal to an analog one.[3] This is a problem frequently encountered in microcomputer applications. Assume that the digital signal consists of 3 bits (binary numbers which can have a value of 1 or 0) and is specified as $b_2 b_1 b_0$. The equivalent analog (decimal) signal is given as

$$(b_2 \times 4) + (b_1 \times 2) + (b_0 \times 1) \tag{2-34}$$

Such an analog signal is readily generated (within a multiplicative constant) from digital input using a resistance network called an R-$2R$ network. The circuit is shown in Fig. 2-6.3(a). Each of the switches in this network is set to the upper position if the corresponding bit is "1," and to the lower position if it is "0." The output current i_{out} is then proportional to the decimal value of the binary number. To see how the circuit operates, assume that the binary number 100 ($b_2 = 1$, $b_1 = 0$, $b_0 = 0$) is to be converted. After the switches are positioned, the equivalent circuit is shown in Fig. 2-6.3(b). The portion of the network from node a to ground may be considered as a

Figure 2-6.3 R-$2R$ resistance network.

ladder network. Analyzing it we find that it may be replaced by a single equivalent resistor of value $2R$. The total resistance seen by the voltage source is thus $3R$, and the current out of the source is thus $V_0/3R$. The rightmost resistor and the equivalent resistor now operate as a current divider to produce $i_{out} = V_0/6R$. Similar analyses show that a digital input of 010 produces an output $i_{out} = V_0/12R$ ($\frac{1}{2}$ as great), and an input of 001 produces an output $i_{out} = V_0/24R$ ($\frac{1}{4}$ as great). Thus, as is required by (2-34), each higher-order bit produces an output current twice as great as the preceding one. The effect of setting more than one switch to the upper position is simply the superposition of the individual effects. Thus i_{out} will be directly proportional to the decimal value of the binary input data.

As a final example of a simple circuit, consider the network shown in Fig. 2-6.4(a), consisting of a voltage source, a linear resistor R_1, and a non-linear resistor R_2. Let R_2 have the non-linear characteristic defined in terms of its branch variables $v_2(t)$ and $i_2(t)$ by the curved locus shown in Fig. 2-6.4(b). If we apply KVL to this

[3]A circuit which performs such a function is called a D/A (Digital-to-Analog) converter.

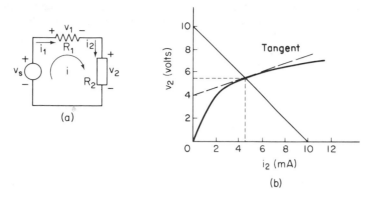

Figure 2-6.4 Analysis of a non-linear circuit.

circuit, we obtain

$$v_s = v_1 + v_2(i_2) \tag{2-35}$$

where v_s is the source voltage and v_1 and $v_2(i_2)$ are the branch voltages for the two resistors. Let us now define a loop current i as shown in Fig. 2-6.4(a). From an examination of the figure we see that $i = i_1 = i_2$, therefore we may now write (2-35) in the form

$$v_s - iR_1 = v_2(i) \tag{2-36}$$

The above equation is a non-linear one since v_2 is a non-linear function of the loop current i. We may, however, solve this equation graphically by drawing two plots, one of the left-hand member of the equation, as was done in Sec. 2-3, and the other of the right-hand member, putting both plots on the same graph. The intersection of the two loci indicates the values of v_2 and i which will satisfy the equality of (2-36). This has been done for the values $v_s = 10$ V and $R_1 = 1$ kΩ, in Fig. 2-6.4(b). The resulting solution is $v_2 = 5.5$ V, $i = 4.5$ mA. Such a solution is frequently referred to as the *operating point* for the non-linear characteristic of R_2. Note that if a specific operating point is desired, then a load line may be drawn through the operating point to find the values of the voltage source v_s and the resistor R_1 which will determine that operating point. Specifically, the intercept with the v_2 axis gives the value of v_s and the intercept with the i_2 axis gives the value v_s/R_1. This is referred to as a solution to the *biasing problem* of a non-linear resistive element. Since the straight line may be drawn with any slope, there are an infinite number of solutions to the problem unless either the value of the source voltage v_s or the value of R_1 is specified. Once an operating point has been found, then, for small variations in the values of the variables of the non-linear element around such an operating point, we can treat the non-linear device as a linear one. This may be justified mathematically by expressing the non-linearity in the form of a Taylor series around the operating point. The value of the small-signal resistance is equal to the slope of the tangent to the non-linear curve at the operating point. For example, in Fig. 2-6.4(b) the tangent has a value of 333 Ω. If, to the circuit in Fig. 2-6.4(a) we now apply a small additional voltage v_δ in series with the bias voltage v_s, we may treat the resulting current as having two components. One of these components we may call i_s; it is the bias current due to v_s and has a value of 4.5 mA. The other component is i_δ and it is due to the excitation voltage v_δ. The circuit is now as shown in Fig. 2-6.5(a). If we delete the bias components, then, assuming that $|v_\delta| \ll |v_s|$, we can construct a *small-signal equivalent circuit* as shown in Fig. 2-6.5(b) by representing R_2 by its small-signal resistance. For example, if $v_\delta = \pm 0.1$, then $i_\delta = \pm 0.1/(R_1 + 333) = \pm 0.1/1333 = \pm 0.075 \times 10^{-3}$ A (± 0.075 mA).

(a)   (b)

Figure 2-6.5 Non-linear circuit and its small-signal equivalent.

2-7 CONCLUSION

In this chapter, we have introduced some simple basic network elements, namely, resistors and sources. For a resistor, we have seen that the terminal variables are related by Ohm's law. Also in this chapter, we applied Kirchhoff's laws to various series and parallel combinations of resistors and sources to determine their properties. The results obtained in this chapter from our study of network elements will be greatly extended in the following chapter, where we will develop methods for analyzing networks containing an arbitrary number of resistors and sources.

PROBLEMS

Problem 2-1 (Sec. 2-1)
    A voltage $10 \sin 2\pi t$ V is applied across a 10-kΩ resistor.
    (a) Find an expression for the current through the resistor.
    (b) Find an expression for the power dissipated in the resistor.
    (c) Draw plots of the waveforms of the three quantities.

Problem 2-2 (Sec. 2-1)
    A voltage $10 \sin 2\pi t$ V is applied across a diode which has the characteristic shown in Fig. P2-2.

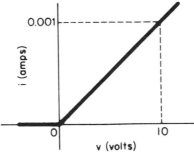

Figure P2-2

Resistors, Sources, and Simple Circuits   Chap. 2

(a) Draw a plot of the waveform of the current which flows through the diode.

(b) Draw a plot of the power dissipated in the diode.

Note: Although the voltage has positive and negative polarity the current only has one polarity. In this case, we say that *rectification* has taken place.

Problem 2-3 (Sec. 2-2)

Two sources and a single resistor are shown in Fig. P2-3. Determine the sign and the value of the power associated with each of the sources.

Figure P2-3   **Figure P2-4**

Problem 2-4 (Sec. 2-2)

The circuit shown in Fig. P2-4 consists of two sources and two resistors. Determine the polarity and the value of the power associated with each of the elements for the case where $v_1 = 10$, $v_2 = -5$ V.

Problem 2-5 (Sec. 2-2)

Repeat Problem 2-4 for the case where v_1 and v_2 have the waveshapes shown in Fig. P2-5.

Figure P2-5

Problem 2-6 (Sec. 2-2)

For a transistor in a common-base configuration as shown in Fig. P2-6(a), an ICIS can be used as a model as shown in Fig. P2-6(b). Find the relation between the parameter α defined in this figure and the parameter β defined in Fig. 2-2.7(b).

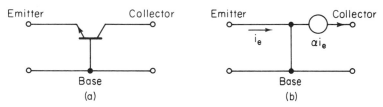

Figure P2-6

Problem 2-7 (Sec. 2-3)

A dry-cell battery has an open-circuit voltage of 1.5 V. When a short circuit is applied across the terminals of the battery, a current of 5 A flows through the short.

(a) Derive two circuit models for the battery.

(b) Which of the models is the "better" one from the viewpoint of matching the internal dissipation in the battery?

Problem 2-8 (Sec. 2-3)

The ac power available at the terminals of a standard household wall outlet may be modeled by an ideal sinusoidal voltage source in series with a very small resistance. Such a model is shown in Fig. P2-8. Determine the expression for the current that would flow if a screwdriver with a resistance of 0.01 Ω is placed across the terminals (from the magnitude of this quantity you should be able to understand why circuit breakers and fuses are extensively used in household wiring circuits). DO NOT try the actual experiment.

Problem 2-9 (Sec. 2-3)

Find the locus of operating points on a v_0-i_0 plane of the terminal characteristics of the network shown in Fig. P2-9 in which a current source in parallel with a resistor is connected to some arbitrary network.

| Figure P2-8 | Figure P2-9 |

Problem 2-10 (Sec. 2-4)

Find an equivalent single resistance which may be used to replace the resistance network shown in Fig. P2-10 at the pair of terminals labeled *a-b*.

Figure P2-10

Problem 2-11 (Sec. 2-4)

Find the input resistance at the terminals *a-b* of the network of resistors shown in Fig. P2-11.

| Figure P2-11 | Figure P2-12 |

Problem 2-12 (Sec. 2-4)

Find the input resistance at the terminals *a-b* of the network containing two time-varying resistors shown in Fig. P2-12.

Problem 2-13 (Sec. 2-4)

For the network shown in Fig. P2-13, find the value of the resistor R_0 such that the input resistance at the pair of terminals labeled *a-b* is 2 Ω.

Figure P2-13

Problem 2-14 (Sec. 2-4)

(a) Find the voltage transfer function v_2/v_1 for the network shown in Fig. P2-11. Define v_1 at the terminals *a-b* (with the + reference at terminal *a*), and v_2 as the voltage across the 5-Ω resistor (with the + reference at the left end of the resistor).

(b) Repeat part (a) for the case where v_2 is similarly defined across the 3-Ω resistor.

Problem 2-15 (Sec. 2-4)

(a) Find the voltage transfer function v_2/v_1 for the network shown in Fig. P2-15(a). Assume that all resistors have unit value.

(b) Can the voltage transfer function for the network of Fig. P2-15(a) be found by first determining the voltage transfer function for the two (1-Ω) resistor network shown in Fig. P2-15(b), and then treating the original network as a cascade of four such voltage dividers? Why or why not?

(a)   (b)

Figure P2-15

Problem 2-16 (Sec. 2-5)

Replace the network of sources shown in each part of Fig. P2-16 with a single equivalent source.

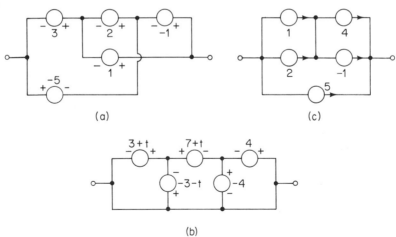

(a)   (c)

(b)

Figure P2-16

Problem 2-17 (Sec. 2-5)

Will the networks shown in Fig. P2-17 produce the same current in the resistance R_0? If not, modify the network shown in part (b) of the figure by making any necessary changes, so that the currents will be the same. Do *not* attempt to solve the network.

(a)　　　　　(b)

Figure P2-17

Problem 2-18 (Sec. 2-5)

Repeat Problem 2-17 for the networks shown in Fig. P2-18.

(a)　　　　　(b)

Figure P2-18

Problem 2-19 (Sec. 2-5)

The network shown in Fig. P2-19(a) may be reduced to the form shown in Fig. P2-19(b) by appropriate source transformations. Find the value of R_e and v_e.

(a)　　　　　(b)　　　　　**Figure P2-19**

Problem 2-20 (Sec. 2-5)

The network shown in Fig. P2-20(a) may be reduced to the form shown in Fig. P2-20(b) by appropriate source transformations. Find the values of G_e and i_e.

(a)　　　　　(b)　　　　　**Figure P2-20**

Problem 2-21 (Sec. 2-6)

For the network shown in Fig. P2-21, find the voltage transfer functions v_1/v_0 and v_2/v_0.

Figure P2-21

Problem 2-22 (Sec. 2-6)

For the network shown in Fig. P2-22, find the current transfer functions i_1/i_0, i_2/i_0, and i_3/i_0.

Figure P2-22

Problem 2-23 (Sec. 2-6)

A current generator of value $i_s(t) = 3 \cos 2t$ A is connected to a resistance network as shown in Fig. P2-23. Find the expression for the current $i_0(t)$.

Figure P2-23

Problem 2-24 (Sec. 2-6)

An ideal voltage source with an output $v_s(t) = 7 + 3e^{-t}$ V is connected to a resistance network as shown in Fig. P2-24. Find an expression for the voltage $v_0(t)$.

Figure P2-24

Problem 2-25 (Sec. 2-6)

A tunnel diode characteristic is shown in Fig. P2-25(a). The diode is connected in series with a source and resistor as shown in Fig. P2-25(b).

(a) Determine the three possible operating points if $v_s = 0.5$ V and $R_1 = 625 \, \Omega$.

(b) Compute the small-signal resistance of the tunnel diode at each of these operating points.

(c) For each operating point, draw a small-signal equivalent circuit and compute the effects of a small-signal voltage $v_\delta = \pm 0.01$ V placed in series with v_s.

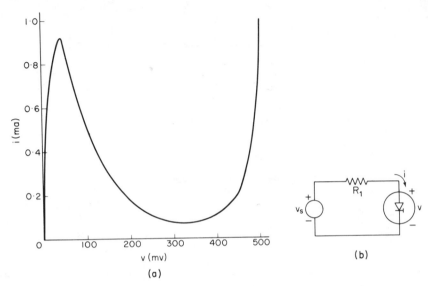

Figure P2-25

Problem 2-26 (Sec. 2-6)

Draw and analyze an *R-2R* D/A converter for a 4-bit digital input.

Problem 2-27 (Sec. 2-6)

(a) Another form of a D/A converter for a 3-bit digital input is shown in Fig. P2-27. Analyze the circuit and show how it operates.

(b) Extend the circuit shown in Fig. P2-27 so that it may be used to convert 4-bit digital input.

Figure P2-27

Resistors, Sources, and Simple Circuits   Chap. 2

3

Resistance Networks

In the preceding chapter, we defined the resistor, and showed how circuits consisting of a single loop (or a single node-pair) could be formulated by connecting such an element to a voltage or a current source. In this chapter, we shall extend our ideas to the case of a general resistance network.

3-1 MESH EQUATIONS FOR A TWO-MESH CIRCUIT

In Chap. 1, in connection with our discussion of KVL, we defined a loop as a connected set of branches which formed a closed path and which had the property that each node in the path was incident to exactly two branches of the set. In this chapter, we shall restrict our attention to a particular type of loop, which we will call a *mesh*. A mesh is defined as a collection of branches which forms a closed path and which has the properties: (1) every node in the path is incident to exactly two branches; and (2) no other branches are enclosed by the collection. For example, in Fig. 3-1.1(a) the closed path indicated by the heavy line forms a loop, while in Fig. 3-1.1(b), the closed paths indicated by the heavy lines form meshes. Since a mesh is simply a loop which satisfies the additional condition given in (2) of the above definition, it should be clear that KVL is satisfied for any mesh.[1]

[1] One of the reasons for the use of meshes to determine the unknown current variables is that the resulting equations describing the network will always be independent. If, on the other hand, an arbitrary set of loops is chosen (without taking some special precautions), the resulting equations may not be independent and thus they will not have a solution.

(a)　　　　　　　　　　　　(b)

Figure 3-1.1 Examples of a loop and meshes.

Now let us consider a resistance network containing two meshes, with a voltage source in each mesh as shown in Fig. 3-1.2(a). For each mesh we may define a mesh current as shown in Fig. 3-1.2(b). Note that the positive reference direction for both

(a)　　　　　　　　　　　　(b)

Figure 3-1.2 Two-mesh resistance network.

meshes is chosen the same, namely, clockwise. Also note that for the resistors R_1 and R_2, the mesh currents $i_1(t)$ and $i_2(t)$ are exactly the same as the defined branch currents $i_{R1}(t)$ and $i_{R2}(t)$. For the resistor R_3, however, the branch current $i_{R3}(t)$ defined in the figure is actually equal to the difference of the mesh currents, i.e., it is equal to $i_1(t) - i_2(t)$. We may now apply KVL around each mesh. In terms of the branch voltages and the source voltages we obtain

$$v_{S1}(t) = v_{R1}(t) + v_{R3}(t) \quad \text{and} \quad -v_{S2}(t) = v_{R2}(t) - v_{R3}(t) \tag{3-1}$$

Note that in both the equations given above we have chosen to put the source voltages on one side of the equation and the branch voltages on the other. In addition, we have set up the equation so that in traveling around the mesh in the direction indicated by the mesh reference current, a source whose reference polarity represents a voltage rise is treated as positive. This is a practice we shall find convenient to follow in all of our future discussions. The branch relations for the three resistors are

$$v_{R1}(t) = R_1 i_{R1}(t), \quad v_{R2}(t) = R_2 i_{R2}(t), \quad \text{and} \quad v_{R3}(t) = R_3 i_{R3}(t) \tag{3-2}$$

Substituting these branch relations in Eq. (3-1), we obtain

$$v_{S1}(t) = R_1 i_{R1}(t) + R_3 i_{R3}(t) \tag{3-3a}$$

$$-v_{S2}(t) = R_2 i_{R2}(t) - R_3 i_{R3}(t) \tag{3-3b}$$

Finally, we may put Eqs. (3-3a) and (3-3b) into a still more useful form by substituting the relations between the branch currents and the mesh currents, namely,

$$i_{R1}(t) = i_1(t), \quad i_{R2}(t) = i_2(t) \quad \text{and} \quad i_{R3}(t) = i_1(t) - i_2(t) \tag{3-4}$$

　　　　　　　　　　　　　　　　　　　　　　　Resistance Networks　　Chap. 3

Substituting the above relations in Eqs. (3-3a) and (3-3b) and rearranging terms, we obtain

$$v_{S1}(t) = (R_1 + R_3)i_1(t) - R_3i_2(t) \tag{3-5a}$$

$$-v_{S2}(t) = -R_3i_1(t) + (R_2 + R_3)i_2(t) \tag{3-5b}$$

The relations given in Eqs. (3-5a) and (3-5b) are usually referred to as a set of *simultaneous equations*. We may now make some observations about these relations. In the first equation (which is the one expressing KVL for mesh 1), note the following points:

1. The reference polarity of the voltage source is such as to produce a reference voltage *rise* when traveling around the mesh in the direction of the mesh reference current $i_1(t)$. Thus, this source voltage appears as a positive quantity in the left member of the equation.

2. The term $(R_1 + R_3)$ multiplying the quantity $i_1(t)$ is simply the sum of the resistances which are present in mesh 1.

3. The term $-R_3$ which multiplies the quantity $i_2(t)$ is simply the resistance which is common to both mesh 1 and mesh 2 (multiplied by a negative sign). Note that the presence of the negative sign in this term is due to the fact that the mesh 2 reference current $i_2(t)$ passes through the branch containing R_3 in the *opposite* direction to that of the mesh 1 reference current $i_1(t)$, i.e., that there is a negative sign in the third equation of (3-4). This, of course, is the result of choosing the same positive reference direction for both meshes.

A similar set of observations may be made for the KVL equation for mesh 2 as given by (3-5b). We note that:

1. The left member, i.e., the term $-v_{S2}(t)$, represents the effect of a voltage source with its reference polarity so oriented that it produces a reference voltage *drop* when traveling around the mesh in the direction of the mesh reference current $i_2(t)$.

2. The term $(R_2 + R_3)$ multiplying the quantity $i_2(t)$ is simply the sum of the resistances which are present in the mesh.

3. The term $-R_3$ which multiplies the quantity $i_1(t)$ is simply the resistance which is common to both mesh 1 and mesh 2 (multiplied by a negative sign). As before, the negative sign in this term is due to the fact that the mesh reference currents $i_1(t)$ and $i_2(t)$ pass through the branch in opposite directions.

Now let us write Eqs. (3-5a) and (3-5b) in matrix format[2]. We obtain

$$\begin{bmatrix} v_{S1}(t) \\ -v_{S2}(t) \end{bmatrix} = \begin{bmatrix} R_1 + R_3 & -R_3 \\ -R_3 & R_2 + R_3 \end{bmatrix} \begin{bmatrix} i_1(t) \\ i_2(t) \end{bmatrix} \tag{3-6}$$

This is a specific example of a general matrix expression for the equations describing a two-mesh resistance network. Such an expression will always have the form

[2]An introductory treatment of matrices is given in Appendix A.

$$\mathbf{v}(t) = \begin{bmatrix} v_1(t) \\ v_2(t) \end{bmatrix} = \begin{bmatrix} r_{11} & r_{12} \\ r_{21} & r_{22} \end{bmatrix} \begin{bmatrix} i_1(t) \\ i_2(t) \end{bmatrix} = \mathbf{R}i(t) \tag{3-7}$$

where $v_1(t)$ and $v_2(t)$ may be considered as the elements of a column matrix $\mathbf{v}(t)$, $i_1(t)$ and $i_2(t)$ may be considered as the elements of a column matrix $\mathbf{i}(t)$, and the quantities r_{ij} may be considered as elements of a square matrix $\mathbf{R}$, with the subscript i being used to represent the row of $\mathbf{R}$ in which the element occurs and the subscript j being used to represent the column of $\mathbf{R}$ in which the element occurs. We may directly compare Eq. (3-6) with Eq. (3-7) by noting that the element $v_1(t)$ represents the source in mesh 1, the element $v_2(t)$ represents the source (the orientation of the reference polarity must be taken into consideration) in mesh 2, the element r_{11} represents the sum of the values of the resistors in mesh 1, the element r_{22} represents the sum of the values of the resistors in mesh 2, and the elements r_{12} and r_{21} (which are equal) represent the value of the resistor which is common to meshes 1 and 2 (multiplied by a negative sign).

In the above paragraphs we have seen that for a two-mesh network, we may develop a matrix equation giving the relations between the voltage sources (or excitations) that are found in each of the meshes (these may be referred to as the "known" variables), and the mesh currents (the "unknown" variables). We would now like to solve these equations. By the word "solve" we mean that we desire to find the values of the mesh currents for some specified values of the voltage sources and the resistors. Such a solution is readily shown to be[3]

$$\mathbf{i}(t) = \begin{bmatrix} i_1(t) \\ i_2(t) \end{bmatrix} = \frac{1}{r_{11}r_{22} - r_{12}r_{21}} \begin{bmatrix} r_{22} & -r_{12} \\ -r_{21} & r_{11} \end{bmatrix} \begin{bmatrix} v_1(t) \\ v_2(t) \end{bmatrix} = \mathbf{R}^{-1}\mathbf{v}(t) \tag{3-8}$$

where $\mathbf{R}^{-1}$ is the inverse of the $\mathbf{R}$ matrix. Note that for this second-order case the inverse is easily found by interchanging the "11" and "22" elements of the original matrix $\mathbf{R}$, multiplying the "12" and "21" elements by a minus sign, and dividing all the elements by the quantity $(r_{11}r_{22} - r_{12}r_{21})$, the *determinant* of the array of elements specified by the matrix $\mathbf{R}$. It is a theorem of linear algebra that the necessary and sufficient condition for a solution to exist, i.e., for the inverse of a matrix $\mathbf{R}$ to exist, is that the determinant of $\mathbf{R}$ be non-zero. Thus the relations of Eq. (3-8) provide a general solution for the two-mesh resistance network. An example follows.

Example 3-1.1

A two-mesh resistance network is shown in Fig. 3-1.3. The voltage source in mesh 1 represents a voltage rise of 8 V, and the voltage source in mesh 2 represents a voltage drop of 9 V. The total resistance in mesh 1 is 3 Ω. The total resistance in mesh 2 is

ohms

Figure 3-1.3 Two-mesh resistance network.

[3]General methods for the solution of a set of simultaneous equations are given in Sec. 3-3.

Resistance Networks   Chap. 3

5 Ω. There is a 2-Ω resistor common to both mesh 1 and mesh 2. From the discussion given above, we see that we may write the KVL equations for this resistance network as

$$\begin{bmatrix} 8 \\ -9 \end{bmatrix} = \begin{bmatrix} 3 & -2 \\ -2 & 5 \end{bmatrix} \begin{bmatrix} i_1 \\ i_2 \end{bmatrix}$$

The solution is found by computing the inverse matrix. Following the relations given in (3-18), we obtain

$$\begin{bmatrix} i_1 \\ i_2 \end{bmatrix} = \frac{1}{11} \begin{bmatrix} 5 & 2 \\ 2 & 3 \end{bmatrix} \begin{bmatrix} 8 \\ -9 \end{bmatrix} = \begin{bmatrix} 2 \\ -1 \end{bmatrix}$$

Thus, the current in mesh 1 is 2 A flowing in the indicated reference direction, and the current in mesh 2 is −1 A, i.e., 1 A flowing in a direction opposite to the reference direction.

The method of analysis described above for the two-mesh resistance network is easily extended to cases where there are additional voltage sources in the meshes. The process is best illustrated by an example.

Example 3-1.2

A two-mesh resistance network is shown in Fig. 3-1.4. The net effect of the two voltage sources that are encountered in mesh 1 is such as to produce a voltage rise of 12 V. The net effect of the two voltage sources that are encountered in mesh 2 is such as to

ohms

Figure 3-1.4 Two-mesh resistance network with three sources.

produce a voltage drop of 2 V. Evaluating the values of the total resistance in both meshes and the resistance common to the two meshes, we may write the KVL equations for the network as

$$\begin{bmatrix} 12 \\ -2 \end{bmatrix} = \begin{bmatrix} 5 & -3 \\ -3 & 7 \end{bmatrix} \begin{bmatrix} i_1 \\ i_2 \end{bmatrix}$$

The inverse set of relations is easily found to be

$$\begin{bmatrix} i_1 \\ i_2 \end{bmatrix} = \frac{1}{26} \begin{bmatrix} 7 & 3 \\ 3 & 5 \end{bmatrix} \begin{bmatrix} 12 \\ -2 \end{bmatrix} = \begin{bmatrix} 3 \\ 1 \end{bmatrix}$$

Thus the current in mesh 1 is 3 A flowing in the indicated reference direction, and the current in mesh 2 is 1 A flowing in the indicated reference direction.

A resistance network such as the two-mesh ones described in the preceding examples is an illustration of a *linear* network. Such a network must have the property that superposition (as defined in Sec. 1-3) applies for any combination of inputs and for any variable considered as an output. For example, if we use the inverse matrix found in solving for the mesh currents of the network in Example 3-1.1, and leave the values of the two voltage sources unspecified, we may write

$$\begin{bmatrix} i_1(t) \\ i_2(t) \end{bmatrix} = \frac{1}{11} \begin{bmatrix} 5 & 2 \\ 2 & 3 \end{bmatrix} \begin{bmatrix} v_1(t) \\ v_2(t) \end{bmatrix} \tag{3-9}$$

If we now apply an excitation to the network only in mesh 1, by letting $v_1 = 8$ V, and setting v_2 to zero, then from Eq. (3-9) we see that $i_1 = \frac{40}{11}$ A and $i_2 = \frac{16}{11}$ A. Now let us change our inputs and apply an excitation to the network only in mesh 2, setting $v_2 = -9$ V, and setting v_1 to zero. In this case, from Eq. (3-9), we find that $i_1 = -\frac{18}{11}$ A and $i_2 = -\frac{27}{11}$ A. The currents for the case where excitations are applied in both of the meshes simultaneously, i.e., where $v_1 = 8$ and $v_2 = -9$ V, are simply found by the superposition of the results of the above two cases, namely $i_1 = (40 - 18)/11 = 2$, and $i_2 = (16 - 27)/11 = -1$ A. These are the results which were obtained in the example.

We could also have made use of the technique of superposition directly in the physical network, without using the inverse matrix relationships given in Eq. (3-9). As an example of this, in Fig. 3-1.5(a), we have redrawn the network used in Example 3-1.1 for the case where $v_1 = 8$ V and $v_2 = 0$. Since the characteristic of an ideal voltage source with an output voltage of zero is exactly that of a short circuit, we have replaced the source in mesh 2 by a short circuit. Such a replacement is always implied when we state that a voltage source is set to zero. The resistance seen by the 8 V source is $\frac{11}{5}$ Ω, thus the current from this source is $\frac{40}{11}$ A (which is i_1). When this current reaches the junction of the 2- and 3-Ω resistors, these resistors will act as a current divider. The current through the 3-Ω resistor will thus be $(\frac{2}{5})(\frac{40}{11}) = \frac{16}{11}$ A (which is i_2). These results, of course, agree with the results found from the inverse matrix relation given in Eq. (3-9). A similar analysis may be made to find the contributions to the currents i_1 and i_2 when the source $v_2 = -9$ V and $v_1 = 0$. The physical situation is shown in Fig. 3-1.5(b). The addition of the results from the two separate cases yields the same results as were given in Example 3-1.1.

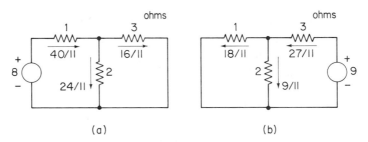

Figure 3-1.5 Application of superposition.

For a system with multiple inputs such as is discussed above, there is a right way and a wrong way to apply the tests for superposition. As an example let us consider the two-mesh resistance network shown in Fig. 3-1.6. The mesh equations for this network have the form

$$\begin{bmatrix} v_1(t) \\ v_2(t) \end{bmatrix} = \begin{bmatrix} 2 & -1 \\ -1 & 1 \end{bmatrix} \begin{bmatrix} i_1(t) \\ i_2(t) \end{bmatrix} \tag{3-10}$$

Figure 3-1.6 Two-mesh resistance network.

The inverse set of equations is easily found to be

$$\begin{bmatrix} i_1(t) \\ i_2(t) \end{bmatrix} = \begin{bmatrix} 1 & 1 \\ 1 & 2 \end{bmatrix} \begin{bmatrix} v_1(t) \\ v_2(t) \end{bmatrix} \tag{3-11}$$

Now let us consider the case where $v_1 = 1$ V and $v_2 = 1$ V. We see that the currents i_1 and i_2 have the values 2 A and 3 A respectively. Suppose that we now double the value of the voltage v_1 (while keeping the value of v_2 the same). Inserting the values $v_1 = 2$ V and $v_2 = 1$ V in Eq. (3-11), we obtain the results that $i_1 = 3$ A and $i_2 = 4$ A. Note that doubling the voltage v_1 did not double the currents i_1 and i_2. This at first seems to violate the concept of superposition which we have used to define the property of linearity. Now, however, let us make a different test. Let us set $v_1 = 1$ V, while setting v_2 to zero. In this case we find from Eq. (3-11) that i_1 and i_2 are both 1 A. If we double the value of v_1 (while keeping v_2 at zero) we see that i_1 and i_2 both have the value of 2 A. In this case, doubling the input to the network has indeed doubled the outputs, i.e., the currents i_1 and i_2. It is easily shown that the same result holds true for the other source, namely, if v_2 is multiplied by some factor while v_1 is held equal to zero, then the currents i_1 and i_2 will also be multiplied by the same factor.

Considering the above results, we see that a linear resistance network may be characterized as one comprised of linear resistors and sources. It has the following properties:

1. The value of any variable may be found as the sum of the values of that variable produced by each of the excitation sources acting separately.
2. If *all the sources except one are set to zero*, then, multiplying the value of the source by a constant multiplies the values of all the network variables by the same constant.

This definition applies to the case of resistance networks excited by either voltage or current sources. The latter case will be discussed in a later section.

3-2 MESH EQUATIONS FOR THE GENERAL CIRCUIT

In this section we shall extend the results obtained for the two-mesh resistance circuit to the general case. First let us consider a network containing three meshes as shown in Fig. 3-2.1. As before, we have assigned the same (clockwise) positive reference directions for all the mesh currents. To determine the network equations for this circuit, we may apply the same procedure as was used for the two-mesh example, namely:

Figure 3-2.1 Three-mesh resistance network.

1. Write the KVL equations for each loop in terms of the branch voltage variables (the voltages across the resistive branches) and the voltage sources.

2. Substitute the branch relationships (Ohm's law for each branch) so that the KVL equations are expressed in terms of the branch currents.

3. Substitute the relationships defining the branch currents in terms of the mesh currents $i_1(t)$, $i_2(t)$, and $i_3(t)$ defined in the figure.

The details are left as an exercise for the reader. The result is the following set of equations:

$$\begin{bmatrix} v_a(t) - v_c(t) \\ -v_b(t) \\ v_c(t) \end{bmatrix} = \begin{bmatrix} R_2 + R_4 & -R_4 & -R_2 \\ -R_4 & R_3 + R_4 & -R_3 \\ -R_2 & -R_3 & R_1 + R_2 + R_3 \end{bmatrix} \begin{bmatrix} i_1(t) \\ i_2(t) \\ i_3(t) \end{bmatrix} \quad (3\text{-}12)$$

The set of equations given above is an example of a set of matrix equations of the form $\mathbf{v}(t) = \mathbf{R}i(t)$, which may be written as

$$\mathbf{v}(t) = \begin{bmatrix} v_1(t) \\ v_2(t) \\ v_3(t) \end{bmatrix} = \begin{bmatrix} r_{11} & r_{12} & r_{13} \\ r_{21} & r_{22} & r_{23} \\ r_{31} & r_{32} & r_{33} \end{bmatrix} \begin{bmatrix} i_1(t) \\ i_2(t) \\ i_3(t) \end{bmatrix} = \mathbf{R}i(t) \quad (3\text{-}13)$$

If we compare the elements of (3-13) with those of Eq. (3-12), we may make a set of conclusions similar to those made for the two-mesh network, namely, we see that the elements $v_i(t)$ of the matrix $\mathbf{v}(t)$ each represent the sum (with appropriate note taken of the signs) of the voltage sources that appear in the ith mesh of the network, the elements r_{ii} on the main diagonal of the matrix $\mathbf{R}$ represent the total resistance of the resistors in the ith mesh, and the elements r_{ij} $(i \neq j)$ represent the negative of the values of the resistors common to both the ith and the jth meshes. An example of determining the elements of such an equation follows.

Example 3-2.1

A three-mesh resistance network is shown in Fig. 3-2.2. If we apply the rules specified above for determining the algebraic summation of the voltage sources in each mesh, we see that $v_1(t)$ of (3-13) is 7 V, $v_2(t) = -9 - 3 = -12$ V, and $v_3(t) = 9$ V. The total resistance in the first mesh is 3 Ω, and this mesh has a 2-Ω resistor which is also in mesh 2, and a 1-Ω resistor which is also in mesh 3. Continuing in this manner, we obtain a set of simultaneous equations for the network which may be written in matrix form as

Figure 3-2.2 Three-mesh resistance network.

$$\begin{bmatrix} 7 \\ -12 \\ 9 \end{bmatrix} = \begin{bmatrix} 3 & -2 & -1 \\ -2 & 5 & -3 \\ -1 & -3 & 8 \end{bmatrix} \begin{bmatrix} i_1 \\ i_2 \\ i_3 \end{bmatrix}$$

We shall defer until the following section a discussion of the methods by means of which such a set of equations may be solved, i.e., the techniques by means of which the values of the currents i_1, i_2, and i_3 may be found.

The development made above may be easily extended to the case where there are an arbitrary number of meshes in the network. Consider a network containing n meshes and an arbitrary number of sources. In addition, let all the mesh currents have the same reference direction (we will assume that this is clockwise). An extension of the development made above shows that we may write the network equations for this network in the form

$$\mathbf{v}(t) = \begin{bmatrix} v_1(t) \\ v_2(t) \\ \cdot \\ \cdot \\ \cdot \\ v_n(t) \end{bmatrix} = \begin{bmatrix} r_{11} & r_{12} & \cdots & r_{1n} \\ r_{21} & r_{22} & \cdots & r_{2n} \\ \cdot & \cdot & & \cdot \\ \cdot & \cdot & & \cdot \\ \cdot & \cdot & & \cdot \\ r_{n1} & r_{n2} & \cdots & r_{nn} \end{bmatrix} \begin{bmatrix} i_1(t) \\ i_2(t) \\ \cdot \\ \cdot \\ \cdot \\ i_n(t) \end{bmatrix} = \mathbf{R}i(t) \qquad (3\text{-}14)$$

The procedure for determining the elements of the matrices of Eq. (3-14) may be summarized as follows:

SUMMARY 3-2.1

Writing the Mesh Equations for a Resistance Network. A set of mesh equations having the form given in Eq. (3-14) may be written for a planar[4] resistance network by defining a set of mesh currents all of which have the same reference direction, and determining the elements of the component matrices as follows:

$v_i(t)$ the summation of all the voltage sources in the ith mesh, with sources whose reference polarity constitutes a voltage rise (in the positive reference

[4]A planar network is one with a planar graph. The planar graph is defined in Sec. 1-4.

direction of the ith mesh current) treated as positive, and with sources whose reference polarity constitutes a voltage drop treated as negative

$r_{ij}(i = j)$ the summation of the values of resistors that occur in the ith mesh

$r_{ij}(i \neq j)$ the negative of the sum of the values of any resistors which are common to the ith and the jth mesh[5]

$i_i(t)$ the unknown mesh currents which are the unknown variables for the network   ■

It is left to the reader as an exercise to determine that these rules lead to a set of equations which are exactly the same as those which will result if KVL is applied around each mesh of the network in terms of the branch voltage variables, and if the relations for the mesh currents are substituted into these relations. These rules may be applied to determine the mesh equations for any planar network.

Example 3-2.2

As an example of the application of the rules given above to directly determine the matrix relations for a multi-mesh resistance network, consider the circuit shown in Fig. 3-2.3. Assume that all resistors have unity value. The matrix relations are easily determined to be

$$
\begin{bmatrix} 1 \\ 1 \\ -1 \\ 1 \\ 0 \\ -1 \\ 1 \\ 1 \\ -1 \end{bmatrix} =
\begin{bmatrix}
3 & -1 & 0 & -1 & 0 & 0 & 0 & 0 & 0 \\
-1 & 3 & -1 & 0 & -1 & 0 & 0 & 0 & 0 \\
0 & -1 & 3 & 0 & 0 & -1 & 0 & 0 & 0 \\
-1 & 0 & 0 & 3 & -1 & 0 & -1 & 0 & 0 \\
0 & -1 & 0 & -1 & 4 & -1 & 0 & -1 & 0 \\
0 & 0 & -1 & 0 & -1 & 3 & 0 & 0 & -1 \\
0 & 0 & 0 & -1 & 0 & 0 & 2 & -1 & 0 \\
0 & 0 & 0 & 0 & -1 & 0 & -1 & 3 & -1 \\
0 & 0 & 0 & 0 & 0 & -1 & 0 & -1 & 2
\end{bmatrix}
\begin{bmatrix} i_1 \\ i_2 \\ i_3 \\ i_4 \\ i_5 \\ i_6 \\ i_7 \\ i_8 \\ i_9 \end{bmatrix}
$$

Note that for a general network such as the one analyzed in the above example, the mesh currents as defined by the unknown variables $i_i(t)$ may actually not be directly, physically measurable in the circuit. For example, the mesh current $i_5(t)$ shown in Fig. 3-2.3 always appears in connection with some other mesh current. Therefore, there is no way in which a single current-measuring instrument such as an ammeter may be used to physically observe the value of this mesh current. This will always be the case when a mesh is completely surrounded by other meshes.

Mesh equations are only one method of applying KVL in such a way as to generate an independent set of equations for a given network. The topological concepts introduced in Sec. 1-4 are easily applied to develop other independent sets of equations. The first step is to draw the network graph and select a tree. The next step is to define a set of loops, each containing one link and with the rest of the loop

---

[5]The quantity $r_{ij}(i = j)$ is sometimes called the *self-resistance* of the ith mesh, while the quantity $r_{ij}(i \neq j)$ is called the *mutual resistance* between the ith and jth mesh.

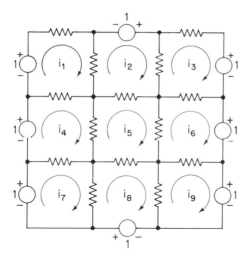

Figure 3-2.3 Nine-mesh resistance network.

branches taken from the tree. The KVL equations for these loops will always be independent since each contains a branch (a link) not in any other. Thus they may be solved for the loop currents. Just as is the case for mesh analysis, there will always be $b - n + 1$ equations, where b is the number of branches and n is the number of nodes. Note that, unlike the mesh methods, the loop method can be applied to non-planar networks as well as to planar ones. As an example of the loop method, consider the network shown in Fig. 3-2.1. Some independent sets of loops (dashed lines) and the trees that they are derived from (bold lines) are shown in Fig. 3-2.4.

(a)

(b)

(c)

Figure 3-2.4 Some independent sets of loops.

3-3 SOLVING THE MESH EQUATIONS

In the preceding section we showed how a resistance network comprised of voltage sources and resistors could be described by a set of simultaneous equations in which the mesh currents were the unknowns. Solving this set of equations, i.e., finding the values of the mesh currents for a given set of values of voltage sources, may be said to give us a *solution for the network*, since, if the mesh currents are known, we may easily find any individual branch current and/or voltage, and thus we may determine the value of any variable in the network.

If we consider a set of simultaneous equations written in matrix form as

$$\mathbf{v}(t) = \mathbf{R}\mathbf{i}(t) \tag{3-15}$$

where the various matrices are defined in (3-14), then the *determinant* of the set of equations has the symbolic representation

$$\det \mathbf{R} = \begin{vmatrix} r_{11} & r_{12} & \cdots & r_{1n} \\ r_{21} & r_{22} & \cdots & r_{2n} \\ \cdot & \cdot & \cdots & \cdot \\ \cdot & \cdot & \cdots & \cdot \\ \cdot & \cdot & \cdots & \cdot \\ r_{n1} & r_{n2} & \cdots & r_{nn} \end{vmatrix} \tag{3-16}$$

This is frequently referred to as a *determinant of order n*. Such a determinant has a definite (scalar) value which is determined by the elements r_{ij}. The rules for finding the value of a determinant are summarized in Appendix A.

A second scalar quantity which may be defined for a square matrix is the *cofactor*. For a matrix $\mathbf{R}$ with elements r_{ij} defined as in Eq. (3-14) it is customary to use the symbol R_{ij} to designate the cofactors. Specifically, the cofactor R_{ij} is the determinant formed by deleting the ith row and jth column from the array of elements, taking the determinant of the remaining array, and multiplying by $(-1)^{i+j}$. A more complete description of cofactors is given in Appendix A.

One of the most important uses of determinants and cofactors is in finding the solutions of sets of simultaneous equations such as arise when the mesh equations are written for a resistance network. An example of such a solution was given for a two-mesh network in Sec. 3-1. Let us now consider again a set of equations describing a three-mesh network. These will always have the form

$$\mathbf{v}(t) = \begin{bmatrix} v_1(t) \\ v_2(t) \\ v_3(t) \end{bmatrix} = \begin{bmatrix} r_{11} & r_{12} & r_{13} \\ r_{21} & r_{22} & r_{23} \\ r_{31} & r_{32} & r_{33} \end{bmatrix} \begin{bmatrix} i_1(t) \\ i_2(t) \\ i_3(t) \end{bmatrix} = \mathbf{R}i(t) \tag{3-17}$$

The solution to such a set of simultaneous equations may be expressed in the form

$$\begin{bmatrix} i_1(t) \\ i_2(t) \\ i_3(t) \end{bmatrix} = \frac{1}{\det \mathbf{R}} \begin{bmatrix} R_{11} & R_{21} & R_{31} \\ R_{12} & R_{22} & R_{32} \\ R_{13} & R_{23} & R_{33} \end{bmatrix} \begin{bmatrix} v_1(t) \\ v_2(t) \\ v_3(t) \end{bmatrix} \tag{3-18}$$

where the elements R_{ij} are cofactors. The square matrix of cofactors is sometimes referred to as the *adjoint* matrix. Note that the indices of the cofactors are reversed, i.e., the element in the ith row and jth column of the square adjoint matrix of Eq. (3-18) is the cofactor formed by deleting the jth row and the ith column from the original array of the elements of $\mathbf{R}$. As an example of the use of Eq. (3-18) to determine the unknown mesh currents, consider the case where it is desired to find only $i_1(t)$ in a three-mesh network specified by the relations of Eq. (3-17). From Eq. (3-18), we obtain

$$i_1(t) = \frac{v_1(t)\begin{vmatrix} r_{22} & r_{23} \\ r_{32} & r_{33} \end{vmatrix} - v_2(t)\begin{vmatrix} r_{12} & r_{13} \\ r_{32} & r_{33} \end{vmatrix} + v_3(t)\begin{vmatrix} r_{12} & r_{13} \\ r_{22} & r_{23} \end{vmatrix}}{\det \mathbf{R}} \tag{3-19}$$

A more convenient form for expressing the above relation is

$$i_1(t) = \frac{\begin{vmatrix} v_1(t) & r_{12} & r_{13} \\ v_2(t) & r_{22} & r_{23} \\ v_3(t) & r_{32} & r_{33} \end{vmatrix}}{\det \mathbf{R}} \tag{3-20}$$

Thus, we may say that the value of the mesh current $i_1(t)$ may be found by replacing the *first* column of the matrix $\mathbf{R}$ by the elements of the matrix $\mathbf{v}$, taking the determinant of the result, and dividing this by the determinant of the matrix $\mathbf{R}$. Similarly, to solve for $i_2(t)$, we replace the *second* column of $\mathbf{R}$ by $\mathbf{v}$ and take the determinant, and to solve for $i_3(t)$ we replace the *third* column of $\mathbf{R}$ by $\mathbf{v}$ and take the determinant. A simplified method for evaluating third-order determinants is given in Sec. A-9 (Appendix A). A necessary and sufficient condition for a solution to exist is that the determinant of $\mathbf{R}$ be non-zero. In this case, the matrix $\mathbf{R}$ is said to be non-singular. An example follows.

Example 3-3.1

We may apply the procedure described above to obtain a solution for the mesh currents of the three-mesh resistance network analyzed in Example 3-2.1. The $\mathbf{R}$ matrix is

$$\mathbf{R} = \begin{bmatrix} 3 & -2 & -1 \\ -2 & 5 & -3 \\ -1 & -3 & 8 \end{bmatrix}$$

The value of the determinant of the resistance matrix for this network is computed in Appendix A and has a value of 44. Inserting the values of the elements of the matrix into Eq. (3-19), we obtain

$$i_1(t) = \frac{v_1(t)\begin{vmatrix} 5 & -3 \\ -3 & 8 \end{vmatrix} - v_2(t)\begin{vmatrix} -2 & -1 \\ -3 & 8 \end{vmatrix} + v_3(t)\begin{vmatrix} -2 & -1 \\ 5 & -3 \end{vmatrix}}{44}$$

Evaluating the second-order determinants, we obtain the result

$$i_1(t) = (\tfrac{31}{44})v_1(t) + (\tfrac{19}{44})v_2(t) + (\tfrac{11}{44})v_3(t)$$

If we now insert the known values of the source voltages specified in Example 3-2.1, namely, $v_1 = 7$ V, $v_2 = -12$ V, and $v_3 = 9$ V, we obtain the result that $i_1 = 2$ A. Following a similar procedure, we see that

$$i_2 = \frac{-7\begin{vmatrix} -2 & -3 \\ -1 & 8 \end{vmatrix} - 12\begin{vmatrix} 3 & -1 \\ -1 & 8 \end{vmatrix} - 9\begin{vmatrix} 3 & -1 \\ -2 & -3 \end{vmatrix}}{44} = -1 \text{ A}$$

$$i_3 = \frac{7\begin{vmatrix} -2 & 5 \\ -1 & -3 \end{vmatrix} + 12\begin{vmatrix} 3 & -2 \\ -1 & -3 \end{vmatrix} + 9\begin{vmatrix} 3 & -2 \\ -2 & 5 \end{vmatrix}}{44} = 1 \text{ A}$$

Thus we have obtained the solution for the mesh currents of a three-mesh resistance network.

The above procedure is frequently referred to as *Cramer's rule*. We may easily generalize this to the case of solving an nth order array produced by writing the mesh equations for an n-mesh resistance network. In such a case, the equations will have the form given in Eq. (3-14). The solution will then have the form

$$\mathbf{i}(t) = \begin{bmatrix} i_1(t) \\ i_2(t) \\ i_3(t) \\ \cdot \\ \cdot \\ \cdot \\ i_n(t) \end{bmatrix} = \begin{bmatrix} a_{11} & a_{12} & a_{13} & \cdots & a_{1n} \\ a_{21} & a_{22} & a_{23} & \cdots & a_{2n} \\ a_{31} & a_{32} & a_{33} & \cdots & a_{3n} \\ \cdot & \cdot & \cdot & \cdots & \cdot \\ \cdot & \cdot & \cdot & \cdots & \cdot \\ \cdot & \cdot & \cdot & \cdots & \cdot \\ a_{n1} & a_{n2} & a_{n3} & \cdots & a_{nn} \end{bmatrix} \begin{bmatrix} v_1(t) \\ v_2(t) \\ v_3(t) \\ \cdot \\ \cdot \\ \cdot \\ v_n(t) \end{bmatrix} = \mathbf{A}\mathbf{v}(t) \qquad (3\text{-}21)$$

where $\mathbf{A} = \mathbf{R}^{-1}$, and where the elements of $\mathbf{A}$ have the dimensions of reciprocal resistance, i.e., conductance. From Cramer's rule, the elements r_{ij} of the matrix $\mathbf{R}$ of Eq. (3-14) are related to the elements a_{ij} of the matrix $\mathbf{A}$ of Eq. (3-21) (which is the inverse of $\mathbf{R}$) by the equation

$$a_{ij} = \frac{R_{ji}}{\det \mathbf{R}} \qquad (3\text{-}22)$$

where R_{ji} is the cofactor formed by taking the determinant of the array $\mathbf{R}$ after the jth row and ith column have been deleted, and multiplying its value by $(-1)^{j+i}$. If we compare Eq. (3-22) with the relations for the inverse of the third-order matrix given in Eq. (3-18) it is readily observed that these relations follow the general rule given above. It may be shown that the necessary and sufficient condition under which a solution for Eq. (3-14) exists, i.e., for which $\mathbf{R}$ has an inverse, is that $\det \mathbf{R} \neq 0$. Note that, in verification of this, if $\det \mathbf{R}$ should be equal to zero, the elements of A as defined by Eq. (3-22) are indeterminate.

The use of Cramer's rule to determine the elements of the adjoint matrix, and thus to provide a solution to a set of simultaneous equations, provides a general means for solving for the mesh currents of a resistance network. The major application of this rule is found in solving two- and three-mesh networks. To solve for the mesh currents of more complicated networks, techniques which are easily implemented on the digital computer are required. Such an implementation is given in Appendix C.

3-4 NODE EQUATIONS FOR A TWO-NODE CIRCUIT

In the early sections of this chapter we described methods by means of which a resistance network could be described using mesh currents as the unknown variables. In this section we shall introduce a second method of describing resistance networks, using node voltages as the unknown variables. To begin, let us consider a resistance network with two node voltages (both the node voltages are defined with respect to the same reference or ground node).[6] If we connect a current source from the reference node to each of the nodes defining the node voltages as shown in Fig. 3-4.1, and apply KCL at node 1 and node 2, we obtain the equations

$$i_{S1}(t) = i_{G1}(t) + i_{G3}(t) \qquad \text{and} \qquad i_{S2}(t) = i_{G2}(t) - i_{G3}(t) \qquad (3\text{-}23)$$

In the above equations we have elected to put the source current on one side of the equation and the branch currents in the resistive elements on the other. Now let us

---

[6]The reference node is also sometimes referred to as a *datum* node; thus, the node voltages are sometimes called node-to-datum voltages.

Figure 3-4.1 Two-node resistance network.

write the branch relations. These are most conveniently written using units of conductance (mhos). Thus, we obtain

$$i_{G1}(t) = G_1 v_{G1}(t), \quad i_{G2}(t) = G_2 v_{G2}(t) \quad \text{and} \quad i_{G3}(t) = G_3 v_{G3}(t) \tag{3-24}$$

Substituting these branch relations in the KCL equations of (3-23), we obtain

$$i_{s1}(t) = G_1 v_{G1}(t) + G_3 v_{G3}(t) \tag{3-25a}$$

$$i_{s2}(t) = G_2 v_{G2}(t) - G_3 v_{G3}(t) \tag{3-25b}$$

Finally, we may put the above equations in a still more useful form by substituting the relations between the node voltages $v_1(t)$ and $v_2(t)$ and the branch voltages. These are easily seen to be

$$v_{G1}(t) = v_1(t), \quad v_{G2}(t) = v_2(t), \quad \text{and} \quad v_{G3}(t) = v_1(t) - v_2(t) \tag{3-26}$$

Substituting the above relations in (3-25) and rearranging terms, we obtain

$$i_{s1}(t) = (G_1 + G_3)v_1(t) - G_3 v_2(t) \tag{3-27a}$$

$$i_{s2}(t) = -G_3 v_1(t) + (G_2 + G_3)v_2(t) \tag{3-27b}$$

These equations may be written as a matrix equation of the form

$$\begin{bmatrix} i_{s1}(t) \\ i_{s2}(t) \end{bmatrix} = \begin{bmatrix} G_1 + G_3 & -G_3 \\ -G_3 & G_2 + G_3 \end{bmatrix} \begin{bmatrix} v_1(t) \\ v_2(t) \end{bmatrix} \tag{3-28}$$

In general, for a two-node network, the matrix equation of (3-28) will have the form

$$\mathbf{i}(t) = \begin{bmatrix} i_1(t) \\ i_2(t) \end{bmatrix} = \begin{bmatrix} g_{11} & g_{12} \\ g_{21} & g_{22} \end{bmatrix} \begin{bmatrix} v_1(t) \\ v_2(t) \end{bmatrix} = \mathbf{Gv}(t) \tag{3-29}$$

For this case we may make the following observations about the elements of the matrices $\mathbf{i}(t)$, $\mathbf{G}$, and $\mathbf{v}(t)$:

1. The element $i_1(t)$ represents the sum of the effects of any current sources connected to node 1. The contribution of an individual source is positive if its reference current flows towards the node and negative if its reference current flows away from the node. The element $i_2(t)$ represents the sum of the effects of any current sources connected to node 2.

2. The element g_{11} of the matrix $\mathbf{G}$ represents the sum of all the conductances which are connected to the first node. The elements g_{12} and g_{21} (which are equal) represent the negative of the value (in mhos) of any resistor which is connected between nodes 1 and 2. The element g_{22} represents the sum of all the conductances which are connected to node 2.

3. The negative sign that appears in the term $-G_3$ is the result of having chosen the negative reference polarity of both the node voltages $v_1(t)$ and $v_2(t)$ at the same node.

The solution to the set of equations given in (3-29) is

$$\mathbf{v}(t) = \begin{bmatrix} v_1(t) \\ v_2(t) \end{bmatrix} = \frac{1}{g_{11}g_{22} - g_{12}g_{21}} \begin{bmatrix} g_{22} & -g_{12} \\ -g_{21} & g_{11} \end{bmatrix} \begin{bmatrix} i_1(t) \\ i_2(t) \end{bmatrix} = \mathbf{G}^{-1}\mathbf{i}(t) \qquad (3\text{-}30)$$

Example 3-4.1

A two-node resistance network is shown in Fig. 3-4.2. The current source connected between the reference node and the first node produces a current of 8 A flowing into the node. The current source connected between the reference node and the second

mhos

Figure 3-4.2 Two-node resistance network.

node represents a current of -9 A flowing into the second node. The total conductance connected to the first node is 3 mhos, and the total of the conductance connected to the second node is 5 mhos. There is a 2-mho conductance connected between the first and the second nodes. From the discussion given above, we see that we may write the KCL equations for this network as

$$\begin{bmatrix} 8 \\ -9 \end{bmatrix} = \begin{bmatrix} 3 & -2 \\ -2 & 5 \end{bmatrix} \begin{bmatrix} v_1 \\ v_2 \end{bmatrix}$$

The solution is given by computing the inverse matrix. From the relations given in (3-30), we obtain

$$\begin{bmatrix} v_1 \\ v_2 \end{bmatrix} = \frac{1}{11} \begin{bmatrix} 5 & 2 \\ 2 & 3 \end{bmatrix} \begin{bmatrix} 8 \\ -9 \end{bmatrix} = \begin{bmatrix} 2 \\ -1 \end{bmatrix}$$

Thus the voltage at the first node is 2 V, while the voltage at the second node is -1 V (both measured with respect to the reference node).

The numerical computations for the example given above exactly parallel the computations for the development given in Example 3-1.1 in which we obtained a solution for the mesh currents of the two-mesh resistance network shown in Fig. 3-1.3. When a pair of circuits, such as the ones shown in Figs. 3-1.3 and 3-4.2, is characterized by a set of equations in which the numerical coefficients have the same values (although the physical significance of the variables may be different), the circuits are said to be *dual* to each other. We will treat dual circuits in more detail in Sec. 6-8.

The method of analysis described above is easily extended to the case where there are additional current sources in the network. An example follows.

Example 3-4.2

A two-node resistance network is shown in Fig. 3-4.3. The net effect of the two current sources that are connected to node 1 is to provide a positive input current to the node of 12 A. Similarly, the effect of the two sources connected to node 2 produces an input

Figure 3-4.3 Two-node resistance network with three sources.

current to the node of −2 A. Evaluating the values of the total conductance connected to each node, and the conductance connected between the nodes, we may write the KCL equations for the two nodes in the form

$$\begin{bmatrix} 12 \\ -2 \end{bmatrix} = \begin{bmatrix} 5 & -3 \\ -3 & 7 \end{bmatrix} \begin{bmatrix} v_1 \\ v_2 \end{bmatrix}$$

The inverse set of relations is easily found to be

$$\begin{bmatrix} v_1 \\ v_2 \end{bmatrix} = \frac{1}{26} \begin{bmatrix} 7 & 3 \\ 3 & 5 \end{bmatrix} \begin{bmatrix} 12 \\ -2 \end{bmatrix} = \begin{bmatrix} 3 \\ 1 \end{bmatrix}$$

Thus, the voltage at the first node is 3 V, while that at the second node is 1 V.

The same conclusions that were made with respect to the linearity of networks described on a mesh basis in Sec. 3-1 apply to networks described on a nodal basis. The only difference is that when a current source is set to zero, it in effect becomes an open circuit, whereas a voltage source set to zero effectively becomes a short circuit. Thus, the definition for a linear resistance network given in Sec. 3-1 applies irrespective of whether the network is described on a mesh basis or on a node basis.

3-5 NODE EQUATIONS FOR THE GENERAL CIRCUIT

In this section we shall extend the discussion given in the preceding section for a two-node resistance network to the general case. First, let us consider a network containing three nodes as shown in Fig. 3-5.1. To determine the node equations for this network, we may follow a procedure similar to that used for the two-node case, namely: (1) write the KCL equations at each node in terms of the branch current variables; (2) substitute the branch relationships for each branch so that the KCL equations are expressed in terms of the branch voltage variables; and (3) substitute the relationships defining the branch voltages in terms of the node voltages $v_1(t)$, $v_2(t)$, and $v_3(t)$ defined in the figure. The result is the following set of equations:

$$\begin{bmatrix} i_a(t) - i_c(t) \\ -i_b(t) \\ i_c(t) \end{bmatrix} = \begin{bmatrix} G_2 + G_4 & -G_4 & -G_2 \\ -G_4 & G_3 + G_4 & -G_3 \\ -G_2 & -G_3 & G_1 + G_2 + G_3 \end{bmatrix} \begin{bmatrix} v_1(t) \\ v_2(t) \\ v_3(t) \end{bmatrix} \qquad (3\text{-}31)$$

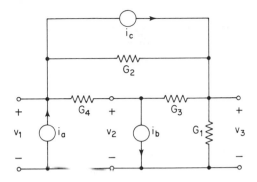

Figure 3-5.1 Three-node resistance network.

This is an example of a set of matrix equations of the general form

$$\mathbf{i}(t) = \begin{bmatrix} i_1(t) \\ i_2(t) \\ i_3(t) \end{bmatrix} = \begin{bmatrix} g_{11} & g_{12} & g_{13} \\ g_{21} & g_{22} & g_{23} \\ g_{31} & g_{32} & g_{33} \end{bmatrix} \begin{bmatrix} v_1(t) \\ v_2(t) \\ v_3(t) \end{bmatrix} = \mathbf{Gv}(t) \qquad (3\text{-}32)$$

If we compare the elements of Eq. (3-32) with those of Eq. (3-31), we may make a set of conclusions similar to those which were made for the two-node network. That is, we see that the elements $i_i(t)$ of Eq. (3-32) each represent the sum (with appropriate note taken of the reference polarity) of the currents incident to node i, the elements g_{ii} on the main diagonal of the conductance matrix $\mathbf{G}$ represent the sum of values of the conductances connected to the ith node, and the elements $g_{ij}(i \neq j)$ represent the negative of the value of any conductance connected between the ith and jth nodes. An example of such a determination follows.

Example 3-5.1

A three-node resistance network is shown in Fig. 3-5.2. If we apply the rules specified above for determining the matrix elements, we see that the KCL equations for the network may be written in the form

$$\begin{bmatrix} 7 \\ -12 \\ 9 \end{bmatrix} = \begin{bmatrix} 3 & -2 & -1 \\ -2 & 5 & -3 \\ -1 & -3 & 8 \end{bmatrix} \begin{bmatrix} v_1 \\ v_2 \\ v_3 \end{bmatrix}$$

mhos

Figure 3-5.2 Three-node resistance network.

The solution to such a set of equations follows exactly the development given in Sec. 3-3. For this example, the known elements of the matrix equation are identical with those given in Example 3-3.1; thus, the numerical values of the solution are also identical, namely, $v_1 = 2$ V, $v_2 = -1$ V, and $v_3 = 1$ V.

The development given above may easily be extended to the case where there are an arbitrary number of nodes in the network. If there are n nodes (plus a reference node), then we may define n node voltage variables, all defined with respect to the same reference node [the $(n + 1)$st node]. An extension of the development made above readily shows that the KCL equations for the various nodes will produce a set of equations of the form

$$\mathbf{i}(t) = \begin{bmatrix} i_1(t) \\ i_2(t) \\ \cdot \\ \cdot \\ \cdot \\ i_n(t) \end{bmatrix} = \begin{bmatrix} g_{11} & g_{12} & \cdots & g_{1n} \\ g_{21} & g_{22} & \cdots & g_{2n} \\ \cdot & \cdot & \cdots & \cdot \\ \cdot & \cdot & \cdots & \cdot \\ \cdot & \cdot & \cdots & \cdot \\ g_{n1} & g_{n2} & \cdots & g_{nn} \end{bmatrix} \begin{bmatrix} v_1(t) \\ v_2(t) \\ \cdot \\ \cdot \\ \cdot \\ v_n(t) \end{bmatrix} = \mathbf{G}\mathbf{v}(t) \qquad (3\text{-}33)$$

The procedure for determining the elements of the matrices of (3-33) may be summarized as follows:

SUMMARY 3-5.1

Writing the Node Equations for a Resistance Network. A set of node equations of the form given in (3-33) may be written for a given resistance network by defining a set of node voltages with their negative reference polarities all at the same common reference node, and determining the elements of the component matrices as follows:

$i_i(t)$ the sum of the current sources connected to the ith node; sources whose positive reference direction is towards the node are treated as positive, and sources whose positive reference direction is away from the node are treated as negative

$g_{ij}(i = j)$ the sum of the values of conductance (in mhos) of all resistors connected to the ith node

$g_{ij}(i \neq j)$ the negative of the value of the conductance (in mhos) of any resistor which is directly connected between the ith and jth nodes[7]

$v_i(t)$ the unknown node voltage variables which are to be found by solving the matrix equation   ■

It should be noted that, unlike the procedure for finding the mesh equations given in Summary 3-2.1, the procedure given above may be applied to any resistance network, irrespective of whether it is planar or non-planar. As an example of the use of this procedure for finding the KCL equations for a network, consider the following.

Example 3-5.2

A resistance network with nine node voltage variables is shown in Fig. 3-5.3. We may apply the rules given above to directly write the KCL equations for the node voltages.

---

[7]The quantity $g_{ij}(i = j)$ is sometimes called the *self conductance* of the ith node, while the quantity $g_{ij}(i \neq j)$ is called the *mutual conductance* between the ith and jth nodes.

Figure 3-5.3 Nine-node resistance network.

For simplicity, we shall assume that all the resistors have unity value. We obtain

$$
\begin{bmatrix} 2 \\ 0 \\ 0 \\ 3 \\ 0 \\ 0 \\ 4 \\ 0 \\ 0 \end{bmatrix}
=
\begin{bmatrix}
2 & -1 & 0 & -1 & 0 & 0 & 0 & 0 & 0 \\
-1 & 3 & -1 & 0 & -1 & 0 & 0 & 0 & 0 \\
0 & -1 & 2 & 0 & 0 & -1 & 0 & 0 & 0 \\
-1 & 0 & 0 & 3 & -1 & 0 & -1 & 0 & 0 \\
0 & -1 & 0 & -1 & 4 & -1 & 0 & -1 & 0 \\
0 & 0 & -1 & 0 & -1 & 3 & 0 & 0 & -1 \\
0 & 0 & 0 & -1 & 0 & 0 & 2 & -1 & 0 \\
0 & 0 & 0 & 0 & -1 & 0 & -1 & 3 & -1 \\
0 & 0 & 0 & 0 & 0 & -1 & 0 & -1 & 2
\end{bmatrix}
\begin{bmatrix} v_1 \\ v_2 \\ v_3 \\ v_4 \\ v_5 \\ v_6 \\ v_7 \\ v_8 \\ v_9 \end{bmatrix}
$$

In Sec. 3-3 we introduced the loop method of analysis which provided a more general way of using KVL to solve resistance networks than did the mesh method. It was based on the topological concepts of Sec. 1-4. The same concepts may be applied to KCL analysis. Rather than using node voltages as the independent variables, we can use the node-pair voltage variables defined across the branches of the tree. The KCL equations are then written for the cutsets formed of a single tree branch and the necessary links. Since each such equation includes a tree branch not included by the others, they will always be independent. Just as is the case for the node equations, the number of equations to be solved will always be $n - 1$, where n is the number of nodes. As an example of the use of node-pair voltages, consider the oriented graph and tree (the heavy lines) shown in Fig. 3-5.4 for the network of Fig. 3-5.2. The one tree-branch cutsets are identified by the letters A, B, and C. For these cutsets the KCL equations are

$$i_1 + i_4 + i_5 = 0, \quad i_3 - i_4 - i_6 - i_7 = 0, \quad \text{and} \quad i_1 + i_2 + i_4 + i_6 + i_7 = 0$$

$$(3\text{-}34)$$

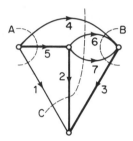

Figure 3-5.4 Set of one-tree-branch cutsets.

3-6 NETWORKS CONTAINING BOTH VOLTAGE AND CURRENT SOURCES

In the preceding sections of this chapter we have discussed methods for writing and solving sets of simultaneous equations which describe networks composed of resistors and sources. These methods have covered two important cases, namely: (1) the case where all the sources are voltage sources (and the equations are written on a mesh basis); and (2) the case where all the sources are current sources (and the equations are written on a node basis). In this section we shall discuss the extension of these methods to the case where voltage *and* current sources are present in a given resistance network.

Let us assume that we desire to write a set of equations describing the resistance network shown in Fig. 3-6.1(a). Note that this network contains two voltage sources

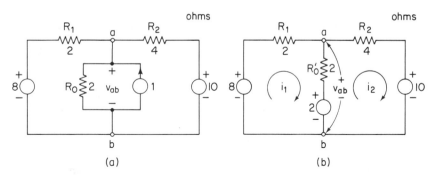

Figure 3-6.1 Resistance network containing voltage and current sources.

and one current source. Note also that the current source has a resistor in parallel with it. If it is desired to write the equations describing such a network on a mesh basis, we may use the concept of equivalent sources demonstrated in Sec. 2-3 to replace the 1-A current source and the 2-Ω shunt resistor R_0 shown in Fig. 3-6.1(a) by an equivalent 2-V voltage source and a series 2-Ω resistor R_0' as shown in Fig. 3-6.1(b). The mesh equations for the circuit of Fig. 3-6.1(b) may be written following the method described in Sec. 3-1. Thus, with the mesh currents defined as shown in Fig. 3-6.1(b), we obtain

$$\begin{bmatrix} 6 \\ -8 \end{bmatrix} = \begin{bmatrix} 4 & -2 \\ -2 & 6 \end{bmatrix} \begin{bmatrix} i_1 \\ i_2 \end{bmatrix} \tag{3-35}$$

The inverse set of equations is

$$\begin{bmatrix} i_1 \\ i_2 \end{bmatrix} = \frac{1}{20} \begin{bmatrix} 6 & 2 \\ 2 & 4 \end{bmatrix} \begin{bmatrix} 6 \\ -8 \end{bmatrix} = \begin{bmatrix} 1 \\ -1 \end{bmatrix} \tag{3-36}$$

Thus we see that $i_1 = 1$ A and $i_2 = -1$ A. The current in the resistor R'_0 in Fig. 3-6.1(b) is thus 2 A (flowing downward). The voltage v_{ab} from node a to node b is 6 V. Since the voltage v_{ab} must be identical for both circuits, the voltage across the resistor R_0 in Fig. 3-6.1(a) must be 6 V. Thus, the current through this resistor is 3 A (flowing downward). From this discussion, we see that the *internal* conditions in the two equivalent (non-ideal) sources are not the same (as was originally pointed out in Sec. 2-3). Note that, since the internal conditions in the sources are not the same, the total power dissipated in the two networks is not the same. For example, in Fig. 3-6.1(a), 2 W is dissipated in R_1, 4 W is dissipated in R_2, and 18 W is dissipated in R_0, making a total of 24 W dissipated in the network. In the circuit shown in Fig. 3-6.1(b), however, the power dissipated in R_1 and R_2 is 2 and 4 W, as before, but the power dissipated in R'_0 is 8 W, making a total of 14 W. This is a conclusion which might have been expected, since equivalent non-ideal sources are only equivalent with respect to their *terminal* behavior, not with respect to their internal conditions. Since the terminal variables of the two sources are identical, however, all variables in the resistance network external to the equivalent sources have the same values in both circuits.

As another example of analyzing resistance networks with both voltage and current sources, consider the resistance network shown in Fig. 3-6.2(a). This network contains two current sources and one voltage source. It may be modified so that it

Figure 3-6.2 Resistance network containing voltage and current sources.

can be described by equations written on a node basis. To do this, the voltage source and the series resistance connected between nodes 1 and 2 may be replaced by an equivalent current source with a shunt resistance, as shown in Fig. 3-6.2(b). The equations describing this circuit are readily found using the methods outlined in Sec. 3-4. We obtain

$$\begin{bmatrix} 4 \\ 1 \end{bmatrix} = \begin{bmatrix} 3 & -2 \\ -2 & 5 \end{bmatrix} \begin{bmatrix} v_1 \\ v_2 \end{bmatrix} \tag{3-37}$$

The inverse set of equations is

$$\begin{bmatrix} v_1 \\ v_2 \end{bmatrix} = \frac{1}{11} \begin{bmatrix} 5 & 2 \\ 2 & 3 \end{bmatrix} \begin{bmatrix} 4 \\ 1 \end{bmatrix} = \begin{bmatrix} 2 \\ 1 \end{bmatrix} \tag{3-38}$$

We see that the nodal voltages in Fig. 3-6.2 are $v_1 = 2$ V and $v_2 = 1$ V. Thus, there is a potential difference of 1 V across the 2-mho resistor G_0' in Fig. 3-6.2(b), and we see that a current of 2 A flows through this resistor (from node 1 to node 2). In Fig. 3-6.2(a), however, a potential of 1 V must also exist between nodes 1 and 2; therefore, the voltage across the 2-mho resistor G_0 must be zero. Thus, the current through the branch is zero. The overall current summation at nodes 1 and 2, however, is the same as it was in the circuit shown in Fig. 3-6.2(b), even though the internal conditions in the equivalent sources connected between nodes 1 and 2 in the two figures are quite different. For example, it is easily shown that a different total amount of power is dissipated in the two networks.

The techniques described in the above paragraphs are obviously applicable to the case where an arbitrary number of current and voltage sources are present in a network as long as every voltage source has a resistor connected in series with it or every current source has a resistor in parallel with it. Such networks may be described by a set of equations written on a node basis or a mesh basis, respectively. We may summarize this by the following statement:

SUMMARY 3-6.1

Analysis of Resistance Networks Containing Voltage and Current Sources. Networks containing both voltage and current sources in which all voltage sources have a series resistor may be described on a node basis by appropriate substitution of equivalent sources. Similarly, networks in which all current sources have a shunt resistor may be described on a mesh basis. ■

Note that the development given above is also applicable to the case of a resistance network containing only voltage sources (with series resistors) for which it is desired to write node equations, or to the case of a network containing only current sources (with shunt resistors) for which it is desired to write mesh equations.

Since an arbitrary resistance network satisfying the characteristics given above may obviously be described either by mesh equations or by node equations, it is desirable to have some sort of a guide as to which formulation will provide the least number of equations, and thus the easier method of solution. To do this, we may proceed as follows.

1. Replace all sources by their internal resistance (a short circuit for a voltage source, and an open circuit for a current source).
2. In the resulting network, count the total number of nodes n, and the total number of branches b.

The number of equations which will be required to describe the network on a node basis is $n - 1$. The number which will be required to describe it on a mesh or loop basis is $b - n + 1$.

Resistance networks which do not fall into the class defined above, i.e., those in which there are voltage sources which do not have a series resistor or current sources without a resistor in parallel, may easily be analyzed through the use óf the techniques described in Sec. 2-5 for adding parallel voltage sources and/or series current sources without changing the network variables. As an example of the application of such techniques, consider the network shown in Fig. 3-6.3(a). This network contains a

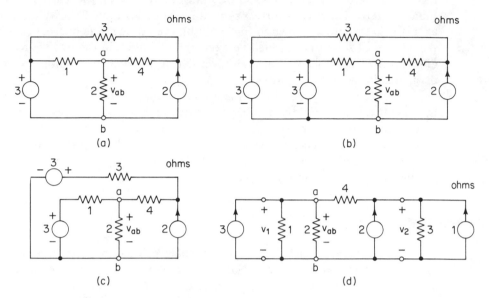

Figure 3-6.3 Resistance network in which source conversion is not directly applicable.

voltage source without a series resistor and also a current source without any shunt resistor. Suppose that it is desired to find the voltage v_{ab} across the 2-Ω resistor. We might proceed in our analysis by the following steps:

1. Replace the 3-V voltage source (which does not have a series resistor) with two 3-V sources connected in parallel as shown in Fig. 3-6.3(b).
2. Since the voltage sources are equal, the connection between them may be broken, leaving each source with a series resistance as shown in Fig. 3-6.3(c).
3. Replace each of the voltage source and series resistor combinations by an equivalent current source and shunt resistor as shown in Fig. 3-6.3(d).

The resulting circuit is now simply a two-node circuit in which the voltage v_{ab} across the resistor is simply v_1. The node equations may now be written as

$$\begin{bmatrix} 3 \\ 3 \end{bmatrix} = \begin{bmatrix} \dfrac{7}{4} & -\dfrac{1}{4} \\ -\dfrac{1}{4} & \dfrac{7}{12} \end{bmatrix} \begin{bmatrix} v_1 \\ v_2 \end{bmatrix} \tag{3-39}$$

The solution for v_1 using the standard techniques for the solution of simultaneous equations developed in Sec. 3-1 provides the desired answer for the value of v_{ab}.

As a second example of the application of the technique of adding additional sources to solve a resistance network, suppose that in the circuit shown in Fig. 3-6.4(a) it is desired to solve for the current i_{ab} in the 1-Ω resistor. The analysis might proceed as follows:

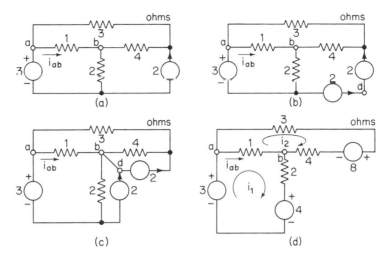

(a) (b) (c) (d)

Figure 3-6.4 Resistance network in which source conversion is not directly applicable.

1. Replace the 2-A current source by a series connection of two 2-A current sources as shown in Fig. 3-6.4(b).
2. Since KCL is satisfied by the two sources at the node d between them, and since the voltage at this node is unspecified, we may connect it to the node b at the junction of the 2-, 1-, and 4-Ω resistors as shown in Fig. 3-6.4(c) without changing any of the network variables.
3. Each of the current sources and its parallel resistor may now by replaced by a voltage source and a series resistor as shown in Fig. 3-6.4(d).

The remaining network is a two-mesh network. The simultaneous equations are easily written in terms of the mesh currents i_1 and i_2 defined in the figure. They are

$$\begin{bmatrix} -1 \\ -8 \end{bmatrix} = \begin{bmatrix} 3 & -1 \\ -1 & 8 \end{bmatrix} \begin{bmatrix} i_1 \\ i_2 \end{bmatrix} \qquad (3\text{-}40)$$

The solution for i_1 and i_2 using standard techniques for the solution of simultaneous equations provides the desired solution for the current i_{ab} using the relation $i_{ab} = i_1 - i_2$. A summary of the various possibilities that may be used in analyzing a resistance network is given in the flow chart shown in Fig. 3-6.5.

To conclude this section, we shall introduce a final technique which is frequently of use as an aid in solving resistance networks containing both current and voltage sources. First, let us give a name to a circuit configuration of three nodes and three resistors, which has the property that each pair of nodes is connected by a single resistor. Such a configuration, consisting of the nodes numbered 1, 2, and 3, and

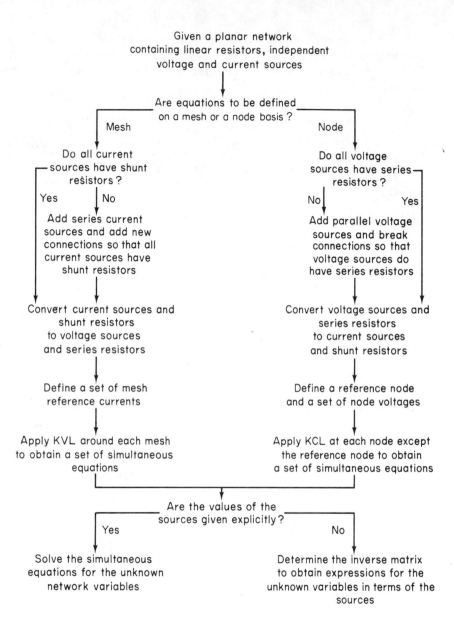

Figure 3-6.5 Flow chart for the analysis of resistance networks containing voltage and current sources.

the resistors labeled R_a, R_b, and R_c, is shown in Fig. 3-6.6(a). This circuit configuration is called a *delta network*, since it has the geometrical appearance of an upside-down Greek letter Δ. We will show that such a network may be replaced by a *wye connection* (which resembles the geometrical appearance of the letter Y) of three resistors R_1, R_2, and R_3 as shown in Fig. 3-6.6(b), which has identical properties at nodes 1, 2, and 3. To find the values of R_1, R_2, and R_3, we need only equate the input resistance seen between the various pairs of terminals in the two networks. For example, equating

Figure 3-6.6 Delta and wye resistance networks.

the resistance measured between terminals 1 and 3 in the two networks, we obtain

$$R_1 + R_3 = \frac{R_b(R_a + R_c)}{R_a + R_b + R_c} \tag{3-41}$$

Similarly, for the input resistance between terminals 1 and 2, and between terminals 2 and 3, to be equal, we must have

$$R_1 + R_2 = \frac{R_c(R_a + R_b)}{R_a + R_b + R_c} \quad \text{and} \quad R_2 + R_3 = \frac{R_a(R_b + R_c)}{R_a + R_b + R_c} \tag{3-42}$$

These equations are readily solved to obtain

$$R_1 = \frac{R_c R_b}{R_a + R_b + R_c} = \frac{G_a}{G_a G_b + G_a G_c + G_b G_c} \tag{3-43a}$$

$$R_2 = \frac{R_a R_c}{R_a + R_b + R_c} = \frac{G_b}{G_a G_b + G_a G_c + G_b G_c} \tag{3-43b}$$

$$R_3 = \frac{R_a R_b}{R_a + R_b + R_c} = \frac{G_c}{G_a G_b + G_a G_c + G_b G_c} \tag{3-43c}$$

Thus, we have found expressions for determining the elements of a wye of resistors in terms of the elements of a delta of resistors in such a manner that the terminal properties of the two resistance networks are identical. Such a process is referred to as a *delta-wye transformation*. By a similar process, we may define the inverse relations, i.e., the ones that determine a *wye-delta transformation*. The relations are

$$G_a = \frac{G_2 G_3}{G_1 + G_2 + G_3} = \frac{R_1}{R_1 R_2 + R_1 R_3 + R_2 R_3} \tag{3-44a}$$

$$G_b = \frac{G_1 G_3}{G_1 + G_2 + G_3} = \frac{R_2}{R_1 R_2 + R_1 R_3 + R_2 R_3} \tag{3-44b}$$

$$G_c = \frac{G_1 G_2}{G_1 + G_2 + G_3} = \frac{R_3}{R_1 R_2 + R_1 R_3 + R_2 R_3} \tag{3-44c}$$

where G_1, G_2, and G_3 are the values of conductance of the resistors labeled R_1, R_2, and R_3 in Fig. 3-6.6(b).

The relations of Eqs. (3-43) and (3-44) may frequently be used to simplify the analysis of resistance networks. Note that the relations involve addition, multiplication, and division, but not subtraction. Thus, if the resistors in one of the circuits shown in Fig. 3-6.6 are positive-valued, then the resistors in the equivalent circuit will also be positive-valued. An example of the use of these relations follows.

Example 3-6.1

As an example of the use of these relations, let us assume that it is desired to find the voltage v_1 across the 5-A current source shown in Fig. 3-6.7(a). The $\frac{3}{11}$-, $\frac{2}{11}$-, and $\frac{1}{11}$-Ω resistors form a wye; thus using (3-44), we find that the elements of the equivalent delta are $G_a = 3$, $G_b = 2$, $G_c = 1$ mhos. Thus, the circuit may be redrawn as shown in Fig. 3-6.7(b). The circuit now has two nodes rather than the original three. Routine node analysis may be applied to the circuit to show that $v_1 = \frac{13}{7}$ V.

(a)   (b)

Figure 3-6.7 Using a delta-wye transformation to analyze a resistance network.

3-7 DEPENDENT SOURCES

Dependent or controlled sources were introduced in Sec. 2-2. As shown in Fig. 3-7.1, there are four types; the VCVS, ICVS, VCIS, and ICIS. To see how these sources are treated in the analysis of resistance networks, let us begin by considering the two-

Figure 3-7.1 Types of dependent sources.

mesh network shown in Fig. 3-7.2. This network contains an ICVS, which has an output voltage $3i_1$, i.e., a voltage numerically equal to three times the value of the mesh reference current i_1. We may use the techniques of Sec. 3-1 to write the mesh

Figure 3-7.2 Two-mesh resistance network with a controlled source.

equations for this network in matrix form, adding the contribution from the two voltage sources in mesh 2, and putting their sum as an element of the voltage matrix **v**. Thus, we obtain

$$\begin{bmatrix} 8 \\ -3 - 3i_1 \end{bmatrix} = \begin{bmatrix} 3 & -2 \\ -2 & 5 \end{bmatrix} \begin{bmatrix} i_1 \\ i_2 \end{bmatrix} \tag{3-45}$$

In the above equation, a term which is a function of the mesh current i_1 appears in the left member. We may easily move this term to the right member of the equation. The resulting set of equations then has the form

$$\begin{bmatrix} 8 \\ -3 \end{bmatrix} = \begin{bmatrix} 3 & -2 \\ 1 & 5 \end{bmatrix} \begin{bmatrix} i_1 \\ i_2 \end{bmatrix} \tag{3-46}$$

The techniques given in Sec. 3-2 are readily applied to find the inverse set of equations, and thus the solution for the mesh currents. The inverse relations are

$$\begin{bmatrix} i_1 \\ i_2 \end{bmatrix} = \frac{1}{17} \begin{bmatrix} 5 & 2 \\ -1 & 3 \end{bmatrix} \begin{bmatrix} 8 \\ -3 \end{bmatrix} = \begin{bmatrix} 2 \\ -1 \end{bmatrix} \tag{3-47}$$

Thus, we have solved the network for the mesh currents, and we find that $i_1 = 2$ A and $i_2 = -1$ A.

The procedure illustrated above is easily generalized to apply to any arbitrary resistance network which is to be defined by mesh equations, and which contains controlled sources of the ICVS type.

The steps are summarized as follows:

SUMMARY 3-7.1

Solution of Resistance Networks Containing Voltage Sources and ICVS's. A set of simultaneous equations of the form $\mathbf{v}(t) = \mathbf{R}i(t)$ may be written for a resistance network containing independent voltage sources and dependent ICVS's by the following steps:

1. Define a set of mesh reference currents.
2. Write the KVL equations for each mesh using the procedure given in Sec. 3-2 and putting all terms representing ICVS's in the appropriate elements of $\mathbf{v}(t)$.
3. Rewrite the simultaneous equations putting all terms representing the contributions from ICVS's into the appropriate elements of the matrix $\mathbf{R}$.
4. Solve the resulting set of simultaneous equations. ∎

An example of this procedure follows.

Example 3-7.1

A three-mesh resistance network is shown in Fig. 3-7.3. In mesh 2, there is an ICVS controlled by the current in mesh 1. In addition, in mesh 3 there is an ICVS controlled

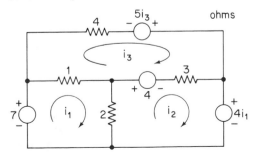

Figure 3-7.3 Three-mesh resistance network with two controlled sources.

by the current in that mesh. Applying the techniques described in Sec. 3-2, we may apply KVL around each of the three meshes to obtain the relations

$$\begin{bmatrix} 7 \\ -4 - 4i_1 \\ 4 + 5i_3 \end{bmatrix} = \begin{bmatrix} 3 & -2 & -1 \\ -2 & 5 & -3 \\ -1 & -3 & 8 \end{bmatrix} \begin{bmatrix} i_1 \\ i_2 \\ i_3 \end{bmatrix}$$

The terms in the left-hand member of this equation which are functions of i_1 and i_3 may be moved into the right-hand member of the equation. If this is done, we obtain the relations

$$\begin{bmatrix} 7 \\ -4 \\ 4 \end{bmatrix} = \begin{bmatrix} 3 & -2 & -1 \\ 2 & 5 & -3 \\ -1 & -3 & 3 \end{bmatrix} \begin{bmatrix} i_1 \\ i_2 \\ i_3 \end{bmatrix}$$

Applying the methods of Sec. 3.3 to solve these equations we obtain

$$i_1 = \frac{7 \begin{vmatrix} 5 & -3 \\ -3 & 3 \end{vmatrix} + 4 \begin{vmatrix} -2 & -1 \\ -3 & 3 \end{vmatrix} + 4 \begin{vmatrix} -2 & -1 \\ 5 & -3 \end{vmatrix}}{25} = 2$$

$$i_2 = \frac{-7 \begin{vmatrix} 2 & -3 \\ -1 & 3 \end{vmatrix} - 4 \begin{vmatrix} 3 & -1 \\ -1 & 3 \end{vmatrix} - 4 \begin{vmatrix} 3 & -1 \\ 2 & -3 \end{vmatrix}}{25} = -1$$

$$i_3 = \frac{7 \begin{vmatrix} 2 & 5 \\ -1 & -3 \end{vmatrix} + 4 \begin{vmatrix} 3 & -2 \\ -1 & -3 \end{vmatrix} + 4 \begin{vmatrix} 3 & -2 \\ 2 & 5 \end{vmatrix}}{25} = 1$$

It should be noted, from the example given above, that if the mesh equations describing a resistance network which contains controlled sources are put in the form $v(t) = \mathbf{R}i(t)$, then the matrix $\mathbf{R}$, in general, will not be symmetric.

Several extensions of the method described above should be readily apparent to the reader. For example, since any branch current can be expressed as a linear combination of the mesh currents, any ICVS which is controlled by a branch current that is not a mesh current can have its output voltage expressed as a function of mesh currents; thus, the development given above applies to this case. As another extension, consider the case where ICIS's are present in the network. If such sources have resistors in parallel with them, they may be converted to ICVS's with series resistors using the source conversion methods described in Sec. 2-3. If such shunt resistors are not present, then the techniques given in Sec. 2-5 for adding additional series current sources may be applied to provide the ICIS's with shunt resistors. Similarly, to treat VCVS's, it is usually possible to express the controlling voltage in terms of some branch current, which, in turn, can be expressed in terms of the unknown variables, the mesh currents. A flow chart of a general procedure to include these various possibilities is given in Fig. 3-7.4. An example follows.

Example 3-7.2

As an example of the extension of the method for including the effects of various types of sources outlined above, consider the network shown in Fig. 3-7.5(a). It contains a VCIS without a paralleling shunt resistor. As a first step in the solution of this network, we may add a second series VCIS as shown in Fig. 3-7.5(b). Since the potential between

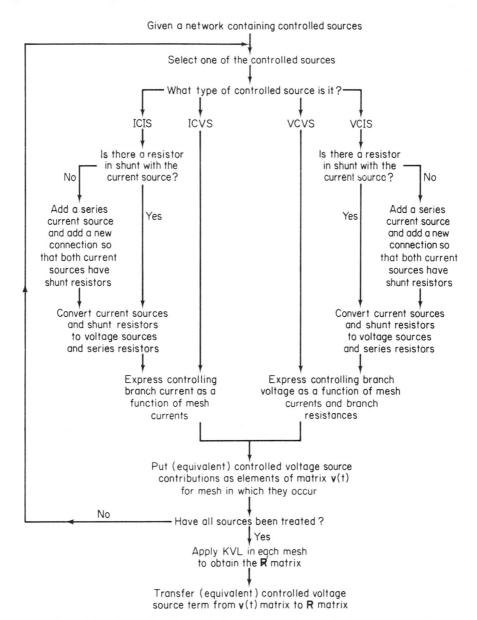

Given a network containing controlled sources

Select one of the controlled sources

What type of controlled source is it?

ICIS   ICVS   VCVS   VCIS

Is there a resistor in shunt with the current source?

No

Is there a resistor in shunt with the current source?

No

Add a series current source and add a new connection so that both current sources have shunt resistors

Yes

Yes

Add a series current source and add a new connection so that both current sources have shunt resistors

Convert current sources and shunt resistors to voltage sources and series resistors

Convert current sources and shunt resistors to voltage sources and series resistors

Express controlling branch current as a function of mesh currents

Express controlling branch voltage as a function of mesh currents and branch resistances

Put (equivalent) controlled voltage source contributions as elements of matrix $\mathbf{v}(t)$ for mesh in which they occur

No

Have all sources been treated?

Yes

Apply KVL in each mesh to obtain the $\mathbf{R}$ matrix

Transfer (equivalent) controlled voltage source term from $\mathbf{v}(t)$ matrix to $\mathbf{R}$ matrix

Figure 3-7.4 Flow chart for including the effect of controlled sources in mesh (or loop) equations.

the VCIS's is undetermined, we may connect this point so that each of the current sources has a shunt resistor as shown in Fig. 3-7.5(c). The VCIS's and their shunt resistors may now be converted to VCVS's with series resistors as shown in Fig. 3-7.5(d). In this latter figure, two mesh reference currents i_1 and i_2 have been defined. In terms of these mesh currents, we see that $v_0 = 3(i_1 - i_2)$. Applying KVL to the two meshes, we obtain the relations

$$\begin{bmatrix} 5 - 2v_0 \\ 6v_0 \end{bmatrix} = \begin{bmatrix} 5 - 6(i_1 - i_2) \\ 18(i_1 - i_2) \end{bmatrix} = \begin{bmatrix} 4 & -4 \\ -4 & 10 \end{bmatrix} \begin{bmatrix} i_1 \\ i_2 \end{bmatrix}$$

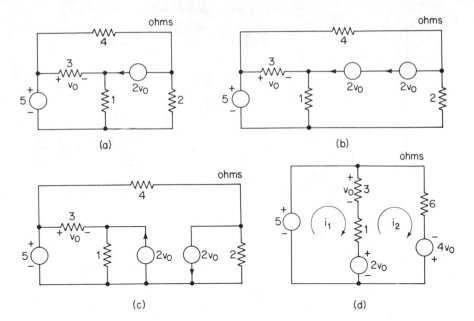

Figure 3-7.5 Resistance network with a controlled source.

These equations may be rewritten in the form

$$\begin{bmatrix} 5 \\ 0 \end{bmatrix} = \begin{bmatrix} 10 & -10 \\ -22 & 28 \end{bmatrix} \begin{bmatrix} i_1 \\ i_2 \end{bmatrix}$$

The solution for the mesh currents is easily accomplished, and is left to the reader as an exercise.

If one desires to write the equations for a resistance network containing controlled sources using node voltages as the unknown variables, he may use a method similar to that given above. If the controlled sources are VCIS's, the network equations may be written directly. An outline of the procedure follows.

SUMMARY 3-7.2

Solution of Resistance Networks Containing Current Sources and VCIS's. A set of simultaneous equations of the form $i(t) = Gv(t)$ may be written for a resistance network containing independent current sources and dependent VCIS's by the following steps:

1. Define a set of node voltages.
2. Write the KCL equations for each node using the procedure given in Sec. 3-5 and putting all terms representing VCIS's in the appropriate elements of the matrix $i(t)$.
3. Rewrite the matrix equation putting all terms representing contributions from the VCIS's into the appropriate elements of the matrix G.
4. Solve the resulting set of simultaneous equations. ■

Resistance Networks   Chap. 3

The procedure given above is easily modified to take account of various other types of controlled sources. The modifications are similar to those used in writing the mesh equations. A flow chart of the procedure to be followed in writing node equations for networks containing various types of controlled sources is given in Fig. 3-7.6. The reader should note the similarity between this flow chart and the one given for the mesh analysis case in Fig. 3-7.4. An example follows.

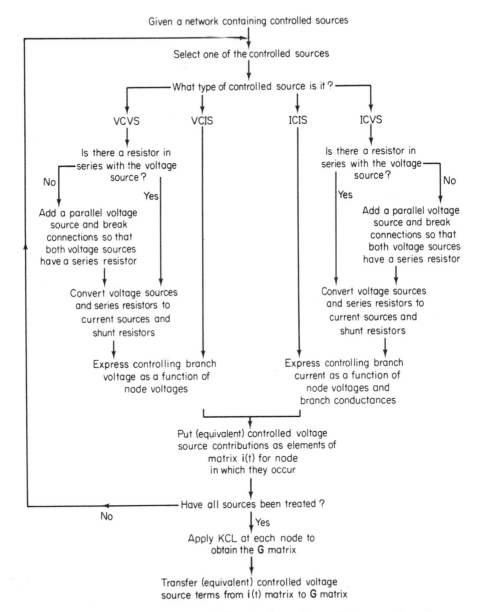

Figure 3-7.6 Flow chart for including the effect of controlled sources in node equations.

Example 3-7.3

The resistance network shown in Fig. 3-7.7(a) contains two controlled sources, a VCIS and a VCVS. To analyze this network on a node basis, we may begin by noting that the VCVS does not have a series resistor, therefore we will parallel this controlled source by a second and then break the connection between them as shown in Fig. 3-7.7(b). The two VCVS's and their series resistors may now be replaced by VCIS's and shunt resistors as shown in Fig. 3-7.7(c). Using the voltage $v_1(t)$ as one

(a)

(b)

(c)

Figure 3-7.7 Resistance network with a pair of controlled sources.

of our node voltages, and defining a second node voltage $v_2(t)$ as shown in the figure, we may apply KCL at each of the nodes to obtain the relations

$$\begin{bmatrix} i_1(t) + 2v_1(t) - 3v_1(t) \\ i_2(t) + 3v_1(t) + 4v_1(t) \end{bmatrix} = \begin{bmatrix} 7 & -2 \\ -2 & 7 \end{bmatrix} \begin{bmatrix} v_1(t) \\ v_2(t) \end{bmatrix}$$

Rearranging the terms of this equation, we obtain a set of equations relating the node voltages $v_1(t)$ and $v_2(t)$ and the input currents from the independent current sources, namely, $i_1(t)$ and $i_2(t)$. These relations have the form

$$\begin{bmatrix} i_1(t) \\ i_2(t) \end{bmatrix} = \begin{bmatrix} 8 & -2 \\ -9 & 7 \end{bmatrix} \begin{bmatrix} v_1(t) \\ v_2(t) \end{bmatrix}$$

The inverse set of relations which gives the values of the node voltages as functions of the input currents is easily obtained to yield a solution for the network for any arbitrary set of input currents.

Another topic concerning controlled sources is of interest for future application. This concerns circuits formed by an interconnection of controlled sources and resistors which do not contain any independent sources. Any such circuit may be replaced at a given pair of terminals by a single equivalent resistor. To see how the value of such a resistor is determined, consider the following example.

Example 3-7.4

The circuit shown in Fig. 3-7.8(a) consists of a single controlled source and three resistors. The source is an ICVS controlled by the current $i_0(t)$ in resistor R_1. It is desired to find a single equivalent resistor to replace the network, so that the relation

(a)   (b)   (c)

Figure 3-7.8 Equivalent resistance for a resistance network with a controlled source.

between the variables at terminals a-b will be unchanged. To find the value of such an equivalent resistor, we may apply a voltage source with a constant output voltage V_0 to the terminals a-b. As indicated in Fig. 3-7.8(b), this in effect defines a two-mesh resistance network with mesh current variables $i_0(t)$ and $i_2(t)$. Applying KVL to the two meshes, we obtain the relations

$$\begin{bmatrix} 0 \\ -V_0 \end{bmatrix} = \begin{bmatrix} 4 & -1 \\ -3 & 2 \end{bmatrix} \begin{bmatrix} i_0(t) \\ i_2(t) \end{bmatrix}$$

Solving for $i_2(t)$, we find that $i_2(t) = -4V_0/5$. The relation between the input voltage V_0 provided by the voltage source and the current flowing out of the positive terminal of this source [which is $-i_2(t)$] provides the value of the equivalent resistance R_{eq}. Thus, we see that

$$R_{eq} = \frac{V_0}{4V_0/5} = \frac{5}{4}$$

We conclude that the entire circuit shown in Fig. 3-7.8(a) may be replaced at terminals a-b by an equivalent resistor of value $\frac{5}{4}$ Ω, as shown in Fig. 3-7.8(c), without any change in the relation between the terminal variables as observed at those terminals.

The procedure described above is easily generalized. It should be especially noted that the value of the source voltage applied cancels when the value of R_{eq} is determined. Thus, any numerical value of voltage may be applied in calculating the value of R_{eq} (unity is usually a convenient value). An alternative method of computation would be to apply a current source to the terminals a-b and determine the voltage appearing across it. The ratio of voltage to current will also be $\frac{5}{4}$ in this case. It should be noted that some connections of controlled sources and resistors may produce values of an equivalent resistance which are negative. Such a circuit will supply energy to any network to which it may be connected.

3-8 THEVENIN AND NORTON'S THEOREMS FOR RESISTANCE NETWORKS

Frequently in our analysis of a resistance network we are not interested in the solution for all of the unknown network variables, but only in one specific variable. Thus, we may wish to calculate a single branch current or a single branch voltage. In addition, many problems call for the calculation of the variables in a given branch for several different values of the resistance of the branch. In such a case, it is desirable to be able to separate the branch from the rest of the network, and apply some type of solution to the rest of the network that will not have to be repeated for each different value of the resistance of the branch. This is especially important when resistance networks containing independent sources (for example, transistor or operational amplifier models) are to be analyzed. In this section we shall introduce Thevenin and Norton circuits and show how they provide a solution to such problems.

First let us consider the case where only independent sources are present. As an example of this type of a problem, consider the network shown in Fig. 3-8.1(a) in which a resistor R_0 is connected to nodes a and b. Let us assume that we desire to

(a)   (b)

Figure 3-8.1 Two-mesh resistance network.

compute the current through the resistor for any specified value of resistance. Let us first attack this problem by conventional mesh analysis techniques. Defining two mesh currents $i_1(t)$ and $i_2(t)$ as shown in Fig. 3-8.1(b), we obtain the relations

$$\begin{bmatrix} v_1(t) \\ 0 \end{bmatrix} = \begin{bmatrix} 2 & -1 \\ -1 & 1 + R_0 \end{bmatrix} \begin{bmatrix} i_1(t) \\ i_2(t) \end{bmatrix} \tag{3-48}$$

The inverse set of relations is

$$\begin{bmatrix} i_1(t) \\ i_2(t) \end{bmatrix} = \frac{1}{1 + 2R_0} \begin{bmatrix} 1 + R_0 & 1 \\ 1 & 2 \end{bmatrix} \begin{bmatrix} v_1(t) \\ 0 \end{bmatrix} \tag{3-49}$$

The current through the resistor R_0 is $i_2(t)$. From the above set of equations, we see that

$$i_2(t) = \frac{v_1(t)}{1 + 2R_0} \qquad (3\text{-}50)$$

This equation obviously gives us a solution for the current through the resistor R_0 for any value of resistance. As an introduction to a second means for solving this problem, let us connect a resistor R_0 to nodes c and d of a circuit consisting of a single voltage source of value $v_1(t)/2$ and a resistor of value $\frac{1}{2}\,\Omega$ as shown in Fig. 3-8.2. Applying KVL to this circuit, we may write

$$i_0(t) = \frac{v_1(t)/2}{\frac{1}{2} + R_0} = \frac{v_1(t)}{1 + 2R_0} \qquad (3\text{-}51)$$

Figure 3-8.2 Thevenin equivalent circuit for the network shown in Fig. 3-8.1.

Comparing the last two equations, we see that, for a given expression for $v_1(t)$, the same current will flow in the resistor R_0 whether it is connected to nodes a and b of the circuit shown in Fig. 3-8.1 [in this case the current is $i_2(t)$], or to nodes c and d of the circuit in Fig. 3-8.2 [in this case the current is $i_0(t)$]. Furthermore, this statement is true for any value of R_0. Thus we may say that the two circuits (not including R_0) shown in the two figures are equivalent with respect to their behavior at the labeled pair of nodes. The circuit shown in Fig. 3-8.2, consisting of a single voltage source in series with a single resistor, is called the *Thevenin equivalent* of the resistance network (not including R_0) shown in Fig. 3-8.1. Such an equivalent circuit may be derived for any network consisting of linear resistors and independent sources. To see this, consider the properties that such an arbitrary network must have if all of its internal sources are set to zero, i.e., if all current sources are replaced by open circuits and all voltage sources are replaced by short circuits. Since the resulting (sourceless) network contains only resistors, it must be able to be replaced at a given pair of nodes by a single resistor, the value of which is the input resistance of the network seen at that pair of nodes. Thus, if we define variables $v(t)$ and $i(t)$ at the terminals of such a sourceless network, as shown in Fig. 3-8.3, the input characteristics must be represented by a straight line passing through the origin of an i-v plane as shown in Fig. 2-1.2. The slope of this line gives the value of the input resistance.

Figure 3-8.3 Voltage and current variables at the terminals of a sourceless network.

Now let us consider again the original network including its independent sources. In general, a voltage will be present across the terminals, due to the effects of the sources, even though no current is flowing at the terminals. If, in addition, we apply a current to the network terminals (from an external current source), then, due to the linearity of the resistance network, this source must produce the same effect as it did when it was used to excite the sourceless network. In this case, however, such an effect is superimposed upon the effect of the network's internal sources. The additional effect is determined by the value of the equivalent resistance of the network, i.e., the slope of the *i-v* characteristic. Thus, we see that the original network must have an overall characteristic consisting of a straight line on the *i-v* plane *which does not pass through the origin* as shown in Fig. 3-8.4. Such a representation was shown in Sec. 2-3 to be equivalent to that of an ideal voltage source in series with a linear resistor.[8]

Figure 3-8.4 Volt-ampere characteristic for a resistance network with sources.

Thus we conclude that our original network may be replaced by an equivalent circuit consisting of a voltage source in series with a single resistor. We may summarize our results as follows:

SUMMARY 3-8.1

Thevenin's Theorem. Any network consisting of linear resistors and independent sources may be replaced at a given pair of nodes by an equivalent circuit consisting of a single voltage source and a series resistor as shown in Fig. 3-8.5. Such a circuit is

Figure 3-8.5 Thevenin equivalent circuit.

called a *Thevenin equivalent circuit*. The value of the resistor is the input resistance seen at the pair of nodes when all the independent sources of the original network are set to zero. The value of the voltage source is that seen at the open-circuited terminals.

■

As an example of the use of a Thevenin equivalent circuit in analyzing a resistance network, consider the following example.

[8]The slope of the locus in Fig. 3-8.4 is reversed from the one shown in Fig. 2-3.5 since the direction of the reference current is reversed in the two figures.

Example 3-8.1

It is desired to find the current i_R in the resistor R in the circuit shown in Fig. 3-8.6(a) for any value of the resistor. We may use Thevenin's theorem to replace the resistance network (excluding the resistor R) by a single source and a resistor to simplify the computations. First let us find the (open-circuit) voltage v_{ab} across the nodes a-b when the resistor R is removed as shown in Fig. 3-8.6(b). The network equations are the

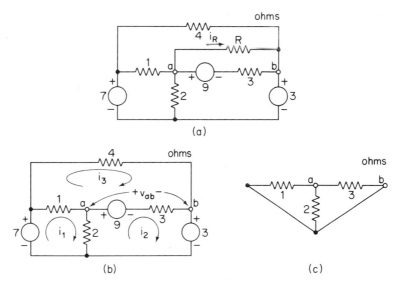

Figure 3-8.6 Analysis of a resistance network.

same as those given in Example 3-3.1, and the solution for the mesh currents i_2 and i_3 was shown in that example to be -1 A and 1 A, respectively. Thus, there is a net current of 2 A flowing through the 3-Ω resistor from right to left and the voltage across it is 6 V with a polarity opposite to that of the 9-V voltage source. Adding the two voltages, we see that the voltage v_{ab} as defined in Fig. 3-8.6(b) is 3 V. This is also the voltage of the voltage source in the Thevenin equivalent circuit. Now let us find the input resistance of the sourceless network. Replacing the voltage sources by short circuits and redrawing the network, as shown in Fig. 3-8.6(c), we see that the equivalent resistance of the network as seen at nodes a-b is $\frac{6}{11}$ Ω. This is the resistance of the Thevenin equivalent circuit. Combining the above results, the Thevenin equivalent circuit for the resistance network is shown in Fig. 3-8.7. Adding the resistance R, as shown in the figure, we see that the current i_R is given by the relation

$$i_R = \frac{3}{R + \frac{6}{11}}$$

Figure 3-8.7 Thevenin equivalent circuit for the resistance network shown in Fig. 3-8.6(a).

The Thevenin equivalent circuit shown in Fig. 3-8.5 is not the only simple circuit that may be used to replace a given resistance network. Using the techniques introduced in Sec. 2-3, the Thevenin circuit may be replaced by one consisting of a single current source and a shunt resistor. The current source has a value which is given by the current flowing through a short circuit connected across the terminals of the original resistance network. The resistance is again the input resistance of the network when the sources are set to zero. The equivalent circuit is named a *Norton equivalent circuit*. We may summarize its description as follows.

SUMMARY 3-8.2

Norton's Theorem. Any network consisting of linear resistors and independent sources may be replaced at a given pair of nodes by an equivalent circuit consisting of a single current source and a shunt resistor as shown in Fig. 3-8.8. The value of the

Figure 3-8.8 Norton equivalent circuit.

resistor is the resistance seen at the pair of nodes when all the independent sources of the original network are set to zero. It is equal to the Thevenin resistance. The value of the current source is equal to the current which flows in a short circuit connecting the nodes. ∎

It should be noted that the Norton equivalent circuit is easily derived directly from the Thevenin equivalent circuit or vice versa using the non-ideal source transformations developed in Sec. 2-3. It should also be noted that, if we let v_{oc} be the voltage measured at a pair of open-circuited terminals and i_{sc} be the current measured when the terminals are shorted (with the relative reference polarities shown in Fig. 3-8.9), then the value of the resistance R_s that is required in either a Thevenin or a

Figure 3-8.9 Reference polarities for Eq. (3-52).

Norton equivalent circuit is given by the relation

$$R_s = \frac{v_{oc}}{i_{sc}} \qquad (3\text{-}52)$$

An example of the direct determination of a Norton equivalent circuit follows.

Example 3-8.2

We may use basic network analysis techniques to derive a Norton equivalent circuit for the network shown in Fig. 3-8.10(a) at the nodes *a-b* shown in the figure. This is the same network that was used in Example 3-8.1. Thus, the input resistance at the terminals *a-b* found when all the internal sources are set to zero will be $\frac{6}{11}$ Ω, just as it was in the preceding example. If we now short-circuit the terminals *a-b* as shown in Fig. 3-8.10(b), we find that the short completely isolates the network branch consisting of the 9-V source and the series 3-Ω resistor from the rest of the network. Thus, this

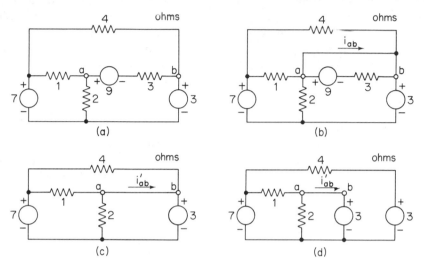

(a) (b) (c) (d)

Figure 3-8.10 Analysis of a resistance network.

branch will produce a contribution to the current i_{ab} of 3 A, independent of the effects of the rest of the network. To determine the effects of the rest of the network, we may redraw the circuit without this branch, as shown in Fig. 3-8.10(c), where i'_{ab} is defined as $i_{ab} - 3$. By using the concept of equivalent paralleled voltage sources developed in Sec. 2-5, the network may be redrawn as shown in Fig. 3-8.10(d). Obviously, the branch consisting of the right 3-V source and the 4-Ω resistor makes no contribution to the current i'_{ab}. We may solve for the contribution to the current i'_{ab} from the remainder of the circuit. Routine mesh analysis applied to this two-mesh circuit shows that this contribution to i'_{ab} is $\frac{5}{2}$ A. The superposition of the above effects gives a total short-circuit current i_{ab} of $\frac{11}{2}$ A. Thus the Norton equivalent network for the resistance network as viewed from the terminals *a-b* is a current source of $\frac{11}{2}$ A and a shunt resistance of $\frac{6}{11}$ Ω as shown in Fig. 3-8.11. Comparison of this Norton equivalent circuit with the Thevenin equivalent circuit shown in Fig. 3-8.7 (without the resistor *R*) readily verifies the equivalence of the two circuits.

ohms

Figure 3-8.11 Norton equivalent circuit for the resistance network shown in Fig. 3-8.10(a).

Now let us consider the case where, in addition to having resistors and independent sources present, as was disucssed in the above development, we also have *dependent* sources present. We may modify the development given above to obtain a Thevenin or Norton equivalent circuit for the entire network, including both types of sources. The value of the Thevenin resistance to be used in such an equivalent circuit is found by using the method given in Example 3-7.4. When applying this method, all *independent* voltage sources must be replaced by short circuits, and all *independent* current sources by open circuits. The controlled sources remain unchanged. To find $V_{\text{Thev.}}$, it is merely necessary to find the open-circuit voltage appearing at the specified terminals of the original network with all the independent sources replaced. The Thevenin equivalent circuit then consists of an independent voltage source, whose value is equal to the open-circuit voltage, in series with the equivalent resistance found when all independent sources are set to zero. An example of this procedure follows.

Example 3-8.3

The network used in Example 3-7.4 and shown in Fig. 3-7.8(a) has been modified by the addition of a voltage source to provide the network shown in Fig. 3-8.12(a). To find a Thevenin equivalent circuit for this network at terminals *a-b*, we first set the

(a)   (b)

Figure 3-8.12 Thevenin equivalent circuit for a resistance network with dependent and independent sources.

(independent) voltage source to zero and find the equivalent resistance seen at terminals *a-b*. From Example 3-7.4, we know that this is $\frac{5}{4}$ Ω. We next find the open-circuit voltage appearing at the terminals *a-b* with the independent voltage source restored to the network. Applying KVL around the mesh containing the two sources, we obtain

$$2 = i_0 + i_0 + 2i_0$$

Solving we obtain $i_0 = \frac{1}{2}$ A. Thus the output of the ICVS is 1 V and the drop across the (vertical) resistor is $\frac{1}{2}$ V. Combining these two voltage drops we find that the open-circuit voltage at terminals *a-b* is $\frac{3}{2}$ V (positive at terminal *a*). Thus, the Thevenin equivalent circuit which includes the effects of both the dependent and the independent sources from the original network consists of the independent voltage source and resistor shown in Fig. 3-8.12(b).

In view of the above development, we may modify the statement of Thevenin's theorem given in Summary 3-8.1 to include the effect of controlled sources. We obtain the following:

SUMMARY 3-8.3

Thevenin's Theorem (for resistance networks containing dependent sources). Any network consisting of linear resistors, independent sources, and linear dependent sources may be replaced at a given pair of nodes by an equivalent circuit consisting of a single voltage source and a series resistor. The value of the resistor is the input resistance seen at the pair of nodes when all the independent sources of the network are set to zero. The value of the voltage source is the voltage seen at the open-circuited terminals. ∎

A parallel development may be made to arrive at a more general statement of Norton's theorem. Thus, we obtain the following:

SUMMARY 3-8.4

Norton's Theorem (for resistance networks containing dependent sources). Any circuit consisting of linear resistors, independent sources, and linear dependent sources may be replaced at a given pair of nodes by an equivalent circuit consisting of a single current source and a shunt resistor. The value of the resistor is the input resistance seen at the pair of nodes when all the independent sources of the network are set to zero. The value of the current source is the current flowing through a short circuit placed across the terminals. ∎

An alternate procedure which may be used in determining the Thevenin equivalent circuit is to apply a test current I_0, but to leave this current as a literal quantity during the analysis procedure. The solution for the resulting output voltage v at the terminals where I_0 is applied will then have the form

$$v = R_0 I_0 + V_0 \tag{3-53}$$

where V_0 gives the open-circuit ($I_0 = 0$) output voltage and R_0 gives the dependence of the output voltage on I_0. If we now apply a test current $i = I_0$ to the Thevenin equivalent circuit shown in Fig. 3-8.5, we obtain

$$v = R_{\text{Thev.}} I_0 + V_{\text{Thev.}} \tag{3-54}$$

Comparing this with Eq. (3-53), we see that $R_0 = R_{\text{Thev.}}$ and $V_0 = V_{\text{Thev.}}$. Thus, putting the equation for the output voltage in the form given by Eq. (3-53) directly determines the Thevenin circuit quantities. Similarly, if a test voltage $v = V_0$ is applied at the output terminals, from Fig. 3-8.8 we see that the resulting current i will be given by an expression having the form

$$i = \frac{V_0}{R_{\text{Nort.}}} - I_{\text{Nort.}} \tag{3-55}$$

An example follows.

Example 3-8.4

For the circuit used in Example 3-8.3, applying a test voltage $v = V_0$ at the terminals a-b (with the positive reference at terminal a), defining two clockwise mesh currents i_1 and i_2 (with i_1 in the left mesh), and noting that $i_0 = i_1$, we obtain the mesh equations

$$\begin{bmatrix} -2i_1 + 2 \\ 2i_1 - V_0 \end{bmatrix} = \begin{bmatrix} 2 & -1 \\ -1 & 2 \end{bmatrix} \begin{bmatrix} i_1 \\ i_2 \end{bmatrix}$$

Solving, we obtain

$$-i_2 = \tfrac{4}{5}V_0 - \tfrac{6}{5}$$

from which we see that $R_{\text{Thev.}} = \tfrac{5}{4}$, $I_{\text{Nort.}} = \tfrac{6}{5}$, and $V_{\text{Thev.}} = \tfrac{3}{2}$, confirming the results given in Example 3-8.3.

3-9 APPLICATIONS—TRANSISTORS AND OPERATIONAL AMPLIFIERS

In this section we consider some applications of the analysis techniques which have been presented in this chapter. As a first application consider the analysis of a circuit which contains a transistor. An example of such a circuit is shown in Fig. 3-9.1(a).[9]

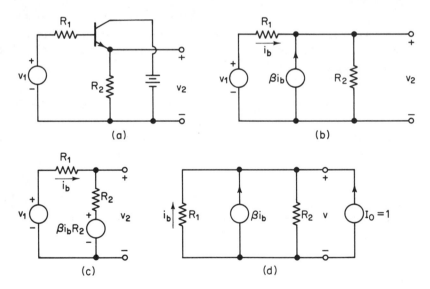

Figure 3-9.1 Emitter follower and its model.

Let us assume that it is desired to find the small signal voltage transfer function $v_2(t)/v_1(t)$. The first step in the analysis is to replace the transistor by the ICIS model defined in Fig. 2-2.7. In addition, the biasing voltage source should be deleted so that

[9]In this circuit an *npn* transistor is shown. The small signal model, however, applies equally well to a *pnp* transistor.

the small signal behavior can be studied (see Sec. 2-6). The resulting circuit is shown in Fig. 3-9.1(b). To analyze the circuit, the shunt-connected ICIS and resistor should be changed to a series-connected ICVS and resistor as shown in Fig. 3-9.1(c). For this circuit

$$v_1(t) = i_b(t)[R_1 + R_2(1 + \beta)] \quad \text{and} \quad v_2(t) = i_b(t)R_2(1 + \beta) \quad (3\text{-}56)$$

Eliminating i_b from the two equations we obtain

$$\frac{v_2(t)}{v_1(t)} = \frac{R_2(1 + \beta)}{R_1 + R_2(1 + \beta)} \quad (3\text{-}57)$$

Note that for large values of β, $v_2(t)/v_1(t) \approx 1$. This circuit is called an *emitter follower*. To find its output resistance, we replace the *independent* source $v_1(t)$ by a short circuit (its internal resistance), and apply the technique illustrated in Example 3-7.4, applying a test current I_0 at the output terminals and determining the resulting voltage v. Using $I_0 = 1$, the circuit is shown in Fig. 3-9.1(d). Using the relation $i_b = -v/R_1$, and applying KCL we obtain

$$-\frac{v}{R_1} - \beta\frac{v}{R_1} + 1 = \frac{v}{R_2} \quad (3\text{-}58)$$

Solving for v we find

$$r_{\text{out}} = \frac{v}{1} = \frac{1}{\dfrac{1 + \beta}{R_1} + \dfrac{1}{R_2}} \quad (3\text{-}59)$$

The input resistance is simply the ratio of $v_1(t)$ to $i_b(t)$ as given by the first equation of (3-56). Thus we see that $r_{\text{in}} = R_1 + R_2(1 + \beta)$.

The determination of the open-circuit voltage $v_2(t)$ given as a function of the input $v_1(t)$ in Eq. (3-57) follows the same procedure as is used to determine the Thevenin voltage. In addition, the output resistance found in Eq. (3-59) is determined in exactly the same manner as is used to find the Thevenin resistance. Thus, the small-signal behavior of the transistor circuit shown in Fig. 3-9.1, as viewed from its output terminals, may be represented by the Thevenin equivalent circuit shown in Fig. 3-8.5, in which

$$V_{\text{Thev.}}(t) = \frac{v_1(t)R_2(1 + \beta)}{R_1 + R_2(1 + \beta)} \quad \text{and} \quad R_{\text{Thev.}} = \frac{1}{\dfrac{1 + \beta}{R_1} + \dfrac{1}{R_2}} \quad (3\text{-}60)$$

A second transistor circuit is shown in Fig. 3-9.2(a). Since the emitter is grounded, this type of circuit is usually called a *grounded-emitter* or a *common-emitter* circuit. A small-signal model for the circuit is given in Fig. 3-9.2(b). For this model

$$v_1(t) = i_b(t)R_1, \quad v_2(t) = -\beta i_b(t)R_2, \quad \text{and} \quad \frac{v_2(t)}{v_1(t)} = -\beta\frac{R_2}{R_1} \quad (3\text{-}61)$$

Since $|v_2(t)/v_1(t)| \gg 1$, this circuit is frequently used as an (inverting) amplifier. The output resistance is readily shown to be R_2 and the input resistance to be R_1. The parameters of the Thevenin equivalent circuit as seen at the terminals where $v_2(t)$ is defined are

$$V_{\text{Thev.}}(t) = -\frac{\beta R_2 v_1(t)}{R_1} \quad \text{and} \quad R_{\text{Thev.}} = R_2 \qquad (3\text{-}62)$$

(a)   (b)

Figure 3-9.2   Common-emitter transistor amplifier.

Example 3-9.1

As another example of the modeling of a transistor, consider the circuit shown in Fig. 3-9.3(a). For this circuit, it is desired to find the voltage v_0. The analysis of the circuit may be simplified by representing the elements to the right of v_0 by their Theve-

(a)   (b)

Figure 3-9.3   Common-emitter transistor amplifier.

nin equivalent. For this portion of the circuit we find $V_{\text{Thev.}} = 5$ V, $R_{\text{Thev.}} = 5$ kΩ. Using the ICIS model for the transistor we obtain the circuit shown in Fig. 3-9.3(b). Analysis of this circuit yields $v_0 = 2.5$ V.

Another dc small-signal equivalent model which is used for transistors in their active region is shown in Fig. 3-9.4. It uses a VCIS as an active element, and thus is

Figure 3-9.4   VCIS model for a transistor.

extremely convenient to use in nodal analysis. As an example of the application of this model, consider the circuit shown in Fig. 3-9.5(a). For this circuit, it is desired to find the Thevenin equivalent at the terminals a-b. Inserting the model shown in Fig. 3-9.4 into this circuit, and transforming the non-ideal input source, we obtain the result shown in Fig. 3-9.5(b). In the figure, for convenience, we have defined $g_\pi = 1/r_\pi$,

Figure 3-9.5 Common-emitter transistor amplifier with feedback.

$G_f = 1/R_f$, and $G_o = 1/R_o$. Applying the test current I_0 as shown by the dashed lines in the figure, we obtain the node equations

$$\begin{bmatrix} V_s(t)G_1 \\ I_0 \end{bmatrix} = \begin{bmatrix} G_1 + g_\pi + G_f & -G_f \\ g_m - G_f & G_o + G_f \end{bmatrix} \begin{bmatrix} v_1(t) \\ v_2(t) \end{bmatrix} \tag{3-63}$$

Solving these equations, and using Eq. (3-54) we find

$$v_2(t) = R_{\text{Thev.}}I_0 + V_{\text{Thev.}}(t) = \frac{(G_1 + g_\pi + G_f)I_0}{D} - \frac{G_1(g_m - G_f)V_s(t)}{D} \tag{3-64}$$

where

$$D = (G_o + G_f)(G_1 + g_\pi + G_f) + G_f(g_m - G_f) \tag{3-65}$$

Note that in Eq. (3-64), $V_{\text{Thev.}}(t)$ is negative, thus the input voltage $V_s(t)$ is inverted.

Another important application of dependent sources is in the modeling of operational amplifiers. An operational amplifier is basically a VCVS with a high value of gain (ideally it approaches infinity). A circuit symbol and a dependent source model for a *differential input operational amplifier* is shown in Fig. 3-9.6. Here the term "differential input" refers to the fact that the output voltage is proportional to the *difference* between the two input voltages $v_+(t)$ and $v_-(t)$. Note that, as shown by the

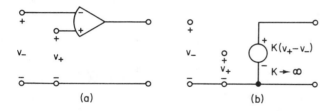

(a)　　　　(b)

Figure 3-9.6 Operational amplifier and its model.

model, no current can flow into either the $v_+(t)$ or the $v_-(t)$ terminal, i.e., the input resistance of these terminals is infinite. Also, the output resistance is that of the dependent voltage source, i.e., zero. Real-world operational amplifiers approach the model quite closely. For example, the μA 741, a popular integrated circuit operational amplifier, typically has an input resistance of 2 MΩ, an output resistance of 75 Ω,[10] and a dc gain of 200,000. Circuits which include operational amplifiers may be analyzed using the techniques developed in the preceding sections of this chapter. As an example, consider finding the voltage transfer function $v_2(t)/v_1(t)$ for the circuit shown in Fig. 3-9.7(a). Using the model of Fig. 3-9.6(b), defining a current $i(t)$ as shown in Fig. 3-9.7(b), and applying KVL we obtain

$$v_1(t) = i(t)R_1 + v_-(t) \quad \text{and} \quad v_2(t) = -i(t)R_2 + v_-(t) \quad (3\text{-}66)$$

Since $v_2(t) = -Kv_-(t)$ and thus $v_-(t) = -v_2(t)/K$, in the limit as K approaches infinity [for a finite value of $v_2(t)$], $v_-(t)$ approaches 0. Thus, taking the ratio of the equations of (3-66) and cancelling i we obtain

$$\frac{v_2(t)}{v_1(t)} = -\frac{R_2}{R_1} \quad (3\text{-}67)$$

Since $v_-(t) = 0$, the input resistance of the circuit is simply R_1. The output resistance is simply that of the dependent VCVS, i.e., zero. Thus, from (3-67), the operational amplifier circuit may be modeled as shown in Fig. 3-9.7(c). Note that if R_1 is large

Figure 3-9.7   Inverting VCVS.

enough, it may be ignored, and the circuit approximates an ideal VCVS of gain $-R_2/R_1$. Because the gain is negative, we refer to this as an *inverting* VCVS. Note that, although the gain of the operational amplifier is infinite, the use of the two resistors has provided a feedback network which allows us to realize low values of gain such as -1, -5.7, etc.

As a second example of a circuit including an operational amplifier, consider the one shown in Fig. 3-9.8(a) and modeled in Fig. 3-9.8(b). Defining a current $i(t)$ as shown, and applying KVL we obtain

$$v_2(t) = i(t)(R_1 + R_2), \quad v_1(t) = v_+(t), \quad \text{and} \quad v_-(t) = i(t)R_1 \quad (3\text{-}68)$$

[10]Although this output resistance seems high compared with the 1-Ω resistors we frequently use in our analyses, in practice, operational amplifiers usually drive loads of 10 kΩ or more. Compared with such values, the output resistance of 75 Ω can be treated as insignificant, i.e., zero.

Resistance Networks   Chap. 3

Figure 3-9.8 Noninverting VCVS.

Since $K[v_+(t) - v_-(t)] = v_2(t)$, then $v_+(t) - v_-(t) = v_2(t)/K$, and in the limit as K approaches infinity, $v_+(t) = v_-(t)$. Using this result in (3-68) we readily derive the result

$$\frac{v_2(t)}{v_1(t)} = \frac{R_1 + R_2}{R_1} \qquad (3\text{-}69)$$

Since the input resistance of the circuit is infinite and the output resistance is zero, this circuit realizes a *non-inverting* VCVS with a gain of $(R_1 + R_2)/R_1$ as shown in Fig. 3-9.8(c). Like the previous circuit, it uses an infinite-gain operational amplifier to realize a VCVS with small values of gain. Note that if R_1 is an open circuit and R_2 is a short circuit, we then realize a VCVS with a gain of unity. Such a circuit is sometimes referred to as a *buffer amplifier* because it can be used to isolate (or buffer) one circuit from another to prevent them from interacting. It is also called a *voltage follower*.

As another example of the use of operational amplifiers, consider the circuit shown in Fig. 3-9.9(a). Two currents are labeled in this figure, namely, $i_1(t)$ and $i_2(t)$. Since no current flows into the minus terminal of the operational amplifier, the entire input current $i_1(t)$ must flow through the resistor R_1. Similarly, all of $i_2(t)$ must flow through the resistor R_2. Now consider the input voltages $v_1(t)$ and $v_2(t)$. From the discussion of the preceding paragraphs, we see that these voltages are equal, namely, $v_1(t) = v_2(t)$. This implies that the voltage drop across R_1 must be the same as that across R_2. Therefore, the relation $R_1 i_1(t) = R_2 i_2(t)$ must hold. Thus, this circuit defines a three-terminal device in which the terminal variables of voltage and current are related by the following equations:

$$v_1(t) = v_2(t) \qquad \text{and} \qquad R_1 i_1(t) = R_2 i_2(t) \qquad (3\text{-}70)$$

Figure 3-9.9 Negative-impedance converter.

In order to determine the properties of such a device, let us connect a resistor R_L from terminal 2 to terminal 3 as shown in Fig. 3-9.9(b). The variables $v_2(t)$ and $i_2(t)$ must now satisfy the relation $v_2(t) = -R_L i_2(t)$. Substituting this in Eq. (3-70), and taking the ratio of the equations we obtain

$$\frac{v_1(t)}{i_1(t)} = -\frac{R_1}{R_2} R_L \qquad (3\text{-}71)$$

The ratio $v_1(t)/i_1(t)$, however, defines the input resistance seen between the terminals 1 and 3 of the circuit shown in Fig. 3-9.9(b). From Eq. (3-71), we see that this input resistance is a negative-valued constant, namely, $-R_1/R_2$ times the resistance R_L. Thus, the circuit shown in Fig. 3-9.9(b) effectively produces a two-terminal, *negative-valued* resistor. This means that the portion of the circuit labeled NIC has *converted* the positive-valued resistor R_L to a negative-valued resistor. Thus, the circuit may be appropriately referred to as a *negative-resistance converter*. In Chap. 7 we shall show that the term *impedance* is a more general term which includes resistance as a special case. Anticipating this result, we may call the device shown in Fig. 3-9.9 a *negative-impedance converter* or NIC for short. It may be shown that stable operation will result only if the resistance connected between terminals 1 and 3 is greater than that connected between terminals 2 and 3.

Another multi-terminal device which is frequently encountered in network studies is the *gyrator*. In terms of the voltage and current variables shown in Fig. 3-9.10(a), it is defined by the equations

$$v_1(t) = -Ri_2(t) \qquad \text{and} \qquad v_2(t) = Ri_1(t) \qquad (3\text{-}72)$$

(a)   (b)   **Figure 3-9.10**   Gyrator.

where R is called the *gyration resistance* of the device. The name "gyrator" comes about since this device may be shown to be the electrical analog of a familiar mechanical device, the gyroscope. The relations of Eq. (3-72) can be modeled using two ICVS's. Such a model is shown in Fig. 3-9.10(b). A circuit which realizes the relations of Eq. (3-72) is shown in Fig. 3-9.11. As illustrated, it requires the use of two NIC's. It is assumed that both of these have unity gain, i.e., in the circuit shown in Fig. 3-9.9(a), $R_1 = R_2$. The details of analyzing this circuit are left as an exercise for the

Figure 3-9.11   Using NIC's to realize a gyrator.

reader. Both the NIC and the gyrator can also be realized by direct use of solid state devices such as transistors. Examples of such realizations may be found in the problems at the end of this chapter. Additional information on these devices and their use may be found in the references given in the Bibliography at the end of this book.

One of the advantages of using operational amplifiers in circuit applications is that these devices are readily available in a wide range of prices, qualities, and sizes, ranging from small, completely integrated units which sell for less than a dollar to precision hybrid units which may have a price of $100 or more. The techniques for analyzing resistance networks containing operational amplifiers can be summarized as follows:

SUMMARY 3-9.1

Operational Amplifiers. An operational amplifier, as shown symbolically in Fig. 3-9.6(a), can be modeled as an ideal VCVS with infinite gain. The input terminals where $v_+(t)$ and $v_-(t)$ are defined have infinite resistance to ground. Thus, no current flows into these terminals. In addition, when the operational amplifier is used in a network such that there is feedback from its output to either of these input terminals, then the voltage difference between them is driven to zero, i.e., $v_+(t) = v_-(t)$.　■

3-10 CONCLUSION

This chapter has presented our first definitive study of the analysis of networks. Although the techniques presented in it have been developed only for resistance networks, we shall see in later chapters that the same techniques apply with only slight modifications to a considerably broader class of circuits, namely, those which include inductors and capacitors, the network elements which are described in the following chapter.

PROBLEMS

Problem 3-1 (Sec. 3-1)
Find the mesh currents i_1 and i_2 in the resistance network shown in Fig. P3-1.

Figure P3-1

Problem 3-2 (Sec. 3-1)

In the circuit shown in Fig. P3-2, it is desired to have $i_1 = 2$ A and $i_2 = -3$ A. Find the voltages v_1 and v_2 which will achieve this.

Figure P3-2   Figure P3-3

Problem 3-3 (Sec. 3-1)

For the circuit shown in Fig. P3-3, find values of R_1 and R_2 such that $i_1 = 1$ A and $i_2 = 2$ A.

Problem 3-4 (Sec. 3-1)

For the circuit shown in Fig. P3-4, find the mesh currents $i_1(t)$ and $i_2(t)$.

Figure P3-4

Problem 3-5 (Sec. 3-1)

(a) Find the value of v_0 in the network shown in Fig. P3-5.

(b) Add 1 V to the value of all three sources and repeat the problem.

Figure P3-5

Problem 3-6 (Sec. 3-2)

Write (but do not solve) the matrix equations for the resistance network hown in Fig. P3-6.

Figure P3-6

Problem 3-7 (Sec. 3-2)

Eliminate the current source in the network shown in Fig. P3-7 in such a way that the resulting network may be described by one additional mesh current (in addition to the two already shown). Write the mesh equations in matrix form for the resulting network. Do not solve the equations.

ohms

Figure P3-7

Problem 3-8 (Sec. 3-2)

Find the mesh currents $i_1(t)$ and $i_2(t)$ for the circuit shown in Fig. P3-8 in which one of the resistors is time-varying.

ohms

Figure P3-8

Problem 3-9 (Sec. 3-2)

(a) For the network shown in Fig. 3-2.1, write the KVL equations using the loops defined in Fig. 3-2.4(a).

(b) Repeat the problem for the loops defined in Fig. 3-2.4(b).

(c) Repeat the problem for the loops defined in Fig. 3-2.4(c).

Problem 3-10 (Sec. 3-3)

In the set of equations which follow, find i_1, i_2, and i_3.

$$2i_1 + 3i_2 - 4i_3 = 1$$
$$i_1 + 2i_2 - i_3 = 12$$
$$4i_1 + i_2 - 3i_3 = 7$$

Problem 3-11 (Sec. 3-3)

Find the inverse of the matrix **A**.

$$\mathbf{A} = \begin{bmatrix} 3 & -1 & 0 \\ 2 & 7 & 1 \\ 1 & -1 & 1 \end{bmatrix}$$

Problem 3-12 (Sec. 3-3)

Find the inverse of the following matrix.

$$\begin{bmatrix} \dfrac{s}{s+1} & -s \\ -s & \dfrac{2s}{s+1} \end{bmatrix}$$

Problem 3-13 (Sec. 3-3)

Find the mesh currents in the circuit shown in Fig. P3-13.

ohms

Figure P3-13

Problem 3-14 (Sec. 3-3)

(a) Solve for the loop currents defined in Problem 3-9(a), and show that your solution agrees with the one obtained in Example 3-3.1.

(b) Repeat the problem for the loop currents defined in Problem 3-9(b).

(c) Repeat the problem for the loop currents defined in Problem 3-9(c).

Problem 3-15 (Sec. 3-3)

Find the elements a, b, c, d, and e in the matrix shown at the right below such that it is the inverse of the matrix shown at the left. Note that the constant $\frac{1}{44}$ multiplies all the elements of the right array.

$$\begin{bmatrix} 3 & -2 & -1 \\ -2 & 5 & -3 \\ -1 & -3 & 8 \end{bmatrix} \qquad \frac{1}{44}\begin{bmatrix} a & d & e \\ b & 23 & 11 \\ c & 11 & 11 \end{bmatrix}$$

Problem 3-16 (Sec. 3-3)

Find the inverse of the following matrix. Verify the result by multiplying the inverse by the original matrix.

$$\begin{bmatrix} 4 & 1 & 1 & 0 \\ 1 & 7 & 1 & 1 \\ 2 & 6 & 4 & 1 \\ -5 & 2 & -1 & 5 \end{bmatrix}$$

Problem 3-17 (Sec. 3-4)

Find the node voltages for the circuit shown in Fig. P3-17.

mhos

Figure P3-17

Problem 3-18 (Sec. 3-4)

In the circuit shown in Fig. P3-18 it is desired to have $v_1 = 3$ and $v_2 = 1$ V. Find i_1 and i_2.

mhos

Figure P3-18

Problem 3-19 (Sec. 3-4)

In the circuit shown in Fig. P3-19, find G_1 and G_2 such that $v_1 = 1$ and $v_2 = 2$ V.

Resistance Networks   Chap. 3

Figure P3-19   **Figure P3-20**

Problem 3-20 (Sec. 3-4)

For the circuit shown in Fig. P3-20, find the values of the node voltages $v_1(t)$ and $v_2(t)$.

Problem 3-21 (Sec. 3-4)

Find the node voltages in the network shown in Fig. P3-21. Compare the results with those obtained in Problem 3-17.

Figure P3-21

Problem 3-22 (Sec. 3-5)

Write (but do not solve) the matrix node equations for the network shown in Fig. P3-22.

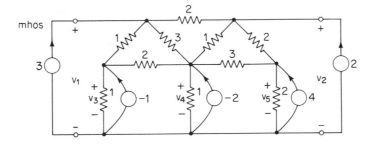

Figure P3-22

Problem 3-23 (Sec. 3-5)

Find the node voltages for the circuit shown in Fig. P3-23.

Figure P3-23

Problem 3-24 (Sec. 3-5)

Solve for the node-pair voltages defined by the graph of Fig. 3-5.4 for the network shown in Fig. 3-5.2, and show that your solution agrees with the one obtained in Example 3-5.1.

Problem 3-25 (Sec. 3-6)

For the circuit shown in Fig. P3-25:
(a) Find i_a using source transformations.
(b) Verify your answer by solving the node equations for v_1 and v_2 and finding the current in the 2-mho resistor.

Figure P3-25

Problem 3-26 (Sec. 3-6)

Solve for i in the network shown in Fig. P3-26.

Figure P3-26

Problem 3-27 (Sec. 3-6)

Find i in the circuit shown in Fig. P3-27.

Problem 3-28 (Sec. 3-6)

For the circuit shown in Fig. P3-28, find i_0. (*Hint:* Use source transformations.)

Figure P3-27

Figure P3-28

Problem 3-29 (Sec. 3-6)

For the circuit shown in Fig. P3-29, find v_0. (*Hint:* Use source transformations.)

Resistance Networks   Chap. 3

Figure P3-29

Problem 3-30 (Sec. 3-6)

In the circuit shown in Fig. P3-30, replace the five sources with a single equivalent source such that the currents and the voltages across resistors R_1 through R_6 remain unchanged.

Figure P3-30

Problem 3-31 (Sec. 3-6)

For the circuit shown in Fig. P3-31, replace the three sources with a single source in such a way that the currents and the voltages across resistors R_1 through R_4 remain the same.

Figure P3-31

Problem 3-32 (Sec. 3-6)

Find delta and wye equivalent circuits for the wye and delta networks shown in Fig. P3-32.

Figure P3-32

Problem 3-33 (Sec. 3-6)

Simplify the expressions given in (3-43) and (3-44) for the case where all the resistors of the wye network (and thus all the resistors of the delta network) have the same value.

Problem 3-34 (Sec. 3-6)

Five solar panels and a conventional (ac line powered) charger are connected to charge a battery as shown in Fig. P3-34. The panels (for a given solar input), diode, battery, and charger may be modeled as shown. Find the charging current into the battery.

Figure P3-34

Problem 3-35 (Sec. 3-7)

Find the mesh currents i_1 and i_2 for the network shown in Fig. P3-35.

Figure P3-35

Problem 3-36 (Sec. 3-7)

Write and solve the matrix equations for the circuit shown in Fig. P3-36.

Figure P3-36   Figure P3-37

Problem 3-37 (Sec. 3-7)

Find the node voltages for the network shown in Fig. P3-37 for the case where $k = -2$.

Problem 3-38 (Sec. 3-7)

For the circuit shown in Fig. P3-38 find i_0.

Figure P3-38

Problem 3-39 (Sec. 3-7)

The three-loop network shown in Fig. P3-39 includes dependent, current-controlled voltage sources. Find the loop currents i_1, i_2, and i_3.

Figure P3-39

Problem 3-40 (Sec. 3-7)

Find the current $i_0(t)$ in the circuit shown in Fig. P3-40.

Figure P3-40

Problem 3-41 (Sec. 3-7)

Find the voltage v_0 in the circuit shown in Fig. P3-41.

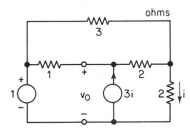

Figure P3-41

Problem 3-42 (Sec. 3-7)

Replace the circuit shown in Fig. P3-42 with a single equivalent resistor connected across terminals *a-b*.

Figure P3-42

Problem 3-43 (Sec. 3-7)

In the circuit shown in Fig. P3-43, find a value of the gain K such that $R_{in} = 0$.

Figure P3-43

Problem 3-44 (Sec. 3-8)

For the circuits shown in Fig. P3-44:
(a) Find the Thevenin equivalent circuit at terminals *a-b*.
(b) Find the Norton equivalent circuit at terminals *a-b*.

(a)   (b)   (c)

Figure P3-44

Problem 3-45 (Sec. 3-8)

For the circuits found in Problem 3-44, find the current that will flow under each of the following conditions:
(a) In a short circuit connected between terminals *a-b*.
(b) In a 1-Ω resistor connected between terminals *a-b*.

Problem 3-46 (Sec. 3-8)

For the circuit shown in Fig. P3-44(a), find the power dissipated in the original circuit, the power dissipated in the Thevenin equivalent, and the power dissipated in the Norton equivalent for the following conditions:
(a) The terminals *a-b* are open-circuited.
(b) The terminals *a-b* are short-circuited.

Problem 3-47 (Sec. 3-8)

Find Thevenin and Norton equivalent circuits for the network shown in Fig. P3-47 at the terminals *a-b*.

Figure P3-47

Problem 3-48 (Sec. 3-8)

For the circuit shown in Fig. P3-48:
(a) Find a Thevenin equivalent circuit.
(b) Find a Norton equivalent circuit.

Figure P3-48

Problem 3-49 (Sec. 3-8)

Find a Thevenin equivalent for the network shown in Fig. P3-49.

Resistance Networks   Chap. 3

Figure P3-49

Problem 3-50 (Sec. 3-8)

Find a Thevenin equivalent for the circuit shown in Fig. P3-50.

Figure P3-50

Problem 3-51 (Sec. 3-8)

Find the Thevenin equivalent of the circuit shown in Fig. P3-51.

Figure P3-51

Problem 3-52 (Sec. 3-9)

Using the transistor model given in Fig. 2-2.7, find the voltage transfer function, input resistance, and output resistance for the circuit shown in Fig. P3-52. Also, at the terminals where v_{out} is defined, find a Thevenin equivalent for the circuit.

Figure P3-52

Problem 3-53 (Sec. 3-9)

Repeat Problem 3-52 for the circuit shown in Fig. P3-53.

Figure P3-53

Problem 3-54 (Sec. 3-9)

Repeat Problem 3-52 for the circuit shown in Fig. P3-54.

Figure P3-54

Problem 3-55 (Sec. 3-9)

A field-effect transistor has the symbol and model shown in Fig. P3-55(a) and (b) (g = gate, d = drain, s = source). Find the voltage transfer function, input resistance, and output resistance for the circuit shown in Fig. P3-55(c). Also, find a Thevenin equivalent for the circuit.

(a)   (b)   (c)   **Figure P3-55**

Problem 3-56 (Sec. 3-9)

Repeat Problem 3-55 for the circuit shown in Fig. P3-56.

Figure P3-56

Problem 3-57 (Sec. 3-9)

Prove that the operational amplifier circuit shown in Fig. P3-57 provides a realization for an ideal inverting ICVS. What is the value of the gain constant?

Figure P3-57

Problem 3-58 (Sec. 3-9)

Prove that the operational amplifier circuit shown in Fig. P3-58 provides a realization for an ideal VCIS with the output current i_2 being taken through the load. What is the value of the gain constant?

Figure P3-58

Problem 3-59 (Sec. 3-9)

Prove that the operational amplifier circuit shown in Fig. P3-59 provides a realization for an ideal ICIS with the output current i_2 being taken through the load. What is the value of the gain constant?

Figure P3-59

Problem 3-60 (Sec. 3-9)

Using the model for a transistor given in Fig. 2-2.7, prove that the transistor circuit shown in Fig. P3-60 provides a realization for an NIC.

Figure P3-60

Problem 3-61 (Sec. 3-9)

Using the transistor model shown in Fig. P2-6, find a set of relations describing the circuit shown in Fig. P3-61 and having the form

$$\begin{bmatrix} i_2 \\ i_1 \end{bmatrix} = \begin{bmatrix} g_{11} & g_{12} \\ g_{21} & g_{22} \end{bmatrix} \begin{bmatrix} v_1 \\ v_2 \end{bmatrix}$$

In the limit as the α's of the transistor approach unity, what device does this circuit realize?

Figure P3-61

Problem 3-62 (Sec. 3-9)

Prove that the circuit shown in Fig. P3-62 containing an ICVS and an ICIS may be used to realize a gyrator.

Figure P3-62

Problem 3-63 (Sec. 3-9)

Prove that the circuit shown in Fig. 3-9.11 provides a realization for a gyrator.

4

Capacitors

and Inductors

In Chap. 2 we introduced some of the basic elements of networks, namely, resistors and sources. We showed that the resistor was a two-terminal element in which the terminal variables of voltage and current were related by a linear algebraic expression called Ohm's law. In Chap. 3 we introduced some methods for analyzing circuits comprised of resistors and sources. We found that such networks were characterized by sets of simultaneous linear *algebraic* equations, and we discussed methods for solving such sets of equations.

In this chapter we shall introduce some two-terminal elements which have properties which are quite different than those of the resistor. These elements are the capacitor and the inductor. The terminal variables of voltage and current for these elements are related by integral and differential equations rather than algebraic equations. The equations describing circuits in which these elements are placed are *integro-differential* equations. The solution of such equations is a task of quite different complexity than the solution of sets of simultaneous algebraic equations covered in the preceding chapter. The rewards for attacking and solving this more complex problem, however, are great. We shall find that the properties of networks which contain capacitors and inductors are so interesting that resistance networks will appear drab and dull by comparison. Surprisingly, however, many of the techniques used to analyze resistance networks may be applied to these more interesting networks with relatively little modification. Such applications will be the subject of several of the following chapters.

4-1 THE CAPACITOR

A capacitor is a two-terminal element which has the property that its branch voltage and current variables are related by the integral and differential equations

$$v(t) = \frac{1}{C} \int_{-\infty}^{t} i(\tau)\, d\tau \qquad \text{and} \qquad i(t) = C\frac{dv}{dt} \tag{4-1}$$

where the constant C is used to specify the value of the capacitor in farads (abbreviated F). The circuit symbol and the reference polarities for a capacitor are shown in Fig. 4-1.1. Note that the reference polarities follow the associated reference directions

Figure 4-1.1 Circuit symbol and reference polarities for a capacitor.

defined in Sec. 1-2. From the first equation in (4-1) we see that the value of the voltage which is present at the terminals of a capacitor depends on the values of current starting at $t = -\infty$. Thus we say that a capacitor has memory. Since it is difficult to maintain a history of a variable such as current over such a long range of time, we usually use the following alternative integral relation for the voltage and current variables associated with a capacitor:

$$v(t) = \frac{1}{C} \int_{0}^{t} i(\tau)\, d\tau + v(0) \tag{4-2}$$

The constant term $v(0)$ in this equation is called an *initial condition*. Note that from Eq. (4-2) we may represent a capacitor with an initial condition as a series connection of a capacitor without an initial condition and a voltage source as shown in Fig. 4-1.2. It is frequently convenient to be able to use reciprocal units to express the

Figure 4-1.2 Use of a voltage source to represent an initial condition.

value of a capacitor. In such a case we shall use units of *darafs* (abbreviated D), which is simply "farad" spelled backwards. We shall refer to such a unit as a unit of *elastance*, and use the symbol S when we desire to refer to capacitors defined on a reciprocal basis. Such usage exactly parallels the use of R (for resistors whose values are given in ohms), and G (for resistors whose values are given in mhos), as introduced in Sec. 2-1.

We may illustrate the properties of a capacitor more fully by substituting the relation between charge and current, namely, $i(t) = dq/dt$, in Eq. (4-1). We obtain

$$q(t) = Cv(t) \qquad (4\text{-}3)$$

Thus, we see that the value of the capacitance determines the relationship between the value of the charge stored in the capacitor and the value of the voltage appearing across its teminals. Thus, to represent a capacitor we may define a *v-q* plane, i.e., a plane in which the voltage variable $v(t)$ is plotted along the abscissa or "*x*" axis and the charge variable $q(t)$ is plotted along the ordinate or "*y*" axis. Such a representation follows closely that used for a resistor in Sec. 2-1. We may now define a *linear* capacitor as one characterized by a straight line passing through the origin of the *v-q* plane, as shown in Fig. 4-1.3. Note that the slope of the characteristic specifies the value of the

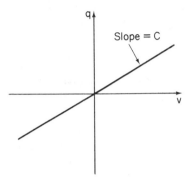

Figure 4-1.3 Voltage-charge characteristic for a capacitor.

capacitor. A non-linear capacitor is thus one whose characteristic is not a straight line and/or does not pass through the origin. As an example of such a characteristic, a solid-state diode, when reverse biased (i.e., with the diode not conducting) exhibits a "space-charge" capacitance which is of the general form $C = Kv^{-1/3}$. For such a diode, we see that Eq. (4-3) must be written in the form $q(t) = C[v(t)]$, i.e., that the relation between charge and voltage is a non-linear one. In addition to the possibility of being nonlinear, physical capacitors may also be time-varying. A linear time-varying capacitor would have a characteristic identical with the one shown in Fig. 4-1.3, except that the slope would vary as a function of time. For a linear time-varying capacitor, the defining relation of Eq. (4-3) becomes

$$q(t) = C(t)v(t) \qquad (4\text{-}4)$$

For such a capacitor, we must modify the terminal relations given at the beginning of this section. For example, differentiating both sides of (4-4), we obtain

$$\frac{dq}{dt} = i(t) = C(t)\frac{dv}{dt} + v(t)\frac{dC}{dt} \qquad (4\text{-}5)$$

This relationship actually includes both the time-varying and the time-invariant cases, since for a time-invariant capacitor $dC/dt = 0$ and Eq. (4-5) simplifies to the derivative relation given in Eq. (4-1). Similarly, for a linear time-varying capacitor, the integral relationship for the terminal variables has the form

$$v(t) = \frac{1}{C(t)}\int_0^t i(\tau)\,d\tau + \frac{q(0)}{C(t)} \qquad (4\text{-}6)$$

Actually, in the real-life or physical world, there is no such thing as a linear time-invariant capacitor. For many applications, however, the terminal properties of physical capacitors so closely approximate the v-q characteristic shown in Fig. 4-1.3 that little is gained by trying to include the non-linear and/or time-varying behavior. Indeed, much may be lost, since the more complicated mathematical treatment required to describe such a capacitor may tend to obscure the fundamental properties of the circuit in which the element is found. To avoid this, we will assume that all physical capacitors are "modeled" by ideal elements obeying the relations given in Eqs. (4-1) through (4-3). Thus, from this point on, when we use the word "capacitor" we imply a *lumped, linear, time-invariant model of some physical capacitor*, unless otherwise specified.

The instantaneous power $p(t)$ which is supplied to a capacitor is easily determined from the relation $p(t) = v(t)i(t)$. Substituting for $i(t)$ from Eq. (4-1), we obtain

$$p(t) = Cv(t)\frac{dv}{dt} \tag{4-7}$$

From this expression we note that, for a positive-valued capacitor, $p(t)$ may be positive or negative, depending on the sign of the quantity $v(t)\,dv/dt$. Thus we see that it is possible for a capacitor to receive energy from the circuit to which it is connected [in which case $p(t)$ will be positive], or to provide energy to the circuit [in which case $p(t)$ will be negative]. These conclusions are based on the associated reference directions developed in Sec. 1-2 for the transferring of energy.

We may readily compute the total energy which has been supplied to a capacitor using the integral relationship given in Eq. (1-3) of Sec. 1-1, and the expression for $p(t)$ derived above. Thus we obtain

$$w(t) = \int_{-\infty}^{t} Cv(\tau)\frac{dv}{d\tau}\,d\tau = \int_{v(-\infty)}^{v(t)} Cv(t)\,dv = \frac{Cv^2(t)}{2} - \frac{Cv^2(-\infty)}{2} \tag{4-8}$$

In this relation, the quantity $v(-\infty)$, i.e., the voltage across the capacitor at $t = -\infty$, must be zero, since at $t = -\infty$ the integral of current supplied to the capacitor must be zero. Therefore, we conclude that the energy $w(t)$ stored in a capacitor is given by the relation

$$w(t) = \frac{Cv^2(t)}{2} \tag{4-9}$$

From the above relation we see that even though the *instantaneous* power $p(t)$ may be positive or negative, the total energy stored in the capacitor as given by (4-9) is always positive (or zero). Thus, we see that a capacitor is a passive network element.[1] It should be noted that the capacitor is a non-dissipative element, i.e., all the energy supplied to a capacitor is stored in the capacitor (actually, more properly, it is stored in the electric field associated with the capacitor). As such, it may be left in storage or it may be reclaimed in the circuit, but it is not dissipated, i.e., it is not converted to heat. Thus, we may expect that circuits containing capacitors will have markedly

---

[1]This, of course, only applies to a linear, time-invariant, positive-valued capacitor. It is interesting to observe that a linear, positive-valued, *time-varying* capacitor may be active! See C. A. Desoer and E. S. Kuh, *Basic Circuit Theory*, Chap. 19, McGraw-Hill Book Company, New York, 1969.

different properties than circuits which contain only dissipative elements such as resistors. The study of such properties will be one of our major concerns in the remainder of this text.

Practical capacitors, such as those used in radios and television sets, may range in value from a few picofarads (10^{-12} F) to hundreds of microfarads (10^{-6} F). As was the case for resistors, however, in our analyses we shall in general use values which are conveniently normalized to lie near unity. In addition to being rated as to their capacitance, a maximum voltage rating, which from Eq. (4-9) determines the maximum energy storage, is also given. Some capacitors, due to the type of materials used in their construction, can only be used in circuit applications in which the polarity of the applied voltage does not change. All real-world capacitors also have some leakage resistance associated with their capacitance. Such a parasitic effect may be modeled by using a shunt resistor in parallel with an ideal capacitor, if the accuracy of the circuit analysis which is being made requires it. In most applications, however, the value of such a resistor is so large as to be negligible.

4-2 SOME FREQUENTLY ENCOUNTERED TIME FUNCTIONS

In our studies of resistance networks in Chaps. 2 and 3 we found that the voltage and current variables could have any desired time dependence since they are always related by *algebraic* equations. A quite different situation exists for networks containing capacitors, however, since from Eq. (4-1) we see that *derivative and integral* relations for the terminal variables replace the algebraic relations which characterize resistors. Thus, we must concern ourselves with the derivatives and integrals of various time functions. In this section we shall define some commonly encountered time functions, and discuss their derivatives and integrals. In the following section we shall investigate what happens when these basic waveshapes are applied as excitations to capacitors.

One of the most frequently encountered time functions used in studies of circuits and systems is the unit step function. It has the waveform shown in Fig. 4-2.1. As shown in the figure, this function has a constant positive value of unity for all time greater than zero, and a value of zero for all time less than zero. Such a function is easily generated physically by a switch. As an example of this, consider the circuit shown in Fig. 4-2.2(a). If the switch is closed at the instant of time when t equals 0, then the waveshape of the voltage across the 1-Ω resistor R has the form shown in Fig. 4-2.2(b), i.e., it is a step function. Similarly, the waveshape of the current in the resistor, as shown in Fig. 4-2.2(c), is also a step function. Although the step function is easily generated physically, it is difficult to describe mathematically. This is the result of the fact that the function does not have a definite value at $t = 0$, even though it does have well-defined values for $t > 0$, and for $t < 0$. Such a function is referred to as a discontinuous function. We may use the symbol $u(t)$ to represent a *unit step function* occurring at $t = 0$. Thus, we may define

$$u(t) = \begin{cases} 0, & t < 0 \\ 1, & t > 0 \end{cases} \tag{4-10}$$

Its value at $t = 0$ may be taken as 0, $\frac{1}{2}$, or 1 without affecting the results that follow. Such a definition is easily generalized to provide a representation for step functions occurring at times other than zero. Thus, for a unit step occurring at time t_0 as shown

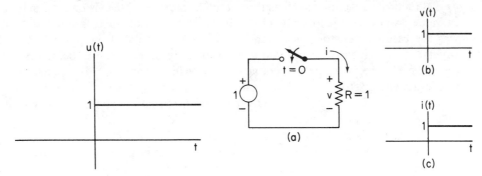

Figure 4-2.1 Unit step function.

Figure 4-2.2 Waveshapes in a simple circuit.

in Fig. 4-2.3, we may define the function $u(t - t_0)$ by the relations

$$u(t - t_0) = \begin{cases} 0, & t - t_0 < 0 \\ 1, & t - t_0 > 0 \end{cases} = \begin{cases} 0, & t < t_0 \\ 1, & t > t_0 \end{cases} \tag{4-11}$$

Note that the significant feature of the definitions given in (4-10) and (4-11) above is the argument of $u(\cdot)$, i.e., the quantity in parentheses following u.[2] Both definitions state the same fact, namely, that when the *quantity or expression* listed as an argument is positive, the function $u(\cdot)$ is unity, and when the quantity is negative, $u(\cdot)$ is zero.

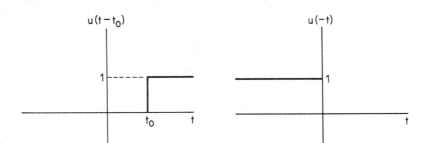

Figure 4-2.3 Delayed unit step function.

Figure 4-2.4 Function $u(-t)$.

Such an interpretation opens the way to extending the definition of the unit step function to many other useful situations. For example, the waveshape shown in Fig. 4-2.4 is simply defined by the expression $u(-t)$.

It is easy to add several unit step functions together to synthesize quite complicated discontinuous waveforms. Some examples follow.

---

[2]The symbolism $u(\cdot)$ is used when it is desired to represent the complete functional relationship, whereas $u(t)$ implies the *value* of the function as determined at some particular value t of its argument.

Example 4-2.1

The pulse $f(t)$ shown in Fig. 4-2.5(a) can be constructed from two step functions, a unit step applied at $t = 0$, as shown in Fig. 4-2.5(b), and a step function with a value of -1 applied at $t = 2$, as shown in Fig. 4-2.5(c). Thus, we may write

$$f(t) = u(t) - u(t - 2)$$

(a) (b) (c)

Figure 4-2.5 Rectangular pulse and its components.

Note that for all values of $t > 2$, the sum of the unit value of $u(t)$ and the -1 value of $-u(t - 2)$ equals zero, as specified for the given function.

Example 4-2.2

The notch waveshape $g(t)$ shown in Fig. 4-2.6(a) can be constructed from two step functions, one an inverted step of value 2 which stops at $t = 1$, as shown in Fig. 4-2.6(b), the other a step function with a value of 2 which is applied at $t = 4$, as shown

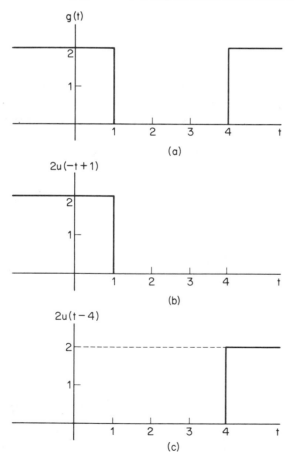

Figure 4-2.6 Notch and its components.

Fig. 4-2.6(c). Thus, we may write

$$g(t) = 2u(-t + 1) + 2u(t - 4)$$

Example 4-2.3

The "staircase" function $h(t)$ shown in Fig. 4-2.7 can be constructed from a series of unit step functions. Thus, we may write

$$h(t) = u(t) + u(t - 2) + u(t - 4) + \cdots$$

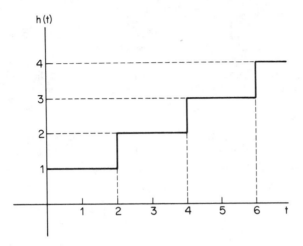

Figure 4-2.7 Staircase function.

Now let us consider a second type of function, the ramp function. A *unit ramp function* $r(t)$ starting at $t = 0$ is shown in Fig. 4-2.8. Such a function is characterized by the properties that it equals t for $t > 0$ and it equals 0 for $t < 0$. This function is easily defined by using the unit step function $u(t)$ to "turn the function off," i.e., set it to zero, for $t < 0$. Thus, we may describe the unit ramp function $r(t)$ shown in Fig. 4-2.8 by the relation

$$r(t) = tu(t) \tag{4-12}$$

More generally, a ramp of slope k which begins at $t = t_0$ as shown in Fig. 4-2.9 may be defined by the expression

$$f(t) = (t - t_0)ku(t - t_0) \tag{4-13}$$

It is easy to show that the unit ramp function $r(t)$ is equal to the integral of the unit

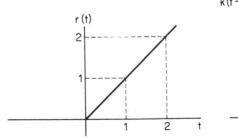

Figure 4-2.8 Unit ramp function.

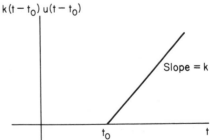

Figure 4-2.9 Delayed ramp function.

Capacitors and Inductors   Chap. 4

step function. Thus we may write

$$r(t) = \int_{-\infty}^{t} u(\tau) \, d\tau \qquad (4\text{-}14)$$

Since a ramp function may be considered as the *integral* of a step function, what sort of function is the *derivative* of a step function? Let us give the symbol $\delta(t)$ to such a function. It must have the property that its integral yields the step function, i.e., that the following relationship holds.

$$u(t) = \int_{-\infty}^{t} \delta(\tau) \, d\tau \qquad (4\text{-}15)$$

This relation defines the *unit impulse function* $\delta(t)$. The opposite relationship may be found by differentiating both sides of (4-15). Thus, we obtain

$$\delta(t) = \frac{du(t)}{dt} \qquad (4\text{-}16)$$

The "slope" of $u(t)$, i.e., the value of its derivative, is zero at every value of t except $t = 0$. At this value of its argument, it is undefined. However, in a non-rigorous sense, we may refer to it as being *infinite*. In such a case, we may define the unit impulse $\delta(t)$ as having a value of zero for all values of its argument t except $t = 0$, and having a value of infinity at $t = 0$. In addition, it must satisfy the constraint

$$\int_{-\epsilon}^{\epsilon} \delta(t) \, dt = 1 \qquad (4\text{-}17)$$

where ϵ is some arbitrary positive constant. The definition of the unit impulse function $\delta(t)$ given above is more easily verified on a limit basis. To do this, let us define a modified unit step function $u_\epsilon(t)$ as shown in Fig. 4-2.10(a). In the limit as the positive

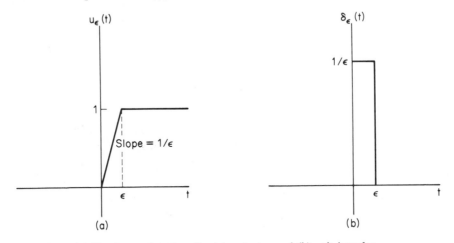

Figure 4-2.10 Approximations for (a) unit step and (b) unit impulse.

constant ϵ approaches 0, this simply becomes the unit step function $u(t)$. Now let us define a modified unit impulse $\delta_\epsilon(t)$ by a relation similar to that given in Eq. (4-16), namely,

$$\delta_\epsilon(t) = \frac{d}{dt} u_\epsilon(t) \qquad (4\text{-}18)$$

The form of $\delta_\epsilon(t)$ is shown in Fig. 4-2.10(b). Note that this function satisfies the relation

$$\int_{-\epsilon}^{\epsilon} \delta_\epsilon(t)\, dt = 1 \qquad (4\text{-}19)$$

As ϵ is made smaller, the function $\delta_\epsilon(t)$ approaches a pulse with very small width, very large height, and an area of unity. Thus, in the limit as $\epsilon \to 0$, we may say that the modified function $\delta_\epsilon(t)$ becomes an impulse function of "zero" width and "infinite" magnitude. Such an interpretation will prove adequate for our applications in this text. The reader who is interested in a more rigorous definition of the properties of the impulse function is referred to the advanced texts listed in the references in the Bibliography.

We may treat impulses occurring at values of time other than zero in a manner similar to that used for unit step functions occurring at other values of time, e.g., the function $\delta(t - t_0)$ represents a unit impulse occurring at $t = t_0$. Similarly, impulses of non-unity value may be thought of as representing the derivative of step functions of non-unity value. Thus, in general we may write

$$k\delta(t - t_0) = k \frac{d}{dt} u(t - t_0) \qquad (4\text{-}20)$$

A symbol which is frequently used to represent the impulse function is shown in Fig. 4-2.11. The number written alongside the arrowhead refers to the area or "strength" of the impulse, i.e., the value of k in Eq. (4-20).

Figure 4-2.11 Delayed unit impulse of strength 4.

In some network situations, we may desire to consider the derivative of an impulse. Such a function is called a *doublet*. Similarly, the derivative of a doublet is referred to as a *triplet*, etc. The treatment of such functions in any detail is beyond the scope of this text.

The impulse function, unlike the step function, is quite difficult to generate physically. Its characterization of near infinite magnitude (even though we say that this occurs for almost zero time) makes it an undesirable waveshape for any variable that might be encountered in a physical circuit. This is so because exceptionally large values of voltages and currents have a disconcerting habit of destroying circuit components long before they reach anything even close to "infinite" magnitude. In many theoretical situations, however, the mathematical use of impulse functions provides considerable insight into the behavior of networks; thus, even though the impulse is physically unattainable, we shall make considerable application of it in the chapters that follow. The reader should be alert to maintain the correct perspective in

Capacitors and Inductors   Chap. 4

such discussions to differentiate between the "mathematical" world and the physical or "real-life" world.

4-3 WAVESHAPES FOR A CAPACITOR

In this section we shall consider the effect of exciting a capacitor by different waveshapes of voltage and current. Such a discussion will provide a clearer picture of the operation of a capacitor in a circuit, and will serve as preparation for a study in the more complicated situations which will be covered in the sections which follow. First of all, let us consider the situation that results when an ideal current source is connected to a capacitor as shown in Fig. 4-3.1. The voltage that appears across the

Figure 4-3.1 Capacitor excited by a current source.

terminals of the capacitor for $t > 0$ is determined by the relation given in Sec. 4-1, namely,

$$v(t) = \frac{1}{C} \int_0^t i(\tau)\, d\tau + v(0) \tag{4-21}$$

Let us now consider the effects of applying various waveforms of current $i(t)$ to the circuit. For example, let $i(t) = u(t)$, a unit step. From the relation given above, we see that in this case

$$v(t) = \frac{1}{C} \int_0^t d\tau + v(0) = \frac{t}{C} + v(0) \qquad t > 0 \tag{4-22}$$

Thus, the voltage that appears across the terminals of the capacitor is a ramp of slope $1/C$ (starting at $t = 0$) plus the constant $v(0)$ representing the value of the initial condition. Clearly, this initial condition must be known before the voltage across the capacitor can be fully defined. If we assume that the initial condition on the voltage is zero, then the current and voltage waveforms for the circuit shown in Fig. 4-3.1 for the case where a unit step of current is applied are shown in Fig. 4-3.2(a).

The presence or absence of an initial voltage on a capacitor has a direct bearing on linearity considerations for circuits which include capacitors. For example, suppose that a 1-F capacitor with an initial condition $v(0)$ of 1 V is excited by a unit step of current $u(t)$ A. The resulting capacitor voltage $v(t)$ is easily seen to be $t + 1$ V ($t > 0$). If we double the input excitation by applying an input current of value $2u(t)$ A, the resulting voltage $v(t)$ is $2t + 1$ V ($t > 0$). Note that by doubling our input we have *not* doubled the resulting voltage, even though the circuit element is a *linear* capacitor. Our conclusion is that superposition with respect to input currents applies *only if the capacitor is uncharged*, i.e., its initial condition is zero. Such considerations will become important when we consider circuits comprised of capacitors and other elements in Chap. 5.

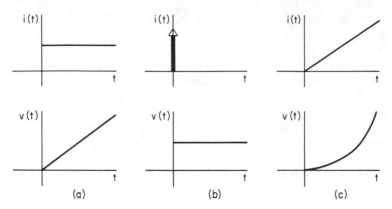

Figure 4-3.2 Waveshapes for a capacitor.

In a similar manner, we may determine the waveshapes of voltage that result for the application of an impulse of current and a ramp of current to a capacitor. These are shown in Figs. 4-3.2(b) and (c), respectively. The waveshapes shown in Fig. 4-3.2(b), representing the situation where an impulse of current is applied to a capacitor, show that the resulting waveshape of voltage is a step function, i.e., a discontinuous wave-shape. This result could also have been predicted from the differential relationship for the voltage and current variables of a capacitor. This has been shown to have the form

$$i(t) = C \frac{dv}{dt} \tag{4-23}$$

Thus, we see that if the voltage waveshape is discontinuous (in which case its derivative must have an infinite magnitude), then the resulting current must be an impulse. This conclusion provides a most important condition on the waveform of the voltage that appears across the terminals of a capacitor. It is known as the *continuity condition*, and it may be summarized as follows:

SUMMARY 4-3.1

Continuity Condition for Capacitors. The voltage appearing across the termi-nals of a linear time-invariant capacitor must always be a continuous function. This assumes that the current through the capacitor is bounded, i.e., that impulses of current are not permitted. ■

We may also express the continuity condition by a mathematical relationship. To do this, let us first define the quantities 0— and 0+ as values of the independent variable t at which some function $f(t)$ is to be evaluated by the relations

$$f(0-) = \lim_{\substack{t \to 0 \\ t < 0}} f(t)$$

$$f(0+) = \lim_{\substack{t \to 0 \\ t > 0}} f(t)$$

Thus $f(0-)$ may be seen to be the limiting value of $f(t)$ as t approaches the value of 0 from "below," i.e., from values of t less than 0. Similarly, $f(0+)$ is the limiting value of $f(t)$ as t approaches 0 from "above." More generally, if we let t_0 be any value of t, then we may define

$$f(t_0-) = \lim_{\substack{t \to t_0 \\ t < t_0}} f(t)$$

$$f(t_0+) = \lim_{\substack{t \to t_0 \\ t > t_0}} f(t)$$

The continuity condition may now be expressed in the form

$$v_C(t_0-) = v_C(t_0) = v_C(t_0+) \tag{4-24}$$

where t_0 is any value of t, and where $v_C(t)$ is the voltage appearing across the terminals of a capacitor. The implication of the above equation is that the voltage $v_C(t)$ is a continuous function of time.

Example 4-3.1

As an example of the effect of the continuity condition, consider the circuit shown in Fig. 4-3.3(a). Assume that the switch has been connected to the 10-V source (switch position 1) for a long time before $t = 0-$, thus allowing the circuit to reach a steady-

(a)   (b)   (c)

Figure 4-3.3 Example of the continuity condition.

state condition. As a result, at $t = 0-$, $i_C(0-) = dv_C(0-)/dt = 0$. Since $i_C(0-) = 0$ independent of the voltage $v_C(0-)$, the capacitor may be treated as an open circuit. Thus the circuit appears as shown in Fig. 4-3.3(b), and we find that $v_R(0-) = 2$ and $v_C(0-) = 8$. At $t = 0$ the switch is changed to position 2. At $t = 0+$, the capacitor voltage $v_C(0+)$ must remain at 8 V, irrespective of any current flowing through it. Thus, it can be modeled by a voltage source as shown in Fig. 4-3.3(c). From this circuit we find that $v_R(0+) = -8$ and $v_C(0+) = 8$. Thus we see that the voltage across a resistor can be discontinuous, even though the capacitor voltage will always be continuous.

Now let us consider again the waveshapes shown for the application of different waveshapes of current to the terminals of a capacitor as illustrated in Fig. 4-3.2. We note that, for the complete specification of the voltage waveshape across a given capacitor, two items of information are necessary. These are: (1) the explicit nature of the waveshape of current applied to the capacitor; and (2) the initial *state* that the capacitor is in, i.e., the initial condition on the value of the voltage appearing across the capacitor. It should be noted that the use of the word "initial" is relative. If we know the value of the capacitor voltage at any given time t_0, then t_0 becomes our starting point for values of the independent variable t, and the "initial" condition is evaluated at t_0. For this case, Eq. (4-21) becomes

$$v(t) = \frac{1}{C} \int_{t_0}^{t} i(\tau)\, d\tau + v(t_0) \tag{4-25}$$

In most of the discussion which follows we shall assume, for simplicity of notation, that our interest in a given network situation begins at $t = 0$, and our initial condition will be specified at that point.

To further illustrate the manner in which voltage and current waveshapes are related in a capacitor, consider the network configuration shown in Fig. 4-3.4 in which a voltage produced by a voltage source is applied across the terminals of a capacitor. The resulting current that flows into the upper terminal of the capacitor is determined by the differential equation given in Eq. (4-23). Some examples of various waveshapes are shown in Fig. 4-3.5. For example, in Fig. 4-3.5(a) we show the effect of applying a parabolic waveshape of voltage $v(t) = t^2 u(t)$. The result is a ramp function of current $i(t)$. The actual value of the slope, of course, depends on the value of the capacitance. Similarly, in Fig. 4-3.5(b), a ramp voltage $v(t) = tu(t)$ is

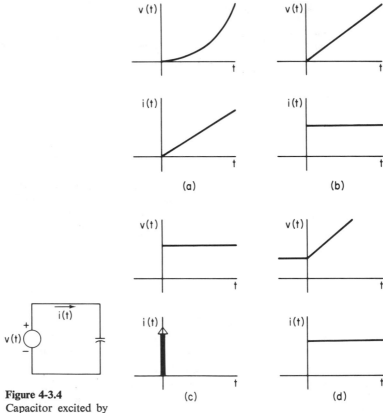

Figure 4-3.4 Capacitor excited by a voltage source.

Figure 4-3.5 Waveshapes for a capacitor.

Capacitors and Inductors   Chap. 4

applied. The result is a step function of current. Finally, in Fig. 4-3.5(c), a step function of voltage is applied. The result, as was discussed above, is an impulse of current. This latter situation, of course, is not possible in physical or real-life situations. It should be noted that all the waveshapes of current shown in Fig. 4-3.5 may apply to many different waveshapes of voltage, all of which represent different initial conditions on the capacitor. As an example of this, in Fig. 4-3.5(d), the effect of applying a ramp of voltage at $t = 0$ in addition to a constant voltage which has been applied at $t = -\infty$ is shown. The waveshape of current (for the same value of capacitance) is indistinguishable from that resulting from a ramp alone as shown in Fig. 4-3.5(b). This, of course, is the result of the fact that differentiation of the constant term in the expression for the voltage applied to the capacitor gives zero. Therefore, such a term has no effect on the value of the current.

4-4 THE INDUCTOR

An inductor is a two-terminal element which has the property that its branch voltage and current variables are related by the expressions

$$i(t) = \frac{1}{L} \int_{-\infty}^{t} v(\tau)\, d\tau \qquad \text{and} \qquad v(t) = L\frac{di}{dt} \qquad (4\text{-}26)$$

where the constant L is used to specify the value of the inductance in henries (abbreviated H). The circuit symbol and the associated reference polarities for an inductor are shown in Fig. 4-4.1. From Eq. (4-26) we see that the value of the current flowing

Figure 4-4.1 Circuit symbol and reference polarities for an inductor.

through an inductor depends on the values of voltage which have been applied to it for values of time starting at $-\infty$. Since it is difficult to maintain such a history, the following alternative definition relating the terminal variables of an inductor is more frequently used

$$i(t) = \frac{1}{L} \int_{0}^{t} v(\tau)\, d\tau + i(0) \qquad (4\text{-}27)$$

The quantity $i(0)$ is referred to as an initial condition. Note that, from Eq. (4-27), we may represent an inductor with an initial condition as a parallel combination of an inductor without an initial condition and a current source as shown in Fig. 4-4.2. It should be noted by the reader that an initial condition for an inductor is preserved in quite a different way than that for a capacitor. If a capacitor has been charged to some voltage, then the open-circuited capacitor will retain this charge. As a matter of fact, high-quality capacitors found in many types of modern electronic apparatus such as radar equipment, television sets, etc., will retain a charge so long that they may

Figure 4-4.2 Use of a current source to represent an initial condition.

actually electrocute the careless worker who touches their terminals. This may happen even though the equipment has been completely disconnected from all sources of power for some time! In an inductor, however, an initial condition or "charge" is preserved by *shorting* the terminals of the inductor. The current (if the inductor is ideal) then simply keeps merrily flowing along. Although such a statement may at first seem strange due to the concept of the continuing "motion" of the charged particles, inductors maintained at cryogenic temperatures (very close to absolute zero) so as to make their resistance exceedingly small have actually been observed to maintain a current for years without any application of external power sources. The integral equations given in (4-26) and (4-27) are quite similar to those defining a capacitor which were presented in Sec. 4-1. As a matter of fact, if we interchange the variables of voltage and current, and substitute the constant C for the constant L, the relations are identical. Network elements with terminal relations whose form is identical if the variables of voltage and current are interchanged are said to be dual to each other. We shall see some interesting applications of the concept of duality in Sec. 6-8.

It is frequently convenient to be able to use reciprocal-valued units when dealing with inductors in a given circuit. In such a case the Greek letter Γ (capital gamma) is usually used. Unlike the resistive and capacitive cases where compact names of "mhos" and "darafs" have been developed for reciprocal-valued elements, the units for Γ are referred to by the clumsy title of "reciprocal henries." If you're looking for an easy chance to become famous, you might try to invent a better name. Obviously, "yrnehs" is not a good choice.

When we described the characteristics of a capacitor, we noted that its properties were based on the relation that this element imposes between the charge stored in the capacitor and the voltage appearing across its terminals. In a similar manner, the properties of an inductor may be defined by the relation that exists between the flux linkages $\phi(t)$ (in weber-turns) and the current flowing in the element. The relation is

$$\phi(t) = Li(t) \tag{4-28}$$

The properties of an inductor are actually a direct result of Faraday's law, introduced in Sec. 1-1, which states that the induced voltage in a circuit is proportional to the time rate of change of flux linkages, namely $v(t) = d\phi/dt$. Combining this expression with the one given in Eq. (4-28), yields the differential relation given in Eq. (4-26).

We may illustrate the properties of an inductor more fully by defining an i-ϕ plane, that is, a plane in which the current variable $i(t)$ is plotted along the abscissa or "x" axis, and the flux-linkage variable $\phi(t)$ is plotted along the ordinate or "y" axis. Such a representation follows closely that used for the resistor in Sec. 2-1 and the capacitor in Sec. 4-1. We may now define a *linear* inductor as one characterized by a straight line passing through the origin of the i-ϕ plane. The slope of the line is the

Capacitors and Inductors   Chap. 4

value of the inductance. An example of such a characteristic is shown in Fig. 4-4.3. Thus, a *time-varying* linear inductor is readily seen to be one with a straight-line characteristic passing through the origin but having a slope which varies as a function of time. For this case the relation of Eq. (4-28) and the differential relation of Eq. (4-26) may be written in the form

$$\phi(t) = L(t)i(t) \quad \text{and} \quad v(t) = L(t)\frac{di}{dt} + i(t)\frac{dL}{dt} \quad (4\text{-}29)$$

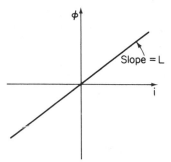

Figure 4-4.3 Current flux-linkage characteristic for an ideal inductor.

A *non-linear* inductor is characterized by a locus on the *i-φ* plane which is not a straight line and/or does not pass through the origin. A type of non-linear behavior which is encountered frequently is that shown in the *i-φ* plane of Fig. 4-4.4. This locus

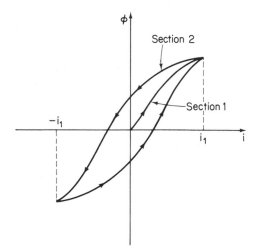

Figure 4-4.4 Current flux-linkage characteristic for a non-linear inductor with hysteresis.

was generated by connecting a current generator to a previously unexcited non-linear inductor, and by increasing the current from 0 to some value i_1. The characteristic generated on the *i-φ* plane is labeled as section 1 of the curve shown in Fig. 4-4.4. If the current is now returned to 0, then as shown in section 2 of the curve, the flux linkages do not go to zero! If this at first seems surprising, the reader should note that this is exactly the type of behavior exhibited by a "permanent" magnet, i.e., it exhibits a flux even though there is no current flow. Additional variations of current between the limits of $-i_1$ and $+i_1$ generate the remaining portions of the characteristic curve in the order indicated by the arrows. This phenomenon is referred to as *hysteresis*. It is of considerable significance in determining the properties of magnetic devices such as solenoids, magnetic cores, magnetic tapes, transformers, and motors.

In the real-life world, all inductors demonstrate non-linear, time-varying behavior. In some applications, such as storage of information in computers, it is the non-linearity itself which provides the mechanism by means of which the desired operation is accomplished. Our primary interest in this text, however, will be with networks whose operation is based on the linear properties of inductors. Thus, we are lead to consider an "ideal" element which obeys the relations given in Eqs. (4-26) through (4-28). From this point on, when we use the word "inductor," we imply a lumped, linear, time-invariant model of some physical inductor, unless otherwise specified.

The power and energy considerations for an inductor are quite similar to those developed for a capacitor. For the power, we obtain

$$p(t) = i(t)v(t) = i(t)L\frac{di}{dt} \tag{4-30}$$

From this relation, we see that the instantaneous power supplied to an inductor may be positive or negative. Thus, at a given instant of time, an inductor may be receiving energy from the circuit to which it is connected, or supplying energy to that circuit. The total energy stored in an inductor is found from the relation

$$w(t) = \int_{-\infty}^{t} p(\tau)\,d\tau = \int_{-\infty}^{t} Li(\tau)\frac{di}{d\tau}\,d\tau = \int_{i(-\infty)}^{i(t)} Li(t)\,di$$

$$= \frac{Li^2(t)}{2} - \frac{Li^2(-\infty)}{2} \tag{4-31}$$

The current though the inductor at $t = -\infty$, however, must be zero. Therefore, we may write $i(-\infty) = 0$. Thus, the development given above simplifies to

$$w(t) = \frac{Li^2(t)}{2} \tag{4-32}$$

From this relation, we see that the energy stored in a positive-valued inductor is always non-negative. Thus, such an inductor is a passive network element.[3] Like the capacitor, the inductor is a non-dissipative element, i.e., all the energy supplied to an inductor may be reclaimed, and none of it is converted to heat.

As a result of the duality which exists for the inductor and the capacitor, we may readily extend the conclusions covering the application of various simple waveshapes for a capacitor to the case of an inductor. For example, if a step input of voltage is applied to an inductor, a ramp of current results. The circuit configuration for such an experiment is shown in Fig. 4-4.5. The waveshapes of voltage and current (assuming that the initial condition is zero) are shown in Fig. 4-4.6(a). Note that these are similar to the waveshapes shown in Fig. 4-3.2(a) for the case where a step input of current was applied to an unexcited capacitor, except for the fact that the voltage and current waveshapes are interchanged. Similarly, the waveshapes that result for the

---

[3]As was pointed out for the capacitor, a positive-valued, linear, *time-varying* inductor, however, may be active.

Figure 4-4.5 Inductor excited by a voltage source.

situations where an impulse of voltage and a ramp of voltage, respectively, are applied to an inductor are shown in Figs. 4-4.6(b) and (c). Note that an impulse of voltage creates a discontinuous function of current. This is readily verified from the derivative relationship given in (4-26). In the real-life world, impulses of voltage cannot exist;

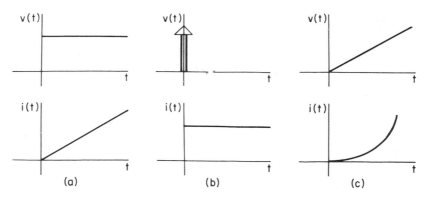

Figure 4-4.6 Waveshapes for an inductor.

thus, we may state a *continuity condition* for inductors exactly paralleling the continuity condition developed for capacitors in Sec. 4-3. The condition may be summarized as follows:

SUMMARY 4-4.1

Continuity Condition for Inductors. The current which flows through a linear time-invariant inductor must always be a continuous function. This assumes that the voltage across the inductor is bounded, i.e., that impulses of voltage are not permitted. ∎

We may also express the continuity condition by the relationship

$$i_L(t_0-) = i_L(t_0) = i_L(t_0+) \tag{4-33}$$

where $i_L(t)$ is the current flowing through an inductor and t_0 is any specific value of t. The notation follows that used for Eq. (4-24). Let us see what happens when we try to violate the continuity condition using the circuit shown in Fig. 4-4.7. We will assume that the battery (a non-ideal source) is producing a constant current flow through the inductor (the details of how such a current flow originates will be treated in the next chapter). Now assume that the switch $S1$ shown in the figure is opened. From the circuit diagram it appears as if this action would instantaneously set the current to

Figure 4-4.7 Violating the continuity condition.

zero, thus producing an impulse of voltage across the terminals of the inductor. If such an experiment is actually attempted, however, an arc will be observed across the switch contacts as they opened. The following is a simplified explanation of the phenomenon: As the switch begins to open and the current begins to change, the voltage induced in the inductor appears across the switch contacts. This voltage is large enough to provide an ionized path which permits the current to continue to flow until the gap between the switch contacts becomes so large that the arc is extinguished. The whole process takes place very quickly, but the net effect is that the current flow is stopped gradually (here "gradually" means in a small fraction of a second); thus the continuity condition is not violated. In effect, the process described above happens every time any switch is opened, since there is some inductance present even in straight wires.

The values of inductance which are frequently encountered in most circuit analyses range from the self inductance of a short length of straight wire (this is some-times called a *lead inductance*), which may be a few microhenries, to the values found in power transformers and large filtering chokes, which may be several henries. As was the case for resistors and capacitors, however, in our analyses we shall, in general, use values which are conveniently normalized to lie near unity. Just as was the case for real-world capacitors, real-world inductors always have some dissipation. This is the result of the resistance of the wire used for winding the coil of the inductor. Also, when an inductor is used in an ac circuit, there is additional dissipation caused by core losses in the magnetic material on which the wire is wound. These effects are sometimes modeled by including resistors in series or parallel with the (ideal) inductor. In general, inductors have proportionately more dissipation than do capacitors, thus, a greater error may be introduced in an analysis due to the neglect of these effects. Successively higher levels of accuracy in a given analysis may be obtained by first including models for the dissipation in the inductors and then including models for the dissipation in capacitors.

4-5 SERIES AND PARALLEL COMBINATIONS OF CAPACITORS AND INDUCTORS

In this section we shall discuss the situations that occur when capacitors or inductors are connected in various series and parallel combinations. There are several interesting situations that arise when such connections are made. First of all, let us consider the case defined by the interconnection of capacitors in which all the initial conditions are zero.

In Fig. 4-5.1 a *parallel* connection of n capacitors of value C_i ($i = 1, 2, \ldots, n$) farads is shown. In terms of the variables defined in the figure, we may write the

Capacitors and Inductors   Chap. 4

Figure 4-5.1 Parallel connection of capacitors.

relations

$$i_i(t) = C_i \frac{dv}{dt} \qquad i = 1, 2, \ldots, n \qquad (4\text{-}34)$$

Application of KCL to the current variables shown in the figure shows that $i(t) = i_1(t) + i_2(t) + \cdots + i_n(t)$. Therefore, using the relations of Eq. (4-34), we may write

$$i(t) = \left(\sum_{i=1}^{n} C_i\right) \frac{dv}{dt} \qquad (4\text{-}35)$$

From the above relation, we may conclude that if n uncharged capacitors of value C_i farads ($i = 1, 2, \ldots, n$) are connected in parallel, they may be replaced by a single equivalent capacitor of value C_{eq}, where

$$C_{eq} = \sum_{i=1}^{n} C_i \qquad (4\text{-}36)$$

Now let us consider the *series* connection of n uncharged capacitors as shown in Fig. 4-5.2. In terms of the variables defined in the figure, we may write

$$v_i(t) = \frac{1}{C_i} \int_0^t i(\tau)\, d\tau \qquad i = 1, 2, \ldots, n \qquad (4\text{-}37)$$

Figure 4-5.2 Series connection of capacitors.

Applying KVL to the circuit we see that $v(t) = v_1(t) + v_2(t) + \cdots + v_n(t)$. Therefore, using the relations of Eq. (4-37), we may write

$$v(t) = \left(\sum_{i=1}^{n} \frac{1}{C_i}\right) \int_0^t i(\tau)\, d\tau \qquad (4\text{-}38)$$

We conclude that if n uncharged capacitors of value C_i farads ($i = 1, 2, \ldots, n$) are connected in series, they may be replaced by a single equivalent capacitor of value C_{eq}, where

$$\frac{1}{C_{eq}} = \sum_{i=1}^{n} \frac{1}{C_i} \qquad (4\text{-}39)$$

This relation is more easily expressed by using values of reciprocal capacitance, i.e., elastance (the units are darafs). Thus, if n uncharged capacitors of value S_i darafs ($i = 1, 2, \ldots, n$) are connected in series, they may be replaced by a single equivalent capacitor of value S_{eq}, where

$$S_{eq} = \sum_{i=1}^{n} S_i \qquad (4\text{-}40)$$

The relations given above are easily combined to determine the value of simple series-parallel combinations of uncharged capacitors. An example follows.

Example 4-5.1

The value C_{eq} of an equivalent capacitor which may be used to replace the four uncharged capacitors shown in Fig. 4-5.3 is easily found by making appropriate replace-

Figure 4-5.3 Network of capacitors.

ments of series and parallel capacitors. Thus, the parallel combination of C_1 and C_2 may be replaced by a single equivalent capacitor of 3 F. The series combination of this capacitor and C_3 is equivalent to a single capacitor of $\frac{6}{5}$ F, and the parallel combination of the latter with C_4 yields an equivalent capacitance for the entire network of $\frac{11}{5}$ F.

The method described in the above example is always applicable to capacitor networks which have the form of a ladder. Some techniques which may be applied to find the equivalent capacitance of more complicated networks may be found in the problems at the end of this chapter.

The discussion given in the above paragraphs is restricted to the case of uncharged capacitors, i.e., ones in which the initial condition is zero. We will now make some remarks about the interconnection of charged capacitors. First let us consider the parallel connection of two capacitors, one charged, the other uncharged, as shown in Fig. 4-5.4(a). The connection is made by means of the switch $S1$ shown in the figure.

Figure 4-5.4 Using one capacitor to charge another.

Before the switch is closed, let us assume that the capacitor labeled C_1 has been charged to V_a volts. Thus, it has $Q_0 = C_1 V_a$ coulombs of charge stored on its plates. When the switch is closed, since the two capacitors are effectively paralleled, the voltage across both must be the same. Let this voltage be V_b as shown in Fig. 4-5.4(b). The total charge that is to be found on the plates of both capacitors must still be Q_0 since charge is always conserved (see Sec. 1-1). Therefore, we may calculate the resulting voltage V_b by the relation

$$V_b = \frac{Q_0}{C_1 + C_2} = \frac{C_1 V_a}{C_1 + C_2} \tag{4-41}$$

It should be noted that the closing of the switch presents a discontinuous waveshape of voltage on capacitor C_2. Thus, from our discussion in Sec. 4-3, an impulse of current must flow into the capacitor. We conclude that the readjustment of voltage takes

Capacitors and Inductors   Chap. 4

place instantaneously if we are considering ideal capacitors. In the case of physical capacitors, there will always be some resistance present, and the transfer of energy will take place in a non-zero (but still very small) interval of time. Some other interesting facts about the values of the variables that result during this operation may be found in the problems at the end of this chapter.

Now let us consider the behavior of inductors when these elements are connected in various series and parallel combinations. We shall see that the results are very similar to those previously discussed for the capacitor. First we will treat the case where the inductors are "uncharged," i.e., the initial condition of current through them is zero. In Fig. 4-5.5, a *series* connection of n inductors of value L_i ($i = 1, 2, \ldots, n$) is shown. In terms of the variables defined in the figure, we may write the relations

$$v_i(t) = L_i \frac{di}{dt} \qquad i = 1, 2, \ldots, n \tag{4-42}$$

Figure 4-5.5 Series connection of inductors.

From KVL, we see that $v(t) = v_1(t) + v_2(t) + \cdots + v_n(t)$. Therefore, using the relations of Eq. (4-42) we may write

$$v(t) = \left(\sum_{i=1}^{n} L_i \right) \frac{di}{dt} \tag{4-43}$$

From Eq. (4-43) we may conclude that if n uncharged inductors of value L_i henries ($i = 1, 2, \ldots, n$) are connected in series, they may be replaced by a single equivalent inductor of value L_{eq}, where

$$L_{eq} = \sum_{i=1}^{n} L_i \tag{4-44}$$

Now let us consider the *parallel* connection of n uncharged inductors as shown in Fig. 4-5.6. In terms of the variables defined in the figure, we may write

$$i_i(t) = \frac{1}{L_i} \int_0^t v(\tau) \, d\tau \qquad i = 1, 2, \ldots, n \tag{4-45}$$

Figure 4-5.6 Parallel connection of inductors.

From KCL we see that $i(t) = i_1(t) + i_2(t) + \cdots + i_n(t)$. Therefore, using the relations of Eq. (4-45), we may write

$$i(t) = \left(\sum_{i=1}^{n} \frac{1}{L_i} \right) \int_0^t v(\tau) \, d\tau \tag{4-46}$$

From Eq. (4-46) we may conclude that if n uncharged inductors are connected in parallel, they may be replaced by a single equivalent inductor of value L_{eq}, where

$$\frac{1}{L_{eq}} = \sum_{i=1}^{n} \frac{1}{L_i} \tag{4-47}$$

Alternately, we see that if n uncharged inductors of value Γ_i reciprocal henries $(i = 1, 2, \ldots, n)$ are connected in parallel, they may be replaced by a single equivalent inductor of value Γ_{eq} reciprocal henries, where

$$\Gamma_{eq} = \sum_{i=1}^{n} \Gamma_i \tag{4-48}$$

The relations given above may be readily applied to find the equivalent inductance of any simple combinations of uncharged inductors in a manner similar to that used for the case of capacitors. Note that the relations for combining inductors are dual to those for combining capacitors, i.e., inductors *in series* add their individual values (in henries) while capacitors in *parallel* add their individual values (in farads). A summary of the properties of series and parallel connections of resistors, capacitors, and inductors is given in Table 4-5.1.

TABLE 4-5.1 Series and Parallel Connections of Resistors, Capacitors, and Inductors

Type of Element	Value of Elements	Units Used for Value	Total Value for a Series Connection	Total Value for a Parallel Connection
resistor	R_i G_i	ohms mhos	$R_{\text{total}} = \sum R_i$	$G_{\text{total}} = \sum G_i$
capacitor	C_i S_i	farads darafs	$S_{\text{total}} = \sum S_i$	$C_{\text{total}} = \sum C_i$
inductor	L_i Γ_i	henries reciprocal henries	$L_{\text{total}} = \sum L_i$	$\Gamma_{\text{total}} = \sum \Gamma_i$

In treating the case where inductors which have non-zero initial conditions are interconnected, we recall first that an inductor with an initial condition must have its terminals shorted, since there must be a closed path for the non-zero current to follow. The voltage appearing across such a short circuit will, of course, be zero. Now consider a circuit consisting of two inductors and a switch $S1$ which is initially closed, as shown in Fig. 4-5.7(a). Let us assume that the inductor of value L_1 henries has a non-zero initial current of value I_a flowing through it. In addition, let us assume that the inductor of value L_2 henries is uncharged. Thus, the current flowing through the (closed) switch $S1$ has the value I_a. From (4-28) we see that there must be $\phi_0 = L_1 I_a$ flux linkages with the coil of the charged inductor. When the switch $S1$ is opened,

Capacitors and Inductors   Chap. 4

Figure 4-5.7 Using one inductor to charge another.

as shown in Fig. 4-5.7(b), the two inductors are connected in series. Thus, the current through both of them must be the same. Let this current be I_b as shown in the figure. The total flux linkages that exist for both of the inductors must still be ϕ_0, since flux linkages are conserved in a magnetic circuit, just as charge is conserved in a capacitive circuit (see Sec. 1-1). Therefore, we may determine the resulting current I_b by the relation

$$I_b = \frac{\phi_0}{L_1 + L_2} = \frac{L_1 L_a}{L_1 + L_2} \tag{4-49}$$

In a manner similar to that explained for the capacitor, the opening of the switch will produce an impulse of voltage across the terminals of the two inductors (since the current through the inductors is discontinuous). Some other interesting properties of this circuit may be found in the problems at the end of this chapter.

4-6 MUTUAL INDUCTANCE—THE TRANSFORMER

In Sec. 3-9 we introduced some multi-terminal network elements, namely, the transistor, the differential-input operational amplifier, the NIC, and the gyrator. In addition to being characterized as having more than two terminals, these devices all have the property that their voltage and current variables are related by *algebraic* equations. In this section, we shall discuss a quite different multi-terminal device in which the voltage and current variables are related by *differential* equations. Such a device is called a *set of coupled coils* or, more generally, a *transformer*. Its principles of operation are a logical extension of the discussions pertaining to the inductor given in Sec. 4-4.

To begin our study of a pair of coupled coils, let us consider first the single coil shown in Fig. 4-6.1. The symbol ϕ_1 has been used for the flux linkages of this coil. The positive reference direction for this quantity follows the usual right-hand rule relating current and flux.[4] If we assume this coil is linear and time-invariant, the relation between the current $i_1(t)$ and the flux linkages $\phi_1(t)$ is

$$\phi_1(t) = L_1 i_1(t) \tag{4-50}$$

where the quantity L_1 is called the *self-inductance* of the coil. Now consider the effect of adding a second coil in the vicinity of the first as shown in Fig. 4-6.2. Some of the lines of flux produced by the current in this new coil will also link the first coil. Thus, the flux linkages $\phi_1(t)$ of the first coil are now determined by the current $i_2(t)$ as well as by the current $i_1(t)$. If we again assume that the relations between flux and current are

---

[4]The right-hand rule, as formulated in basic physics courses, specifies that if the fingers of the right hand are closed around a coil in a direction which follows the positive reference direction of current in the winding, the thumb will point in the positive reference direction for the flux.

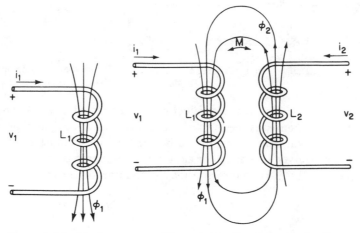

Figure 4-6.1 Single coil.

Figure 4-6.2 Coupled pair of coils.

linear and time-invariant, we may write an expression for $\phi_1(t)$ for this circuit as follows:

$$\phi_1(t) = L_1 i_1(t) + M_{12} i_2(t) \qquad (4\text{-}51)$$

where L_1 is the self-inductance of coil 1 and the quantity M_{12} is called the *mutual inductance* of coil 1 with respect to coil 2. The unit for both the self-inductance and the mutual inductance is the henry. The voltage $v_1(t)$ appearing across the terminals of coil 1 is easily found by applying Faraday's law. Thus, we have

$$v_1(t) = \frac{d\phi_1}{dt} = L_1 \frac{di_1}{dt} + M_{12} \frac{di_2}{dt} \qquad (4\text{-}52)$$

Now let us consider the second coil shown in Fig. 4-6.2. The flux $\phi_2(t)$ linking this coil is produced by the current $i_1(t)$ as well as the current $i_2(t)$. Thus, we may write

$$\phi_2(t) = M_{21} i_1(t) + L_2 i_2(t) \qquad (4\text{-}53)$$

where L_2 is the self-inductance of coil 2 and M_{21} is the mutual inductance of coil 2 with respect to coil 1. The voltage $v_2(t)$ appearing across the terminals of coil 2 is given by the expression

$$v_2(t) = \frac{d\phi_2}{dt} = M_{21} \frac{di_1}{dt} + L_2 \frac{di_2}{dt} \qquad (4\text{-}54)$$

The equations of (4-52) and (4-54) determine the relations between the voltage and current variables for the multi-terminal network element comprised of the pair of coupled coils shown in Fig. 4-6.2. We shall examine these relations in more detail in the following paragraphs.

As a first investigation of the properties of the pair of coupled coils shown in Fig. 4-6.2, we shall use energy considerations to prove that the mutual inductance terms M_{12} and M_{21} are equal. To show this, let us connect a current source to the terminals of coil 1 and raise the current $i_1(t)$ from 0 to I_1 over the period $0 \le t \le t_1$ as shown in Fig. 4-6.3(a). It is assumed that the change of current is made without

Capacitors and Inductors   Chap. 4

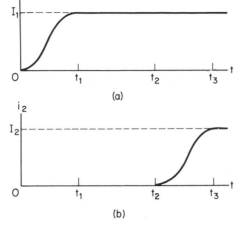

(a)

(b)

Figure 4-6.3 Waveforms in a pair of coupled coils.

producing any discontinuities in $i_1(t)$. The reason for this is to avoid any resultant impulses in $v_1(t)$. If we assume, as shown in Fig. 4-6.3(b), that the current $i_2(t)$ is zero over this time interval, we may compute the total energy W_a stored in coil 1 at time t_1 as follows.

$$W_a = \int_0^{t_1} v_1(t)i_1(t)\,dt = \int_0^{I_1} L_1 i_1\,di_1 = \tfrac{1}{2}L_1 I_1^2 \tag{4-55}$$

Now consider the time interval $i_1 \le t \le i_2$. During this period, both $v_1(t)$ and $v_2(t)$ are zero, so no energy is supplied to the coils. Now, leaving $i_1(t)$ constant at the value I_1, let us raise $i_2(t)$ from 0 to some value I_2 over a time period $t_2 \le t \le t_3$, as shown in Fig. 4-6.3(b). During this interval, since $i_2(t)$ is changing and $i_1(t)$ is constant, the voltages $v_1(t)$ and $v_2(t)$ are found from (4-52) and (4-54) to be

$$v_1(t) = M_{12}\frac{di_2}{dt} \qquad v_2(t) = L_2\frac{di_2}{dt} \tag{4-56}$$

Thus, the energy W_b from both current sources supplied to the coils during this period is

$$W_b = \int_{t_2}^{t_3} v_1(t)i_1(t)\,dt + \int_{t_2}^{t_3} v_2(t)i_2(t)\,dt$$

$$= \int_0^{I_2} I_1 M_{12}\,di_2 + \int_0^{I_2} L_2 i_2\,di_2$$

$$= M_{12}I_1 I_2 + \tfrac{1}{2}L_2 I_2^2 \tag{4-57}$$

The total energy W supplied to the pair of coupled coils over the time range $0 \le t \le t_3$ is simply the sum of (4-55) and (4-57). Thus, we may write

$$W = W_a + W_b = \tfrac{1}{2}L_1 I_1^2 + M_{12}I_1 I_2 + \tfrac{1}{2}L_2 I_2^2 \tag{4-58}$$

Now let us conduct an experiment similar to the one described above, again starting from a zero energy state. This time we will first raise the current $i_2(t)$ to a value I_2 and then raise the current $i_1(t)$ to the value I_1. Typical waveforms for such a process are shown in Fig. 4-6.4. An energy analysis directly parallel to that given above shows that in this case the total energy W supplied over the time period $0 \le t \le t_3$ is

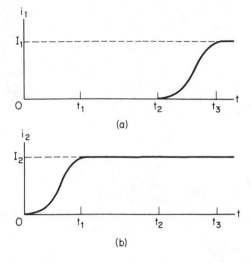

(a)

(b)

Figure 4-6.4 Waveforms in a pair of coupled coils.

$$W = \tfrac{1}{2}L_1 I_1^2 + M_{21}I_1 I_2 + \tfrac{1}{2}L_2 I_2^2 \qquad (4\text{-}59)$$

In both of the cases described above, the final energy supplied to the network must be the same since the final values of $i_1(t)$ and $i_2(t)$ are the same, namely, I_1 and I_2. Thus, the expressions given in Eqs. (4-58) and (4-59) must be equal. This can only be true if the quantities M_{12} and M_{21} are equal. Although we have arrived at this conclusion from a particular example, it may be shown to be a general property of coupled coils. Therefore, from this point in the text we will drop the subscript notation and simply refer to the quantity M as the mutual inductance of a pair of coupled coils with respect to each other. Thus, an expression for the energy stored in such a pair of coils may be written in the form

$$w(t) = \tfrac{1}{2}L_1 i_1^2(t) + M i_1(t) i_2(t) + \tfrac{1}{2}L_2 i_2^2(t) \qquad (4\text{-}60)$$

The mutual inductance M discussed above may be positive or negative. To see this, consider the circuit shown in Fig. 4-6.5(a). To emphasize the path followed by the lines of magnetic flux in this circuit, we have shown the coils as being wound on a core of some high-permeability magnetic material. The usual associated reference directions have been used for the voltage and current variables of each winding, and the reference directions for the fluxes follow the right-hand rule. First observe that in coil 1, if di_1/dt is positive, $d\phi_1/dt$ will also be positive. In addition, of course, $v_1(t)$

(a)

(b)

Figure 4-6.5 Pair of coupled coils with positive mutual inductance.

Capacitors and Inductors   Chap. 4

will be positive because the self-inductance L_1 of coil 1 is always positive. Now consider the effect of the quantity di_2/dt on $d\phi_1/dt$. Because of the winding directions chosen for the two coils, when di_1/dt is positive, $d\phi_1/dt$ as well as $d\phi_2/dt$ will be positive. Therefore, a positive contribution to the voltage $v_1(t)$ will result. Similar reasoning shows that a positive di_1/dt or di_2/dt will produce a positive $v_2(t)$. To describe such a case we say that *the mutual inductance is positive*. The relations for the voltage and current variables for this case are

$$v_1(t) = L_1 \frac{di_1}{dt} + |M| \frac{di_2}{dt} \quad \text{and} \quad v_2(t) = |M| \frac{di_1}{dt} + L_2 \frac{di_2}{dt} \quad (4\text{-}61)$$

where we have emphasized the positive nature of the mutual inductance M by using magnitude signs. It should be noted that, in order for the mutual inductance to be positive, the positive reference directions for $\phi_1(t)$ and $\phi_2(t)$ must be such as to aid each other.

Frequently, coupled coils are completely encapsulated or are constructed in such a way as to make it very difficult to actually observe the winding directions of the coils. In such a case, if we apply a positive increasing current $i_1(t)$, i.e., if we make di_1/dt positive and if we then find that $v_2(t)$ is also positive (assuming di_2/dt is zero), we can conclude that the mutual inductance M is positive. To indicate this, it is customary to use the symbol shown in Fig. 4-6.5(b) in which dots are used to indicate the relative coil winding directions. The significance of these dots is that a positive increasing current into the dot of one coil produces a voltage at the terminals of the other coil such that the dotted terminal is positive.

To illustrate the effect of a negative mutual inductance, consider the circuit shown in Fig. 4-6.6(a). In this circuit, if di_1/dt is positive, a positive $d\phi_1/dt$ and, thus, a positive $v_1(t)$ will be produced as a result. However, applying the right-hand rule, we see that a positive di_2/dt will produce a negative $d\phi_1/dt$ and thus a negative contribution to the voltage $v_1(t)$. Similarly, a positive di_1/dt will produce a negative contribution to $v_2(t)$. To describe this situation, we say that the mutual inductance is negative. The equations relating the voltage and current variables now have the form

$$v_1(t) = L_1 \frac{di_1}{dt} - |M| \frac{di_2}{dt} \quad \text{and} \quad v_2(t) = -|M| \frac{di_1}{dt} + L_2 \frac{di_2}{dt} \quad (4\text{-}62)$$

It should be noted that for negative mutual inductance, the positive reference directions for the two fluxes $\phi_1(t)$ and $\phi_2(t)$ oppose each other in the magnetic path. The

(a)　　　　　　　　　　　　　　　(b)

Figure 4-6.6 Pair of coupled coils with negative mutual inductance.

circuit symbol for the case of negative mutual inductance is shown in Fig. 4-6.6(b). The dot convention used here has the identical significance as it had in Fig. 4-6.5(a). In general, it should be noted that *the dot positions are determined by the actual winding direction of the coil* and thus they are invariant for a given set of coupled coils. The sign of the mutual inductance, however, *is determined by the relative reference directions chosen for the voltage and current variables at the terminals of the two coils*. Thus, we may reverse the sign of the mutual inductance by reversing the voltage and current reference directions at the terminals of either coil. As an example of this, note that the coupled coils shown in Fig. 4-6.7(a) have the same winding configuration as those

(a)   (b)

Figure 4-6.7   Pair of coupled coils with negative mutual inductance.

shown in Fig. 4-6.5(a). However, the reference directions for the variables $v_2(t)$ and $i_2(t)$ have been reversed. In this case, the positive reference direction for the two fluxes are such that the two fluxes oppose each other; thus, the mutual inductance is negative. The dot positions, however, are still the same and the coil may, therefore, be represented by the circuit symbol in Fig. 4-6.7(b), which is the same as that used in Fig. 4-6.5(b). The results given above may be summarized as follows:

SUMMARY 4-6.1

Coupled Coils.   The properties of a pair of coupled coils as shown in Fig. 4-6.2 are defined by specifying the values of the self-inductance of each coil and the mutual inductance which couples them. The relations between the voltage and current variables are

$$v_1(t) = L_1 \frac{di_1}{dt} + M \frac{di_2}{dt} \quad \text{and} \quad v_2(t) = M \frac{di_1}{dt} + L_2 \frac{di_2}{dt}$$

The mutual inductance M may be positive or negative. A pair of dots is used to identify the relative winding directions. If the positive reference current directions both flow into (or both out of) the dotted terminals of the two coils, the mutual inductance is positive; otherwise it is negative.   ■

The magnitude of the value of the mutual inductance found for a given set of coils has a definite upper bound which is determined by the values of the self inductances L_1 and L_2. This is a result of the fact that the total energy stored in a set of coupled

Capacitors and Inductors   Chap. 4

coils can never be negative. To investigate this upper bound, let us write the expression for the energy $w(t)$ given in Eq. (4-60) in a different form by completing the square in the variable $i_1(t)$. We obtain

$$w(t) = \frac{1}{2} \left\{ \left[\sqrt{L_1}\, i_1(t) + \frac{M}{\sqrt{L_1}}\, i_2(t) \right]^2 + \left(L_2 - \frac{M^2}{L_1} \right) i_2^2(t) \right\} \qquad (4\text{-}63)$$

To insure that $w(t)$ be positive for all values of $i_1(t)$, it is sufficient to require that the term $L_2 - (M^2/L_1)$ be positive (or zero), i.e., that[5]

$$|M| \leq \sqrt{L_1 L_2} \qquad (4\text{-}64)$$

To indicate the relative amount by which this inequality is satisfied, we may define a constant k by the relation

$$k = \frac{|M|}{\sqrt{L_1 L_2}} \qquad (4\text{-}65)$$

The constant k is called the *coefficient of coupling*. It is a measure of the amount of flux generated by a current flowing in one coil which links the turns of the other coil. If this amount is small, k is small and we say that the coils are *loosely coupled*. On the other hand, if *all* the flux generated by one coil links the turns of the other, we say that the coils are *perfectly coupled*. In this case, the coefficient of coupling k is unity and the mutual inductance M is equal to the geometric mean of the self-inductances L_1 and L_2. This case is covered in more detail in the following paragraphs.

To determine the properties of a perfectly coupled set of coils, consider a set of coils in which $\phi(t)$ is the flux that would link a one-turn coil located anywhere in the flux path. Let us also assume that the mutual inductance is positive. If we let n_1 be the number of turns in coil 1 and n_2 be the number of turns in coil 2, then $\phi_1(t)$ and $\phi_2(t)$, the flux linkages with coil 1 and coil 2, respectively, are given as

$$\phi_1(t) = n_1 \phi(t) \qquad \phi_2(t) = n_2 \phi(t) \qquad (4\text{-}66)$$

The voltages $v_1(t)$ and $v_2(t)$ are determined by Faraday's law and have the form

$$v_1(t) = \frac{d\phi_1}{dt} = n_1 \frac{d\phi}{dt} \qquad v_2(t) = \frac{d\phi_2}{dt} = n_2 \frac{d\phi}{dt} \qquad (4\text{-}67)$$

Taking the ratio of these two latter equations, we obtain

$$\frac{v_2(t)}{v_1(t)} = \frac{n_2}{n_1} \qquad (4\text{-}68)$$

Thus, we see that for a perfectly coupled set of coils, and for a positive mutual inductance, the ratio of the voltages at the terminals of the two coils is directly proportional to the ratio of the turns in the windings of the coils. Now let us assume that the magnetic permeability of the core on which the two coils are wound is infinite. This, of course, implies that the self inductances of the two coils as defined in Sec. 4-4 are also infinite. Since the magnetic reluctance $\mathcal{R}$ is inversely proportional to the permeability, it will be zero. This means that the "Ohm's law" for a magnetic circuit becomes

$$\text{mmf} = \phi \mathcal{R} = 0 \qquad (4\text{-}69)$$

---

[5] Actually this condition is necessary as well as sufficient, as may be seen by putting (4-63) in the standard form of a parabola.

The magnetomotive force or mmf, however, is simply the sum (if the mutual inductance was negative, it would be the difference) of the mmf's which are applied to the two coils. This has the form

$$\text{mmf} = n_1 i_1(t) + n_2 i_2(t) = 0 \tag{4-70}$$

Thus, for the infinite permeability case, we may write the ratio of the two coil currents as

$$\frac{i_2(t)}{i_1(t)} = -\frac{n_1}{n_2} \tag{4-71}$$

Thus, the ratio of the magnitudes of the coil currents is inversely proportional to the magnitude of the turns ratio. The expressions given in Eqs. (4-68) and (4-71) determine the relations between the terminal variables of voltage and current for a perfectly coupled set of coils with infinite magnetic permeability and positive mutual inductance. Such a multi-terminal device is called an *ideal transformer*. Its circuit symbol is shown in Fig. 4-6.8(a). In this figure use have defined $n = n_1/n_2$. If the mutual inductance is negative, similar relations apply. However, a minus sign is introduced into each of Eqs. (4-68) and (4-71). Thus Eq. (4-71) ends up with a plus sign. The circuit symbol for this case is shown in Fig. 4-6.8(b).

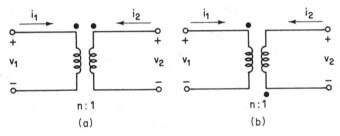

Figure 4-6.8 Circuit symbol for an ideal transformer with (a) positive mutual and (b) negative mutual inductance.

SUMMARY 4-6.2

The Ideal Transformer. An ideal transformer is represented as shown in Fig. 4-6.8. It may be considered as a pair of coupled coils in which the coefficient of coupling is unity and in which the self- and mutual-inductances are infinite. As a circuit device, it may be treated as a linear, time-invariant, multi-terminal element in which the relations between the terminal variables of voltage and current are defined by

$$\frac{v_2(t)}{v_1(t)} = \pm\frac{1}{n} \quad \text{and} \quad \frac{i_2(t)}{i_1(t)} = \mp n \tag{4-72}$$

where the upper polarity signs apply if the mutual inductance is positive, and the lower ones if it is negative. ∎

One of the important properties possessed by an ideal transformer is that of *resistance multiplication*. For example, consider the circuit shown in Fig. 4-6.9(a) in

Figure 4-6.9   Resistance multiplication effect of an ideal transformer.

which a resistor R_L is connected through an ideal transformer of turns ratio $n:1$ to another resistor R_s and a source $v_s(t)$. The resistor R_L imposes a constraint on the transformer variables $v_2(t)$ and $i_2(t)$ such that

$$\frac{v_2(t)}{-i_2(t)} = R_L \tag{4-73}$$

Substituting for $v_2(t)$ and $i_2(t)$ from Eq. (4-72) we obtain

$$\frac{\pm v_1(t)/n}{\pm n i_1(t)} = R_L \quad \text{or} \quad \frac{v_1(t)}{i_1(t)} = n^2 R_L \tag{4-74}$$

Thus, at the input to the ideal transformer we see a resistance of $n^2 R_L$. Effectively, the transformer has multiplied the value of the resistor by n^2. Note that this result is independent of the sign of the mutual inductance. This effect is frequently indicated by drawing an arrow as shown in Fig. 4-6.9(b). The arrow and the accompanying notation "$n^2 R_L$" specify the resistance of the portion of the circuit pointed to by the arrowhead, *ignoring any components of the circuit which are located on the other side of the arrowhead*. Thus, the transformer and resistor R_L may be represented by an equivalent resistor of value $n^2 R_L$ as shown in Fig. 4-6.9(c). It should be noted that the resistance multiplication effect of an ideal transformer depends on whether the resistor is on the right or the left of the transformer. As an example of this, consider the combination of ideal transformer and a Thevenin equivalent circuit shown in Fig. 4-6.10(a). The resulting overall Thevenin equivalent circuit as seen at the terminals $a - b$ is shown in Fig. 4-6.10(b). Note that in this case the value of the resistance is multiplied by $1/n^2$.

(a)

(b)

Figure 4-6.10   Resistance multiplication applied to a Thevenin equivalent circuit.

The ideal transformer is a very important multi-terminal network element since it models the actual performance of many physical transformers quite closely. Such a model can be made even more accurate by adding parasitic elements to represent some of the additional physical phenomena which are associated with an actual transformer.

Frequently it is desirable, for analysis purposes, to replace a pair of coupled coils by an equivalent circuit. As an example of such a circuit, consider the "T" of inductors shown in Fig. 4-6.11(a). In terms of the indicated variables, applying KVL we may write

$$v_1(t) = (L_1 - M)\frac{di_1}{dt} + M\frac{d}{dt}[i_1(t) + i_2(t)]$$

$$v_2(t) = (L_2 - M)\frac{di_2}{dt} + M\frac{d}{dt}[i_2(t) + i_1(t)]$$

(4-75)

(a)

(b)

Figure 4-6.11 T-equivalent circuit for a pair of coupled coils.

Rearranging terms we obtain

$$v_1(t) = L_1\frac{di_1}{dt} + M\frac{di_2}{dt} \quad \text{and} \quad v_2(t) = M\frac{di_1}{dt} + L_2\frac{di_2}{dt}$$

(4-76)

If M is positive, then these equations are the same as those given in (4-61), thus we conclude that this T of inductors may be used to replace a set of coupled coils. Since the mutual inductance M may be greater than L_1 or L_2 (but not greater than both), it is possible for one of the inductors in such an equivalent circuit to be negative; however, this is no way affects the validity of the method. A similar T of inductors which may be used to model a set of coupled coils in which the mutual inductance is negative is shown in Fig. 4-6.11(b). In the figure we have used magnitude signs to emphasize the parity of the various terms. Analysis of this circuit yields a set of equations identical to those given in Eq. (4-62). It should be noted that the T-equivalent circuits described above fail to model the behavior of a set of coupled coils in one important respect, i.e., they do not provide the physical isolation which the coupled coils do. Such an isolation may, however, be added by using an ideal 1:1 turns ratio transformer at either end of the network. Some other examples of circuits which provide isolation may be found in the problems.

4-7 CONCLUSION

The material contained in this section opens a whole new series of challenges in our study of circuit theory. The integral and differential equations that relate the terminal variables of capacitors and inductors require the development of an entirely new set of techniques for the analysis of circuits containing such elements. These new techniques, however, will not replace the analysis methods that we presented for resistance networks in Chap. 3. Rather, they will supplement and generalize them. Thus, in the chapters that follow, we shall make frequent applications of Kirchhoff's laws to enable us to write network equations. The results of such applications, however, will be simultaneous sets of integro-differential equations rather than the simultaneous sets of algebraic equations which were encountered for resistance networks. In general, some entirely new techniques will have to be developed to enable us to solve such sets of integro-differential equations. For certain special situations, however, such as the sinusoidal steady-state one discussed in Chap. 7, we shall see that the solution methods for sets of simultaneous equations applied in Chap. 3 may be directly applied to analyze networks containing capacitors and inductors.

PROBLEMS

Problem 4-1 (Sec. 4-1)

The voltage across a capacitor of value C farads is described by the waveform shown in Fig. P4-1. Sketch the current and charge as functions of time.

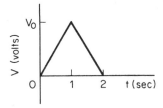

Figure P4-1

Problem 4-2 (Sec. 4-1)

If the voltage across a 5-F capacitor has the waveform drawn in Fig. P4-2, find the current through the capacitor.

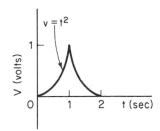

Figure P4-2

Problem 4-3 (Sec. 4-1)

If the current through a $\frac{1}{4}$-F capacitor is described by the waveform shown in Fig. P4-3, find the voltage across the capacitor.

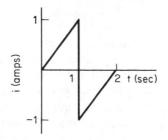

Figure P4-3

Problem 4-4 (Sec. 4-1)

Prove that the energy stored in a capacitor is equal to $w(t) = \frac{1}{2}[q^2(t)/C]$ by using the relation

$$w(t) = \int_{-\infty}^{t} v(t)i(t)\, dt.$$

Problem 4-5 (Sec. 4-1)

Let the voltage across a particular capacitor be equal to $v(t) = \frac{1}{4}q^4(t)$, where $q(t)$ is the charge. Find an expression for $w(q)$, the energy stored.

Problem 4-6 (Sec. 4-1)

The circuit shown in Fig. P4-6 has reached steady-state conditions (the current to the capacitor is zero) with the switch in position 1. If the switch is thrown to position 2, and the circuit again reaches a steady-state condition, find the value of the total energy dissipated in all the intervening time in the circuit at the right (consisting of the 500-kΩ resistor and the 5-V source).

Problem 4-7 (Sec. 4-1)

A time-varying capacitor of value $C(t) = 3 + t$ F is excited by a voltage source having the value $v(t) = \sin 2t$ V. Find an expression for the input current $i(t)$ to the capacitor.

Problem 4-8 (Sec. 4-1)

For the network shown in Fig. P4-8:
(a) Apply KVL and write a set of mesh current equations.
(b) Apply KCL and write a set of node equations.

Figure P4-6

Figure P4-8

Problem 4-9 (Sec. 4-2)

Write a mathematical expression describing each of the functions shown in Fig. P4-9.

Problem 4-10 (Sec. 4-2)

Write mathematical expressions which describe the derivatives of the functions shown in Fig. P4-9.

Problem 4-11 (Sec. 4-2)

Write mathematical expressions which describe the integrals of the functions shown in Fig. P4-9(b), (d), (e), and (f).

Capacitors and Inductors   Chap. 4

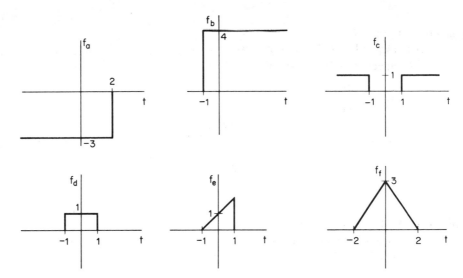

Figure P4-9

Problem 4-12 (Sec. 4-2)

Construct a sketch of each of the following functions.

(a) $f_a(t) = 2u(t + 3)$

(b) $f_b(t) = 3u(3 - t)$

(c) $f_c(t) = u(t - 2) + 2(t - 1)u(t - 1)$

(d) $f_d(t) = u(t^2 - 1)$

Problem 4-13 (Sec. 4-3)

If the waveshapes shown in Fig. P4-13 are applied as input currents to an uncharged 1-F capacitor, plot the resulting voltage $v(t)$ that appears across the terminals of the capacitor for each of the cases.

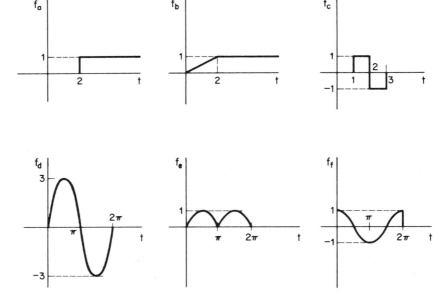

Figure P4-13

Problem 4-14 (Sec. 4-3)

If the waveshapes shown in Fig. P4-14 are applied as input voltages to the terminals of a 2-F capacitor, plot the resulting current $i(t)$ that flows into the capacitor in each case.

Figure P4-14

Problem 4-15 (Sec. 4-3)

The capacitor in the circuit shown in Fig. P4-15 varies with time as shown, but the voltage across the capacitor remains constant at 1 V. Sketch $i(t)$.

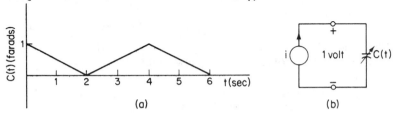

Figure P4-15

Problem 4-16 (Sec. 4-3)

The switch in the circuit shown in Fig. P4-16 is closed at $t = 0$. It is found that $v_c(0+) = 0$ and $(dv_c/dt)(0+) = 10$. What is the value of C?

Figure P4-16

Problem 4-17 (Sec. 4-4)

The waveshapes shown in Fig. P4-13 are applied as input voltages to a 1-H inductor. Draw the waveshapes of the resulting current. It is assumed that the inductor is initially uncharged.

Problem 4-18 (Sec. 4-4)

The waveshapes shown in Fig. P4-14 are applied as input currents to a 2-H inductor. Draw the resulting waveshapes of voltage that will appear at the terminals of the inductor for each of the cases shown.

Problem 4-19 (Sec. 4-4)

The current $i(t)$ in the circuit shown in Fig. P4-19 has the variation indicated. Find:

(a) $q(10)$

(b) $\phi(10)$

(c) $v(10)$

(a)   (b)

Figure P4-19

Problem 4-20 (Sec. 4-4)

The current through an inductor has the form $i(t) = 10 \sin [t(\pi/2)]$ A. Sketch $v(t)$ if $L = \frac{1}{2}$. Does the voltage sinusoid lead or lag the current?

Problem 4-21 (Sec. 4-4)

Let the current through a $\frac{1}{2}$-H inductor, as provided by a current source, be equal to 2A. Find the energy stored in the inductor. If the terminals of the inductor are suddenly short-circuited, what happens to the stored energy?

Problem 4-22 (Sec. 4-4)

Prove that the energy stored in an inductor is equal to $w(t) = \frac{1}{2}[\phi^2(t)/L]$ by using the relation

$$w(t) = \int_{-\infty}^{t} v(t)i(t)\, dt.$$

Problem 4-23 (Sec. 4-4)

Switch $S1$ in the circuit shown in Fig. P4-23 closes at $t = 0$, and switch $S2$ simultaneously opens. Find $i(0+)$ and $(di/dt)(0+)$.

ohms, henries

Figure P4-23

Problem 4-24 (Sec. 4-4)

For the network shown in Fig. P4-24:
(a) Apply KVL and write a set of mesh current equations.
(b) Apply KCL and write a set of node equations.

Figure P4-24

Problem 4-25 (Sec. 4-4)

For the network shown in Fig. P4-25:
(a) Apply KVL and write a set of mesh current equations.
(b) Apply KCL and write a set of node equations.

Figure P4-25

Problem 4-26 (Sec. 4-4)

In the circuit shown in Fig. P4-26, a 10-V dc source is connected as shown. It is assumed that the source has been connected to the circuit for a long time, so that all variables in the circuit have reached a constant (steady-state) value at $t = 0-$. At $t = 0$ the source is switched off, leaving a short circuit to ground. Find the following initial conditions:

(a) $i_1(0+)$ (b) $i_2(0+)$
(c) $v_1(0+)$ (d) $i'_1(0+)$
(e) $i'_2(0+)$ (f) $v'_1(0+)$
(g) $v''_1(0+)$

ohms, henries, farads

Figure P4-26

Problem 4-27 (Sec. 4-5)

In the networks shown in Fig. P4-27, all of the capacitors are assumed to be uncharged. Find equivalent capacitors which may be used to represent the capacitor networks at the terminals a and b for each of the two networks.

farads

(a)   (b)   Figure P4-27

Problem 4-28 (Sec. 4-5)

In the networks shown in Fig. P4-28, all of the inductors are assumed to be uncharged. Find

henries

(a)   (b)

Figure P4-28

                                          Capacitors and Inductors   Chap. 4

equivalent inductors which may be used to represent the inductor network at the terminals a and b for each of the two networks.

Problem 4-29 (Sec. 4-5)

In the circuit shown in Fig. P4-29, the capacitor C_1 is initially charged to 10 V and the capacitor C_2 is uncharged. The switch is closed at $t = 0$. Find:

(a) The voltage in the circuit after the switch is closed.

(b) The energy stored in C_1 before the switch is closed.

(c) The total energy stored in C_1 and C_2 after the switch is closed.

Explain any difference in the values found in parts (b) and (c).

Figure P4-29

Problem 4-30 (Sec. 4-5)

The inductor L_1 shown in Fig. P4-30 is "charged" in the sense that a current of 1 A is flowing in the loop formed by it and the closed switch for values of $t < 0$. No current is flowing through L_2. The switch is opened at $t = 0$. Find:

(a) The current in the circuit after the switch is opened.

(b) The energy stored in the inductor L_1 before the switch is opened.

(c) The total energy stored in both L_1 and L_2 after the switch is opened.

Explain any difference in the values computed in parts (b) and (c).

Figure P4-30

Problem 4-31 (Sec. 4-6)

In Fig. P4-31 two possible interconnections of a pair of coupled coils are shown. Find the inductance of a single equivalent inductor as seen from terminals a and b for each of the circuit connections.

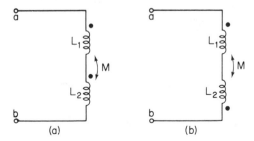

(a)   (b)   Figure P4-31

Problem 4-32 (Sec. 4-6)

For each of the circuits shown in Fig. P4-32, do the following:

(a) Write the equations indicating the sign of the mutual inductance between the various coils by using plus and minus signs and magnitude symbols.

(b) Identify the winding sense between coils 1 and 2 by dots, between coils 2 and 3 by squares, and between coils 1 and 3 by triangles.

Figure P4-32

Problem 4-33 (Sec. 4-6)

Find an equivalent circuit representation for an ideal transformer which uses two controlled sources, one a VCVS, the other an ICIS.

Problem 4-34 (Sec. 4-6)

It is possible to replace a set of coupled coils with a "T" of inductors and an ideal transformer in such a way that the individual inductors are always positive-valued. Two such configurations are shown in Fig. P4-34. The circuits are drawn assuming a positive value of M. For a negative value of M it is only necessary to change the position of the dot on one of the windings of the ideal transformer. Verify the validity of the circuits.

Figure P4-34

Problem 4-35 (Sec. 4-6)

For the network shown in Fig. P4-35, find an expression for the voltage $v_2(t)$ as a function of the current $i(t)$ and the resistance R.

Figure P4-35

Problem 4-36 (Sec. 4-6)

Find the Thevenin equivalent at the terminals *a-b* for the circuit shown in Fig. P4-36.

4:1   **Figure P4-36**

Problem 4-37 (Sec. 4-6)

Find Thevenin equivalents at the terminals *a-b* for each of the circuits shown in Fig. P4-37.

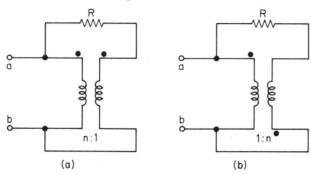

(a)   (b)   **Figure P4-37**

5

First-Order Circuits

In the last chapter we showed that capacitors and inductors both have the property that they are capable of storing energy. In this chapter we shall introduce the study of circuits characterized by a single energy-storage element. Thus, the type of circuit to be discussed in this chapter is one which contains a single capacitor or a single inductor. In addition, the circuit may contain any number of resistors and sources. It will be shown that the equations describing such a circuit may be put in a form involving an unknown variable and its first derivative. Such an equation is referred to as a *first-order* differential equation, thus we shall refer to circuits which contain only a single energy-storage element as *first-order circuits*. The techniques which we shall discuss in this chapter will be ones that may readily be adapted to the study of higher-order circuits, i.e., circuits containing more than one energy-storage element. This will be done in succeeding chapters.

5-1 EXCITATION BY INITIAL CONDITIONS

If we consider a circuit in which all the variables are identically zero, we may say that such a circuit is *unexcited*. Conversely, circuits in which at least some of the variables are non-zero for some range of time are referred to as circuits which have been *excited*. In this chapter we shall consider the ways in which circuits can be excited and the results of such excitation. There are two basic ways of exciting a circuit. The first is by initial conditions, the second is by sources. In this section we shall study the effects of using an initial condition to excite a first-order circuit.

As an example of a circuit which is excited only by an initial condition, consider the circuit consisting of a capacitor, a resistor, and a battery (a dc voltage source) shown in Fig. 5-1.1(a). If switch $S1$ is closed for all values of $t < 0$, then an initial

$$(a) \qquad (b)$$

Figure 5-1.1 First-order RC circuit excited by an initial condition.

voltage V_0 is established on the capacitor by the battery. During this same period, it is assumed that switch $S2$ is open. If we open switch $S1$ just before $t = 0$, and close switch $S2$ at $t = 0$, then for all $t \geq 0$, we may ignore the battery shown in Fig. 5-1.1(a). Thus, for $t \geq 0$ we need only consider the circuit comprised of a resistor and a capacitor shown in Fig. 5-1.1(b). The initial condition on the capacitor voltage $v(t)$ for this circuit is $v(0) = V_0$. For the range of time $t \geq 0$, such a circuit may be considered as excited only by the initial condition which is present on the capacitor at $t = 0$. Applying KCL to this circuit, we obtain the differential equation

$$C\frac{dv}{dt} + \frac{1}{R}v(t) = 0 \qquad t \geq 0 \tag{5-1}$$

We may use the following words and phrases to describe this equation:

Ordinary This term applies since the differentiation is made with respect to only one independent variable, namely t. *Note:* A differential equation which has derivatives taken with respect to more than one independent variable is known as a *partial* differential equation rather than an ordinary differential equation.

First-order This term applies since the highest-order derivative found in the equation is the first.

Linear This descriptive term tells us that the dependent variable and its derivatives appear only as terms of the first degree, i.e., that there are no terms of the type $v^2(t)$ or $v(t) \times dv/dt$ (these are both second-degree terms).

Constant coefficients This term applies since the coefficients C and $1/R$ are constants, i.e., they are *not* functions of the variable t. *Note:* When the coefficients are functions of the variable t, they are referred to as *time-varying* coefficients.

Homogeneous This term applies since there is no term in the equation which does not include $v(t)$ or its derivative dv/dt. *Note:* An equation in $v(t)$ and its derivatives which includes a term such as $g(t)$, where $g(t)$ may be a constant or some known function of t, is said to be *non-homogeneous*.

The complete description of Eq. (5-1) is thus: *an ordinary, first-order, linear, homogeneous, differential equation with constant coefficients.* This is too much for an author to write more than once (or for a reader to read more than once), so we shall assume that all the differential equations to be discussed in this text are ordinary, linear, and have constant coefficients unless otherwise stated. Thus we shall refer to Eq. (5-1) as a *first-order, homogeneous, differential equation.* Such an equation will always result when KVL or KCL is applied to a first-order circuit excited only by initial conditions. The first-order nature of the resulting differential equation comes from the fact that a single-storage element introduces only a single differential relation between the various branch variables of the circuit. The homogeneous nature of the resulting differential equation comes from the fact that, without sources, there can be no term which does not include some network variable or its derivative. Thus, if KCL and/or KVL is applied to a first-order circuit excited only by initial conditions, the equation that must be solved to determine any variable is a first-order homogeneous, differential equation.

The solution of a first-order, homogeneous, differential equation of the type given in Eq. (5-1) is easily found. First, for convenience, let us divide Eq. (5-1) by the quantity C. We obtain

$$\frac{dv}{dt} + \frac{1}{RC}v(t) = 0 \qquad t \geq 0 \tag{5-2}$$

We may employ a mathematical technique called *separation of variables* to solve the above equation. To do this, we first write Eq. (5-2) in the form

$$\frac{dv}{v} = \frac{-1}{RC}\,dt \qquad t \geq 0 \tag{5-3}$$

Integrating both sides of (5-3) we obtain

$$\ln v(t) = \frac{-1}{RC}t + K \qquad t \geq 0 \tag{5-4}$$

where ln is the natural logarithm. Now let us use both sides of (5-4) as an exponent for the quantity e, where e is the base of the natural logarithm. We obtain[1]

$$e^{\ln v(t)} = v(t) = e^{(K-t/RC)} = K'e^{-t/RC} \qquad t \geq 0 \tag{5-5}$$

where K' is a new constant equal, by definition, to e^K. This constant is easily given a physical significance by noting that when $t = 0$, $v(0) = K'$. Thus, we may write (5-5) in the form

$$v(t) = v(0)e^{-t/RC} \qquad t \geq 0 \tag{5-6}$$

This last equation provides the solution for the first-order, homogeneous, differential equation given in Eq. (5-1). The fact that it is a solution is easily verified by substituting Eq. (5-6) into (5-1). This is left as an exercise for the reader.

The steps given above to find the solution of (5-1) are exactly the same for the solution of any other first-order, homogeneous, differential equation. Thus, we may generalize as follows:

---

[1] In this equation, we use the relation $e^{\ln x} = \ln(e^x) = x$.

Solution of a First-Order Circuit Excited Only by Initial Conditions. A first-order circuit excited only by initial conditions is characterized by a first-order, homogeneous, differential equation having the form

$$\frac{df}{dt} + af(t) = 0 \tag{5-7}$$

The solution to this equation is

$$f(t) = f(0)e^{-at} \qquad t \geq 0 \tag{5-8}$$

∎

This result allows us to readily solve any first-order, homogeneous, differential equation which may occur in our studies of circuits. Two examples follow.

Example 5-1.1

Assume that the circuit components shown in Fig. 5-1.1 have the following values: $C = 1 \ \mu F$, $R = 10 \ k\Omega$; and that $V_0 = v(0) = 10$ V. From (5-6) we see that the solution for the voltage $v(t)$ across the capacitor for $t \geq 0$ is $v(t) = 10e^{-100t}$ V (where t is in seconds). The expression for the current $i(t)$ through the resistor is easily found using Ohm's law. Thus we see that $i(t) = 0.001e^{-100t}$ A. Plots of the waveshapes for $v(t)$ and $i(t)$ are given in Fig. 5-1.2. The scales for these plots have been chosen so as to make them collinear.

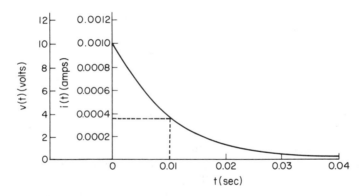

Figure 5-1.2 Waveforms for Example 5-1.1.

Example 5-1.2

The circuit shown in Fig. 5-1.3 consists of an inductor with a resistor connected across its terminals. A switch $S1$ is also connected directly across the two terminals. It is assumed that the switch is closed for all $t < 0$. Thus, for this range of time the node-

Figure 5-1.3 First-order RL circuit excited by an initial condition.

pair voltage $v(t)$ between the terminals must be zero, and consequently, no current will flow through the resistor. (From Ohm's law $v(t) = Ri_R(t)$, therefore if $v(t) = 0$ and $R \neq 0$, $i_R(t)$ must equal 0.) Let us assume that the inductor current, however, is non-zero.[2] Let this current be I_0. If we open the switch $S1$ at $t = 0$, then, for $t \geq 0$, we may apply KVL around the mesh consisting of the inductor and the resistor to obtain the first-order, homogeneous, differential equation. After dividing through by L, we obtain

$$\frac{di}{dt} + \frac{R}{L}i(t) = 0 \qquad t \geq 0$$

The continuity condition for an inductor tells us that the current in the inductor at the instant the switch opens must be the same as it was just before the switch opened; thus, we see that $i(0) = I_0$. From the preceding summary, we find that the solution of the differential equation is

$$i(t) = I_0 e^{-Rt/L} \qquad t \geq 0$$

To make this example more concrete, suppose that $L = 1$ mH, $R = 10$ kΩ, and $I_0 = 0.001$ A. The solution for the inductor current is $i(t) = 0.001e^{-10^7 t}$ A (where t is in seconds). The voltage across the resistor can be found (from Ohm's law) to be $v(t) = 10e^{-10^7 t}$ V. This is also the voltage which appears across the inductor. An alternative means of finding the voltage across the inductor is to use the relation $v(t) = -L\, di/dt$. (The minus sign is required because the variables $i(t)$ and $v(t)$, shown in Fig. 5-1.3, have relative reference directions which are opposite to those used for defining the terminal relations for the inductor.) The waveshapes for $v(t)$ and $i(t)$ for this example are similar to those shown in Fig. 5-1.2.

An interesting and most important fact may be observed in the above examples. In Example 5-1.1, the waveshapes for both the voltage and the current variables have the same exponential form, that is, they both have the form Ke^{-100t} (where the value of K is different for the two variables). It may be easily verified that any other variable, such as the charge on the capacitor or the current through the capacitor, also has the same form. Similarly, in Example 5-1.2 the current and the voltage variable both have the form $Ke^{-10^7 t}$, as do the other variables in the circuit. Thus, we conclude that in each of these two circuits *all variables have an exponential behavior characterized by the same exponential term*, and they differ only in the magnitude (and sign) of the constant specifying the initial value of the variable. We may easily see that this must be true in general. Recall that all circuit variables are related by either of two types of relations: (1) algebraic relations such as those characterizing the resistor; and (2) differential or integral relations such as those characterizing the inductor and the capacitor. The derivative or integral of an exponential function Ke^{-at}, however, is still an exponential function $K'e^{-at}$, where the multiplicative constant K' is equal to $-aK$ or $-K/a$. Therefore, once we have found the exponential waveshape for any variable (in a first-order circuit excited only by initial conditions), we know the waveshape for all other variables. We need merely find the initial values of the other variables to completely characterize them. To emphasize this point, we state the following:

[2] Remember that an initial current in an ideal inductor will flow indefinitely through a short circuit such as is provided by $S1$. If this is not clear, the student should review the properties of an inductor presented in Sec. 4-4.

that, from a practical viewpoint, for values of time greater than four or five time constants, the values of the variables in a first-order circuit may be thought of as being zero, although in a mathematical sense, zero value is reached only when the variable t goes to infinity.

The techniques described in this section may also be applied to the case where the application of KCL or KVL produces an integral equation rather than a differential equation. The following example will serve to illustrate the process.

Example 5-1.3

We may solve directly for the voltage $v(t)$ across the resistor for the network shown in Fig. 5-1.3 and discussed in Example 5-1.2. For values of $t \geq 0$, we note that $v(t)$ is also the voltage across the inductor. Thus, we may apply KCL to either of the nodes connecting the resistor and the inductor to obtain

$$\frac{1}{R}v(t) + \frac{1}{L}\int_0^t v(\tau)\,d\tau + i_L(0) = 0 \qquad t \geq 0$$

Differentiating the above equation and multiplying both sides by R, we obtain

$$\frac{dv}{dt} + \frac{R}{L}v(t) = 0 \qquad t \geq 0$$

This equation should be compared with the first equation given in Example 5-1.2. The coefficients are the same, but the variable is different. The solution is directly seen to be

$$v(t) = v(0)e^{-Rt/L} \qquad t \geq 0$$

The constant $v(0)$ is easily evaluated by noting that when the switch $S1$ is opened at $t = 0$, all of the current I_0 formerly flowing through the switch must now flow through the resistor. Thus we see that $v(0) = I_0 R$. Using the numerical values given in Example 5-1.2, we obtain the same result, namely, $v(t) = 10e^{-10^7 t}$.

The techniques given in this section are easily extended to the case where more than one resistor is present in the circuit or where dependent sources are used. These situations are illustrated by the following examples:

Example 5-1.4

A first-order circuit consisting of a capacitor and three resistors is shown in Fig. 5-1.4(a). An initial condition $i(0) = 4$ A is specified for the $\frac{1}{2}$-Ω resistor. It is desired to find an expression for the current through the $\frac{1}{2}$-Ω resistor for all $t \geq 0$. This may be done by first replacing the network of three resistors connected to the capacitor by a single equivalent resistor, R_{eq}. Such a resistor has a value of 2 Ω. The circuit is now as shown in Fig. 5-1.4(b). From KCL we may write for this circuit,

$$C\frac{dv}{dt} + \frac{1}{R_{eq}}v(t) = 0 \qquad t \geq 0$$

Substituting the values $C = \frac{1}{6}$ F, $R_{eq} = 2\,\Omega$, we find that the solution for the voltage

(a)   (b)

Figure 5-1.4 Circuit for Example 5-1.4.

Exponential Form of the Variables in a First-Order Circuit Excited Only by Initial Conditions. In a circuit characterized by a first-order, homogeneous, differential equation, all variables have an exponential behavior characterized by the same exponential term. The initial magnitude and polarity of each variable will, in general, be different.[3] ∎

Since the variables of first-order circuits excited only by initial conditions are all exponential in form, it is worth our time to note some useful properties of the exponential waveshape. These properties apply to all variables in first-order circuits. First consider the circuit shown in Fig. 5-1.1. The solution for $v(t)$ is

$$v(t) = v(0)e^{-t/RC}$$

When t has the value RC, this becomes

$$v(RC) = v(0)e^{-1} = 0.368v(0) \tag{5-9}$$

The quantity $(-t/RC)$, which is the exponent of e, must be dimensionless; therefore, the quantity RC must have the dimensions of time. This is readily verified by direct dimensional analysis. Such a quantity is referred to as the *time constant* of the given circuit. From (5-9) we see that we may define the time constant of a given circuit as the time required for *any* variable to decay to 36.8 percent of its initial value when the circuit is excited only by initial conditions. It is equal to the reciprocal of the quantity a in Eq. (5-7).

The time constant may be expressed either in literal form or in numerical form, whichever is appropriate. Thus, for the numerical values given in Example 5-1.1 the time constant for the circuit is 0.01 s. This is indicated by the dashed lines in Fig. 5-1.2. For the circuit shown in Example 5-1.2, the time constant is L/R, and, for the numerical values used in the example, the time constant is 10^{-7} S. Note that, for a given circuit, all variables have the same time constant. Since the time constant is a property of the circuit itself, it is convenient to refer to it as a *natural* property of the circuit and its variables (as opposed to a property *forced* on the circuit by the excitation from some source). Thus, we may refer to the *natural time constant* of a first-order circuit. Further, since frequency and time are reciprocal quantities, we may say that the circuit shown in Fig. 5-1.1 has a *natural frequency* of value $1/RC$, while the one shown in Fig. 5-1.3 has a natural frequency of value R/L. We shall see considerable application of the concept of natural frequencies in a later chapter. Note that the use of the word "frequency" here does not imply any connection with the usual "sinusoidal frequency."

In the last paragraph it was shown that an exponential function decays to about 36.8 percent of its initial value in one time constant. Similar calculations show that in two time constants the exponential function decays to about 13.5 percent of its initial value; in three time constants to about 5 percent, in four time constants to about 1.8 percent, and in five time constants to about 0.67 percent. Thus, we see

---

[3] When we refer to "all variables" here, we exclude quantities such as power and energy which are found as a product of basic variables such as voltage and current.

across the capacitor is $v(t) = v(0)e^{-3t}$. The exponential behavior of all variables in the circuit must be the same. Therefore, we conclude that the expression for the current in the $\frac{1}{2}$-Ω resistor is $i(t) = 4e^{-3t}$.

Example 5-1.5

In the circuit shown in Fig. 5-1.5(a) we assume that the switch $S1$ has been in position 1 for a long time, allowing the 10-V source to establish an initial condition $v(0)$ on the capacitor. To determine this initial condition, we note that the capacitor voltage is

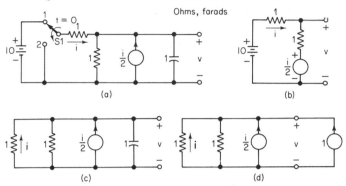

Ohms, farads

(a)

(b)

(c)

(d)

Figure 5-1.5 Circuit for Example 5-1.5.

constant, and thus there will be no current flow into or out of it. Thus, at $t = 0-$ the circuit may be analyzed as shown in Fig. 5-1.5(b) in which a Thevenin equivalent has been used for the ICIS and shunt resistor. For this circuit

$$10 = i + i + \frac{i}{2} \qquad \text{and} \qquad v = i + \frac{i}{2}$$

Solving these relations we obtain $i = 4$ and $v = 6$. Thus, our initial condition is $v(0) = 6$ V. At $t = 0$, the switch is moved to position 2. The circuit now appears as shown in Fig. 5-1.5(c). The two resistors and the ICIS may be replaced (see Sec. 3-7) by a single resistor R_{eq}. To find the value of this resistor we apply a 1-A current source as shown in Fig. 5-1.5(d). In this circuit $v = -i$. Thus we have

$$i - v + \frac{i}{2} + 1 = 0 \qquad \text{and} \qquad -v - v - \frac{v}{2} + 1 = 0$$

Solving we obtain $R_{eq} = v/1 = \frac{2}{5}\,\Omega$, thus the RC time constant is $\frac{2}{5}$ s, and $v(t) = 6e^{-5t/2}$ for $t \geq 0$.

We conclude this section by briefly considering the linearity aspects of first-order circuits. For our purposes in this section, a linear first-order circuit may be considered as one comprised of linear two-terminal elements, namely, resistors, and a single energy-storage element. From the preceding discussion in this section, it should be clear that the response of such a circuit is linearly related to the initial condition used as an excitation, i.e., if the initial condition is multiplied by some constant, the response is multiplied by the same constant. As an example of this result, consider again the circuit analyzed in Example 5-1.1. Since an initial voltage $v(0) = 10$ V resulted in a current $i(t) = 0.001e^{-100t}$ A, and since the circuit is linear, we may directly conclude that an initial voltage $v(0) = 20$ V will produce a current $i(t) = 0.002e^{-100t}$ A. We shall explore some other considerations of the property of linearity when we discuss the excitation of a first-order circuit by sources in the following section.

5-2 EXCITATION BY SOURCES

In the last section we discussed the excitation of first-order circuits by means of initial conditions. The resulting variation of the circuit variables is sometimes referred to as the *zero-input response* of the network, since the variables respond to the excitation provided by initial conditions rather than responding to some input excitation from a source. In this section we shall discuss the excitation of a first-order circuit by an independent source (or sources), assuming the initial conditions are zero. A term which is frequently used to imply that the initial conditions of a circuit are zero is to say that the circuit is in the "zero state." The variations in the network variables which result from excitation by sources in the absence of initial conditions may correspondingly be referred to as the *zero-state response*. We shall make additional usage of this terminology when we discuss the excitation of a network by initial conditions *and* sources. This will be done in the following section.

As an example of a first-order circuit which is excited by a source, consider the circuit shown in Fig. 5-2.1. From KCL we may write

$$C\frac{dv}{dt} + \frac{1}{R}v(t) = i(t) \tag{5-10}$$

Figure 5-2.1 First-order *RC* circuit excited by a source.

If we refer to the descriptive phrases given for differential equations in the last section, we see that Eq. (5-10) may be characterized as an ordinary first-order, linear, non-homogeneous, differential equation with constant coefficients. Thus, the difference between this equation and the ones treated in the last section is that this equation is *non-homogeneous*, i.e., a term $i(t)$ appears which is not a function of the dependent variable $v(t)$ or its derivative. In general, we shall refer to an equation such as Eq. (5-10) more compactly as a *first-order, non-homogeneous, differential equation*. The solution to such an equation consists of two parts. The first of these parts is the solution to the related homogeneous equation, i.e., the solution to Eq. (5-10) when $i(t)$ is set to zero. We will refer to this part of the solution as $v_h(t)$ (the *h* stands for homogeneous). This component of the solution is given the mathematical appellation of *complementary solution* or complementary function. To determine it, we simply apply the methods of the last section. Thus, we see that

$$v_h(t) = Ke^{-t/RC} \tag{5-11}$$

Note that in the above equation we have written the multiplicative constant as K rather than writing it as $v(0)$. The reason for this, of course, is that we must take both parts of the solution of (5-10) into account when evaluating the constant K so as to determine the required value of $v(0)$. We may generalize our discussion to this point by considering a differential equation of the form

$$\frac{df}{dt} + af(t) = g(t) \qquad (5\text{-}12)$$

The complementary solution for this differential equation is defined as follows:

SUMMARY 5-2.1

Complementary Solution of a First-Order Differential Equation. The complementary solution is one of the parts of the overall solution of a non-homogeneous, first-order, differential equation. It is the solution to the related homogeneous equation. If the differential equation has the form given in Eq. (5-12), then the complementary solution $f_h(t)$ has the form Ke^{-at}, where K is a constant which must be evaluated in the complete solution. ∎

The complementary solution given in Eq. (5-11) for the non-homogeneous differential equation of Eq. (5-10) is independent of the particular form of the excitation function $i(t)$. However, the second part of the overall solution, which we will now discuss, has a form which is quite similar to the particular form of the excitation function being used. Thus, it is logically called the *particular solution*. For example, suppose that $i(t)$ is a constant I_0 applied at $t = 0$. If we use $v_p(t)$ to designate the particular solution for $v(t)$, then $v_p(t)$ will also be a constant. Let us call the constant A. Thus, we may write $v_p(t) = A$. To show that this is indeed a solution, and also to determine the value of A in terms of the coefficients of the differential equation and the constant I_0, we need merely substitute A for $v(t)$ in Eq. (5-10). The derivative of A is, of course, zero. Thus, the first term in (5-10) disappears, and we obtain

$$\frac{1}{R}A = I_0 \qquad (5\text{-}13)$$

The solution for A is thus RI_0, and we see that the particular solution is given as

$$v_p(t) = RI_0 \qquad t \geq 0 \qquad (5\text{-}14)$$

As another example, suppose that $i(t) = 2 + t$ for all values of $t \geq 0$. In this case, $v_p(t)$ will have the form $A + Bt$, where A and B are constants. To show this and to find values of A and B, we need merely substitute this expression for $v(t)$ in Eq. (5-10). Thus, we obtain

$$CB + \frac{1}{R}(A + Bt) = 2 + t \qquad (5\text{-}15)$$

For this equation to hold for all values of t, the coefficients of corresponding powers of t on both sides of the equation must be the same. This requires that

$$CB + \frac{A}{R} = 2 \qquad \frac{B}{R} = 1 \qquad (5\text{-}16)$$

Solving the two equations, we obtain

$$B = R \qquad A = R(2 - RC) \qquad (5\text{-}17)$$

Thus, the particular solution is

$$v_p(t) = R(2 - RC) + Rt \qquad t \geq 0 \tag{5-18}$$

The method of determining the value of $v_p(t)$ given in the preceding examples is called the method of *undetermined coefficients*. It is easily extended to a large range of excitation functions. The discussion given above may be summarized as follows:

SUMMARY 5-2.2

Particular Solution of a First-Order Differential Equation. If a non-homogeneous differential equation has the form $(df/dt) + af(t) = g(t)$, given in Eq. (5-12), the particular solution $f_p(t)$ may be found by substituting an assumed form for $f(t)$ and solving for any unknown constants. A listing of the assumed forms to use for $f(t)$ for various types of excitation functions $g(t)$ is given in Table 5-2.1. ■

Once the complementary solution and the particular solution have been found for a given differential equation, then the overall or *complete solution* is easily found as the sum of the two solutions. For the circuit shown in Fig. 5-2.1, for the case where

TABLE 5-2.1 Forms for the Particular Solution for a Non-Homogeneous Differential Equation $(df/dt) + af(t) = g(t)$†

Form of the Excitation Function $g(t)$	Form of the Particular Solution $f_p(t)$
K_0	A
$K_0 t$	$A + Bt$
$K_0 + K_1 t$	$A + Bt$
$K_0 + K_1 t + K_2 t^2$	$A + Bt + Ct^2$
$K_0 e^{-bt} \ (b \neq a)$	Ae^{-bt}
$K_0 e^{-at}$	Ate^{-at}
$K_0 \sin bt$	$A \sin bt + B \cos bt$
$K_0 \cos bt$	$A \sin bt + B \cos bt$

†Note that, if $a = 0$, i.e., if the network has a natural frequency at 0, then an excitation function $g(t)$ will have a particular solution $f_p(t) = A \int g(t) \, dt$. For example, if $g(t)$ is a constant, then $f_p(t)$ will have the form At.

$i(t)$ is a constant current I_0, the solutions for $v_h(t)$ and $v_p(t)$ are given in (5-11) and (5-14), respectively. Thus we see that the complete solution for $v(t)$ is

$$v(t) = v_h(t) + v_p(t) = Ke^{-t/RC} + RI_0 \qquad t \geq 0 \tag{5-19}$$

The constant K may now be evaluated by inserting the initial condition on $v(t)$. For this case, let us assume that $v(0) = 0$, i.e., that there is no initial condition. From (5-19) we see that

$$v(0) = K + RI_0 = 0 \tag{5-20}$$

Thus, $K = -RI_0$, and the complete solution for $v(t)$ is

$$v(t) = RI_0(1 - e^{-t/RC}) \qquad t \geq 0 \tag{5-21}$$

A plot of this waveshape is shown in Fig. 5-2.2. The value of $v(t)$ at t equals one time constant is shown by the dashed lines. It is equal to $(1 - 0.368)RI_0$ or $0.632RI_0$.

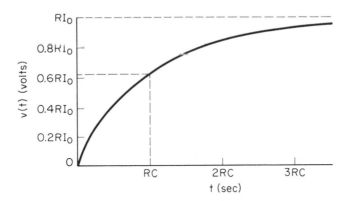

Figure 5-2.2 Waveshapes for the circuit shown in Fig. 5-2.1 for $i(t) = I_0$.

The above steps are easily applied to solve any first-order non-homogeneous differential equation. The procedure is summarized as follows:

SUMMARY 5-2.3

Solution of a First-Order Circuit Excited by Sources. A first-order circuit which is excited by sources is characterized by a first-order, non-homogeneous, differential equation. The complete solution of such an equation is found by adding the complementary solution and the particular solution. The constant (which is part of the complementary solution) is evaluated in the complete solution to match the specified initial condition. ∎

Example 5-2.1

A first-order RL circuit is excited by a voltage source as shown in Fig. 5-2.3. If we assume that $v(t)$, the output of the voltage source, is zero for $t < 0$, and is equal to $V_0 \sin \omega t$ for $t \geq 0$, then, applying KVL, we obtain the first-order, non-homogeneous, differential equation

$$L\frac{di}{dt} + Ri(t) = V_0 \sin \omega t \qquad t \geq 0$$

From Table 5-2.1, we see that the correct assumed form for the particular solution $i_p(t)$ is

Figure 5-2.3 First-order RL circuit excited by a source.

$$i_p(t) = A \sin \omega t + B \cos \omega t$$

Substituting this expression into the non-homogeneous differential equation describing the circuit, we obtain

$$L\omega[A \cos \omega t - B \sin \omega t] + R[A \sin \omega t + B \cos \omega t] = V_0 \sin \omega t$$

Equating corresponding coefficients of $\sin \omega t$ and $\cos \omega t$ on both sides of the above equation, we see that

$$-L\omega B + RA = V_0 \quad \text{and} \quad L\omega A + RB = 0$$

The two equations given above may be written in a matrix form as follows:

$$\begin{bmatrix} R & -\omega L \\ \omega L & R \end{bmatrix} \begin{bmatrix} A \\ B \end{bmatrix} = \begin{bmatrix} V_0 \\ 0 \end{bmatrix}$$

This is simply a set of simultaneous equations which may be solved to determine the constants A and B. Thus, we obtain

$$A = \frac{V_0 R}{R^2 + (\omega L)^2} \quad \text{and} \quad B = \frac{-\omega L V_0}{R^2 + (\omega L)^2}$$

Let us assume that the circuit parameters have the following values: $R = 2 \, \Omega$, $L = 1 \, \text{H}$, $\omega = 1 \, \text{rad/s}$, $V_0 = 10 \, \text{V}$. Then, the constants have the values $A = 4$, $B = -2$. Thus, the particular solution $i_p(t)$ becomes

$$i_p(t) = 4 \sin t - 2 \cos t \quad t \geq 0$$

The complementary solution $i_h(t)$ is easily found from the homogeneous equation. It is

$$i_h(t) = Ke^{-Rt/L} = Ke^{-2t} \quad t \geq 0$$

The complete solution for $i(t)$ is simply written as the sum of the complementary solution and the particular solution. Thus, we obtain

$$i(t) = Ke^{-2t} + 4 \sin t - 2 \cos t \quad t \geq 0$$

To determine the value of the constant K, let us assume that there is no initial condition present, i.e., that $i(0) = 0$. From the preceding equation, we see that $i(0) = K - 2 = 0$; therefore, $K = 2$. The complete solution for $i(t)$ is thus found to be[4]

$$i(t) = 2e^{-2t} + 4 \sin t - 2 \cos t \quad t \geq 0$$

To verify the entries in Table 5-2.1 (or to determine particular solutions for functions not listed in the table), we may note that multiplying both sides of (5-12) by an appropriate "integrating factor" makes it possible to readily integrate both members of the equation. Specifically, multiplying both sides of (5-12) by e^{at}, we obtain

$$e^{at} \frac{df}{dt} + af(t)e^{at} = g(t)e^{at} \tag{5-22}$$

We now observe that the integrating factor was chosen so as to make the left member of (5-22) expressible as the derivative of the product of two functions. Thus, we may write

$$\frac{d}{dt}[f(t)e^{at}] = e^{at}\frac{df}{dt} + af(t)e^{at} = g(t)e^{at} \tag{5-23}$$

---

[4]Any expression involving both a sine term and a cosine term (with the same argument) can be rewritten using a single term of the form $A \sin (\omega t + \phi)$ or $B \cos (\omega t + \theta)$. A method for doing this will be given in Sec. 7-1.

Integrating both sides of the above equation and multiplying by the factor e^{-at}, we obtain

$$f(t) = e^{-at} \int g(t) e^{at} \, dt + K e^{-at} \tag{5-24}$$

The integral term in the right member of this result is called the particular integral. Evaluating the integral and multiplying by e^{-at} gives us the particular solution. The second term in the right member of (5-24) is the complementary solution. The constant K is determined so as to match the specified initial condition. Thus, this is the general form of the complete solution to the first-order, homogeneous, differential equation given in (5-12).

The techniques defined in this section are easily applied to a circuit containing several voltage and current sources and/or several resistors. In such a case, we need merely reduce the portion of the circuit which contains the sources and the resistors to a single source and a single resistor, that is, we need simply use the techniques of Sec. 3-8 to find a Thevenin or Norton equivalent for the portion of the network consisting of the sources and resistors. This equivalent network may then be used to solve for the network variable associated with the energy-storage element. KCL and KVL may then be applied to find any other network variable. The procedure is illustrated by the following examples.

Example 5-2.2

The first-order circuit shown in Fig. 5-2.4(a) contains a voltage source, a current source, two resistors, and a capacitor. It is desired to find the current $i(t)$ flowing from the voltage source for $t \geq 0$. It is assumed that the circuit is initially in the zero state, i.e., that the initial voltage $v(0)$ across the capacitor is zero. To solve this problem, it is convenient to first solve for the voltage $v(t)$ rather than the current $i(t)$. To do this we must first reduce the circuit to the form where a single source and a single resistor are used to replace the two sources and the two resistors. First let us replace the voltage source and the resistor in series with it by a Norton equivalent source. Thus we obtain the circuit shown in Fig. 5-2.4(b). The two sources and the two resistors may now be combined to yield the circuit shown in Fig. 5-2.4(c). For this circuit, using KCL, we may write

Figure 5-2.4 First-order circuit with two sources.

$$\frac{1}{4}\frac{dv}{dt} + v(t) = \frac{1}{2} + t \qquad t \geq 0$$

From Table 5-2.1, the assumed form for the particular solution $v_p(t)$ is

$$v_p(t) = A + Bt$$

Substituting this expression in the above differential equation and equating like coefficients of t, we readily obtain the values $A = \frac{1}{4}$, $B = 1$. Thus the particular solution is known. The complementary solution $v_h(t)$ is easily seen to be Ke^{-4t}. The complete solution, with the value of K determined such that $v(0) = 0$, is thus

$$v(t) = \tfrac{1}{4}(1 - e^{-4t}) + t \qquad t \geq 0$$

We may now solve for the current $i(t)$ by noting that this variable may be expressed as a function of the capacitor voltage $v(t)$ and the source voltage $2t$. Applying KVL around the right-hand mesh of the circuit shown in Fig. 5-2.4(a) we obtain

$$i(t) = \frac{2t - v(t)}{R_2}$$

Substituting the result for $v(t)$ determined above and the value of R_2 we obtain the result

$$i(t) = \tfrac{1}{2}t - \tfrac{1}{8}(1 - e^{-4t}) \qquad t \geq 0$$

Example 5-2.3

The first-order circuit shown in Fig. 5-2.5(a) contains both a dependent source and an independent source. It is desired to find the current $i_1(t)$ flowing out of the independent source for all values of $t \geq 0$. It is assumed that the initial condition $v(0)$ on the

Figure 5-2.5 First-order circuit with a dependent and an independent source.

capacitor is zero. The first step in solving this problem is to replace the portion of the network containing the resistors and the sources by a Thevenin equivalent circuit, following the procedure outlined in Summary 3-8.3. This portion of the network is the same as the one used in Example 3-8.3. Using the results of that example, we obtain the equivalent circuit shown in Fig. 5-2.5(b). Using KVL, for this circuit, we may write

$$\tfrac{3}{2} = \tfrac{5}{4}i(t) + v(t)$$

Substituting the relation $i(t) = \frac{4}{5}\,dv/dt$ for the terminal variables of the capacitor, the above relation may be written

$$\frac{dv}{dt} + v(t) = \frac{3}{2} \qquad t \geq 0$$

where $v(t)$ is the voltage across the capacitor. The solution for $v(t)$ is

$$v(t) = \tfrac{3}{2}(1 - e^{-t}) \qquad t \geq 0$$

Since the voltage across the capacitor is fixed by this expression, we may replace the capacitor by a voltage source of value $\tfrac{3}{2}(1 - e^{-t})$ V in the original circuit. Thus we obtain the circuit shown in Fig. 5-2.5(c). Using the techniques for solving resistance networks with controlled sources given in Sec. 3-7, we see that the equations describing this circuit may be put in the form

$$\begin{bmatrix} 2 \\ -\dfrac{3}{2}(1 - e^{-t}) \end{bmatrix} = \begin{bmatrix} 4 & -1 \\ -3 & 2 \end{bmatrix} \begin{bmatrix} i_1(t) \\ i_2(t) \end{bmatrix}$$

The solution for this set of equations is

$$\begin{bmatrix} i_1(t) \\ i_2(t) \end{bmatrix} = \frac{1}{5} \begin{bmatrix} 2 & 1 \\ 3 & 4 \end{bmatrix} \begin{bmatrix} 2 \\ -\dfrac{3}{2}(1 - e^{-t}) \end{bmatrix}$$

Thus, we find that

$$i_1(t) = \tfrac{1}{2} + \tfrac{3}{10}e^{-t} \qquad t \geq 0$$

This is the solution for the current flowing out of the independent source.

It has been pointed out in the discussion of this section and in the examples that the complementary portion of the complete solution to a non-homogeneous differential equation is found by solving the related homogeneous equation. If the differential equation is actually known and has the form given in (5-12), this is readily accomplished by setting the right member $g(t)$ to zero. It is also possible, however, to find the homogeneous equation directly from the physical network and in many cases this provides a simpler procedure. Since the complementary part of the complete solution does not include the effect of the independent sources, to find the homogeneous equation we need only set all such sources to zero, replace all the remaining resistors and dependent sources with a single equivalent resistance, and apply KVL or KCL. It should be recalled from Sec. 2-2 that when a voltage source is set to zero, it may be replaced by a short circuit. Similarly, when a current source is set to zero, it may be replaced by an open circuit. Thus, if we set the sources in the circuit shown in Fig. 5-2.4(a) to zero, we obtain the circuit shown in Fig. 5-2.4(d). Replacing the two resistors shown in this circuit with a single equivalent resistor, we see that the homogeneous differential equation has the form $\tfrac{1}{4}\,dv/dt + v(t) = 0$. This is the same expression as was found for this circuit in the discussion given in Example 5-2.2, although in the example we reduced the entire network, including the sources, to find the equation.

The linearity aspects of circuits excited only by sources are similar to those discussed in the preceding section for circuits excited only by initial conditions. For example, consider the problem discussed in Example 5-2.1. If the source excitation $v(t)$ is changed to $20 \sin \omega t$ (twice the value given in the example), then it is readily verified that the resulting response for $i(t)$ is $4e^{-2t} + 8 \sin t - 4 \cos t$, namely, twice the value found in the example. More generally, for a first-order circuit with a zero initial condition, it may be shown that if a single source is used to excite the circuit, and if an excitation $e_1(t)$ produces a response $r_1(t)$ and an excitation $e_2(t)$ produces a response $r_2(t)$, then an excitation $e_1(t) + e_2(t)$ will produce a response $r_1(t) + r_2(t)$. If there is more than one (independent) source in the network, then considerations similar to those developed in Sec. 3-1 for the resistance network with multiple inputs apply. That is, the relation between an excitation provided by any one of the sources and the response of any network variable to it, when all the other sources are set to zero, will be a linear relation. By a linear relation, we mean that the properties of superposition and homogeneity defined in Sec. 1-3 apply.

As an example of this result, consider again the circuit discussed in Example 5-2.2. If we let $i_i(t)$ be the complete solution for $i(t)$ when only the current source is used as an excitation, i.e., when the voltage source is set to zero, we obtain $i_i(t) = (e^{-4t} - 1)/4$ A. Similarly, if we let $i_v(t)$ be the complete solution for $i(t)$ when only the voltage source is used as an excitation, we find that $i_v(t) = (t/2) + (1 - e^{-4t})/8$ A. The sum of $i_i(t)$ and $i_v(t)$ obviously gives the answer for $i(t)$ found in the example. If we now consider a new problem where the current source excitation is $\frac{3}{2}$ A (three times the value used in the example), the voltage source output is $4t$ V (two times the value used in the example), and both sources are applied simultaneously, then, since the network is linear, the resulting current $i(t) = 3i_i(t) + 2i_v(t)$. This result is readily verified by direct solution of the network.

5-3 EXCITATION BY INITIAL CONDITIONS AND SOURCES

In the first section of this chapter we discussed the procedure to be used in solving for the network variables of a first-order circuit which was excited *only by initial conditions*. One of the major steps in this procedure was the solution of a first-order, homogeneous, differential equation. In the following section we showed how to solve for the variables of a network excited *only by sources*. In this case, it was necessary to solve a first-order, non-homogeneous, differential equation. The next topic to be discussed follows very logically, namely, what happens when sources *and* initial conditions are present. This will be the subject of our discussion in this section. In addition, we shall present some further properties of first-order circuits which follow from our discussion.

In the last section we showed that the solution for a non-homogeneous, first-order, differential equation could be found by the following steps:

1. Find the complementary solution (with an unspecified constant).
2. Find the particular solution (based on the form of the excitation function).
3. Add the complementary solution and the particular solution to obtain the complete solution.
4. Evaluate the unspecified constant so as to satisfy some known initial condition.

These same steps, which were used in the last section for problems in which the initial condition was zero, may be directly applied to circuits in which the initial condition is non-zero. The only change in the form of the solution will be that the value of the constant in the complementary portion of the complete solution will be different. An example follows.

Example 5-3.1

The first-order circuit shown in Fig. 5-2.3 and discussed in Example 5-2.1 was shown to have a complete solution for the current $i(t)$ which was of the form

$$i(t) = Ke^{-2t} + 4\sin t - 2\cos t \qquad t \geq 0$$

The example stipulated that the initial condition on $i(t)$ was zero. It was found that $K = 2$ satisfied this condition. Suppose that we now specify a non-zero initial condition; for example, let $i(0) = I_0$. We may evaluate the equation given above at $t = 0$ to find the new value of K. We obtain $i(0) = K - 2 = I_0$. Thus we find that in this case $K = 2 + I_0$. The complete solution for the current $i(t)$, including the effect of the initial condition, now becomes

$$i(t) = (2 + I_0)e^{-2t} + 4\sin t - 2\cos t \qquad t \geq 0$$

Summary 5-2.3 may now be repeated in the following form:

SUMMARY 5-3.1

Solution of a First-Order Circuit Excited by Initial Conditions and Sources. The complete solution for a first-order circuit excited by both initial conditions and sources may be found by solving the non-homogeneous differential equation describing the circuit. The complete solution is the sum of the complementary solution and the particular solution. The constant (which is part of the complementary solution) is evaluated in the complete solution to match the specified initial condition. ∎

The expression for $i(t)$ given in the above example may be broken into separate parts to illustrate some of the properties of first-order circuits. For example, we may write

$$i(t) = \underbrace{I_0 e^{-2t}}_{\substack{\text{Zero-input response} \\ \text{(term due to the} \\ \text{excitation of the} \\ \text{network by a non-} \\ \text{zero initial condition)}}} + \underbrace{2e^{-2t} + 4\sin t - 2\cos t}_{\substack{\text{Zero-state response} \\ \text{(terms due to the excitation} \\ \text{of the network by a source)}}} \qquad t \geq 0 \qquad (5\text{-}25)$$

As indicated by the above decomposition, we note that the complete response is the sum of the response produced by the initial condition and the response produced by the source. It should be noted, however, that the quantity characterizing the *natural response* of the network, namely, the exponential e^{-2t}, appears both in the term giving the response from the initial condition (the zero-input response) and in the term giving the response from the source (the zero-state response). Another way of expressing this fact is to say that the network's natural frequency is excited both by the initial condition and by the source.

Now let us consider a different decomposition of the complete solution for $i(t)$. This is

$$i(t) = \underbrace{I_0 e^{-2t}}_{\text{Transient response}} + \underbrace{2e^{-2t} + 4\sin t - 2\cos t}_{\text{Forced response}} \qquad (5\text{-}26)$$

Transient response
(term due to the natural
response of the circuit
which disappears
for large values of time)

Forced response
(term which is independent
of the natural response
of the circuit and which
depends only on the
form of the excitation
from the source)

As indicated by Eq. (5-26) we note that a second way of expressing the complete response is as the sum of the transient response and the forced response. If the excitation is a constant or a sinusoid, then, after the transient response has disappeared, the remaining portion of the complete solution has the value of a constant (if the excitation is a constant) or a sinusoid of constant magnitude (if the excitation is such a sinusoid). In either of these cases, we say that the circuit has reached "steady-state" conditions, and thus the forced response term on the right in Eq. (5-26) may also be referred to as the *steady-state response*. This term is only used for the case of excitations which are either constant or sinusoidal.

The conclusions reached above for Example 5-3.1 may easily be shown to be true for first-order circuits in general. Let us consider a first-order circuit described by a non-homogeneous differential equation of the form

$$\frac{df}{dt} + af(t) = g(t) \qquad t \geq 0 \qquad (5\text{-}27)$$

In the following summary, we describe the complete solution of such a differential equation in terms of its component parts based on the manner in which the circuit is excited [see (5-25) as an example]:

SUMMARY 5-3.2

Solution of a First-Order Circuit (zero-input response and zero-state response). The complete solution of a non-homogeneous differential equation of the form given in (5-27) describing a first-order circuit will consist of a term resulting from the initial condition (the zero-input response) which has the form $K_0 e^{-at}$, and a term produced by the excitation function (the zero-state response) which has the form $K_1 e^{-at} + g_0(t)$, where $g_0(t)$ is a function having the same general form as the excitation function $g(t)$, and may be determined from Table 5-2.1. ∎

Similarly, in the following summary, we describe the complete solution of a first-order circuit in terms of its component parts based on the natural and forced behavior of the network [see (5-26) as an example]:

SUMMARY 5-3.3

Complete Solution of a First-Order Circuit (transient response and forced response). The complete solution of a non-homogeneous differential equation of the form given in (5-27) describing a first-order circuit will consist of a transient response term related to the natural behavior of the network and having the form Ke^{-at}, where the value of K is the result of the excitation from both the initial condition and the application of some excitation function $g(t)$.[5] A second term in the complete response will also be produced by the excitation function $g(t)$. It will have a form similar to $g(t)$ and may be determined from Table 5-2.1. ■

The linearity considerations given in Sec. 5-1 for the case of a first-order circuit excited only by initial conditions (the zero-input case) and in Sec. 5-2 for a network excited only by sources (the zero-state case) are easily extended to the case where such a circuit is excited by both initial conditions and sources. As an illustration of the principles involved, consider the following example.

Example 5-3.2

The linear first-order circuit shown in Fig. 5-3.1 is excited by three separate effects, namely, the initial condition $v(0) = 1$ V on the capacitor, the voltage source with output voltage $v_S(t) = t$ V, and the current source with output current $i_S(t) = 1$ A (both applied at $t = 0$). Let us consider the following cases of reduced excitation:

Figure 5-3.1 First-order circuit.

Case 1 (example of zero-input response): If both sources are set to zero, then the solution for the voltage $v(t)$ across the capacitor yields $v(t) = 1e^{-2t}$. Now suppose the initial condition on the capacitor is doubled, so that $v(0) = 2$. We find in this case that $v(t) = 2e^{-2t}$. We conclude that the variable $v(t)$ is linearly related to the excitation provided by the initial condition. This is the same conclusion given in Summary 5-1.1.

Case 2 (example of zero-state response): If the initial condition and the voltage source are both set to zero but $i_S(t) = 1$ A (applied at $t = 0$), then solving for $v(t)$, we find that $v(t) = -\frac{1}{2}e^{-2t} + \frac{1}{2}$ V. If we double the excitation by changing $i_S(t)$ from 1 to 2 A, then we find that $v(t) = -1e^{-2t} + 1$. We conclude that for a single source, the variable $v(t)$ is linearly related to the excitation if the initial condition and the other source are set to zero.

Case 3 (another example of zero-state response): If the initial condition and the current source are set to zero but $v_S(t) = t$ V (starting at $t = 0$), then, solving for $v(t)$, we find that $v(t) = \frac{1}{4}e^{-2t} - \frac{1}{4} + \frac{1}{2}t$ V. If the excitation from the voltage source is changed from t V to $2t$ V, we find that the response is also doubled, namely, $v(t) = \frac{1}{2}e^{-2t} - \frac{1}{2} + t$. Just as in case 2, we conclude that the response $v(t)$ is linearly related to the excitation. The results for these two cases are the same as given in Summary 5-2.3.

[5] In some cases, this term may be suppressed. As an illustration of this, note that in Example 5-3.1, if we choose the value of -2 for the constant I_0, then K will be zero and the exponential term will not be present. For a detailed discussion of how specific natural frequencies may be excited or suppressed, see L. P. Huelsman, *Circuits, Matrices, and Linear Vector Spaces*, Chap. 6, McGraw-Hill Book Company, New York, 1963.

Case 4 (example of general response): If the initial condition and both sources are used to excite the network, then we find that $v(t) = \frac{3}{4}e^{-2t} + \frac{1}{4} + \frac{1}{2}t$. Note that superposition applies, since this response is the sum of the three responses obtained for cases 1, 2, and 3 (for the original excitations). If we now double any of the excitations, however, we find that our response has not doubled. For example, if the initial condition is changed from 1 to 2 V, we find that $v(t) = \frac{7}{4}e^{-2t} + \frac{1}{4} + \frac{1}{2}t$. Of course, if all excitations are doubled, then the resulting expression for $v(t)$ will also be doubled.

Generalizing the results given in the above example, we see that the response of any variable in a first-order circuit consisting of linear elements is linearly related to the following excitations: (1) the excitation from an initial condition when all sources are set to zero; and (2) the excitation from any source when the initial condition and all other sources are set to zero. In general, the response is linearly related to any single excitation, and is not linearly related to a given excitation if other excitations are present.

In this and the preceding sections we have presented techniques for finding the solution of a first-order circuit which is excited by sources and initial conditions. The overall procedure is illustrated by the flow chart shown in Fig. 5-3.2. It is strongly

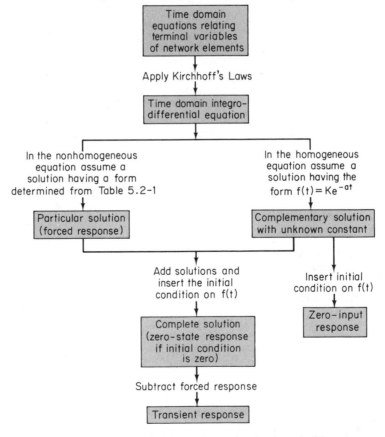

Figure 5-3.2 Flow chart for solving a first-order circuit excited by sources and initial conditions.

recommended that the student, as a review, restudy the examples given to this point in this chapter and identify each step in their solution with a corresponding step in the flow chart.

5-4 RESPONSE TO SOURCES WITH CONSTANT EXCITATION

A simplified method may be applied to solve first-order circuit problems in which *all independent sources have a constant output.* For such a case, let us assume that the first-order differential equation describing the circuit has been put in the form

$$\frac{df}{dt} + af(t) = G \qquad t \geq 0 \tag{5-28}$$

where $a > 0$ and G is a constant. From the preceding discussion, we know that the solution must have the form

$$f(t) = Ke^{-at} + G_0 \tag{5-29}$$

where G_0 is also a constant. If we let the initial and final values of $f(t)$ be $f(0)$ and $f(\infty)$, respectively, then, from (5-29), the following relations must hold.

$$f(0) = K + G_0 \qquad f(\infty) = G_0 \tag{5-30}$$

From these two equations, we may express the constant K of (5-29) as

$$K = f(0) - f(\infty) \tag{5-31}$$

Substituting the values of K and G_0 in (5-29), we obtain the following general solution for (5-28):

$$f(t) = [f(0) - f(\infty)]e^{-at} + f(\infty) \tag{5-32}$$

To emphasize this method of solution we may state it in the following form:

SUMMARY 5-4.1

Solution of a First-Order Circuit with Constant Excitation. If a first-order circuit is excited by initial conditions and constant-valued sources, and if the complementary solution for the circuit has the form Ke^{-at} $(a > 0)$, then the complete solution for $f(t)$ for $t > 0$ is given by

$$f(t) = [f(0) - f(\infty)]e^{-at} + f(\infty)$$

where $f(t)$ is any variable in the circuit. ∎

The utility of this method lies in the fact that it merely requires that we know the initial and final values of any circuit variable (and the natural exponential behavior of the circuit) to completely determine the value of that variable. Thus, we conclude that the circuit variables all change exponentially from their value when the network is in some initial state to their value when it is in some final state.

This method is easily implemented directly from the circuit diagram for a given network. We have already seen how to find the form of the complementary solution by setting the independent sources to zero. Now let us see how to determine the initial and final values of the network variables. First we will consider the initial values. The continuity conditions developed in Chap. 4 tell us that capacitor voltages and inductor currents cannot be discontinuous. Therefore, for a network with zero initial conditions at $t = 0$, we need simply replace a capacitor by a voltage source of zero voltage (this is simply a short circuit) or an inductor by a current source of zero output current (this is an open circuit), and solve for the initial values of any network variable in which we are interested. Note that if some initial condition happens to be present on either of these elements, then the replacement source will simply have the appropriate value of the initial voltage or current. Two examples follow.

Example 5-4.1

The method described above is easily applied to find the initial value of the current through the capacitor in the circuit used in Example 5-2.2 and shown in Fig. 5-2.4(a). Since the initial voltage across the capacitor is zero, at $t = 0$ the capacitor may be replaced by a voltage source of zero voltage, namely, a short circuit. The current source provides $\frac{1}{2}$ A through this short, and the independent voltage source, since it has zero output voltage at $t = 0$, provides zero output voltage. Thus, we see that the initial value of current through the capacitor is $\frac{1}{2}$ A.

Example 5-4.2

It is assumed that the initial condition at $t = 0$ for the inductor current $i_L(t)$ in the circuit shown in Fig. 5-4.1(a) is -5 A. It is desired to find the initial voltage $v_L(0)$ across the inductor. Since the initial current in the inductor is -5 A, at $t = 0$ the inductor may be replaced by a current source of value 5 A. Thus, at $t = 0$, the circuit appears as shown in Fig. 5-4.1(b). From the figure we see that the voltage across the 2-Ω resistor is 8 V, and the voltage across the 1-Ω resistor is 5 V. From KVL, we find that the voltage across the 5-A current source is 3 V. This is also the initial voltage $v_L(0)$ across the inductor.

Figure 5-4.1 Circuit for Example 5-4.2.

The method described above for finding the initial values of network variables is quite general. It may be applied to a circuit which is excited by sources having any time dependence, not just constant-valued sources. In addition, we will see in later chapters that this method is applicable to circuits of higher than first-order, i.e., circuits containing more than one energy-storage element. To emphasize this, we present the following:

SUMMARY 5-4.2

Initial Values of Circuit Variables. The initial values of any variables in a circuit may be found by replacing the capacitors with voltage sources of value equal to the initial voltage across them, replacing the inductors with current sources of value equal to the initial current through them, and solving the resistance network. If the circuit was unexcited for $t \leq 0$, then all the initial conditions on the energy-storage elements are zero. In this case, capacitors may be replaced by short circuits, and inductors by open circuits. ∎

Now let us consider finding the *final* values of circuit variables. If the excitations provided by the independent sources are all constant in value, then as t approaches infinity, all the circuit variables must reach some constant value. This means that all the derivatives on the circuit variables must go to zero. Since the voltage across an inductor depends on the derivative of the current, any such voltage must also go to zero. The current through the inductor, however, may have any value. We conclude that as t approaches infinity, every inductor will be characterized as having zero voltage across it independent of the current flowing through it. This is exactly the property possessed by a short circuit. Therefore, if we desire to evaluate the circuit variables at $t = \infty$, we may replace all inductors by short circuits. Similar reasoning tells us that at $t = \infty$ we may replace capacitors by open circuits. Having done this, the values of any of the network variables are easily found. This technique is applicable to circuits containing any number of inductors and capacitors, as long as the sources exciting the circuit are all constant-valued. To emphasize this, we state the following:

SUMMARY 5-4.3

Final Values of Circuit Variables. If all the independent sources in a circuit are constant-valued, then the final values of any network variables may be found by replacing inductors with short circuits and capacitors with open circuits.[6] ∎

Note that the replacement operations in the two cases described above are opposite, i.e., to determine *initial* values we replace inductors by open circuits and capacitors by short circuits (assuming that the initial conditions are zero), but to determine *final* values we replace inductors by short circuits and capacitors by open circuits. An example of finding the final value of a network variable follows.

Example 5-4.3

The final value of the voltage $v_C(t)$ across the capacitor in the circuit shown in Fig. 5-4.2(a) is easily found by the method described above. To simplify our visualization

[6] It is also necessary to require that the magnitude of the complementary solution decays exponentially. The reason for this stipulation will become clear when circuits containing inductors and capacitors are studied in the following chapter.

of the process, the circuit has been redrawn in Fig. 5-4.2(b), with the inductor replaced by a short circuit and the capacitor replaced by an open circuit. This schematic represents the conditions that are found in the circuit as t approaches infinity. By routine analysis of the resulting resistance network, we see that $v_C(\infty) = -4$ V.

(a)   (b)

Figure 5-4.2 Circuit for Example 5-4.3.

By the methods described above, we may easily find the initial and final values of any variable in a first-order circuit excited by constant-valued sources. Thus, we may apply the techniques specified in Summary 5-4.1 to determine the time variation of any variable directly. An example follows.

Example 5-4.4

As an example of the solution of circuits excited by constant-valued sources, consider the transistor circuit shown in Fig. 5-4.3(a). In this circuit, the voltage source shown as a battery on the right is a power supply providing a bias voltage for the transistor, while the voltage source on the left provides an input signal consisting of a 10-V step

(a)   (b)   (c)

Figure 5-4.3 Transistor stage excited by a step of voltage.

function applied at $t = 0$. Using the transistor model shown in Fig. 2-2.7 with $\beta = 100$, it is desired to find $v_o(t)$ for $t \geq 0$. The first step is to find the initial condition $v_o(0)$. For this we may use the circuit shown in Fig. 5-4.3(b). In this circuit we assume that the power supply voltage has been applied to the circuit for a long enough time so that the circuit variables have reached steady state ($t < 0$). In this circuit, $i_b(0) = 0$, thus $v_o(0) = 10$ V. The next step in the analysis of this circuit is to find the complementary solution. In this case, in Fig. 5-4.3(b), we set the (independent) power supply source to zero, and, using the relation $i_b(t) = C \, dv_o/dt$, applying KCL at the + terminal where $v_o(t)$ is defined, we obtain

$$i_b(t) + 100i_b(t) + \frac{v_o(t)}{R} = 101C\frac{dv_o}{dt} + \frac{v_o(t)}{R} = 0$$

Thus, the time constant for the circuit is $101RC = 101 \times 10^{-6}$ s, i.e., the value of the constant a in the general solution expression given in Eq. (5-32) is (approximately) 10^4. The final step in the analysis is to find $v_o(\infty)$. For this we may use the circuit shown in Fig. 5-4.3(c), representing the situation when both sources are energized. In this circuit, under steady-state conditions, we see that $i_b = 10/(200 \times 10^3) = 50 \times 10^{-6}$ A. Applying KVL we obtain

$$v_o(\infty) = 10 - R(100i_b) = 5 \text{ V}$$

If we now insert the results obtained above in Eq. (5-32), we find

$$v_o(t) = 5e^{-10^4 t} + 5$$

Thus, the output voltage of the transistor stage changes from 10 V to 5 V in about 400 μs (four time constants) when a 10-V input step is applied to the circuit.

5-5 INITIAL CONDITIONS APPLIED AT $t \neq 0$

In all our discussions of first-order circuits to this point, we have used initial conditions which are specified at $t = 0$. In this section we shall extend our treatment to cover the case of "initial" conditions which are specified for values of t not equal to zero. As a first example of such a case, consider the circuit shown in Fig. 5-5.1. Let us

Figure 5-5.1 First-order RC circuit.

assume that this circuit is excited only by an initial condition on the capacitor. In Sec. 5-1 we pointed out that the time constant for this circuit, i.e., the time required for a variable to decay to 36.8 percent of its initial value, is equal to RC. Since the circuit is time-invariant, this fact must hold true no matter at what instant we start our count of time. Therefore, if the voltage $v(t)$ across the capacitor is equal to V_0 at some time t_0, it will be equal to $0.368V_0$ at time $t_0 + RC$. In addition, all following values of $v(t)$ will exhibit a similar exponential behavior. For example, $v(t)$ will be equal to $0.018V_0$ at a value of t equal to four time constants after t_0, i.e., when $t = t_0 + 4RC$. To illustrate this, the waveshape of $v(t)$ for an initial condition V_0 applied at $t = t_0$ has been sketched in Fig. 5-5.2 using a solid line. For convenience, the curve

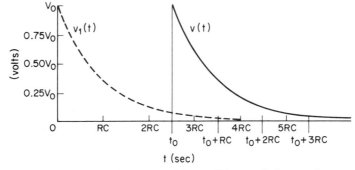

Figure 5-5.2 Effect of an initial condition applied at $t \neq 0$.

$V_0 e^{-t/RC} u(t)$ has also been sketched using a dashed line. It should be apparent, both from the sketch and from the discussion given above, that the curves are identical in shape, although they are displaced along the abscissa by the quantity t_0. If we call the dashed curve $v_1(t)$ and the solid curve $v(t)$, then we may mathematically define such a displacement by the relation

$$v(t) = v_1(t - t_0) \tag{5-33}$$

In other words, if we substitute the argument $t - t_0$ for the argument t in the expression for the dashed curve of Fig. 5-5.2, we obtain the correct expression (as a function of t) for the solid curve. The expression for the dashed curve is[7]

$$v_1(t) = V_0 e^{-t/RC} u(t) \tag{5-34}$$

thus the expression for the solid curve $v(t)$ is

$$v(t) = V_0 e^{-(t-t_0)/RC} u(t - t_0) \tag{5-35}$$

To check the validity of this expression, we note that when $t = t_0$, $v(t_0) = V_0$, as is to be expected. Similarly, one time constant later, $v(t_0 + RC) = V_0 e^{-1} = 0.368 V_0$, etc.

An operation of the type defined by Eq. (5-33) is usually referred to mathematically as *shifting* or *translation*. It is readily applied to a wide variety of problems in which elements are added to or deleted from a circuit by opening or closing switches at different values of time. An example of such a situation follows.

Example 5-5.1

The circuit shown in Fig. 5-5.3(a), at $t = 0$, has an initial condition $v(0)$ of 10 V across the capacitor, and switch $S1$ is open. At $t = 5$ s, switch $S1$ is closed. It is desired to find the voltage $v(t)$ for all time $t \geq 0$. The problem may be attacked in two parts. For the first part, consider the circuit during the period $0 \leq t \leq 5$. For this range

Figure 5-5.3 Circuit for Example 5-5.1.

of time, the original circuit has the electrical appearance of the circuit shown in Fig. 5-5.3(b). The time constant for this circuit is 5 s. Thus, the solution for the voltage $v(t)$ is readily seen to be

$$v(t) = 10e^{-t/5} \qquad 0 \leq t \leq 5$$

Note that $v(5) = 10e^{-1} = 3.68$ V. We may now consider the second part of this problem, namely, the range of time $t \geq 5$, during which the switch $S1$ is closed. For this range of time, the original circuit has the electrical appearance of the one shown in Fig. 5-5.3(c). The time constant for this circuit is 2.5 s. In addition, since the voltage $v(t)$ must be continuous at the instant switch $S1$ is closed, we have the "initial" condition $v(5) = 3.68$ V. Thus, the solution for $v(t)$ is

[7] The expressions $u(t)$ and $u(t - t_0)$ are defined in Sec. 4-2.

$$v(t) = 3.68e^{-(t-5)/2.5} \qquad t \geq 5$$

A plot of the waveshape of $v(t)$ for a range of time of 0 to 15 s is shown in Fig. 5-5.4.

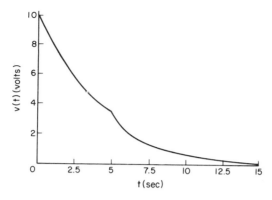

Figure 5-5.4 Output waveform for the circuit shown in Fig. 5-5.3.

The changing of circuit parameters by means of switches illustrated in the preceding example may introduce transients into a circuit. Thus, although we know that capacitor voltages and inductor currents must be continuous in a physical network, other variables may exhibit discontinuities. To evaluate these discontinuities, and thus to determine the necessary values of initial conditions which occur when elements are switched into or out of a circuit, we may use the techniques for finding initial values of variables described in Sec. 5-4. This type of situation is illustrated by the following example.

Example 5-5.2

The circuit shown in Fig. 5-5.5 has an initial condition of 10 V on the capacitor. The switch $S1$ is initially open, but is closed at $t = 5$ s. It is desired to find the current $i(t)$ through the indicated resistor. The problem may be solved in two parts in a manner

Figure 5-5.5 Circuit for Example 5-5.2.

similar to that used in the previous example. For the range of time $0 \leq t < 5$ s, the time constant of the circuit is 5 s, and the initial current through the capacitor is readily seen to be 2 A. Thus, we see that

$$i(t) = 2e^{-t/5} \qquad 0 \leq t < 5$$

Now let us consider the circuit variables before and after the switch $S1$ is closed. For the voltage $v(t)$ across the capacitor, we may write[8]

$$v(5-) = v(5) = v(5+) = 3.68$$

---

[8]Using the notation introduced in Sec. 4-3, the + and − signs in the arguments are used to indicate the limit of the function $v(t)$ as the variable t approaches the value of 5 from above and below, respectively.

For the resistor current $i(t)$, we note that $i(5-) = 0.736$ A. After the switch $S1$ is closed, however, we find that the capacitor voltage of 3.68 V now appears across a single 2.5 Ω resistor. Thus, $i(5+) = 1.472$ A. This value provides the initial condition for the second part of the problem, namely, the range of time $t > 5$. Since the time constant of the circuit for this range of time is 2.5 s, we may write

$$i(t) = 1.472e^{-(t-5)/2.5} \qquad t > 5$$

A plot of $i(t)$ for a range of 0 to 10 s is shown in Fig. 5-5.6. The discontinuous nature of this variable is readily apparent in the figure.

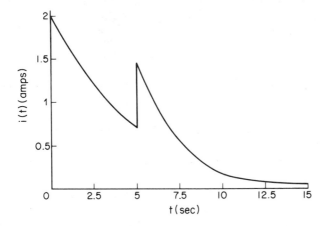

Figure 5-5.6 Output waveform for the circuit shown in Fig. 5-5.5.

We may summarize the development given above as follows:

SUMMARY 5-5.1

Solution of a First-Order Circuit Excited by Initial Conditions Applied at $t \neq 0$. A first-order circuit which has an initial condition specified on a variable for some time $t_0 \neq 0$ will have a zero-input response of the form

$$f(t) = f(t_0)e^{-(t-t_0)/\tau} \qquad t > t_0$$

where $f(t)$ is the expression for the variable, $f(t_0)$ is the initial condition, and τ is the time constant of the circuit. ∎

The techniques described above are readily applicable to the case where a first-order circuit is excited by sources which are "turned on" (or off) at times other than $t = 0$. The approach is illustrated by the following examples.

Example 5-5.3

The circuit shown in Fig. 5-5.7 contains a current source with an output $i_S(t) = 3u(t - 4)$ A. Thus, for all values of $t < 4$, the source current is zero, while for values of $t > 4$, the source current is a constant 3 A. It is desired to find the current $i(t)$ in the inductor, assuming that there is no initial condition. First let us consider what would happen if the step of current was applied at $t = 0$. The techniques given in Sec. 5-4 may be directly applied in this case. Since the initial inductor current must be zero, and the final inductor current must be 3 A, the expression for the current may be

ohms,
henries

Figure 5-5.7 Circuit for Example
5-5.3.

directly determined from Summary 5-4.1. It has the form $3(1 - e^{-2t})u(t)$. To solve the
original problem, since the circuit is linear and time-invariant, we need merely "shift"
both the excitation and the response by replacing t by $t - 4$ in the above expression.
Thus, we see that the answer for the original problem is

$$i(t) = 3(1 - e^{-2(t-4)})u(t - 4)$$

Example 5-5.4

The circuit shown in Fig. 5-5.8(a) is excited by two sources, namely a current source
which provides a constant excitation of 2 A starting at $t = 0$, and a voltage source
which puts out zero voltage for $t \leq 2$, and a voltage $3e^{-(t-2)}$ V for $t \geq 2$. To find the
voltage $v(t)$ across the capacitor assuming a zero initial condition, we may consider the

ohms, farads

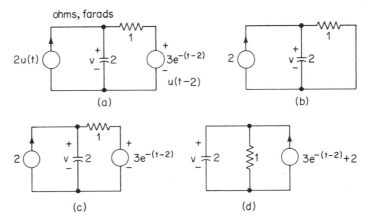

(a)

(b)

(c)

(d)

Figure 5-5.8 Circuit for Example 5-5.4.

circuit as it appears for two different time ranges. In Fig. 5-5.8(b) the circuit is shown as
it appears for the period $0 \leq t \leq 2$. The solution for this circuit is easily found from the
techniques of Sec. 5-4. Thus, we see that

$$v(t) = 2(1 - e^{-t/2}) \qquad 0 \leq t \leq 2$$

From this relation we note that $v(2) = 1.264$. At this time, the second source is "turned
on" by the step function term $u(t - 2)$. Thus, for the range of time $t \geq 2$, the circuit
appears as shown in Fig. 5-5.8(c). By making source transformations, we obtain the
circuit shown in Fig. 5-5.8(d). The non-homogeneous differential equation describing
the circuit (after multiplying through by the factor $\frac{1}{2}$) is

$$\frac{dv}{dt} + \frac{1}{2}v(t) = \frac{3}{2}e^{-(t-2)} + 1 = \frac{3}{2}e^2 e^{-t} + 1 \qquad t \geq 2$$

The form of the particular solution $v_p(t)$ for this circuit is found from Table 5-2.1. It is
$Ae^{-t} + B$. Solving for the constants, we find that

$$v_p(t) = -3e^2 e^{-t} + 2$$

The complementary solution to the related homogeneous equation has the form

$$v_h(t) = Ke^{-(t-2)/2}$$

Adding the two components of the solution, and evaluating the constant K of the complementary solution so that the initial condition $v(2) = 1.264$ is satisfied, we obtain

$$v(t) = 2.264e^{-(t-2)/2} - 3e^2e^{-t} + 2 \qquad t \geq 2$$

This equation, together with the one found for the range of time $0 \leq t \leq 2$, specifies the solution for the capacitor voltage $v(t)$ for this problem.

The examples given above treated cases in which the initial conditions on the circuit were zero. These results, however, are readily extended to the case where initial conditions are present and where various sources are connected to a network at times other than $t = 0$. The details of this extension are left to the reader as an exercise. We conclude this section by summarizing the development given above as follows:

SUMMARY 5-5.2

Solution of a First-Order Circuit Excited by Initial Conditions and Sources Applied at $t \neq 0$. If sources are used to excite a first-order circuit at some time $t \neq 0$, the particular solution which is valid for values of time following the application of the excitation may be found using the method outlined in Summary 5-2.2 and Table 5-2.1. The complementary solution will have the form specified in Summary 5-5.1, and will contain a constant whose value is determined so that the correct initial conditions at the time the excitation is applied are obtained. ∎

5-6 THE USE OF IMPULSES TO GENERATE INITIAL CONDITIONS

In Sec. 4-2, we introduced the impulse function which was characterized (in the limit) by infinite magnitude and zero duration. Such a function has many applications in the investigation of the properties of networks from a theoretical viewpoint. In this section we shall present one such application, namely, the use of impulse functions to establish initial conditions in first-order circuits. As an example of such a case, consider the circuit shown in Fig. 5-6.1. Applying KCL, we obtain for this circuit the relation

$$C\frac{dv}{dt} + Gv(t) = i(t) = K\delta(t) \qquad (5\text{-}36)$$

where $K\delta(t)$ is an impulse of strength K applied at $t = 0$. Let us assume that there is

Figure 5-6.1 *RC* network excited by an impulse of current.

no charge present on the capacitor for $t < 0$. If we now integrate (5-36) between the limits $-\infty$ and $0+$, we obtain

$$Cv(0+) + G \int_{-\infty}^{0+} v(\tau)\, d\tau = \int_{-\infty}^{0+} i(\tau)\, d\tau = K \qquad (5\text{-}37)$$

In the left member of this equation, the integral of $v(t)$ from $-\infty$ to $0+$ must be zero, since no excitation is applied before $t = 0$. Therefore, we see that

$$v(0+) = \frac{K}{C} \qquad (5\text{-}38)$$

We conclude that the impulse of current applied by the current source has established an initial condition on the previously unexcited capacitor. The response of the capacitor voltage $v(t)$ to the application of such an impulse of current at $t = 0$ is thus easily found using the methods of Sec. 5-1. We obtain

$$v(t) = \frac{K}{C} e^{-t/RC} \qquad t > 0 \qquad (5\text{-}39)$$

A similar (dual) situation holds true for the circuit shown in Fig. 5-6.2, in which an

Figure 5-6.2 *RL* network excited by an impulse of voltage.

impulse of voltage of strength K is applied to the circuit by the voltage source at $t = 0$. We assume that the initial condition at $t = 0-$ is zero. From KVL, we obtain for this circuit

$$L\frac{di}{dt} + Ri(t) = v(t) = K\delta(t) \qquad (5\text{-}40)$$

Integrating from $-\infty$ to $0+$, we obtain

$$Li(0+) + R \int_{-\infty}^{0+} i(\tau)\, d\tau = \int_{-\infty}^{0+} v(\tau)\, d\tau = K \qquad (5\text{-}41)$$

As before, the integral in the left member of this equation is zero, and we see that the impulse of voltage has produced an initial condition of current in the circuit which has the value

$$i(0+) = \frac{K}{L} \qquad (5\text{-}42)$$

Thus, the solution for $i(t)$ is

$$i(t) = \frac{K}{L} e^{-Rt/L} \qquad t > 0 \qquad (5\text{-}43)$$

It should be noted that in both the examples given above, the use of impulses has violated the continuity condition for the energy-storage element concerned. We may summarize the results given in the above examples as follows:

The Use of Impulses to Generate Initial Conditions. An impulse of current applied by a current source in parallel with a capacitor will generate an initial condition of voltage on the capacitor. An impulse of voltage applied by a voltage source in series with an inductor will generate an initial condition of current in the inductor.

∎

The response of some network variable to an impulse of excitation as illustrated in the examples given above is logically referred to as the *impulse response*. Since, from the above discussion, impulses can be thought of as establishing initial conditions in a circuit, it should be apparent that the impulse response is identical to the response which results from initial conditions which have been established by any other means. Thus, the treatment of circuits containing sources which apply impulses as well as other excitations is the same as the treatment of circuits excited by initial conditions and sources with non-impulsive outputs. This means that all the techniques of analysis which have been developed so far in this chapter can be applied to such circuits. An example of such a situation follows.

Example 5-6.1

The circuit shown in Fig. 5-6.3 is excited by two voltage sources connected in series. Source 1 applies an impulse of strength I_0 (volts) to the circuit at $t = 0$. Source 2 applies an excitation $(10 \sin t)u(t)$ V to the circuit, i.e., an excitation of $10 \sin t$ V starting at

Figure 5-6.3 Circuit for Example 5-6.1.

$t = 0$. It is assumed that there is no initial condition present in the circuit for $t < 0$. From Eq. (5-42) we see that the effect of Source 1 is to establish an initial condition $i(0+) = I_0$ A in the inductor. The solution for the current $i(t)$ for $t > 0$ follows exactly the procedure given in Example 5-3.1. Thus we see that

$$i(t) = (2 + I_0)e^{-2t} + 4 \sin t - 2 \cos t \qquad t > 0$$

Note that we have specified this answer for the strict inequality $t > 0$, rather than the weak inequality $t \geq 0$ used in Example 5-3.1, since the conditions in the circuit discussed here are actually indeterminate at $t = 0$. In addition, we note that the two voltage sources could be combined into a single voltage source with an output voltage $v(t) = I_0\delta(t) + 10 \sin t$ V to obtain the same response for the variable $i(t)$.

Since the impulse response described above is identical with the response found when a given network is excited only by initial conditions, its form is determined solely by the properties of the network, rather than the form of the excitation. Thus it is an intrinsic or *natural* property of the network itself. We may express this in another manner by saying that the impulse response of a given network is the result of exciting the *natural frequencies* of the network. This concept will become of great utility when we discuss the Laplace transformation in Chap. 8.

In linear circuits there is a direct relation between the impulse response of a given network and the response obtained by the application of any other waveform by the same source. We shall discuss this relation in more detail in Sec. 5-7. Here, as an introduction to the topic, consider the indefinite integral of an impulse $\delta(t)$. In Sec. 4-2 this was shown to be the unit step $u(t)$. Thus, we may write

$$u(t) = \int_{-\infty}^{t} \delta(\tau) \, d\tau \qquad (5\text{-}44)$$

Not only is the unit step the integral of the unit impulse, but, if we apply a unit step $u(t)$ as an excitation to a given network, using the same source which was used to apply an impulse, the resulting response will be the integral of the impulse response. As an example of this, consider again the circuit shown in Fig. 5-6.1. If we let $v_\delta(t)$ be the impulse response, then, from Eq. (5-39), we may write

$$v_\delta(t) = \frac{K}{C} e^{-t/RC} u(t) \qquad (5\text{-}45)$$

where we have used the symbol $u(t)$ to avoid having to consider the limitation $t > 0$ in connection with the equation. To find the response to a unit step, we need merely integrate the impulse response given in Eq. (5-45). Let $v_u(t)$ be the step response. We obtain

$$v_u(t) = \int_{-\infty}^{t} v_\delta(\tau) \, d\tau$$

$$= \int_{-\infty}^{t} \frac{K}{C} e^{-\tau/RC} u(\tau) \, d\tau = KR(1 - e^{-t/RC}) u(t) \qquad (5\text{-}46)$$

It is readily verified by direct solution of the problem that the expression for $v_u(t)$ given in Eq. (5-46) is the same as the voltage resulting from the excitation of the circuit by a source current $Ku(t)$, i.e., a step of magnitude K A applied at $t = 0$. A similar conclusion is readily established for the *RL* circuit shown in Fig. 5-6.2. We may summarize the development given above by the following:

SUMMARY 5-6.2

Relation Between the Impulse Response and the Step Response. The response of a given network variable to the application of a step input is the integral of the response of that same variable to the application of an impulse. It is assumed that the initial conditions in the network are zero before the application of the excitations. ∎

As a further example of the importance of the impulse response, we leave the reader with the following conclusion, which will be developed further in Sec. 5-7: If the impulse response relating a given network variable to a specific source is known, then this expression may be used to find the response for any arbitrary excitation. Thus, the behavior of the network is completely characterized. For such a property to apply, of course, the network must be linear, i.e., comprised of linear elements.

5-7 CONVOLUTION AND SUPERPOSITION INTEGRALS

One of the limitations in the methods of analyzing first-order circuits which have been presented in this chapter is the difficulty of finding the forced response when the excitation function is defined by a complicated mathematical expression. Indeed, if the excitation is specified by a graphical waveform that has been experimentally determined, the methods are not applicable at all. In this section we shall introduce a most important general analysis method. The technique makes use of the impulse response and step response introduced in Sec. 5-6. It may be used to find the zero-state response of a network (see Summary 5-3.2) for an arbitrary waveform, even one which is graphically defined.

Let us begin our discussion by assuming that some arbitrary excitation function $g(t)$ is defined over the period $t_0 < t < t_n$ as shown in Fig. 5-7.1(a). We may approxi-

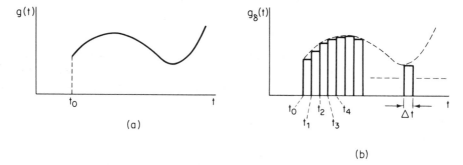

(a)

(b)

Figure 5-7.1   Arbitrary excitation function and its approximation.

mate this function by a series of n pulses of duration Δt and of varying amplitude as shown in Fig. 5-7.1(b). Such a series of pulses may be represented analytically by defining a unit-area pulse function $p_i(t)$ by the relationship

$$p_i(t) = \begin{cases} 0 & t < t_i \\ \dfrac{1}{\Delta t} & t_i < t < t_{i+1} \\ 0 & t > t_{i+1} \end{cases} \qquad (5\text{-}47)$$

where $\Delta t = t_{i+1} - t_i$. If we now define the quantity $g_\delta(t)$ as the pulse representation of $g(t)$, we may write

$$g_\delta(t) = \sum_{i=0}^{n-1} p_i(t)g(t_i)\,\Delta t \qquad (5\text{-}48)$$

This expression defines a function over all values of t in terms of a set of discrete values, namely, the quantities $g(t_i)$. Since all the pulses p_i have been defined with the same width, an alternative series representation for $g_\delta(t)$ can be made which will be more useful in the following development. To do this, let us define a single rectangular pulse $p_\delta(t)$ of width, Δt, beginning at $t = 0$, and having unit area. Thus, we may write

$$p_\delta(t) = \begin{cases} 0 & t < 0 \\ \dfrac{1}{\Delta t} & 0 < t < \Delta t \\ 0 & t > \Delta t \end{cases} \qquad (5\text{-}49)$$

To use this function to represent any of the pulses $p_i(t)$ defined in Eq. (5-47), we need only shift $p_\delta(t)$ so that it begins at time t_i by replacing the argument t by $t - t_i$. Thus, we obtain

$$p_\delta(t - t_i) = \begin{cases} 0 & t - t_i < 0 \\ \dfrac{1}{\Delta t} & 0 < t - t_i < \Delta t \\ 0 & t - t_i > \Delta t \end{cases} = \begin{cases} 0 & t < t_i \\ \dfrac{1}{\Delta t} & t_i < t < t_i + \Delta t \\ 0 & t > t_i + \Delta t \end{cases} = p_i(t) \qquad (5\text{-}50)$$

The series of pulses $g_\delta(t)$ of Eq. (5-48) may now be written

$$g_\delta(t) = \sum_{i=0}^{n-1} p_\delta(t - t_i)g(t_i)\,\Delta t \qquad (5\text{-}51)$$

Let us now assume that the excitation waveform discussed above is applied as an input to a network which has a response $h_\delta(t)$ when excited by a pulse $p_\delta(t)$. If we assume that the network is linear and time-invariant, and that it is in the zero state, the response due to a shifted pulse $p_\delta(t - t_i)$ must be $h_\delta(t - t_i)$. Since homogeneity applies to linear networks, the response to a pulse $p_\delta(t - t_i)g(t_i)\,\Delta t$ must therefore be $h_\delta(t - t_i)g(t_i)\,\Delta t$, since the term $g(t_i)\,\Delta t$ only changes the amplitude of the pulse. We now have an expression for the response for any one of the pulses in the summation given in Eq. (5-51). Thus, we may apply superposition to find the response to the complete sum of pulses, i.e., to the function $g_\delta(t)$. For convenience, let us define $f_\delta(t)$ as the resulting response. For the discussion above, we see that we may write

$$f_\delta(t_n) = \sum_{i=0}^{n-1} h_\delta(t - t_i)g(t_i)\,\Delta t \qquad (5\text{-}52)$$

Now let us see what happens as we increase n, the number of pulses used to represent $g(t)$ over the indicated time span from t_0 to t_n, where t_0 and t_n are fixed values. As n is increased, the quantity Δt becomes smaller, and in the limit as n approaches infinity, several items of interest occur. First of all, the pulse train $g_\delta(t)$ becomes equal to the actual input $g(t)$. Correspondingly, in the limit, $f_\delta(t)$ may be written as $f(t)$, the actual zero-state network response to the excitation $g(t)$. Now consider the pulse $p_\delta(t)$. In the limit, this simply becomes the impulse $\delta(t)$. Thus, the quantity $h_\delta(t)$ in the limit becomes the impulse response which we will designate as $h(t)$. Finally, we may note that in the limit the summation of (5-52) becomes an integral, the sequence of variables t_i becomes a continuous variable of integration which we may designate as τ, and the incremental quantity Δt becomes $d\tau$. Inserting these results in (5-52), we obtain

$$f(t) = \int_{t_0}^{t} h(t - \tau)g(\tau)\,d\tau \qquad (5\text{-}53)$$

where we have replaced t_n, an arbitrary value of time which is greater than t_0, by t. This relation defines the *convolution integral*. It allows us to find the response of a network to any excitation $g(t)$ just so long as we know the impulse response $h(t)$ of the network. An example illustrating the application of the convolution integral follows.

Example 5-7.1

A first-order RL network is shown in Fig. 5-7.2. The impulse response $i_\delta(t)$ of this network to an excitation $v(t) = \delta(t)$ is given as $i_\delta(t) = e^{-t}u(t)$ and has the form shown in Fig. 5-7.3(a). If a unit step $v(t) = u(t)$ is applied as input, the resulting response $i(t)$ may be determined using Eq. (5-53). Thus, we obtain

$$i(t) = \int_0^t i_\delta(t - \tau)v(\tau)\, d\tau = \int_0^t e^{-(t-\tau)}\, d\tau = e^{-t}\int_0^t e^{\tau}\, d\tau = (1 - e^{-t})u(t)$$

A plot of the resulting output $i(t)$ is shown in Fig. 5-7.3(b).

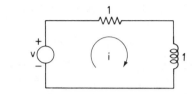

Figure 5-7.2 First-order RL network.

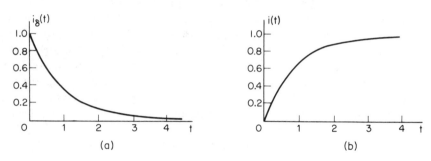

Figure 5-7.3 Waveforms for the network shown in Fig. 5-7.2.

The results given above may be summarized as follows:

SUMMARY 5-7.1

The Convolution Integral. If an arbitrary excitation $g(t)$ which is zero for $t < t_0$ is applied to a linear time-invariant network whose response to a unit impulse applied at $t = 0$ is $h(t)$, and if the network is in the zero state, the response $f(t)$ is given by the convolution integral

$$f(t) = \int_{t_0}^t h(t - \tau)g(\tau)\, d\tau$$

More specifically, this integral is said to define the convolution of two time-domain functions $h(t)$ and $g(t)$. ∎

There are several pertinent observations that can be made regarding convolution. First of all, we note that the quantity t appears as part of the integrand in Eq. (5-53). This means that if numerical or graphical methods of integration are used, then for each new value of t at which we desire to find $f(t)$ we must perform a new integration starting from t_0. This point will be illustrated further in connection with the presentation of techniques for graphical evaluation of the convolution integral. A second interesting fact is that the convolution process indicated in Eq. (5-53) may be applied to circuits which include distributed elements as well as those consisting only of lumped and active elements. This occurs because the convolution integral does not require the use of a specific differential equation representation for a network. This property is one of significant importance since the other time-domain techniques that we present in this text are not, in general, applicable to circuits including distributed components. As a final property of the convolution integral, let us consider the case where the input function begins at $t = 0$. Letting $t_0 = 0$ in (5-53), we obtain

$$f(t) = \int_0^t h(t - \tau)g(\tau)\,d\tau \tag{5-54}$$

Now let us make a change of variable $t - \tau = x$. Thus, $\tau = t - x$, $d\tau = -dx$, and the integration limits of 0 and t become t and 0, respectively. Substituting these results in Eq. (5-54), we obtain

$$f(t) = \int_t^0 -h(x)g(t - x)\,dx = \int_0^t h(x)g(t - x)\,dx \tag{5-55}$$

Changing the variable of integration in the right member of the above equation from x to τ, we obtain

$$f(t) = \int_0^t h(\tau)g(t - \tau)\,d\tau \tag{5-56}$$

Comparing Eqs. (5-54) and (5-56), we see that the role of the impulse response $h(t)$ and the excitation function $g(t)$ is an interchangeable one in the convolution process. Such an interchange is frequently useful in evaluating convolution integrals.

The evaluation of the convolution integral for simple functions and for functions for which no explicit mathematical expression is known is conveniently done by employing graphical techniques. As an example of this, let us assume that a network has a triangular impulse response $h(\tau)$ as shown in Fig. 5-7.4(a),[9] and that the input is a unit step $u(\tau)$. First let us determine $f(0.5)$, i.e., the value of the integral given in Eq. (5-54) for $t = 0.5$. The graphical procedure is illustrated in Fig. 5-7.4. In part (b) of this figure we show $h(-\tau)$, i.e., the triangular function $h(\tau)$ in which the direction of increasing argument τ has been inverted. In part (c) we show the function $h(0.5 - \tau)$ obtained by shifting $h(-\tau)$ one-half unit to the right. In part (d) we show the product $h(0.5 - \tau)u(\tau)$. The value of the integral is equal to the shaded area. This is readily seen to be 0.875. Thus, we may write $f(0.5) = 0.875$. Repeating this process for $t = 1$, we find $f(1) = 1.5$. Similarly, $f(1.5) = 1.875$, $f(2) = 2$, and $f(t)$ for all $t > 2$ is 2. Thus, we obtain the waveform for $f(t)$ shown in Fig. 5-7.5.

The convolution integral is only one of many methods in which the response of a network to some arbitrary excitation can be specified in terms of its known response

[9]Such an impulse response is physically unrealizable but provides a useful mathematical example.

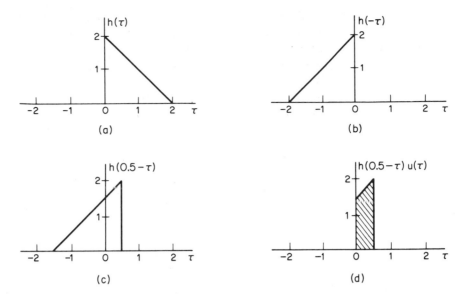

(a)

(b)

(c)

(d)

Figure 5-7.4 Waveforms for graphical convolution example.

to some other excitation. Although the convolution integral, which is based on the impulse response of the network, is by far the most important of these methods, there is one other one, known as the superposition integral, which is also of importance. In this case, the known response of the network is assumed to be the response $w(t)$ to a unit step input $u(t)$ applied at $t = 0$. We may begin our discussion by assuming that some arbitrary input function $g(t)$ is defined over the period $t_0 < t < t_n$, as shown in Fig. 5-7.1(a). We may approximate this function by a series of n step functions $g_i(t)$ displaced in time by Δt and of varying amplitude as shown in Fig. 5-7.6. Such an

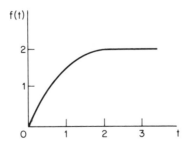

Figure 5-7.5 Output waveform from graphical convolution example.

approximation may be conveniently accomplished by using a shifted unit step function $u(t - t_i)$ defined as

$$u(t - t_i) = \begin{cases} 0 & t < t_i \\ 1 & t > t_i \end{cases} \tag{5-57}$$

The sequence of step functions in the approximation shown in Fig. 5-7.6 may now be written explicitly by first defining the quantities g_i as

$$g_0 = g(t_0)u(t - t_0)$$
$$g_1 = [g(t_1) - g(t_0)]u(t - t_1)$$
$$\begin{matrix} \cdot & \cdot \\ \cdot & \cdot \\ \cdot & \cdot \end{matrix}$$
$$g_i = [g(t_i) - g(t_{i-1})]u(t - t_i)$$

(5-58)

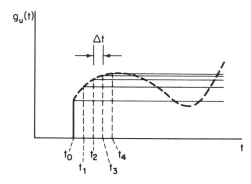

Figure 5-7.6 Arbitrary excitation function and its approximation.

where the last equation holds for all i from 0 to n if we define $g(t_{-1}) = 0$. If we now define $g_u(t)$ as the step function approximation of $g(t)$, we may write

$$g_u(t) = \sum_{i=0}^{n} g_i = \sum_{i=0}^{n} [g(t_i) - g(t_{i-1})]u(t - t_i) \tag{5-59}$$

Now let us assume that the approximate excitation waveform $g_u(t)$ is applied as an input to a network which has the property that a unit step input $u(t)$ applied at $t = 0$ produces a response $w(t)$. If the network is linear, time-invariant, and in the zero state, then the response to any of the steps $g_i(t)$ defined in Eq. (5-58) is $[g(t_i) - g(t_{i-1})]w(t - t_i)$. We may now define $f_u(t)$ as the response to the series of n step functions $g_u(t)$ defined in (5-59). Applying superposition, we obtain

$$f_u(t) = \sum_{i=0}^{n} [g(t_i) - g(t_{i-1})]w(t - t_i) \tag{5-60}$$

As a preliminary step for our next development, let us multiply and divide each member of the summation in the above equation by the quantity Δt. Thus, we obtain

$$f_u(t) = \sum_{i=0}^{n} \left[\frac{g(t_i) - g(t_{i-1})}{\Delta t} \right] w(t - t_i) \Delta t \tag{5-61}$$

Now let us investigate what happens as we increase n, the number of step functions which approximate the input excitation $g(t)$ over the time span t_0 to t_n. As n increases, the quantity Δt becomes smaller and, in the limit as n approaches infinity, the series of step functions $g_u(t)$ becomes equal to the actual input $g(t)$. Similarly, in the limit, $f_u(t)$ becomes equal to $f(t)$, the actual zero-state network response to the excitation $g(t)$. Now consider the summation given in Eq. (5-61). In the limit, the quantity enclosed in brackets in the right member simply becomes equal to the derivative of the excitation function. In addition, the sum becomes an integral, the sequence of variables t_i becomes a continuous variable which we may designate as τ, and the incremental quantity Δt thus becomes $d\tau$. Therefore, we may write

$$f(t) = \int_{t_0}^{t} \frac{dg(\tau)}{d\tau} w(t - \tau)\, d\tau \qquad (5\text{-}62)$$

where we have represented the quantity t_n, an arbitrary value of time, by t. This relation defines the *superposition integral*.[10] It enables us to find the response of the network to any excitation function $g(t)$ as long as we know the unit step response $w(t)$ of the network. An example illustrating the application of the superposition integral follows.

Example 5-7.2

A first-order RL network is shown in Fig. 5-7.2. The response of this network to an input unit step $v(t) = u(t)$ was shown in Example 5-7.1 to be $i_u(t) = (1 - e^{-t})u(t)$. We may now apply the superposition integral given in (5-62) to determine the response of this network to a unit ramp function $v(t) = tu(t)$. For such an input, the resulting response $i(t)$ may be determined as:

$$i(t) = \int_{0}^{t} i_u(t - \tau) \frac{dv(\tau)}{d\tau}\, d\tau = \int_{0}^{t} [1 - e^{-(t-\tau)}]\, d\tau$$

$$= \int_{0}^{t} d\tau - e^{-t} \int_{0}^{t} e^{\tau}\, d\tau = t - 1 + e^{-t} \qquad t > 0$$

The results given above may be summarized as follows:

SUMMARY 5-7.2

The Superposition Integral. If an arbitrary excitation $g(t)$ which is zero for $t < t_0$ is applied to a linear time-invariant network whose response to a unit step function applied at $t = 0$ is $w(t)$, and if the network is in the zero state, the response $f(t)$ is given by the superposition integral

$$f(t) = \int_{t_0}^{t} \frac{dg(\tau)}{d\tau} w(t - \tau)\, d\tau \qquad \blacksquare$$

The superposition integral may be evaluated graphically as well as analytically. The process is similar to that presented for the convolution integral in the first part of this section. Some alternative equivalent representations of the superposition integral are frequently useful in such a case. For example, if we let t_0 be 0 so that the super-

[10]This integral is also referred to as Duhamel's integral.

position integral of Eq. (5-62) has the form

$$f(t) = \int_0^t \frac{dg(\tau)}{d\tau} w(t - \tau)\, d\tau \tag{5-63}$$

then the following form is equivalent.

$$f(t) = \int_0^t \frac{dw(\tau)}{d\tau} g(t - \tau)\, d\tau \tag{5-64}$$

Other equivalent forms of the superposition integral which may be used are

$$f(t) = \frac{d}{dt} \int_0^t w(t - \tau)g(\tau)\, d\tau$$

$$f(t) = \frac{d}{dt} \int_0^t g(t - \tau)w(\tau)\, d\tau \tag{5-65}$$

The integral portions of these last two relations are similar in form to those given for the convolution integrals in Eqs. (5-54) and (5-56). Thus, we see that the superposition integral may be considered as the derivative of the convolution of an input excitation function and the step response of a given network.

We shall again refer to the results obtained in this section when we consider the Laplace transformation in Chap. 8. There we shall show that the integration process defined by the convolution and superposition integrals can be represented by a simple multiplication operation through the use of this transformation.

5-8 CONCLUSION

In this chapter we have made a detailed presentation of various methods which may be used to solve a first-order circuit, i.e., one containing a single energy-storage element (a capacitor or an inductor). The first method we discussed was the one presented in Sec. 5-1 for the case where the circuit was excited only by an initial condition. The procedure is so simple that the student should be able to apply it almost by inspection. The method for solving a first-order network which is excited by sources, or by sources and initial conditions, however, becomes somewhat more complex. The approaches to be followed in these cases are covered in Secs. 5-2 and 5-3. For the special case where only constant-valued sources are used, simplified methods may be applied. These are described in Sec. 5-4. All the techniques referred to above require that the first-order circuit be linear and time-invariant. A final topic which was presented in this chapter was the use of convolution and superposition integrals (in Sec. 5-7) to obtain the response of a given first-order circuit to some specified excitation. Such a technique is considerably different than the ones referred to earlier in this section since it does *not* require that an equation describing the system be known, only that the network's response to an impulse or a step excitation be given. In addition, such a response may be displayed as a plot, i.e., a mathematical expression for it is not required. Thus, and most important, it may be obtained by direct experimental measurement. The convolution and superposition techniques discussed in Sec. 5-7 are applicable to linear systems of any order.

PROBLEMS

Problem 5-1 (Sec. 5-1)

For the circuit shown in Fig. P5-1, for $t \geq 0$, find $i(t)$, $q(t)$ (the charge in the capacitor), $v(t)$, $w_C(t)$ (the energy stored in the capacitor), and $p_R(t)$ (the power dissipated in the resistor). Assume that $v(0) = 2$ V.

Figure P5-1

Problem 5-2 (Sec. 5-1)

If the switch in the circuit shown in Fig. P5-2 is opened at $t = 0$, find V_0 such that $v(0.5) = 0.3$ V.

Figure P5-2

Problem 5-3 (Sec. 5-1)

The switch in the circuit shown in Fig. P5-3 is closed at $t = 0$. Find V_0 such that $i(0.003) = 0.001$ A.

Figure P5-3

Problem 5-4 (Sec. 5-1)

For the circuit shown in Fig. P5-4, find $i(t)$ for $t > 0$.

Figure P5-4

Problem 5-5 (Sec. 5-1)

For the circuit shown in Fig. P5-5, find $v(t)$ for $t > 0$.

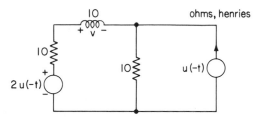

Figure P5-5

Problem 5-6 (Sec. 5-1)

The circuit shown in Fig. P5-6 has the initial conditions $v_C(0) = 1$ and $i_L(0) = 2$. Find the initial conditions $i_R(0)$ and $i_R'(0)$.

Figure P5-6

Problem 5-7 (Sec. 5-1)

Show that the tangent to $f(t) = e^{-t/\tau}u(t)$ at any point t_0 intersects the t axis exactly τ seconds later as shown in Fig. P5-7.

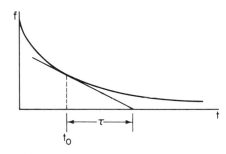

Figure P5-7

Problem 5-8 (Sec. 5-2)

Find expressions for $t \geq 0$ for the following quantities for the circuit shown in Fig. P5-8.
(a) $i(t)$
(b) $\phi(t)$
(c) $v(t)$
(d) $w_L(t)$ (the energy stored in the inductor)
(e) $v_R(t)$
(f) $p_L(t)$ (the power delivered to the inductor)
(g) $p_R(t)$ (the power dissipated in the resistor)
(h) $p(t)$ (the power supplied by the source)

Figure P5-8

Problem 5-9 (Sec. 5-2)

For the circuit shown in Fig. P5-9, find $v(t)$ for $t \geq 0$, if $v(0) = 0$ and $i(t) = 1 + t + t^2$ A. Then find $v(t)$ for the three cases:

(a) $i(t) = 1$
(b) $i(t) = t$
(c) $i(t) = t^2$

Is the sum of the last three solutions equal to the first solution? Is the sum of the *particular* solutions for the last three cases equal to the particular solution of the first case?

Figure P5-9

Problem 5-10 (Sec. 5-2)

In the circuit shown in Fig. P5-10, if $v(t) = 10e^{-4t} + 5$ V, find $i(t)$ for $t \geq 0$ for the initial condition $i(0) = 0$. Then find $i(t)$ for the two cases $v(t) = 10e^{-4t}$ and $v(t) = 5$. Is the sum of the last two solutions equal to the solution for the first case? Is the sum of the *particular* solutions for the last two cases equal to the particular solution for the first case?

ohms, henries

Figure P5-10

Problem 5-11 (Sec. 5-2)

For the circuit shown in Fig. P5-11, find $v(t)$ for $t \geq 0$ if:

(a) $i(t) = \cos(t)u(t)$
(b) $i(t) = \sin(t)u(t)$ A

Assume that $v(0) = 0$.

ohms, farads

Figure P5-11

Problem 5-12 (Sec. 5-2)

In the circuit shown in Fig. P5-12, find $v(t)$ for $t \geq 0$ if $i(0) = 0$.

ohms, henries

Figure P5-12

Problem 5-13 (Sec. 5-2)

In the circuit shown in Fig. P5-13, find $v(t)$ for $t \geq 0$ if the initial current in the inductor is zero.

ohms, henries

Figure P5-13

Problem 5-14 (Sec. 5-2)

The switch in the circuit shown in Fig. P5-14 changes from position a to position b at $t = 0$. If $i(t) - \frac{2}{3}t + 1$ A, find $v(t)$. Assume that $v(0) = 0$.

ohms, farads

Figure P5-14

Problem 5-15 (Sec. 5-2)

In the circuit shown in Fig. P5-15, the switch closes at $t = 0$. If $v(t) = e^{-2t}$ V, find $i(t)$ for $t \geq 0$. Assume that the capacitor is initially uncharged.

ohms, farads

Figure P5-15

Problem 5-16 (Sec. 5-2)

For the network shown in Fig. P5-16, assume that the initial condition is $v(0) = 0$.
(a) Find the voltage $v(t)$ for $t \geq 0$.
(b) Construct a Norton equivalent for the circuit by replacing all the elements to the left of the capacitor by a single resistor and a single (independent) current source.

ohms, farads

Figure P5-16

Problem 5-17 (Sec. 5-2)

The circuit shown in Fig. P5-17 is unexcited for $t < 0$. Find $v(t)$ for $t \geq 0$.

ohms, farads

Figure P5-17

Problem 5-18 (Sec. 5-2)

In the circuit shown in Fig. P5-18, the initial condition is $i(0) = 0$. Find an expression for $i(t)$ for $t \geq 0$.

ohms, farads

Figure P5-18

Problem 5-19 (Sec. 5-3)

Repeat Problem 5-9 for the case where the initial condition is $v(0) = 1$ V.

Problem 5-20 (Sec. 5-3)

Repeat Problem 5-10 for the case where the initial condition is $i(0) = 1$ A.

Problem 5-21 (Sec. 5-3)

For the circuit shown in Fig. P5-21, find $i(0)$ such that the *transient* part of the complete solution is equal to zero.

ohms, henries

Figure P5-21

Problem 5-22 (Sec. 5-3)

In the circuit shown in Fig. P5-22, find R_1, R_2, and C such that:

(a) $v(t) = u(t)$, $i(t) = 0$, and $v_C(t) = 2e^{-2t} + \frac{1}{2}$ when $t \geq 0$.

(b) $i(t) = u(t)$, $v(t) = 0$, and $v_C(t) = \frac{1}{2}e^{-2t} + 2$ when $t \geq 0$.

Figure P5-22

Problem 5-23 (Sec. 5-3)

In the circuit shown in Fig. P5-23, it is found that $i_L(t) = 0.001 + 0.005e^{-at}$ A for $t \geq 0$ when $v(t) = u(t)$ V. What is $i_L(t)$ if $v(t) = 2u(t)$ V?

ohms, henries

Figure P5-23

Problem 5-24 (Sec. 5-3)

For the circuit shown in Fig. P5-24, at a given time t_0, $v_2(t_0) = 2$ V and $dv_2(t_0)/dt = -10$ V/s. Find the value of C.

Figure P5-24

Problem 5-25 *(Sec. 5-3)*

For the circuit and the waveshape of $v_0(t)$ shown in Fig. P5-25, find $v_s(t)$.

(a)

(b)

Figure P5-25

Problem 5-26 *(Sec. 5-3)*

In the circuit shown in Fig. P5-26, the switch opens at $t = 0$. Find $v(t)$ for all $t \geq 0$.

Figure P5-26

Problem 5-27 *(Sec. 5-3)*

In the circuit shown in Fig. P5-27, the switch has been in position a for a "long time." It is suddenly switched to position b. Find the complete solution for $i(t)$.

Figure P5-27

Problem 5-28 *(Sec. 5-3)*

In the circuit shown in Fig. P5-28, at $t = 0$ the switch is suddenly changed from position a to position b. Find the complete solution for $v(t)$.

Figure P5-28

Problem 5-29 (Sec. 5-3)
In the circuit shown in Fig. P5-29, the switch closes at $t = 0$. Find $i(t)$ and $v(t)$ for $t \geq 0$.

Figure P5-29

Problem 5-30 (Sec. 5-3)
In the circuit shown in Fig. P5-30, the switch opens at $t = 0$. Find $v_1(0+)$, $v_1'(0+)$, $v_1(\infty)$, and $v_1'(\infty)$.

Figure P5-30

Problem 5-31 (Sec. 5-3)
The circuit shown in Fig. P5-31 is valid for $t \geq 0$ only. It has an initial condition $i_L(0) = 3$.
(a) Draw a new circuit adding switches and constant-valued sources as necessary to establish the initial condition. The new circuit should be valid for all values of time.
(b) Repeat part (a), but instead of using switches, use sources whose output is a step function such as $u(t)$ or $u(-t)$.

Figure P5-31

Problem 5-32 (Sec. 5-4)
The switch in the circuit shown in Fig. P5-32 is closed at $t = 0$. If $v(0) = 0$, find $i(0+)$, $(di/dt)(0+)$, $(d^2i/dt^2)(0+)$, $i(\infty)$, and $(di/dt)(\infty)$.

Figure P5-32

Problem 5-33 (Sec. 5-4)
In the circuit shown in Fig. P5-33, find $v(0+)$, $(dv/dt)(0+)$, $(d^2v/dt^2)(0+)$, $v(\infty)$, and $(dv/dt)(\infty)$. Assume that the initial current in the inductor is zero.

First-Order Circuits   Chap. 5

Figure P5-33

Problem 5-34 (Sec. 5-4)

In the circuit shown in Fig. P5-34, the voltage applied by the voltage source is changed from 1 V to 2 V at t equals 0. Assume that the 1-V value has been applied for a long time before t equals 0.

(a) Find an expression for $v_C(t)$, the voltage across the capacitor for $t \geq 0$.

(b) Find an expression for $i(t)$ for $t \geq 0$.

(a)

(b)

Figure P5-34

Problem 5-35 (Sec. 5-4)

For the circuit shown in Fig. P5-35, find $v_{L1}(0+)$, $v_{L2}(0+)$, $i_C(0+)$, $i_{L1}(\infty)$, $i_{L2}(\infty)$, and $v_C(\infty)$. Assume that all circuit variables are zero for $t < 0$.

Figure P5-35

Problem 5-36 (Sec. 5-4)

In the circuit shown in Fig. P5-36, for $t < 0$ the circuit is in a steady state. The switch is closed at $t = 0$. Find $i_L(t)$ for $t \geq 0$.

Figure P5-36

Problem 5-37 (Sec. 5-4)

For the circuit shown in Fig. P5-37, find $v(t)$ for $t \geq 0$. Assume that the initial condition is zero.

ohms, farads

Figure P5-37

Problem 5-38 (Sec. 5-4)

For the transistor circuit shown in Fig. 5-4.3, find the output $v_o(t)$ if a signal $[5 + \sin (2\pi \times 100t)]u(t)$ is applied at the voltage source at the left instead of the indicated $10u(t)$ signal. Assume that the initial condition is the same as found in Example 5-4.4.

Problem 5-39 (Sec. 5-4)

For the transistor circuit shown in Fig. 5-4.3, find the output $v_o(t)$ if a pulse $10[u(t) - u(t - t_0)]$ is applied at the voltage source at the left instead of the indicated $10u(t)$ signal. Assume that the initial condition is the same as found in Example 5-4.4, and let $t_0 = 100 \ \mu s$.

Problem 5-40 (Sec. 5-5)

For the circuit shown in Fig. P5-40, find $v(t)$ and $i(t)$ for $t \geq 0$. Assume that $v(0) = 0$.

Figure P5-40

Problem 5-41 (Sec. 5-5)

In the circuit shown in Fig. P5-41, find $i(t)$ and $v(t)$ for $t \geq 0$. Assume that $i(0) = 0$.

Figure P5-41

Problem 5-42 (Sec. 5-5)

In the circuit shown in Fig. P5-42, switch $S1$ is closed at $t = 0$ and switch $S2$ is closed at $t = 3$ s. Find $v(t)$ and $i(t)$ for $t \geq 0$. Assume that $v(0) = 0$.

Figure P5-42

Problem 5-43 (Sec. 5-6)

For $t < 0$, the circuit shown in Fig. P5-43 is in a steady-state condition. At $t = 0$, the switch is moved from position a to position b. Draw a circuit without a switch and with a capacitor

206

which is uncharged for $t < 0$, which will have the same solution for $v_C(t)$ for $t > 0$ as the original circuit.

Figure P5-43

Problem 5-44 (Sec. 5-6)

For $t < 0$, the circuit shown in Fig. P5-44 is in a steady-state condition. At $t = 0$, the switch is closed. Draw a circuit without a switch, and with an inductor which is uncharged for $t < 0$, which will have the same solution for $i_L(t)$ for $t > 0$.

Figure P5-44

Problem 5-45 (Sec. 5-7)

Apply the convolution integral to find the response $v(t)$ for the circuit shown in Fig. P5-45 for the case where the excitation $i(t)$ has the following values:

(a) $e^{-t}u(t)$
(b) $\sin t[u(t) - u(t - \pi)]$
(c) $2(1 - t)[u(t) - u(t - 3)]$

Figure P5-45

Problem 5-46 (Sec. 5-7)

Use graphical integration to solve Problem 5-45(c).

Problem 5-47 (Sec. 5-7)

Repeat Problem 5-45 using the superposition integral.

Problem 5-48 (Sec. 5-7)

Use graphical integration to solve Problem 5-47(c).

6

Second-Order

and Higher-Order Circuits

In Chap. 5 we studied the properties of first-order circuits, i.e., circuits containing a single energy-storage element, either an inductor or a capacitor. In this chapter we shall study the properties of second-order and higher-order circuits, i.e., circuits containing two or more energy-storage elements. First we will consider second-order circuits. Such circuits will, in general, be characterized by second-order differential equations. There are three possible types of such circuits, namely, circuits with two inductors, circuits with two capacitors, and circuits with one inductor and one capacitor. Just as was the case for the first-order circuit, the second-order circuit may also contain an arbitrary number of resistors, independent sources, and controlled sources. In our study of second-order circuits we shall find that many of the same techniques which were used to analyze first-order circuits may be applied with minor modifications to second-order circuits. The properties of the second-order circuit, however, are considerably more varied than those of the first-order circuit, as we shall discover.

6-1 EXCITATION BY INITIAL CONDITIONS—CASES I AND II

In this section we shall consider the effect of exciting a second-order circuit by initial conditions. We shall begin our discussion by considering a second-order circuit characterized by an inductor and a capacitor. Second-order circuits containing two inductors or two capacitors will be covered in a subsequent section. As an example of a second-order *RLC* circuit, consider the one-mesh circuit shown in Fig. 6-1.1.[1] Applying KVL to this circuit, we may write

$$L\frac{di}{dt} + Ri(t) + \frac{1}{C}\int_{-\infty}^{t} i(\tau)\,d\tau = 0 \qquad (6\text{-}1)$$

Figure 6-1.1 Second-order *RLC* circuit.

If we differentiate this equation and divide all terms by L, so as to set the coefficient of the highest-order derivative term to unity, we obtain

$$\frac{d^2 i}{dt^2} + \frac{R}{L}\frac{di}{dt} + \frac{1}{LC}i(t) = 0 \qquad (6\text{-}2)$$

Following the terminology conventions introduced in Sec. 5-1, we see that this equation is an ordinary, second-order, linear, homogeneous differential equation with constant coefficients, or, more briefly, *a second-order, homogeneous, differential equation*. In Sec. 5-1 we found that a simple exponential provided a solution for the first-order homogeneous differential equation. Let us try that approach here by assuming a solution for $i(t)$ of the form

$$i(t) = Ke^{st} \qquad (6\text{-}3)$$

where, for the moment, K and s are undefined. Substituting this assumed solution in Eq. (6-2), we obtain

$$s^2 Ke^{st} + s\frac{R}{L}Ke^{st} + \frac{1}{LC}Ke^{st} = 0 \qquad (6\text{-}4)$$

This may be written in the form

$$\left(s^2 + s\frac{R}{L} + \frac{1}{LC}\right)Ke^{st} = 0 \qquad (6\text{-}5)$$

In this last equation we see that the product of the polynomial $s^2 + (sR/L) + (1/LC)$ and the term Ke^{st} must be 0 for the equation to be satisfied. If the term Ke^{st} is 0, however, the solution is trivial, since from Eq. (6-3) this requires that $i(t) = 0$. Therefore, for a non-trivial solution we must have

$$s^2 + s\frac{R}{L} + \frac{1}{LC} = 0 \qquad (6\text{-}6)$$

This equation is called the *characteristic equation* for the differential equation given in (6-2). Correspondingly, the polynomial which is the left member of Eq. (6-6) is called

[1] All the conclusions developed in this and the following section for a one-mesh *RLC* circuit apply equally well to a one-node-pair *RLC* circuit in which the elements are connected in parallel. Examples of such a circuit may be found in the problems.

the *characteristic polynomial.* Since Eq. (6-6) is a second-degree equation in s, there will be two values of s which will satisfy it. Let these be s_1 and s_2. We find that

$$s_1 = -\frac{R}{2L} + \sqrt{\left(\frac{R}{2L}\right)^2 - \frac{1}{LC}} \qquad s_2 = -\frac{R}{2L} - \sqrt{\left(\frac{R}{2L}\right)^2 - \frac{1}{LC}} \qquad (6\text{-}7)$$

There are three possible cases for the resulting values of s_1 and s_2, based on the relative values of the quantities R, L, and C. More specifically, these cases are specified by the polarity of the quantity inside the radical in Eq. (6-7), namely, $(R/2L)^2 - 1/LC$. This quantity is known as the *discriminant* of the second-degree equation. The first case (Case I) occurs when $(R/2L)^2 > 1/LC$, i.e., when the discriminant is greater than zero. For this case, the expressions given in Eq. (6-7) directly define two real roots of the characteristic equation. Since both s_1 and s_2 satisfy the original differential equation given in Eq. (6-2), the solution for $i(t)$ must be of the form

$$i(t) = K_1 e^{s_1 t} + K_2 e^{s_2 t} \qquad (6\text{-}8)$$

where the constants K_1 and K_2 are determined from the initial conditions. Thus, we see that the roots of the characteristic equation are the natural frequencies of the circuit.[2] Note that in Eq. (6-8) there are *two* constants, thus *two* initial conditions must be specified. This is always the case for a second-order differential equation. Suppose that the initial conditions are given as $i(0)$ and $i'(0)$. Evaluating Eq. (6-8) for $t = 0$, we obtain

$$i(0) = K_1 + K_2 \qquad (6\text{-}9)$$

This is the first of two equations needed to determine the constants K_1 and K_2. To find a second equation, we may first differentiate Eq. (6-8). We obtain

$$i'(t) = s_1 K_1 e^{s_1 t} + s_2 K_2 e^{s_2 t} \qquad (6\text{-}10)$$

Evaluating this equation for $t = 0$, we obtain

$$i'(0) = s_1 K_1 + s_2 K_2 \qquad (6\text{-}11)$$

Equations (6-9) and (6-11) may be written as a set of simultaneous equations, with K_1 and K_2 as unknowns. Using the methods of Chap. 3, we find that the solution for K_1 and K_2 is

$$\begin{bmatrix} K_1 \\ K_2 \end{bmatrix} = \frac{1}{s_2 - s_1} \begin{bmatrix} s_2 & -1 \\ -s_1 & 1 \end{bmatrix} \begin{bmatrix} i(0) \\ i'(0) \end{bmatrix} \qquad (6\text{-}12)$$

Thus, we have obtained the values of the unknown constants K_1 and K_2 as functions of the known quantities s_1, s_2, $i(0)$, and $i'(0)$. The process is readily illustrated by a numerical example.

Example 6-1.1

In the series *RLC* circuit shown in Fig. 6-1.1, assume that the elements have the following values: $R = 6\,\Omega$, $L = 1$ H, $C = \frac{1}{5}$ F. From (6-7), the resulting roots of the characteristic equation are: $s_1 = -1$, $s_2 = -5$. Thus, from (6-8) the solution for $i(t)$ has the form

$$i(t) = K_1 e^{-t} + K_2 e^{-5t}$$

[2]See Sec. 5-1 for a definition of "natural frequency."

The values of the constants K_1 and K_2 may be expressed in terms of the initial conditions $i(0)$ and $i'(0)$. From (6-12) we see that

$$K_1 = \frac{s_2 i(0) - i'(0)}{s_2 - s_1} = \frac{5}{4} i(0) + \frac{1}{4} i'(0)$$

$$K_2 = \frac{-s_1 i(0) + i'(0)}{s_2 - s_1} = -\frac{1}{4} i(0) - \frac{1}{4} i'(0)$$

As a more specific example, suppose that the initial conditions are specified as $i(0) = 1$, $i'(0) = 0$. Then, from the above relations, $K_1 = \frac{5}{4}$, $K_2 = -\frac{1}{4}$, and the solution for $i(t)$ is

$$i(t) = \tfrac{5}{4} e^{-t} - \tfrac{1}{4} e^{-5t} \qquad t \geq 0$$

A plot of this function is shown in Fig. 6-1.2.

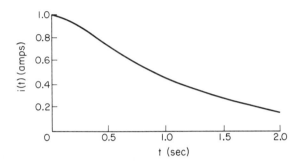

Figure 6-1.2 Waveform of $i(t)$ for Example 6-1.1.

SUMMARY 6-1.1

Solution of a Second-Order Circuit Excited Only by Initial Conditions—Case I.
A passive second-order circuit excited only by initial conditions is described by a second-order, homogeneous, differential equation having the form

$$\frac{d^2 f}{dt^2} + a_1 \frac{df}{dt} + a_0 f(t) = 0 \tag{6-13}$$

where $f(t)$ is any network variable, and in which the coefficients a_1 and a_0 are positive. If $(a_1/2)^2 > a_0$, this equation has a solution of the form

$$f(t) = K_1 e^{s_1 t} + K_2 e^{s_2 t} \tag{6-14}$$

in which s_1 and s_2 are the real roots of the characteristic equation $s^2 + a_1 s + a_0 = 0$, and they are not equal. The two constants K_1 and K_2 are found from the two initial conditions $f(0)$ and $f'(0)$ by the relation

$$\begin{bmatrix} K_1 \\ K_2 \end{bmatrix} = \frac{-1/2}{\sqrt{(a_1/2)^2 - a_0}} \begin{bmatrix} s_2 & -1 \\ -s_1 & 1 \end{bmatrix} \begin{bmatrix} f(0) \\ f'(0) \end{bmatrix} \tag{6-15}$$

■

There are several points of interest concerning the above development. First let us note that the roots of the characteristic equation for (6-13) are of the form

$$s_1 = \alpha + \beta \qquad s_2 = \alpha - \beta \qquad (6\text{-}16)$$

where

$$\alpha = -\frac{a_1}{2} \qquad \beta = \sqrt{\left(\frac{a_1}{2}\right)^2 - a_0} \qquad (6\text{-}17)$$

and where $(a_1/2)^2 - a_0$ (the discriminant) is assumed to be greater than zero. Now we may observe that:

1. From Eq. (6-17), if the coefficient a_1 is greater than zero, then α is negative.
2. From Eq. (6-17), if the coefficients a_1 and a_0 are both greater than zero, then the magnitude of β is smaller than the magnitude of α since $\beta = \sqrt{\alpha^2 - a_0}$.
3. From Eq. (6-16), if α is negative (see item 1), and if β is smaller in magnitude than α (see item 2), then both s_1 and s_2 are negative. In addition (for this case) they are unequal.
4. From Eq. (6-14), if s_1 and s_2 are negative (see item 3), then $f(t)$ disappears as t approaches infinity, i.e., $f(\infty) = 0$.

Now let us use the preceding observations to develop a conclusion relating the parity of the coefficients a_1 and a_0 to the passivity of a given second-order circuit. If such a circuit is comprised only of passive, i.e., positive-valued, resistors and energy-storage elements, then, assuming that some dissipation is present in the circuit, the energy provided by the initial conditions must eventually disappear, i.e., the variables must go to zero as t approaches infinity. From the above observations, this will only be true if a_1 and a_0 are positive. We thus conclude that for a second-order circuit, requiring the circuit to be passive guarantees that the coefficients a_1 and a_0 will be positive.

Now let us consider a second case (Case II) for the second-order, homogeneous, differential equation. We shall define this case by the condition that the discriminant be identically zero. For the circuit shown in Fig. 6-1.1, which has solutions to its characteristic equation given by (6-7), this requires that $(R/2L)^2 = 1/LC$. For such a restriction, we see that the roots of the characteristic equation are

$$s_1 = s_2 = -\frac{R}{2L} \qquad (6\text{-}18)$$

Thus, this case is characterized by the fact that the characteristic equation has two equal roots. In this case, we say that the circuit has a second-order natural frequency at s_1. A solution for $i(t)$ which satisfies the original differential equation is obviously

$$i(t) = K_a e^{s_1 t} \qquad (6\text{-}19)$$

where $s_1 = -R/2L$. The general solution to a second-order, homogeneous, differential equation, however, requires two arbitrary constants, not just one; therefore, we must find a second solution for $i(t)$. Let us assume that this is somewhat like the solution given in Eq. (6-19), but that it differs by having a multiplicative factor of the form $y(t)$. Thus, we assume a second solution to the differential equation which has the form

$$i(t) = y(t) e^{s_1 t} \qquad (6\text{-}20)$$

Now let us see if we can find what form $y(t)$ will take. To do this, we may proceed by substituting Eq. (6-20) into the original differential equation given in Eq. (6-2). We obtain (the details are left as an exercise for the reader)

$$e^{-Rt/2L}y''(t) = 0 \qquad (6\text{-}21)$$

where $y''(t)$ is the second derivative of $y(t)$ with respect to t. Since $e^{-Rt/2L}$ is obviously not equal to 0, $y''(t)$ must equal 0 for all t. Integrating the equation $y''(t) = 0$ two times, we obtain

$$y(t) = K_b + K_2 t \qquad (6\text{-}22)$$

where K_h and K_1 are two arbitrary constants. Substituting this result in Eq. (6-20), we obtain for the second solution for $i(t)$,

$$i(t) = K_b e^{s_1 t} + K_2 t e^{s_1 t} \qquad (6\text{-}23)$$

If we now add the two solutions for $i(t)$ given in (6-19) and (6-23), we obtain

$$i(t) = K_1 e^{s_1 t} + K_2 t e^{s_1 t} \qquad (6\text{-}24)$$

where we have summed the constants K_a of (6-19) and K_b of (6-23) to define a new constant K_1. This equation gives the general solution for the homogeneous differential equation given in (6-2) for the case in which the two roots of the characteristic equation are equal. If we evaluate Eq. (6-24) for $t = 0$, we obtain

$$i(0) = K_1 \qquad (6\text{-}25)$$

This is the first of two equations needed to determine the constants K_1 and K_2. To find a second equation, let us first differentiate (6-24). We obtain

$$i'(t) = s_1 K_1 e^{s_1 t} + K_2 e^{s_1 t}(1 + s_1 t) \qquad (6\text{-}26)$$

Evaluating this result for $t = 0$, we see that

$$i'(0) = s_1 K_1 + K_2 \qquad (6\text{-}27)$$

From Eqs. (6-25) and (6-27), we obtain the following expressions for K_1 and K_2.

$$K_1 = i(0) \qquad K_2 = i'(0) - s_1 i(0) \qquad (6\text{-}28)$$

Thus, we have obtained expressions giving the values of the constants K_1 and K_2 in terms of the initial conditions $i(0)$ and $i'(0)$ and the natural frequency of the circuit s_1. The process is readily illustrated by a numerical example.

Example 6-1.2

In the series RLC circuit shown in Fig. 6-1.1, assume that the elements have the following values: $R = 2\sqrt{5}\ \Omega$, $L = 1$ H, $C = \frac{1}{5}$ F. (These are the same values as were used in Example 6-1.1 for Case I, except for the resistor, which has been lowered in value.) The characteristic equation has a second-order root at $s_1 = -\sqrt{5}$, thus the form of the solution for the current is

$$i(t) = K_1 e^{-\sqrt{5}t} + K_2 t e^{-\sqrt{5}t}$$

If we assume that the initial conditions are $i(0) = 1$, $i'(0) = 0$, then from (6-28) we see that $K_1 = 1$, $K_2 = \sqrt{5}$. Thus, the solution for $i(t)$ is

$$i(t) = e^{-\sqrt{5}t} + \sqrt{5}\, t e^{-\sqrt{5}t}$$

Solution of a Second-Order Circuit Excited Only by Initial Conditions—Case II. A passive second-order circuit excited only by initial conditions is described by a second-order, homogeneous, differential equation having the form

$$\frac{d^2f}{dt^2} + a_1 \frac{df}{dt} + a_0 f(t) = 0 \qquad (6\text{-}29)$$

where $f(t)$ is any network variable, and in which the coefficients a_1 and a_0 are positive. If $(a_1/2)^2 = a_0$, this equation has a solution of the form

$$f(t) = K_1 e^{s_1 t} + K_2 t e^{s_1 t} \qquad (6\text{-}30)$$

where s_1 is the root of the characteristic equation $s^2 + a_1 s + a_0 = 0$. The two constants K_1 and K_2 are found from the two initial conditions $f(0)$ and $f'(0)$ by the relations

$$K_1 = f(0) \qquad K_2 = f'(0) - s_1 f(0) \qquad (6\text{-}31)$$

The conclusions made with respect to the passivity of a second-order circuit and the related positive polarity of the coefficients a_1 and a_0 for Case I also hold true for this case. ∎

In this section we have discussed two of the three possible cases for the second-order, homogeneous, differential equation, namely, the cases where the discriminant is positive (Case I) and zero (Case II). The third case, in which the discriminant is negative, will be covered in the following section.

6-2 EXCITATION BY INITIAL CONDITIONS—CASE III

In this section we continue our study of second-order circuits which are excited only by initial conditions. The case we shall consider here (Case III) is the one which occurs when the discriminant is negative.

To begin, let us again consider the circuit shown in Fig. 6-1.1, with the homogeneous differential equation

$$\frac{d^2i}{dt^2} + \frac{R}{L} \frac{di}{dt} + \frac{1}{LC} i(t) = 0 \qquad (6\text{-}32)$$

The characteristic equation for this differential equation is

$$s^2 + \frac{R}{L} s + \frac{1}{LC} = 0 \qquad (6\text{-}33)$$

If the circuit has values of R, L, and C such that the discriminant $(R/2L)^2 - 1/LC$ is negative (Case III), the roots of this equation may be written in the form

$$s_1 = -\frac{R}{2L} + j \sqrt{\frac{1}{LC} - \left(\frac{R}{2L}\right)^2} \qquad s_2 = -\frac{R}{2L} - j \sqrt{\frac{1}{LC} - \left(\frac{R}{2L}\right)^2} \qquad (6\text{-}34)$$

where $j = \sqrt{-1}$ is the usual complex operator.[3] To simplify the resulting development,

---

[3]A review of complex algebra is given in Appendix B.

Second-Order and Higher-Order Circuits   Chap. 6

let us define

$$\sigma = -\frac{R}{2L} \qquad \omega_d = \sqrt{\frac{1}{LC} - \left(\frac{R}{2L}\right)^2} \tag{6-35}$$

Thus, σ is the real part of the roots s_1 and s_2, and ω_d is the magnitude of the imaginary part of these roots. We may now write s_1 and s_2 in the form

$$s_1 = \sigma + j\omega_d \qquad s_2 = \sigma - j\omega_d \tag{6-36}$$

The solution for $i(t)$ must now have the form

$$i(t) = K_1 e^{s_1 t} + K_2 e^{s_2 t} \tag{6-37}$$

The equantity $i(t)$ in the left member of the above equation represents a physical variable. Therefore, it must be real. The two terms in the right member, however, involve complex quantities. In order for the sum of these terms to be real, they must be complex conjugates, i.e., their real parts must be equal and their imaginary parts must be opposite in sign but equal in magnitude. From Eq. (6-36) we see that s_1 and s_2 are already complex conjugates; therefore, $e^{s_1 t}$ and $e^{s_2 t}$ will also be complex conjugates. It is necessary, therefore, that the constants K_1 and K_2 be complex conjugates to ensure that $i(t)$ will be real. To emphasize this, we may define

$$K_1 = \frac{K_r}{2} - j\frac{K_i}{2} \qquad K_2 = \frac{K_r}{2} + j\frac{K_i}{2} \tag{6-38}$$

where we have introduced the factor of $\frac{1}{2}$ for convenience in future calculations. If we substitute the relations of Eqs. (6-36) and (6-38) in (6-37), we obtain

$$i(t) = \tfrac{1}{2}(K_r - jK_i)e^{\sigma t + j\omega_d t} + \tfrac{1}{2}(K_r + jK_i)e^{\sigma t - j\omega_d t} \tag{6-39}$$

Using Euler's identity,[4] we may write

$$e^{\sigma t + j\omega_d t} = e^{\sigma t}(\cos \omega_d t + j \sin \omega_d t)$$
$$e^{\sigma t - j\omega_d t} = e^{\sigma t}(\cos \omega_d t - j \sin \omega_d t) \tag{6-40}$$

Substituting these relations in (6-39) and rearranging terms, we obtain

$$i(t) = e^{\sigma t}(K_r \cos \omega_d t + K_i \sin \omega_d t) \tag{6-41}$$

which is as predicted, purely real. We will show in Sec. 7-1 that such an expression may be put in the form

$$i(t) = I_0 e^{\sigma t} \cos (\omega_d t + \varphi) \tag{6-42}$$

where ϕ is an angle (measured in radians). Thus, it is convenient to refer to the quantity ω_d as the *frequency of oscillation* of the circuit. For our purposes here, the form shown in Eq. (6-41) is more convenient. Equation (6-41) contains two unknown constants, namely, K_r and K_i. The first of the two equations needed to determine these constants may be found by setting $t = 0$ in (6-41). Thus, we see that

$$i(0) = K_r \tag{6-43}$$

If we differentiate Eq. (6-41), we obtain

$$i'(t) = e^{\sigma t}[K_r(\sigma \cos \omega_d t - \omega_d \sin \omega_d t) + K_i(\omega_d \cos \omega_d t + \sigma \sin \omega_d t)] \tag{6-44}$$

[4]See Appendix B.

Evaluating this expression for $t = 0$, we obtain

$$i'(0) = \sigma K_r + \omega_d K_i \qquad (6\text{-}45)$$

Thus, from Eqs. (6-43) and (6-45), the solutions for K_r and K_i are readily seen to be

$$K_r = i(0) \qquad K_i = \frac{i'(0) - \sigma i(0)}{\omega_d} \qquad (6\text{-}46)$$

Example 6-2.1

The circuit shown in Fig. 6-1.1 is assumed to have the following values for the components: $R = 2\,\Omega$, $L = 1$ H, $C = \frac{1}{5}$ F. (These are the same values used in Examples 6-1.1 and 6-1.2, except that the value of the resistor has been reduced.) The characteristic equation for these element values is

$$s^2 + 2s + 5 = 0$$

The roots of the characteristic equation, i.e., the natural frequencies of the network, are $s_1 = -1 + j2$ and $s_2 = -1 - j2$. Thus, $\sigma = -1$ and $\omega_d = 2$. The general form of the solution for $i(t)$ is thus

$$i(t) = e^{-t}(K_r \cos 2t + K_i \sin 2t)$$

If we assume that the initial conditions are $i(0) = 1$, $i'(0) = 0$, then, from (6-46) we see that $K_r = 1$, $K_i = \frac{1}{2}$. Thus, the solution for $i(t)$ is

$$i(t) = e^{-t}(\cos 2t + \tfrac{1}{2} \sin 2t) \qquad t \geq 0$$

A plot of the waveform of $i(t)$ is shown in Fig. 6-2.1.

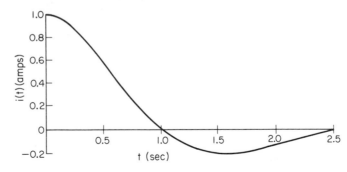

Figure 6-2.1 Waveform of $i(t)$ for Example 6-2.1.

SUMMARY 6-2.1

Solution of a Second-Order Circuit Excited Only by Initial Conditions—Case III. A passive second-order circuit excited only by initial conditions is described by a second-order, homogeneous, differential equation having the form

$$\frac{d^2 f}{dt^2} + a_1 \frac{df}{dt} + a_0 f(t) = 0 \qquad (6\text{-}47)$$

where $f(t)$ is any network variable, and in which the coefficients a_1 and a_0 are positive. If $(a_1/2)^2 < a_0$, this equation has a solution of the form

$$f(t) = e^{\sigma t}(K_1 \cos \omega_d t + K_2 \sin \omega_d t) \qquad (6\text{-}48)$$

Second-Order and Higher-Order Circuits   Chap. 6

where σ and ω_d are the real and imaginary parts of the complex roots of the characteristic equation $s^2 + a_1 s + a_0 = 0$. The two constants K_1 and K_2 are found from the initial conditions $f(0)$ and $f'(0)$ using the relations

$$K_1 = f(0) \qquad K_2 = \frac{f'(0) - \sigma f(0)}{\omega_d} \tag{6-49}$$

■

There are several points of interest concerning the above development. First let us note that the roots of the characteristic equation are of the form

$$s_1 = \sigma + j\omega_d \qquad s_2 = \sigma - j\omega_d \tag{6-50}$$

where

$$\sigma = -\frac{a_1}{2} \qquad \omega_d = \sqrt{a_0 - \left(\frac{a_1}{2}\right)^2} \tag{6-51}$$

and where $(a_1/2)^2 - a_0$ (the discriminant) is less than zero. Now we may observe that:

1. Since $(a_1/2)^2$ is always positive, the coefficient a_0 must be positive, or this case will not apply, i.e., the discriminant will not be negative.
2. From (6-51), if the coefficient a_1 is greater than zero, then σ is negative.
3. From Eq. (6-48), the waveshape of $f(t)$ will have a sinusoidal variation with a peak magnitude or envelope proportional to $e^{\sigma t}$. If σ is negative (see item 2), then from Eq. (6-48), $f(t)$ disappears as t approaches infinity, i.e., $f(\infty) = 0$.

An interesting subcase of Case III occurs when the coefficient a_1 of Eq. (6-47) is zero. If this is true, then from (6-51) we see that $\sigma = 0$ in the general form of the solution given in Eq. (6-48). Thus, a sinusoidal variation of the variable $f(t)$ will continue for all time. Physically, this represents a situation in which no dissipation (resistance) is present in the network, since in such a case any energy orginally provided by the initial conditions as excitation for the network remains in the network indefinitely. For the network discussed in Example 6-2.1, this occurs when $R = 0$. An example of such a case follows.

Example 6-2.2

It is assumed that the circuit shown in Fig. 6-1.1 has a value of resistance which is identically zero. In this case, the differential equation for the circuit becomes

$$\frac{d^2 i}{dt^2} + \frac{1}{LC} i(t) = 0$$

The characteristic equation for this differential equation is

$$s^2 + \frac{1}{LC} = 0$$

Thus, the roots of the characteristic equation and, correspondingly, the natural frequencies of the network are purely imaginary and are equal to $\pm j\sqrt{1/LC}$. For the values $L = 1$ H and $C = \frac{1}{5}$ F (the same values used in the preceding examples), $a_1 = 0$ and $a_0 = 5$ in (6-47) and from Eq. (6-48) we see that the solution is

$$i(t) = K_1 \cos \sqrt{5}\, t + K_2 \sin \sqrt{5}\, t$$

If we assume that the initial conditions are $i(0) = 1$, $i'(0) = 0$, then $K_1 = 1$, $K_2 = 0$, and our solution is

$$i(t) = \cos \sqrt{5}\,t \qquad t \geq 0$$

The circuit described in the preceding example may be said to be *oscillating* (at a frequency of $\sqrt{5}$ rad/s). Some additional insight into this type of circuit may be gained by considering the energy relationships that exist in it. Let $w_L(t)$ be the energy stored in the inductor. Then

$$w_L(t) = \tfrac{1}{2}Li^2(t) = \tfrac{1}{2}\cos^2 \sqrt{5}\,t = \tfrac{1}{4}(1 + \cos 2\sqrt{5}\,t) \qquad (6\text{-}52)$$

The energy stored in the capacitor is readily found by first noting that the voltage across the capacitor $v(t)$ is the same as the voltage across the inductor, thus

$$v(t) = L\frac{di}{dt} = -\sqrt{5}\,\sin \sqrt{5}\,t \qquad (6\text{-}53)$$

If we let $w_C(t)$ be the instantaneous value of the energy stored in the capacitor, we may write

$$w_C(t) = \tfrac{1}{2}Cv^2(t) = \tfrac{1}{2}\sin^2 \sqrt{5}\,t = \tfrac{1}{4}(1 - \cos 2\sqrt{5}\,t) \qquad (6\text{-}54)$$

It is readily shown that $w_L(t) + w_C(t)$ is constant and equal to $\tfrac{1}{2}$. Thus, the total energy in the circuit is constant. Plots of $i(t)$, $v(t)$, $w_L(t)$, and $w_C(t)$ are shown in Fig. 6-2.2. From these plots we see that the total energy present in the circuit is transferred back and forth from the inductor to the capacitor at a frequency equal to two times the frequency at which the circuit is oscillating. Such a transfer of energy always occurs

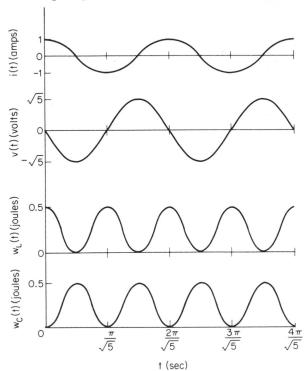

Figure 6-2.2 Waveforms for Example 6-2.2.

Second-Order and Higher-Order Circuits   Chap. 6

in circuits characterized by Case III, i.e., circuits in which the response has a sinusoidal component.

From Example 6-2.2 we note that the a_0 coefficient of the characteristic equation has a special significance, namely, it is the square of the frequency (in radians per second) at which sinusoidal oscillations will occur when the dissipation of the circuit has been reduced to zero. Thus, it is determined only by the energy-storage elements of the circuit, i.e., the inductor and the capacitor. Such a frequency is referred to as the *undamped natural frequency*. Using the symbol ω_0 for this frequency, we may write

$$\omega_0 = \sqrt{a_0} \tag{6-55}$$

The undamped natural frequency ω_0 is equal to the frequency of oscillation ω_d only when the dissipation is zero, otherwise ω_d is less in value than ω_0. More explicitly, from the definitions for σ and ω_d given in Eq. (6-51), we see that we may write

$$\omega_d = \sqrt{a_0 - \left(\frac{a_1}{2}\right)^2} = \sqrt{\omega_0^2 - \sigma^2} \tag{6-56}$$

This may also be written in the form

$$\omega_0 = \sqrt{\sigma^2 + \omega_d^2} \tag{6-57}$$

Thus, it is convenient to think of the relation between the magnitudes of σ, ω_d, and ω_0 as one characterized by the sides of a right triangle in which the sides adjacent to the right angle are of length σ and ω_d and the hypotenuse is of length ω_0 as shown in Fig. 6-2.3. To emphasize the above discussion, we present the following summary:

Figure 6-2.3   Relation between ω_0, ω_d, and σ.

SUMMARY 6-2.2

Sinusoidal Oscillations in Second-Order Circuits.   The sinusoidal component of the response of second-order circuits characterized by Case III (see Summary 6-2.1) has a frequency of ω_d radians per second. If the circuit contains any dissipative, i.e., resistive elements, this frequency will be lower than ω_0, the undamped natural frequency. The latter is determined only by the energy-storage elements of the circuit, i.e., the inductor and the capacitor, and is always equal to $\sqrt{a_0}$, where a_0 is defined in Eq. (6-47) as the zero-order coefficient of the differential equation. ■

We may correlate the separate cases for the solution of the second-order, homogeneous, differential equation presented in this and the preceding section by indicating the permissible locations of the roots of the characteristic equation on a complex "s plane," where s is the variable of the characteristic equation, for the three cases. For passive circuits, in which the coefficients a_1 and a_0 of Eq. (6-47) are positive (and non-zero), Case I is characterized by a pair of unequal roots on the negative real axis (excluding the origin), Case II is characterized by a single second-order root on the negative real axis (excluding the origin), and Case III is characterized by complex conjugate roots in the left-half plane (and not on the imaginary axis) symmetrically located with respect to the real axis. Note that if the coefficient a_0 is zero, then one of the negative real axis zeros of Case I will be at the origin. Similarly, as has been discussed, if the coefficient a_1 is zero, the complex zeros will occur on the imaginary axis. The relative locations of the three cases are shown in Fig. 6-2.4. For example, for the network shown in Fig. 6-1.1 and used as an illustration of Cases I, II, and III in Examples 6-1.1, 6-1.2, and 6-2.1, respectively, the locus of the roots of the characteristic equation as R is varied from 6 to 0 Ω, for $L = 1$ H and $C = \frac{1}{5}$ F, is shown in Fig. 6-2.5. Note that the form of the locus for Case III is a portion of a circle

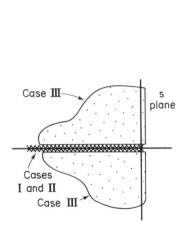

Figure 6-2.4 Location of the zeros of the characteristic equation for Cases I, II, and III.

Figure 6-2.5 Natural frequencies for the circuit of Fig. 6-1.1.

centered at the origin. The radius of this circle is the value of the undamped natural frequency ω_0. For any point on this circle, the real part is σ and the imaginary part is ω_d. Thus, the right triangle relating σ, ω_d, and ω_0 may be inscribed at any point on this portion of the locus as shown in the figure. A plot of the locus of the roots of polynomial (in this case, the *characteristic* polynomial) as a function of some parameter (in this case, R) is given the appropriate name of a *root-locus plot*. Such plots are of considerable importance in studies of control systems as well as in studies of networks. We shall see more of the root-locus plot in future chapters. It should be noted that the positions of the zeros of the characteristic equation on the s plane can be directly correlated to the form of the time response when the network is excited only by initial conditions. Thus, Cases I and II are characterized by exponential waveforms, while Case III is characterized by a sinusoidal waveform with an exponentially decreasing magnitude.

Second-Order and Higher-Order Circuits   Chap. 6

It has been pointed out that the roots of the characteristic equation are the natural frequencies of the network. This being the case, the complex s plane shown in Fig. 6-2.4 is frequently referred to as the *complex frequency plane*. We shall have considerable discussion in future chapters concerning the properties and usage of the complex frequency plane.

When the general form of the solution for any one variable of a second-order circuit has been determined, then, just as was the case for the first-order circuit (see Sec. 5-1), the general form of the solution for any other variable can be found by appropriate application of Kirchhoff's laws. Such an application, however, involves only the operations of differentiation, integration, addition, and subtraction. It is readily verified that when such operations are made to a general solution of the form given in Eqs. (6-14) or (6-30) (Cases I and II, respectively) or Eq. (6-48) (Case III), the result is an expression which has the same form, although it will, in general, have different constants. This is the same conclusion that was reached in Summary 5-1.2 for the first-order circuit. It is a direct consequence of the fact that the behavior of any variable in a given network has a zero-input response which is determined only by the natural frequencies of the network.

One other set of terms is frequently applied to second-order circuits. This has to do with the manner in which any sinusoidal oscillation of network variables disappears, i.e., is damped out. Thus, a circuit characterized by the special subcase of Case III in which the zeros of the characteristic equation are on the $j\omega$ axis is called an *undamped* circuit. Other Case III circuits are referred to as *underdamped*. Circuits characterized by Case II behavior are referred to as *critically damped* in the sense that if there is any less damping, i.e., any less amount of dissipation, then the circuit will exhibit an oscillatory component in its response. Finally, circuits characterized by Case I behavior are referred to as *overdamped*.

6-3 SECOND-ORDER CIRCUITS WITH MORE THAN ONE INDEPENDENT VARIABLE

In this section we shall extend the conclusions which have been reached to this point for the second-order one-mesh and one node-pair RLC circuit to circuits with greater numbers of meshes and node-pairs, and to RL and RC second-order circuits, i.e., circuits containing two inductors or two capacitors. We begin by noting that second-order RL and RC circuits must have more than one mesh or more than one pair of nodes. If this is not the case, there will be two parallel- or series-connected energy-storage elements in the circuit. These may obviously be replaced by a single equivalent energy-storage element, and the resulting circuit can then be treated as a first-order one. An example of a circuit containing two inductors, in which the inductors cannot be combined into a single inductor, is shown in Fig. 6-3.1. To characterize this circuit, it is necessary to write two differential equations. These are found by applying KVL around the two meshes of the circuit. Thus, we obtain

$$R_1 i_1(t) + L_1 \frac{d(i_1 - i_2)}{dt} = 0$$

$$R_2 i_2(t) + L_1 \frac{d(i_2 - i_1)}{dt} + L_2 \frac{di_2}{dt} = 0$$

(6-58)

Figure 6-3.1 Second-order RL circuit with two independent variables.

These equations may be solved by assuming solutions for $i_1(t)$ and $i_2(t)$ as follows:

$$i_1(t) = K_a e^{st} \qquad i_2(t) = K_b e^{st} \tag{6-59}$$

Substituting these values into (6-58) and putting the result into matrix format, we obtain

$$\begin{bmatrix} R_1 + sL_1 & -sL_1 \\ -sL_1 & R_2 + s(L_1 + L_2) \end{bmatrix} \begin{bmatrix} K_a e^{st} \\ K_b e^{st} \end{bmatrix} = \begin{bmatrix} 0 \\ 0 \end{bmatrix} \tag{6-60}$$

We may ignore the trivial solution $K_a e^{st} = K_b e^{st} = 0$ for this simultaneous set of equations, since from Eq. (6-59) this requires that $i_1(t) = i_2(t) = 0$, which is not very interesting. Thus, a solution to the set of equations exists only if the determinant of the square matrix in the left member of Eq. (6-60) is zero.[5] This requires

$$s^2 L_1 L_2 + s(R_1 L_1 + R_1 L_2 + R_2 L_1) + R_1 R_2 = 0 \tag{6-61}$$

This is the characteristic equation for the two-loop circuit shown in Fig. 6-3.1. The discriminant for this equation is

$$\left[\frac{R_1 + R_2}{2L_2} + \frac{R_1}{2L_1} \right]^2 - \frac{R_1 R_2}{L_1 L_2} \tag{6-62}$$

It may be shown that this discriminant is always positive (for positive-valued elements). Since the discriminant is positive, Case I, as presented in Sec. 6-1, applies. Thus, the values of s which satisfy the characteristic equation will be negative and not equal. We may define s_1 and s_2, the roots of the characteristic equation, by the relations

$$s_1 = \alpha + \beta \qquad s_2 = \alpha - \beta \tag{6-63}$$

where β is the positive square root of the discriminant given in Eq. (6-62), and α is defined as

$$\alpha = -\left[\frac{R_1 + R_2}{2L_2} + \frac{R_1}{2L_1} \right] \tag{6-64}$$

Thus, the solution for either $i_1(t)$ or $i_2(t)$ is of the form

$$K_1 e^{s_1 t} + K_2 e^{s_2 t} \tag{6-65}$$

where the constants K_1 and K_2 may be evaluated from the specified initial conditions. A numerical example follows.

[5] A simultaneous set of equations having the matrix form $\mathbf{Ax} = \mathbf{y}$, where the elements of $\mathbf{y}$ are *not* all zero, has a unique solution if $\det \mathbf{A} \neq 0$. If the elements of $\mathbf{y}$ are all zero, however, a non-trivial solution exists only if $\det \mathbf{A} = 0$. This is the case we have here.

Example 6-3.1

It is desired to find the solution for $i_2(t)$ for the circuit shown in Fig.6-3.1 for the case where the circuit elements have the values: $R_1 = 2\,\Omega$, $R_2 = 3\,\Omega$, $L_1 = L_2 = 1$ H. For these values, the square matrix given in the left member of (6-60) becomes

$$\begin{bmatrix} 2+s & -s \\ -s & 3+2s \end{bmatrix}$$

The characteristic equation is found by taking the determinant of this matrix and setting it to 0. Thus, we obtain

$$s^2 + 7s + 6 = 0$$

The roots of this equation are $s_1 = -1$, $s_2 = -6$. We may now write the general form that the solution for $i_2(t)$ (or any other network variable) must have. It is

$$i_2(t) = K_1 e^{-t} + K_2 e^{-6t}$$

Let us assume that the initial conditions are $i_2(0) = 1$ and $i'_2(0) = 0$. To use these conditions to determine expressions for K_1 and K_2, evaluating the above equation for $t = 0$, we find

$$i_2(0) = K_1 + K_2 = 1$$

A second equation involving K_1 and K_2 may be found by first differentiating the general solution for $i_2(t)$, and then evaluating it for $t = 0$. We thus obtain

$$i'_2(0) = -K_1 - 6K_2 = 0$$

Solving the two equations, we see that $K_1 = \frac{6}{5}$ and $K_2 = -\frac{1}{5}$. Thus, the solution for $i_2(t)$ is

$$i_2(t) = \tfrac{6}{5}e^{-t} - \tfrac{1}{5}e^{-6t} \qquad t > 0$$

A plot of the solution for $i_2(t)$ is shown in Fig. 6-3.2.

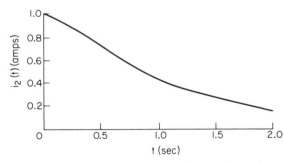

Figure 6-3.2 Waveform of $i_2(t)$ for Example 6-3.1.

An alternate technique for finding the characteristic equation is to use the analysis method developed for ladder networks in Sec. 2-4. As an example of this, consider the circuit shown in Fig. 6-3.3. Starting with the variable $v_2(t)$, and following the approach used in Sec. 2-4 we find that (the functional notation has been deleted for conciseness):

1. $i_{L2} = i_{R2} = v_2/R_2 = 2v_2$
2. $v_{L2} = L_2 i'_{L2} = L_2(2v'_2) = 6v'_2$
3. $v_{R1} = v_{L2} + v_2 = 6v'_2 + v_2$

Figure 6-3.3 Use of ladder network analysis to determine the characteristic equation.

4. $i_{R1} = v_{R1}/R_1 = 6v_2' + v_2$
5. $i_{L1} = i_{R1} + i_{L2} = 6v_2' + 3v_2$
6. $v_{L1} = L_1 i_{L1}' = 6v_2'' + 3v_2'$
7. $v_{L1} + v_{R1} = 6v_2'' + 9v_2' + v_2 = 0$

From the last result we readily find the characteristic equation to be $6s^2 + 9s + 1 = 0$. This equation, of course, applies to any of the variables in the circuit.

It was pointed out in the preceding discussion that the discriminant for the *RL* circuit shown in Fig. 6-3.1 is positive for all positive values of the network elements. This will be true for any second-order *RL* circuit, i.e., Cases II and III do not apply to circuits comprised of resistors and inductors. The same result holds true for second-order *RC* circuits, i.e., the discriminant is always positive for positive-valued circuit elements. It should be noted that even though oscillatory behavior (Case III) does not occur in such circuits, a reversal of the polarity of the variables is still possible. As an example of this, consider the circuit shown in Fig. 6-3.4. If we assume that

Figure 6-3.4 Alternation of the polarity of a variable in an *RC* circuit.

the capacitor C_1 is initially charged and the capacitor C_2 is initially uncharged, then the first action of the circuit will be to charge the capacitor C_2. Eventually, however, the circuit will become completely discharged; thus the current $i(t)$ will first be positive, then negative. The difference between this behavior and the oscillatory behavior of *RLC* circuits is, of course, the fact that with Case III behavior the reversals of polarity continue indefinitely and always have the same time period. Thus, although polarity reversals of variables are possible in *RL* and *RC* circuits, these circuits cannot exhibit true oscillatory behavior.

In Example 6-3.1 we showed how a set of simultaneous equations could be used to determine the characteristic equation for an *RL* (or an *RC*) second-order network. The same procedure must also be used for *RLC* circuits in which more than a single mesh or a single nodal voltage is required to define the network variables. An example follows.

Example 6-3.2

The *RLC* network shown in Fig. 6-3.5 contains two independent nodal voltage variables $v_1(t)$ and $v_2(t)$. It is desired to determine the expression for $v_1(t)$ when the circuit

mhos,
henries,
farads

Figure 6-3.5 Second-order *RLC* circuit with two independent variables.

has the initial conditions $v_1(0) = 1$, $v_1'(0) = 0$. We may do this by first writing the KCL equations at each of the two nodes. Thus, we obtain (note that the resistance values shown in the figure are given in units of mhos)

$$\frac{dv_1}{dt} + 2v_1(t) - 2v_2(t) = 0$$

$$4 \int_{-\infty}^{t} v_2(\tau)\, d\tau + 4v_2(t) - 2v_1(t) = 0$$

If we now assume that $v_1(t) = K_a e^{st}$ and $v_2(t) = K_b e^{st}$ and substitute these expressions into the KCL equations, then, after rearranging terms, we obtain[6]

$$\begin{bmatrix} s+2 & -2 \\ -2 & \dfrac{4}{s}+4 \end{bmatrix} \begin{bmatrix} K_a e^{st} \\ K_b e^{st} \end{bmatrix} = \begin{bmatrix} 0 \\ 0 \end{bmatrix}$$

Setting the determinant of the square matrix in the left member of the above equation to zero yields the characteristic equation

$$s^2 + 2s + 2 = 0$$

The discriminant is negative, thus Case III applies. The natural frequencies of the network are

$$s_1 = -1 + j1 \qquad s_2 = -1 - j1$$

The general form of the solution for $v_1(t)$ is

$$v_1(t) = e^{-t}(K_1 \cos t + K_2 \sin t)$$

Evaluating this equation and its derivative at $t = 0$ to match the specified initial conditions, we find that $K_1 = K_2 = 1$. Thus, the solution for $v_1(t)$ is

$$v_1(t) = e^{-t}(\cos t + \sin t) \qquad t \geq 0$$

[6]It is convenient in this and the following discussion to assume that

$$\int_{-\infty}^{t} K e^{s\tau}\, d\tau = \frac{K e^{st}}{s}$$

This produces the same characteristic equation as would be obtained by first differentiating a given KVL or KCL integro-differential equation, and then making the substitution $K_i e^{st}$ for each of the variables. This conclusion is readily verified for this example by differentiating the KCL equation written for node 2.

The techniques described in this section for solving for the variables of second-order circuits containing two independent variables are easily extended to include the case where second-order circuits contain more than two independent variables. Such a situation would occur for a circuit containing two energy-storage elements and an arbitrary number of resistors and dependent, i.e., controlled, sources. In such a case, if the resistors and the dependent sources are replaced by equivalent resistors and/or Thevenin and Norton equivalent sources (using the techniques of Sec. 3-7), the circuit may usually be reduced to one containing only two independent variables, and the techniques of this section may be applied to determine the solution for the network variables. The process is best illustrated by an example.

Example 6-3.3

The circuit shown in Fig. 6-3.6(a) contains three independent nodal voltage variables. In addition to the two energy-storage elements and the resistors, it also contains a dependent current source whose output current depends on the voltage $v_1(t)$. It is

Figure 6-3.6 Second-order RLC circuit with three independent variables.

desired to solve for the voltage $v_1(t)$ using as initial conditions $v_1(0) = 1$, $v_1'(0) = 0$. To do this, we may first reduce the portion of the circuit at the right which consists of two resistors and the dependent source by replacing it with a Norton equivalent circuit. The result is shown in Fig. 6-3.6(b). For this circuit, from KCL we obtain

$$\frac{dv_1}{dt} + 2v_1(t) - v_2(t) = 0$$

$$4 \int_{-\infty}^{t} v_2(\tau)\, d\tau + 4v_2(t) - v_1(t) = 3v_1(t)$$

Substituting $v_1(t) = K_a e^{st}$ and $v_2(t) = K_b e^{st}$ into these equations and rearranging terms, we obtain

$$\begin{bmatrix} s + 2 & -1 \\ -4 & \frac{4}{s} + 4 \end{bmatrix} \begin{bmatrix} K_a e^{st} \\ K_b e^{st} \end{bmatrix} = \begin{bmatrix} 0 \\ 0 \end{bmatrix}$$

The form of these equations is similar to those found for Example 6-3.2, the only difference being in the diagonal elements of the square matrix in the left member of the equation. The characteristic equation, however, is exactly the same, namely,

$$s^2 + 2s + 2 = 0$$

Thus, since the initial conditions are the same, the details of finding the solution for $v_1(t)$ may be taken directly from that example. The result is

$$v_1(t) = e^{-t}(\cos t + \sin t)$$

It does not, of course, follow that the solutions for all the other variables of this circuit will be the same as the variables of the circuit used in Example 6-3.2.

In the discussion of second-order circuits which we have made to this point, in solving for a given variable it has been assumed that the initial conditions were given for the variable and its first derivative. In other words, if $f(t)$ was the variable, then we assumed that values for $f(0)$ and $f'(0)$ were given to permit us to determine the two arbitrary constants in the solution of the homogeneous equation. In practice, it is more convenient to be able to specify the initial conditions in terms of network variables for which the continuity conditions apply. In Secs. 4-3 and 4-4 such variables were shown to be the voltage across a capacitor and the current through an inductor. If such initial conditions are given, then, to solve the differential equation, we may use the basic circuit equations to relate these initial conditions to $f(0)$ and $f'(0)$. As an example of this type of situation, let us consider the RLC circuit shown in Fig. 6-3.7.

Figure 6-3.7 Second-order RLC circuit.

The differential equation for this circuit is

$$L\frac{di}{dt} + Ri(t) + \frac{1}{C}\int_{-\infty}^{t} i(\tau)\,d\tau = 0 \tag{6-66}$$

If we assume that the values of R, L, and C are such that Case I applies, then the solution for $i(t)$ will have the form

$$i(t) = K_1 e^{s_1 t} + K_2 e^{s_2 t} \tag{6-67}$$

Now let us assume that the initial conditions are given as the current $i(0)$ through the inductor and the voltage $v(0)$ across the capacitor. To determine the two constants K_1 and K_2 in Eq. (6-67), we need to use our knowledge of $v(0)$ and $i(0)$ to find $i'(0)$. To do this, we note that

$$\frac{1}{C}\int_{-\infty}^{t} i(\tau)\,d\tau = v(t) \tag{6-68}$$

Substituting this in (6-66) and evaluating the equation at $t = 0$, we obtain

$$Li'(0) + Ri(0) + v(0) = 0 \tag{6-69}$$

Rearranging terms, we obtain an expression for $i'(0)$ in the form

$$i'(0) = \frac{-Ri(0) - v(0)}{L} \tag{6-70}$$

where all the quantities in the right member of the equation are known. Thus, we have found an expression for the value of $i'(0)$.

The same general procedure may be applied to the case where more than one network variable is involved. For example, consider the network shown in Fig. 6-3.8. If it is desired to solve for the voltage $v_1(t)$, then we require two initial conditions, namely, $v_1(0)$ and $v_1'(0)$. To find $v_1'(0)$ in terms of the more readily specified initial condition on the value of the inductor current $i(0)$, we may proceed by first applying

Figure 6-3.8 Second-order *RLC* circuit.

KCL to the nodes and writing the set of equations

$$C\frac{dv_1}{dt} + G_1v_1(t) - G_1v_2(t) = 0$$

$$\frac{1}{L}\int_{-\infty}^{t} v_2(\tau)\,d\tau + (G_1 + G_2)v_2(t) - G_1v_1(t) = 0$$

$(6\text{-}71)$

We note that the inductor current $i(t)$ may be defined as

$$i(t) = \frac{1}{L}\int_{-\infty}^{t} v_2(\tau)\,d\tau \qquad (6\text{-}72)$$

Now let us substitute this relation in the set of equations of (6-71) and evaluate the resulting set of equations at $t = 0$. We obtain

$$cv'_1(0) + G_1v_1(0) - G_1v_2(0) = 0$$

$$i(0) + (G_1 + G_2)v_2(0) - G_1v_1(0) = 0$$

$(6\text{-}73)$

The above set of equations contains two quantities which may be treated as known, i.e., the initial conditions $i(0)$ and $v_1(0)$. The quantities $v'_1(0)$ and $v_2(0)$, however, are "unknowns." We may solve the second equation of (6-73) for $v_2(0)$. Thus, we obtain

$$v_2(0) = \frac{G_1v_1(0) - i(0)}{G_1 + G_2} \qquad (6\text{-}74)$$

Substituting this value in the first equation and rearranging terms, we obtain

$$v'_1(0) = \frac{-G_1}{C}v_1(0) + \frac{G_1}{C}\left[\frac{G_1v_1(0) - i(0)}{G_1 + G_2}\right] \qquad (6\text{-}75)$$

Thus, the value of $v'_1(0)$ may be found in terms of the given initial conditions $v_1(0)$ and $i(0)$ and the values of the network elements. A similar procedure is readily applied to the problem of finding the desired initial conditions for other networks.

In the discussion of the second-order circuits which we have considered this far, we have assumed that the initial conditions were specified at $t = 0$. It is relatively easy to adapt the solution methods to the case in which initial conditions are specified for values of $t \neq 0$. The techniques are identical with those presented in Sec. 5-5. For example, assume that the values of the network elements are such that Case III applies, and thus that the solution for some network variable $f(t)$ for initial conditions applied at $t = 0$ would normally be of the form

$$f(t) = e^{\sigma t}(K_1 \cos \omega_d t + K_2 \sin \omega_d t) \qquad t \geq 0$$

For the same network, if the initial conditions were specified at t_0, i.e., if $f(t_0)$ and $f'(t_0)$ are given, then the solution will have the form

$$f(t) = e^{\sigma(t-t_0)}[K_1 \cos \omega_d(t - t_0) + K_2 \sin \omega_d(t - t_0)] \qquad t \geq t_0$$

Similar comments apply to Cases I and II, and some examples of such situations will be found in the exercises at the end of this chapter.

6-4 EXCITATION BY INITIAL CONDITIONS AND SOURCES

In the preceding sections of this chapter we have discussed the solution of second-order circuits which were excited only by initial conditions. In this section we shall extend our discussion to include the effects of excitation by sources. Thus, we shall consider the solution of *non-homogeneous* differential equations. As an example of such a situation, consider the circuit shown in Fig. 6-4 1. The KVL equation for this circuit is

$$L\frac{di}{dt} + Ri(t) + \frac{1}{C}\int_{-\infty}^{t} i(\tau)\,d\tau = v(t) \tag{6-76}$$

Figure 6-4.1 Second-order *RLC* circuit.

This is an integro-differential equation. It is more convenient, however, to deal with differential equations than with integro-differential equations, so our first step in finding the solution of Eq. (6-76) will be to change it to a differential equation. One way of achieving this is to replace $i(t)$ in Eq. (6-76) by a different variable. To determine the variable, note that, for the network shown in Fig. 6-4.1, the voltage $v_C(t)$ across the capacitor is defined as

$$v_C(t) = \frac{1}{C}\int_{-\infty}^{t} i(\tau)\,d\tau \tag{6-77}$$

If we differentiate both sides of Eq. (6-77) two times, we obtain

$$i(t) = C\frac{dv_C}{dt} \quad \text{and} \quad \frac{di}{dt} = C\frac{d^2v_C}{dt^2} \tag{6-78}$$

Now let us substitute Eqs. (6-77) and (6-78) into (6-76). For convenience we may also divide both members of the resulting equation by the factor LC. Thus, we obtain

$$\frac{d^2v_C}{dt^2} + \frac{R}{L}\frac{dv_C}{dt} + \frac{1}{LC}v_C(t) = \frac{1}{LC}v(t) \tag{6-79}$$

This is a non-homogeneous, second-order, differential equation. (It is also ordinary, linear, and has constant coefficients.) The techniques used for solving such a differential equation exactly parallel those presented in Sec. 5-2 for the first-order, non-homogeneous, differential equation. It was shown there that the complete solution consists of two parts. The first part is the complementary solution, i.e., the solution to the related homogeneous differential equation. As such, for a second-order circuit, it may have the form of any of the three cases described in the preceding sections of this chapter, depending on whether the discriminant of the characteristic equation is positive, zero, or negative (Cases I, II, and III, respectively). This part of the solution will contain two unspecified constants. The second part of the answer is the particular solution. The form of the particular solution is determined by the form of the excitation term. Table 5-2.1 may be used to determine the correct form to use. The complete solution is then given as the sum of the complementary solution and the particular solution. Finally, the two constants in the complementary solution are evaluated in the complete solution to match the specified initial conditions. The procedure is best illustrated by an example.

Example 6-4.1

In the circuit shown in Fig. 6-4.1 the excitation provided by the voltage source is $v(t) = t/3$ V applied at $t = 0$. The network elements have the same values as were used in Example 6-1.1, namely, $R = 6\,\Omega$, $L = 1$ H, and $C = \frac{1}{5}$ F. For these values, the differential equation of (6-79) becomes

$$\frac{d^2 v_C}{dt^2} + 6\frac{dv_C}{dt} + 5v_C(t) = \frac{5}{3}t \qquad t \geq 0$$

The characteristic equation and its roots are found as follows:

$$s^2 + 6s + 5 = (s + 1)(s + 5) = 0$$

If we let $v_h(t)$ be the complementary function, i.e., the solution to the homogeneous equation, then, noting that all variables in a given circuit have the same natural behavior, we may write

$$v_h(t) = K_1 e^{-t} + K_2 e^{-5t}$$

To find the particular solution $v_p(t)$, we note that the function in the right member of the differential equation is $5t/3$. From Table 5-2.1, the form of the particular solution is seen to be

$$v_p(t) = A + Bt \qquad v_p'(t) = B \qquad v_p''(t) = 0$$

Substituting these relations into the original differential equation we obtain

$$0 + 6B + 5(A + Bt) = (5A + 6B) + 5Bt = \frac{5}{3}t$$

Matching coefficients of like powers of t we find that $A = -\frac{2}{5}$ and $B = \frac{1}{3}$. The complete solution for $v_C(t)$ is now the sum of the complementary solution and the particular solution. Thus, we may write

$$v_C(t) = v_h(t) + v_p(t) = K_1 e^{-t} + K_2 e^{-5t} - \frac{2}{5} + \frac{t}{3} \qquad t \geq 0$$

To determine the initial conditions, let us assume that the circuit is unexcited for $t < 0$. Thus, $v_C(t)$ and $i(t)$ are both zero for $t < 0$. Continuity conditions, however, apply to these variables. Therefore, $i(0) = v_C(0) = 0$. In addition, since $v_C'(t)$ is proportional to $i(t)$, we see that $v_C'(0)$ is also zero. The first equation needed to determine the constants K_1 and K_2 is found by evaluating the complete solution at $t = 0$. Thus, we obtain

$$v_C(0) = K_1 + K_2 - \frac{2}{5} = 0$$

Differentiating the complete solution and evaluating it at $t = 0$ gives us the following second equation

$$v_C'(0) = -K_1 - 5K_2 + \frac{1}{3} = 0$$

Solving the two equations for K_1 and K_2 and inserting the values in the complete solution, we obtain

$$v_C(t) = \frac{5}{12}e^{-t} - \frac{1}{60}e^{-5t} - \frac{2}{5} + \frac{1}{3}t \qquad t \geq 0$$

From Eq. (6-78) we find that the solution for $i(t)$ is

$$i(t) = -\tfrac{1}{12}e^{-t} + \tfrac{1}{60}e^{-5t} + \tfrac{1}{15} \qquad t \geq 0$$

SUMMARY 6-4.1

Solution of a Second-Order Circuit Excited by Initial Conditions and Sources.
A second-order circuit excited by initial conditions and sources is described by a
second-order, non-homogeneous, differential equation having the form

$$\frac{d^2f}{dt^2} + a_1 \frac{df}{dt} + a_0 f(t) = g(t) \qquad (6-80)$$

where $g(t)$ is the excitation function representing the sources and $f(t)$ is any network
variable. The solution to this equation may be found as follows:

1. Find the complementary solution $f_h(t)$, i.e., the solution of the related homogeneous differential equation, using the methods of Secs. 6-1 and 6-2. This
   function will have two unspecified constants.
2. For the specified function $g(t)$ find the form of the particular solution $f_p(t)$ from
   Table 5-2.1. Substitute this expression into the original differential equation of
   (6-80) to find the values of any constants in $f_p(t)$.
3. Form the complete solution $f(t) = f_h(t) + f_p(t)$. The two constants in $f_h(t)$
   may now be determined to match the specified initial conditions $f(0)$ and $f'(0)$. ∎

There are other methods of changing an integro-differential equation of the
form given in Eq. (6-76) to a differential equation than the change of variable technique
which was used in the preceding example. One such method is simply to differentiate
the entire equation. If we do this for Eq. (6-76) and divide the entire equation by L,
we obtain

$$\frac{d^2i}{dt^2} + \frac{R}{L}\frac{di}{dt} + \frac{1}{LC}i(t) = \frac{1}{L}\frac{dv}{dt} \qquad (6-81)$$

The techniques described in Summary 6-4.1 are readily applied to solve this equation.
To illustrate the procedure, we will rework the problem presented in Example 6-4.1
as follows.

Example 6-4.2
    It is desired to find the current $i(t)$ in the circuit shown in Fig. 6-4.1 and used in Example
6-4.1 using the method given above. The differential equation given in (6-81) for the
specified element values and for $v(t) = tu(t)/3$ is

$$\frac{d^2i}{dt^2} + 6\frac{di}{dt} + 5i(t) = \frac{1}{3}u(t)$$

The solution of the homogeneous equation, $i_h(t)$, has the same form as $v_h(t)$ had in
Example 6-4.1, namely,

$$i_h(t) = K_1 e^{-t} + K_2 e^{-5t}$$

The particular solution $i_p(t)$ will simply be a constant. If we let $i_p(t) = A$, then $i'_p(t) = i''_p(t) = 0$. Substituting these expressions in the original differential equation we obtain $5A = \frac{1}{3}$. Thus $A = i_p(t) = \frac{1}{15}$. The complete solution for $i(t)$ is now

$$i(t) = K_1 e^{-t} + K_2 e^{-5t} + \tfrac{1}{15} \qquad t \geq 0$$

Evaluating this solution and its derivative at $t = 0$ we obtain

$$i(0) = K_1 + K_2 + \tfrac{1}{15}$$

$$i'(0) = -K_1 - 5K_2$$

Solving these equations for $i(0) = i'(0) = 0$, we find that $K_1 = -\frac{1}{12}$ and $K_2 = \frac{1}{60}$. Thus our complete solution is

$$i(t) = -\tfrac{1}{12} e^{-t} + \tfrac{1}{60} e^{-5t} + \tfrac{1}{15} \qquad t \geq 0$$

This is, of course, the same result that was obtained in Example 6-4.1.

It should be noted that the solution process illustrated in the preceding example, in which the entire integro-differential equation is differentiated, is somewhat shorter than the method illustrated in Example 6-4.1, in which a change of variable is used to convert the integro-differential equation to a differential equation. Thus, this latter method might appear to be a more desirable one. It should also be noted, however, that the latter method requires that we differentiate the expression for the excitation function. If the waveform of the excitation is discontinuous, as would be the case for a step function, then the resulting derivative will contain an impulse. The solution of a differential equation with an impulse excitation is an advanced topic which will not be treated in this chapter. However, techniques for obtaining such a solution will be covered in connection with the development of the Laplace transformation in Chap. 8.

The procedures for solving second-order differential equations excited by independent sources described in the preceding paragraphs are readily extended to the situation where the second-order circuit has more than one mesh or node-pair. For example, consider the circuit shown in Fig. 6-4.2. Applying KCL to the two nodes of

Figure 6-4.2 Second-order *RLC* circuit with two independent variables.

this circuit, we obtain the equations

$$C\frac{dv_1}{dt} + G_1 v_1(t) - G_1 v_2(t) = i(t)$$

$$-G_1 v_1(t) + (G_1 + G_2) v_2(t) + \frac{1}{L}\int_{-\infty}^{t} v_2(\tau)\, d\tau = 0 \qquad (6\text{-}82)$$

This is a simultaneous set of integro-differential equations. To begin the solution, let us change this set of equations to a set of differential equations by using the variable $i_L(t)$ in place of the variable $v_2(t)$. The necessary relations may be found from the

Second-Order and Higher-Order Circuits   Chap. 6

terminal relations for the variables of the inductor, namely,

$$i_L(t) = \frac{1}{L} \int_{-\infty}^{t} v_2(\tau)\, d\tau \qquad \text{and} \qquad L\frac{di_L}{dt} = v_2(t) \qquad (6\text{-}83)$$

Substituting these relations in (6-82), we obtain

$$C\frac{dv_1}{dt} + G_1 v_1(t) - G_1 L\frac{di_L}{dt} = i(t)$$

$$-G_1 v_1(t) + L(G_1 + G_2)\frac{di_L}{dt} + i_L(t) = 0 \qquad (6\text{-}84)$$

This set of equations may be solved by first assuming an exponential solution for the variables $v_1(t)$ and $i_L(t)$. Thus, if we let

$$v_1(t) = K_a e^{st}$$

$$i_L(t) = K_b e^{st} \qquad (6\text{-}85)$$

and substitute these values in (6-84), we obtain

$$\begin{bmatrix} sC + G_1 & -sLG_1 \\ -G_1 & sL(G_1 + G_2) + 1 \end{bmatrix} \begin{bmatrix} K_a e^{st} \\ K_e e^{st} \end{bmatrix} = \begin{bmatrix} i(t) \\ 0 \end{bmatrix} \qquad (6\text{-}86)$$

The form of the complementary solution for $v_1(t)$ and $i_L(t)$ (and for any other network variable) is found by setting the determinant of the square matrix in the left member of Eq. (6-86) to zero. The roots of the resulting characteristic equation are the natural frequencies of the network. The particular solutions for the network variables $v_1(t)$ and $i_L(t)$ are found by substituting an assumed form for these variables into (6-84). The necessary form is determined from Table 5-2.1 for the specific expression given for the current source. The process is illustrated by the following example.

Example 6-4.3

The circuit shown in Fig. 6-4.2 is assumed to have the following values for its elements: $C = 1$ F, $L = \frac{1}{4}$ H, $G_1 = G_2 = 2$ mhos. (These are the same values which were used for this circuit in Example 6-3.2.) The current source is assumed to have an excitation $i(t) = 4$ A applied at $t = 0$. The resulting set of simultaneous equations given in (6-86) may now be written

$$\begin{bmatrix} s + 2 & -\frac{s}{2} \\ -2 & s + 1 \end{bmatrix} \begin{bmatrix} K_a e^{st} \\ K_b e^{st} \end{bmatrix} = \begin{bmatrix} 4 \\ 0 \end{bmatrix} \qquad t > 0$$

The characteristic equation found by setting the determinant of the square matrix to zero is $s^2 + 2s + 2 = 0$. Thus, the natural frequencies of the network are located at $-1 \pm j1$, and the form of the complementary solution for either $v_1(t)$ or $i_L(t)$ is $e^{-t}(K_1 \cos t + K_2 \sin t)$, where K_1 and K_2 will, of course, be different for the two variables. The particular solution is easily found by noting from Table 5-2.1 that, for a constant-valued excitation function $i(t)$, the particular solution $v_p(t)$ for $v_1(t)$ and the particular solution $i_p(t)$ for $i_L(t)$ will both be constants. If we let $v_p(t) = A$, and $i_p(t) = B$, and substitute these values in the simultaneous differential equations of (6-84), we obtain

$$v_1'(t) + 2v_1(t) - \tfrac{1}{4}i_L'(t) = 2A = 4$$

$$-2v_1(t) + i_L'(t) + i_L(t) = -2A + B = 0$$

Thus, $A = 2$ and $B = 4$. The complete solution for $v_1(t)$ is of the form

$$v_1(t) = e^{-t}(K_1 \cos t + K_2 \sin t) + 2 \qquad t > 0$$

The constants K_1 and K_2 are readily determined for any specified set of initial conditions.

It should be noted that the natural frequencies found in the above example are the same as those found in Example 6-3.2. This is to be expected, since the network is identical in the two examples. Note, however, that the square matrix found in the above example is considerably different from that found in Example 6-3.2. It should especially be noted that the matrix is not symmetric.

Almost all of the techniques and conclusions derived for a first-order circuit in Secs. 5-3 to 5-7 apply directly to the second-order circuit (and also to higher-order circuits). For example, in addition to separating the complete solution into a complementary solution and a particular solution, as we did in the preceding examples, we may also separate it into a zero-input response term (the portion of the solution that results from the presence of initial conditions), and a zero-state response term (the portion of the solution that is produced by excitation from sources). The general forms of such decompositions for the three cases of the second-order differential equation are tabulated in Table 6-4.1. It should be noted that the natural frequencies

TABLE 6-4.1

Case[†]	First Decomposition of Response		Second Decomposition of Response	
	Transient response or complementary solution (terms due to the natural frequencies of the network)	Forced response or particular solution (terms having the form of the excitation)	Zero-input response (terms due to the excitation of the network by initial conditions)	Zero-state response (terms due to the excitation of the network by sources)
I $f(t) =$	$K_1 e^{s_1 t} + K_2 e^{s_2 t}$	$+ \quad f_p(t)$	$= \quad K_{1a} e^{s_1 t} + K_{2a} e^{s_2 t} \ +$	$K_{1b} e^{s_1 t} + K_{2b} e^{s_2 t} + f_p(t)$
II $f(t) =$	$K_1 e^{s_1 t} + K_2 t e^{s_1 t}$	$+ \quad f_p(t)$	$= \quad K_{1a} e^{s_1 t} + K_{2a} t e^{s_1 t} \ +$	$K_{1b} e^{s_1 t} + K_{2b} t e^{s_1 t} + f_p(t)$
III $f(t) =$	$e^{\sigma t}(K_1 \cos \omega_d t \ + K_2 \sin \omega_d t)$	$+ \quad f_p(t)$	$= \quad e^{\sigma t}(K_{1a} \cos \omega_d t \ + K_{2a} \sin \omega_d t) \ +$	$e^{\sigma t}(K_{1a} \cos \omega_d t \ + K_{2b} \sin \omega_d t) + f_p(t)$

[†] $K_{1a} + K_{1b} = K_1$ (for all cases), $K_{2a} + K_{2b} = K_2$ (for all cases).

of the network are, in general, always excited when an input from a source or sources is applied to a network (even though the initial conditions may be zero). Thus, the familiar exponential quantities always appear in the zero-state response term as well

Second-Order and Higher-Order Circuits   Chap. 6

as in the zero-input response term.[7] It should be noted that the forced response may also be referred to as the steady-state response if the excitations provided by the sources are all constant or are all sinusoidal.

Another example of the extension of the conclusions of Chap. 5 to the second-order network is the examination of linearity considerations. Thus we note that the response of any variable in a second-order circuit consisting of linear elements is linearly related to the following excitations: (1) The excitation from any initial condition when all other initial conditions are zero and when all sources are set to zero; and (2) the excitation from any source when all other sources are set to zero and when all initial conditions are zero. In general, the response is linearly related to any single excitation, and is not linearly related to a given excitation when other excitations are present.

6-5 EXAMPLES OF SECOND-ORDER CIRCUITS

In this section we shall present some examples of second-order circuits and show how they may be analyzed using the techniques discussed in the previous sections of this chapter. The first two examples also illustrate some modifications of the ladder analysis technique, originally presented in Sec. 2-4, to take account of controlled sources.

Example 6-5.1

Our first example is the circuit shown in Fig. 6-5.1. This circuit is referred to as an *active-RC* one since it contains only R and C elements plus an active element, namely a VCVS of gain K represented by the triangle. It is desired to find the output voltage

Figure 6-5.1 Active-*RC* circuit.

$v_2(t)$ when the input voltage $v_1(t) = u(t)$. For convenience we will assume that all elements have unity value. Starting at the right end of the circuit, and dropping the functional notation, we find:

1. $v_{C4} = v_2/K$
2. $i_{C4} = C_4 v'_{C4} = v'_2/K$

[7]In Case III, the term involving σ and ω_d may be considered as another way of writing a "complex exponential" term, i.e., e raised to a complex power.

3. $v_{R3} = R_3 i_{R3} = R_3 i_{C4} = v_2'/K$
4. $v_{C2} = v_{R3} + v_{C4} - v_2 = (v_2'/K) + (v_2/K) - v_2$
5. $i_{C2} = C_2 v_{C2}' = (v_2''/K) + (v_2'/K) - v_2'$
6. $v_{R1} = R_1 i_{R1} = R_1 (i_{C2} + i_{R3}) = (v_2''/K) + (2v_2'/K) - v_2'$
7. $v_1 = v_{R1} + v_{R3} + v_{C4} = (v_2''/K) + (3v_2'/K) - v_2' + v_2/K$

Inserting the value $K = 2$, we find that the characteristic polynomial is $s^2 + s + 1$ (Case III). The roots are at $s = -0.5 + j\sqrt{3}/2$. Thus, the complementary solution is $v_{2h}(t) = e^{-0.5t}[K_1 \cos(\sqrt{3}/2)t + K_2 \sin(\sqrt{3}/2)t]$. The particular solution is easily shown to be $v_{2p}(t) = K = 2$. Thus the complete solution is

$$v_2(t) = 2 + e^{-0.5t}\left[K_1 \cos\left(\frac{\sqrt{3}}{2}\right)t + K_2 \sin\left(\frac{\sqrt{3}}{2}\right)t\right]$$

Now let us consider the initial conditions. We will assume that the capacitors are both uncharged at $t = 0$. Thus $v_{C_4}(0+) = 0$ and, from the circuit, we see that $v_2(0+) = Kv_{C4}(0+) = 0$. This is the first of the two required initial conditions on $v_2(t)$. The second initial condition can be found from the equation in step 4 of the analysis procedure given above. Evaluating this equation at $t = 0+$ we obtain

$$v_{C2}(0+) = \frac{v_2'(0+)}{K} + \frac{v_2(0+)}{K} - v_2(0+)$$

Since $v_{C2}(0+) = 0$, we readily find $v_2'(0+) = 0$. This is the second initial condition. Inserting these results in the complete solution given above, we obtain the values of the constants $K_1 = -2$ and $K_2 = -2/\sqrt{3}$. Thus, the complete solution is

$$v_2(t) = 2 + e^{-0.5t}\left[-2 \cos\left(\frac{\sqrt{3}}{2}\right)t - \left(\frac{2}{\sqrt{3}}\right)\sin\left(\frac{\sqrt{3}}{2}\right)t\right]$$

Example 6-5.2

As another example of a second-order *active-RC* circuit, consider the one shown in Fig. 6-5.2. As indicated by the symbol, this circuit uses an operational amplifier as the active element. This circuit may be analyzed, using the ladder analysis method, by

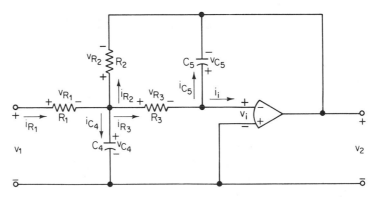

Figure 6-5.2 Active-*RC* circuit.

introducing the usual operational amplifier constraints (see Sec. 3-9) that $v_i = i_t = 0$. For this circuit the analysis steps are as follows (for convenience we have used unity-valued elements):

1. $v_{C5} = -v_2$
2. $i_{R3} = i_{C5} = C_5 v'_{C5} = -v'_2$
3. $v_{C4} = v_{R3} = R_3 i_{R3} = -v'_2$
4. $i_{C4} = C_4 v'_{C4} = -v''_2$
5. $v_{R2} = v_{R3} + v_{C5} = -v'_2 - v_2$
6. $i_{R2} = v_{R2}/R_2 = -v'_2 - v_2$
7. $i_{R1} = i_{R2} + i_{R3} + i_{C4} = -v''_2 - 2v'_2 - v_2$
8. $v_{R1} = R_1 i_{R1} = -v''_2 - 2v'_2 - v_2$
9. $v_1 = v_{R1} + v_{C4} = -v''_2 - 3v'_2 - v_2$

Thus we see that the characteristic equation is $s^2 + 3s + 1 = 0$ (Case I). The determination of the initial conditions and the solution of the differential equation are left as an exercise.

Example 6-5.3

As another example of a second-order circuit, consider the simplified model of an automotive ignition system shown in Fig. 6-5.3. The battery and alternator provide a 12-V dc supply of very low internal resistance that is modeled by the voltage source

Figure 6-5.3   Automotive ignition system.

shown at the left of the figure. This is connected to an ignition coil, a transformer which has a 6-mH self-inductance in its primary (low-voltage) winding and a turns ratio of 1 : 400. The resistance of the wiring and of the transformer winding is modeled by the 3-Ω resistor. The primary winding of the ignition coil is connected to ground by the distributor points, modeled as a switch, and also by the 1-μF capacitor. The high voltage produced by the secondary winding of the ignition coil is applied across the gap of one of the spark plugs selected by the distributor (not shown). Thus, the secondary winding may be considered as being open-circuited until its voltage is sufficient to initiate an arc discharge across the spark plug gap. This voltage will be assumed to be 10 kV. The operation of the ignition system can be divided into two phases. These are described as follows:

Phase 1—Charging the System: Assume that the system has reached a steady-state condition with the points open at $t = 0-$. From the circuit of Fig. 6-5.3, the values of the variables at this time are $i(0-) = 0$, $v_L(0-) = 0$, and $v_C(0-) = 12$. If the points are closed at $t = 0$, the primary winding and the capacitor are shorted to ground, and the primary circuit now appears as shown in Fig. 6-5.4(a).[8] This first-order

---

[8]Actually, this causes an impulse of current to flow from the capacitor through the switch, but this does not affect the other circuit variables.

Figure 6-5.4 Operation of the automotive ignition system.

(a) (b)

circuit can be analyzed using the techniques outlined in Sec. 5-4 to show that $i(t) = 4 - 4e^{-t/(2 \times 10^{-3})}$. Thus, the time constant of the circuit is 2 ms. Since $i(\infty) = 4$ A, after approximately 10 ms the ignition coil is fully charged with a stored energy of

$$W_L = \tfrac{1}{2}Li^2 = \tfrac{1}{2} \times (6 \times 10^{-3}) \times (4)^2 = 48 \times 10^{-3} \text{ J}$$

Phase 2—Generating the Voltage: Let us now assume that the system has been charged as described above, and has reached a steady-state condition with the points closed. We assume that this has happened at a new time reference of $t = 0-$. The values of the variables are $i(0-) = 4$, $v_R(0-) = 12$, $v_L(0-) = v_C(0-) = 0$. If the points are now opened at $t = 0$, the primary circuit appears as shown in Fig. 6-5.4(b). We will now solve this circuit for $v_L(t)$. First we note that, from the continuity conditions, we have $i(0+) = 4$ [thus $v_R(0+) = 12$] and $v_C(0+) = 0$. If we write the KVL equation for the circuit in the form

$$12 = v_R(t) + L\frac{di}{dt} + v_C(t)$$

and evaluate this at $t = 0+$, we find $i'(0+) = 0$. Thus $v_L(0+) = Li'(0+) = 0$. This is our first initial condition for $v_L(t)$. If we now differentiate the KVL equation given above, we obtain

$$0 = \frac{dv_R}{dt} + L\frac{d^2i}{dt^2} + \frac{dv_C}{dt}$$

Evaluating this at $t = 0+$, we may use the relations $v_R'(0+) = Ri'(0+) = 0$ and $v_C'(0+) = i(0+)/C = 4/C$. Thus we obtain the result $v_L'(0+) = Li''(0+) = -v_C'(0+) = -4/C$. This is the second of the two required initial conditions. The characteristic polynomial for the circuit is readily found to be $s^2 + s(R/L) + 1/LC$. Inserting the numerical values, the roots of the characteristic equation are

$$s = \sigma \pm j\omega_d = -0.25 \times 10^3 \pm j1.29 \times 10^4$$

Since $v_L(\infty) = 0$, the particular solution for $v_L(t)$ is zero, and the complete solution has the form

$$v_L(t) = e^{\sigma t}(K_1 \cos \omega_d t + K_2 \sin \omega_d t)$$

Inserting the initial conditions $v_L(0+) = 0$ and $v_L'(0+) = -4/C$, we obtain $K_1 = 0$ and $K_2 = -4/C\omega_d$. Thus, the complete solution is

$$v_L(t) = -310e^{-0.25 \times 10^3 t} \sin (1.29 \times 10^4 t)$$

The voltage $v_g(t)$ across the gap of the spark plug is now found by using the relation $v_g(t) = 400v_L(t)$. To find the time t_f when the spark plug fires, we solve for t_f in the equation $|v_g(t_f)| = 10^4$. The mathematics of the solution may be simplified by neglecting the exponential term, since its value will be nearly constant during the first half-cycle of the sinusoidal term, the half-cycle in which the spark plug is fired, and thus the end of the time range for which the solution given above is valid. Thus we can solve

$$|v_g(t_f)| = 10^4 \approx 400 \times 310 \times \sin (1.29 \times 10^4 t_f)$$

From this relation we find $t_f = 6.26 \times 10^{-6}$ s. It should be noted that, at higher engine speeds, there may not be sufficient time for the charging cycle described in Phase 1 above, to fully charge the inductor. As a result, a longer time period may occur before the spark plug fires. Some details on this situation may be found in the problems.

6-6 CIRCUITS OF HIGHER THAN SECOND ORDER

In Chap. 5 and in the preceding sections of this chapter we have developed methods for analyzing first- and second-order circuits. These methods may be directly extended to permit us to study nth-order circuits, where the value of n is restricted only by our patience in carrying out the mathematical manipulations and by the accuracy of our computations, rather than by any limit of theory. Thus, we may look upon this section as providing techniques to obtain the solution for the variables of a circuit comprised of any combination of the network elements which we have discussed to this point. Despite the magnitude of such an attainment, this section will not be an excessively long one. The reason for this is that it will consist mostly of an extension of previously developed techniques.

As a first example of a circuit of higher than second order, consider the RC network shown in Fig. 6-6.1. Let us assume that we desire to find the voltage $v_3(t)$ for the case where the current source is set to zero, i.e., $i_s(t) = 0$. The initial conditions will

Figure 6-6.1 Third-order RC network.

be specified as $v_3(0) = 1$, $v_3'(0) = 1$, and $v_3''(0) = 1$. The equations found by applying KCL at the three nodes are

$$v_1'(t) + v_1(t) - v_2(t) = 0$$
$$3v_2'(t) + 3v_2(t) - v_1(t) - 2v_3(t) = 0 \qquad (6\text{-}87)$$
$$2v_3'(t) + 2v_3(t) - 2v_2(t) = 0$$

Substituting $v_1(t) = K_a e^{st}$, $v_2(t) = K_b e^{st}$, and $v_3(t) = K_c e^{st}$ in these equations, we obtain

$$\begin{bmatrix} s+1 & -1 & 0 \\ -1 & 3s+3 & -2 \\ 0 & -2 & 2s+2 \end{bmatrix} \begin{bmatrix} K_a e^{st} \\ K_b e^{st} \\ K_c e^{st} \end{bmatrix} = \begin{bmatrix} 0 \\ 0 \\ 0 \end{bmatrix} \qquad (6\text{-}88)$$

Taking the determinant, we find that the characteristic equation for the network is

$$s^3 + 3s^2 + 2s = s(s+1)(s+2) = 0 \qquad (6\text{-}89)$$

Thus, we see that the roots of the characteristic equation are $s_1 = 0$, $s_2 = -1$, and $s_3 = -2$. The complementary solution for $v_3(t)$ (or for any other variable in the network) is

$$v_3(t) = K_1 + K_2 e^{-t} + K_3 e^{-2t} \qquad (6\text{-}90)$$

If we evaluate at $t = 0$ the equation for $v_3(t)$ given above, the equation for $v'_3(t)$, and the equation for $v''_3(t)$, we find that the constants K_1, K_2, and K_3 are related to the specified initial conditions by the equations

$$\begin{bmatrix} 1 & 1 & 1 \\ 0 & -1 & -2 \\ 0 & 1 & 4 \end{bmatrix} \begin{bmatrix} K_1 \\ K_2 \\ K_3 \end{bmatrix} = \begin{bmatrix} 1 \\ 1 \\ 1 \end{bmatrix} \tag{6-91}$$

Solving this set of equations, we obtain $K_1 = 3$, $K_2 = -3$, and $K_3 = 1$. Thus, the solution for $v_3(t)$ is[9]

$$v_3(t) = 3 - 3e^{-t} + e^{-2t} \tag{6-92}$$

The techniques illustrated above are readily applied to an RC circuit of arbitrary complexity. It should be noted that, although the order of the characteristic equation for an RC circuit cannot be greater than the number of independent network variables which is needed to write the KCL equations for the circuit, it may be less. As an example of this, in the circuit shown in Fig. 6-6.2 there are three capacitors. If we determine the characteristic equation for this circuit, we obtain

$$[s(C_1 + C_3) + (G_1 + G_3)][s(C_2 + C_3) + (G_2 + G_3)] - (sC_3 + G_3)^2 = 0 \tag{6-93}$$

Figure 6-6.2   Second-order RC network.

This equation is obviously of second order. Thus, we see that there are only two natural frequencies and that the circuit is a second-order one. The reason for this may be seen by applying KVL to the circuit. Thus we obtain

$$v_{C1} - v_{C2} - v_{C3} = 0 \tag{6-94}$$

Obviously, any one capacitor voltage can be expressed as a linear combination of the other two. As a result, the three capacitor voltages (and their initial conditions) are not independent. In general, any circuit which contains a purely capacitive loop, or a purely capacitive cutset, will have one or more capacitors which are not independent.

Now let us consider the regions in the complex s plane where the roots of the characteristic equations (i.e., the natural frequencies) of an RC circuit may be located. In Sec. 6-3 we showed that, for a second-order RC circuit, the natural frequencies

[9]Note that, since the network has a natural frequency at $s = 0$, if a constant excitation is applied by the current source, the particular solution will be of the form $v_{3p} = Bt$. See also the footnote for Table 5-2.1.

were always on the negative real axis (or the origin) of the complex frequency or s plane. The same conclusion applies to nth-order RC circuits. It may also be shown that the natural frequencies are simple.

In the above paragraphs we have discussed the solution of an nth-order RC network which is excited only by initial conditions. Only a slight extension of the method is necessary to include the case where excitation is also furnished by sources. In such a case, we must find the particular solution for the network variable that we wish to solve for. To do this we need merely substitute an assumed solution for each of the independent variables in the original KCL equations describing the circuit. The correct form of assumed solution to use is determined by the form of the excitation function, and may be taken from Table 5-2.1. The unspecified constants in the expressions given in the table must be determined separately for each of the independent variables. Once the particular solution has been determined, the complete solution may be found by evaluating the constants in the complementary solution (the part of the solution which is determined by the characteristic equation) in such a way as to match the specified initial conditions for the network. The procedure is illustrated by the following example.

Example 6-6.1

The circuit shown in Fig. 6-6.1 is excited by a current source at the left node as illustrated. The excitation $i_s(t)$ provided by the current source is $18e^{-3t}$ A starting at t equals zero. It is desired to find an expression for the voltage $v_3(t)$. It is assumed that the initial conditions are $v_3(0) = 1$, $v_3'(0) = 0$, and $v_3''(0) = 0$. From Table 5-2.1 we find that the assumed form for each of the node voltages $v_i(t)$ ($i = 1, 2, 3$) may be written as $A_i e^{-3t}$ ($i = 1, 2, 3$), where the A_i are quantities whose values are to be determined. Substituting these relations into the original KCL equations of (6-87) and setting the first equation equal to $i_s(t)$, we obtain

$$\begin{bmatrix} -2 & -1 & 0 \\ -1 & -6 & -2 \\ 0 & -2 & -4 \end{bmatrix} \begin{bmatrix} A_1 \\ A_2 \\ A_3 \end{bmatrix} e^{-3t} = \begin{bmatrix} 18 \\ 0 \\ 0 \end{bmatrix} e^{-3t}$$

Since we are only interested in obtaining a solution for $v_3(t)$, we need solve only for A_3. Using determinants, we find that $A_3 = -1$. Thus, the particular solution for $v_3(t)$ is $-e^{-3t}$. The complementary solution is given by Eq. (6-90). Thus, the complete solution for $v_3(t)$ is

$$v_3(t) = K_1 + K_2 e^{-t} + K_3 e^{-2t} - e^{-3t} \qquad t \geq 0$$

Evaluating the equation for $v_3(t)$, the equation for $v_3'(t)$, and the equation for $v_3''(t)$ at $t = 0$, and inserting the specified initial conditions, we obtain the equations

$$\begin{bmatrix} 1 & 1 & 1 \\ 0 & -1 & -2 \\ 0 & 1 & 4 \end{bmatrix} \begin{bmatrix} K_1 \\ K_2 \\ K_3 \end{bmatrix} = \begin{bmatrix} 2 \\ -3 \\ 9 \end{bmatrix}$$

The solution of these equations yields the values $K_1 = 2$, $K_2 = -3$, and $K_3 = 3$. Thus, the solution for $v_3(t)$ is

$$v_3(t) = 2 - 3e^{-t} + 3e^{-2t} - e^{-3t} \qquad t \geq 0$$

There are many fairly obvious variations of the above procedure. For example, if the differential equation can be determined using the ladder analysis technique illustrated in Sec. 6-3, then the particular solution can be found by inserting the assumed particular solution form directly into the differential equation. As another example, if it is desired to find a solution for some network variable which is not one of the independent variables used to determine the KCL equations, then, using the particular solutions for the independent variables, and applying KCL, KVL, and the branch terminal relations, we may determine the particular solution for any other network variable. The complete solution is then found as the sum of the particular solution and the complementary solution. Another variation in the procedure occurs for the case where more than one excitation is present, and the form of the excitations is not the same. In such a case, particular solutions for each of the excitations may be separately determined. The sum of the particular solutions and the complementary solution then gives the complete solution, and the unspecified constants of the complementary solution are evaluated to match the given initial conditions.

A procedure similar to that described in the above paragraphs may be applied to find a solution for an nth-order RL circuit. As an example of this, consider the three-mesh RL circuit shown in Fig. 6-6.3. Let us assume that it is desired to find a

Figure 6-6.3   Third-order RL network.

solution for the network variable $i_3(t)$ for the case where the network is excited only by initial conditions. Applying KVL to each of the meshes, we obtain the equations

$$i'_1(t) + i_1(t) - i_2(t) = 0$$
$$3i'_2(t) + 3i_2(t) - i_1(t) - 2i_3(t) = 0 \tag{6-95}$$
$$2i'_3(t) + 2i_3(t) - 2i_2(t) = 0$$

Substituting the expressions $i_1(t) = K_a e^{st}$, $i_2(t) = K_b e^{st}$, and $i_3(t) = K_c e^{st}$ in the above equations, we obtain

$$\begin{bmatrix} s+1 & -1 & 0 \\ -1 & 3s+3 & -2 \\ 0 & -2 & 2s+2 \end{bmatrix} \begin{bmatrix} K_a e^{st} \\ K_b e^{st} \\ K_c e^{st} \end{bmatrix} = \begin{bmatrix} 0 \\ 0 \\ 0 \end{bmatrix} \tag{6-96}$$

Calculating the determinant, we find that the characteristic equation is

$$s^3 + 3s^2 + 2s = s(s+1)(s+2) = 0 \tag{6-97}$$

Thus we see that the expression for $i_3(t)$ may be written in the form

$$i_3(t) = K_1 + K_2 e^{-t} + K_3 e^{-2t} \tag{6-98}$$

The values of the constants K_i are readily found for any specified set of initial conditions. Similar to the situation for capacitors, there may be non-independent inductors which do not contribute to the order of the circuit. In general these will occur when there are loops or cutsets of inductors.

Second-Order and Higher-Order Circuits   Chap. 6

The procedure illustrated above is readily applied to an *RL* circuit of arbitrary complexity. The extension of the procedure to the case where sources are present is identical with that discussed for the *RC* network. The natural frequencies will always lie on the negative real axis (or at the origin) of the complex frequency plane just as they did for the *RC* case. Although in the examples given above we used KCL to determine the network equations for an *RC* circuit and KVL to find the equations for an *RL* circuit, the procedures are readily interchanged, i.e., we may use KCL to find the equations for an *RL* circuit and KVL to find the equations for an *RC* circuit. In doing this, we need simply use the relation

$$\int Ke^{st} = \frac{K}{s}e^{st} \tag{6-99}$$

Note that here it is not necessary to insert the initial condition for the integration of the exponential, since this term disappears when the equation is converted to a differential equation. Thus, in effect, we use an "operator" notation which replaces the process of integration by the operator $1/s$. Similarly, the process of differentiation may be viewed as represented by the operator s. A more rigorous justification of this process will be provided in connection with the development of the Laplace transformation in Chap. 8.

Now let us consider the determination of the solution for an *RLC* network which is of higher than second order. As an example of such a situation, consider the network shown in Fig. 6-6.4. Let us assume that we desire to find an expression

Figure 6-6.4 Third-order *RLC* network.

for the network variable $v_2(t)$ for the case where the circuit is excited only by initial conditions, and that these are $v_2(0) = 1$, $v_2'(0) = 0$, and $v_2''(0) = 0$. We may begin by applying KCL to the network. Assuming that the output of the current source $i_s(t) = 0$, we obtain the equations

$$v_1'(t) + v_1(t) + \tfrac{5}{8}\int [v_1(t) - v_2(t)]\,dt = 0$$

$$v_2'(t) + v_2(t) + \tfrac{5}{8}\int [v_2(t) - v_1(t)]\,dt = 0 \tag{6-100}$$

If we assume the exponential solutions $v_1(t) = A_1 e^{st}$ and $v_2(t) = A_2 e^{st}$ for the node voltage variables, the above equation may be written in the form

$$\begin{bmatrix} s + 1 + \dfrac{5}{8s} & -\dfrac{5}{8s} \\[2ex] -\dfrac{5}{8s} & s + 1 + \dfrac{5}{8s} \end{bmatrix} \begin{bmatrix} A_1 e^{st} \\[2ex] A_2 e^{st} \end{bmatrix} = \begin{bmatrix} 0 \\[2ex] 0 \end{bmatrix} \tag{6-101}$$

The characteristic equation found by setting the determinant of the square matrix in the above relation to zero may be written as

$$s^3 + 2s^2 + \tfrac{9}{4}s + \tfrac{5}{4} = (s + 1)(s + \tfrac{1}{2} + j1)$$
$$(s + \tfrac{1}{2} - j1) = 0 \tag{6-102}$$

Thus, the natural frequencies of the network are $s_1 = -1$, $s_2 = -\tfrac{1}{2} + j1$, and $s_3 = -\tfrac{1}{2} - j1$. The solution for $v_2(t)$ will have the form

$$v_2(t) = K_1 e^{-t} + K_{2a} e^{[-(1/2)+j1]t} + K_{3a} e^{[-(1/2)-j1]t} \tag{6-103}$$

where K_{2a} and K_{3a} are complex conjugates. To facilitate computations, it is usually desirable to combine complex conjugate quantities of the type shown as the second and third terms in the right member of the above expression using the method given in Sec. 6-2. If this is done, we obtain

$$v_2(t) = K_1 e^{-t} + e^{-(1/2)t}(K_2 \cos t + K_3 \sin t) \tag{6-104}$$

The advantage of this type of formulation is that whereas the constants K_{2a} and K_{3a} of (6-103) are complex, the constants K_2 and K_3 of (6-104) are real. To determine the constants K_1, K_2, and K_3, we note that

$$v_2(0) = K_1 + K_2 = 1 \tag{6-105}$$

This is the first of three equations needed to determine the three constants. Successively differentiating the equation for $v_2(t)$ and evaluating it at t equals zero, we obtain two more equations, namely,

$$v_2'(0) = -K_1 - \tfrac{1}{2}K_2 + K_3 = 0$$
$$v_2''(0) = K_1 - \tfrac{3}{4}K_2 - K_3 = 0 \tag{6-106}$$

If we write these three equations in matrix form, we obtain

$$\begin{bmatrix} 1 & 1 & 0 \\ -1 & -\tfrac{1}{2} & 1 \\ 1 & -\tfrac{3}{4} & -1 \end{bmatrix} \begin{bmatrix} K_1 \\ K_2 \\ K_3 \end{bmatrix} = \begin{bmatrix} 1 \\ 0 \\ 0 \end{bmatrix} \tag{6-107}$$

The solution of these equations is readily found to be $K_1 = 1$, $K_2 = 0$, and $K_3 = 1$. Thus, the solution for $v_2(t)$ is

$$v_2(t) = e^{-t} + e^{-(1/2)t} \sin t \qquad t \geq 0 \tag{6-108}$$

The solution for any other variable in the network may be found by a similar process.

The procedure illustrated above is easily applied to *RLC* networks of any order. The network equations may be written by applying either KCL (if the independent variables are chosen as node voltages) or KVL (if the independent variables are chosen as mesh currents). In general, the order of an *RLC* network will be less than or equal to the number of energy-storage elements in the network. Just as was true for the second-order *RLC* case, the natural frequencies of an *n*th-order *RLC* network may be located anywhere in the left half of the complex frequency plane (including the $j\omega$ axis).

6-7 STATE-VARIABLE EQUATIONS FOR SECOND-ORDER CIRCUITS

In this section we shall introduce methods by means of which any second-order network may be described by a first-order matrix differential equation, i.e., by a set of simultaneous first-order differential equations. For example, consider the circuit shown in Fig. 6-7.1. Applying KVL to the two meshes of this circuit, we obtain

$$L_1 \frac{di_1}{dt} + R_1 i_1(t) - R_1 i_2(t) = v(t)$$

$$L_2 \frac{di_2}{dt} + (R_1 + R_2) i_2(t) - R_1 i_1(t) = 0$$

(6-109)

Figure 6-7.1 Second-order *RL* circuit.

If we divide the first equation by L_1, the second by L_2, and rearrange terms, the equations may be written in the following matrix format:

$$
\begin{bmatrix} \dfrac{di_1}{dt} \\ \dfrac{di_2}{dt} \end{bmatrix} = \begin{bmatrix} -\dfrac{R_1}{L_1} & \dfrac{R_1}{L_1} \\ \dfrac{R_1}{L_2} & -\dfrac{R_1 + R_2}{L_2} \end{bmatrix} \begin{bmatrix} i_1(t) \\ i_2(t) \end{bmatrix} + \begin{bmatrix} \dfrac{v(t)}{L_1} \\ 0 \end{bmatrix}
$$

(6-110)

This is an example of a set of simultaneous first-order differential equations having the form

$$\mathbf{f}'(t) = \mathbf{A}\mathbf{f}(t) + \mathbf{u}(t)$$

(6-111)

where, for this example,

$$
\mathbf{f}'(t) = \begin{bmatrix} \dfrac{di_1}{dt} \\ \dfrac{di_2}{dt} \end{bmatrix} \quad \mathbf{A} = \begin{bmatrix} -\dfrac{R_1}{L_1} & \dfrac{R_1}{L_1} \\ \dfrac{R_1}{L_2} & -\dfrac{R_1 + R_2}{L_2} \end{bmatrix} \quad \mathbf{f}(t) = \begin{bmatrix} i_1(t) \\ i_2(t) \end{bmatrix} \quad \mathbf{u}(t) = \begin{bmatrix} \dfrac{v(t)}{L_1} \\ 0 \end{bmatrix}
$$

(6-112)

The relation given in (6-111) is usually referred to as a *first-order matrix differential equation*. Equations of this type play a very important role in the analysis of linear systems. In addition, they are readily written for time-varying and non-linear systems. If the equations describing a network are put in the form given in Eq. (6-111), then the network variables which are the elements of the column matrix $\mathbf{f}(t)$ are referred to as the *state variables*. Correspondingly, the equations are called the *state-variable equations*, or more simply, the *state equations*. In this section we will explore the process for writing the state equations in more detail. For the second-order circuits, which we are considering in this chapter, there are three possible ways in which the state variables may be chosen, depending on the type of energy-storage elements which are present in the circuit. The first of these ways occurs if the second-order circuit is an *RL* circuit, i.e., a circuit which has two inductors as its energy-storage elements. In this case, the state variables will be the inductor currents. The circuit described in the beginning of this section is an example of such a case. In some *RL* circuits it may happen that the conventional mesh currents do not coincide with the inductor currents. In such a case, we may use loops rather than meshes to define the independent current variables.[10] This should be done in such a manner that each loop-current variable coincides with a single inductor current. To do this, we simply select a tree (see Sec. 3-2) such that the inductors are links. The desired loops are then the one-link ones. The process is illustrated by the following example:

Example 6-7.1

It is desired to determine the state equations for the network shown in Fig. 6-7.2(a). The graph for the network is shown in Fig. 6-7.2(b). The tree is indicated by heavy lines. The loops formed by using a single inductor and the appropriate branches are shown

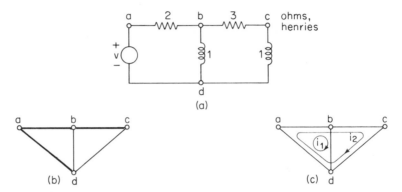

Figure 6-7.2 Determining the state equations for a second-order *RL* network.

in Fig. 6-7.2(c). The direction of the loop currents is, of course, arbitrary. Using KVL and the indicated loop-current reference directions, we may write the equations

$$
\left.\begin{aligned}
2i_1(t) + 2i_2(t) + \frac{di_1}{dt} &= v(t) \\
2i_1(t) + 5i_2(t) + \frac{di_2}{dt} &= v(t)
\end{aligned}\right\} \longrightarrow
\begin{bmatrix} \dfrac{di_1}{dt} \\ \dfrac{di_2}{dt} \end{bmatrix} =
\begin{bmatrix} -2 & -2 \\ -2 & -5 \end{bmatrix}
\begin{bmatrix} i_1(t) \\ i_2(t) \end{bmatrix} +
\begin{bmatrix} v(t) \\ v(t) \end{bmatrix}
$$

This is the desired state-variable formulation.

A procedure similar to that described above may be used to obtain a state-variable formulation for second-order *RC* circuits, i.e., circuits which contain two capacitors as energy-storage elements. For such a network, we may select node-pair voltages in such a manner that each node-pair voltage coincides with the voltage across a capacitor. These node-pair voltage variables are then the state variables. To determine the resulting equations, we need merely draw the graph for the network and form cutsets which each contain a single capacitive branch. To do this, we select

[10]It should be recalled from Chap. 3 that a mesh is a restricted type of loop which has the property that no branches are enclosed by it which are not part of the loop.

a tree (see Sec. 3-5) which contains the capacitors. The one-tree-branch cutsets will then each contain a single capacitor.[11] These cutsets lead directly to a set of KCL equations which contain the capacitor voltages and their derivatives as variables. These are the state equations. The procedure is illustrated by the following example.

Example 6-7.2

To determine the state equations for the network shown in Fig. 6-7.3(a), we first draw the circuit graph as shown in Fig. 6-7.3(b). The two branches which contain capacitors and two node-pair voltage variables are shown in Fig. 6-7.3(c). Using these variables,

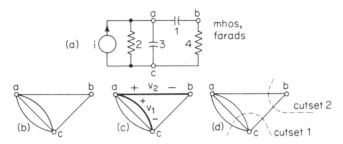

Figure 6-7.3 Determining the state equations for a second-order RC network.

the cutsets defining the state equations are indicated by the dashed lines in Fig. 6-7.3(d). For these cutsets, we obtain the equations

$$\left. \begin{array}{l} 6v_1(t) - 4v_2(t) + 3\dfrac{dv_1}{dt} = i(t) \\[2mm] 4v_2(t) - 4v_1(t) + \dfrac{dv_2}{dt} = 0 \end{array} \right\} \longrightarrow \begin{bmatrix} \dfrac{dv_1}{dt} \\[2mm] \dfrac{dv_2}{dt} \end{bmatrix} = \begin{bmatrix} -2 & \dfrac{4}{3} \\[2mm] 4 & -4 \end{bmatrix} \begin{bmatrix} v_1(t) \\[2mm] v_2(t) \end{bmatrix} + \begin{bmatrix} \dfrac{i(t)}{3} \\[2mm] 0 \end{bmatrix}$$

This is the desired state-variable formulation.

To determine the state equations for the third type of second-order network, namely, the type in which one of the energy-storage elements is a capacitor and the other is an inductor, we may separately treat two cases. The first of these occurs when the second-order circuit consists of a single mesh or a single node-pair. In such a case, the state variables are readily chosen as the voltage across the capacitor and the current through the inductor. If the circuit contains a single mesh, as shown in Fig. 6-7.4, then the first state equation is found by applying KVL. The result is

$$v_s(t) = L\frac{di}{dt} + Ri(t) + v_C(t) \tag{6-113}$$

Figure 6-7.4 Second-order RLC network with a single mesh.

[11]It may also be necessary to use equivalent sources and to replace groups of resistors by a single equivalent resistor.

The second state equation is found from the relation for the terminal variables of the capacitor. It is

$$i(t) = C \frac{dv_C}{dt} \qquad (6\text{-}114)$$

Solving each of the above equations for their first-order derivative term, they may be written in the following matrix form.

$$\begin{bmatrix} \dfrac{di}{dt} \\ \dfrac{dv_C}{dt} \end{bmatrix} = \begin{bmatrix} -\dfrac{R}{L} & -\dfrac{1}{L} \\ \dfrac{1}{C} & 0 \end{bmatrix} \begin{bmatrix} i(t) \\ v_C(t) \end{bmatrix} + \begin{bmatrix} \dfrac{v_s(t)}{L} \\ 0 \end{bmatrix} \qquad (6\text{-}115)$$

Similarly, if the circuit contains a single node-pair, as shown in Fig. 6-7.5, we may

Figure 6-7.5 Second-order RLC network with a single node pair.

apply KCL to find one of the state equations, and we may use the terminal relation for the inductor as the other. For this circuit, after rearranging terms, we readily find the state equations to be

$$\begin{bmatrix} \dfrac{dv}{dt} \\ \dfrac{di_L}{dt} \end{bmatrix} = \begin{bmatrix} -\dfrac{G}{C} & -\dfrac{1}{C} \\ \dfrac{1}{L} & 0 \end{bmatrix} \begin{bmatrix} v(t) \\ i_L(t) \end{bmatrix} + \begin{bmatrix} \dfrac{i_s(t)}{C} \\ 0 \end{bmatrix} \qquad (6\text{-}116)$$

The second case that occurs for second-order circuits, in which one of the energy-storage elements is a capacitor and the other is an inductor, is the case in which more than one variable must be used to write the network equations, i.e., the situation in which the network consists of more than a single loop or more than a single node-pair. In such a case, one of the state variables will be the current through the inductor, while the other will be the voltage across the capacitor. To achieve a state-variable formulation for such a network, we must apply KCL to a cutset which contains the capacitor but which does not contain the inductor, and apply KVL to a loop which includes the inductor but does not include the capacitor. Other KCL and KVL equations and the terminal relations for the non-energy storage elements of the circuit may then be used to eliminate any undesired variables. The procedure is illustrated by the following example.

Example 6-7.3

It is desired to write the state equations for the circuit shown in Fig. 6-7.6(a). This circuit has the graph shown in Fig. 6-7.6(b). If we define node-pair voltages for the branches numbered 2, 3, and 4 in Fig. 6-7.6(b), as shown in Fig. 6-7.6(c), we may apply KCL to the cutset (cutset 1) indicated by the dashed line in this figure. Thus we obtain

$$\frac{1}{4} \frac{dv_1}{dt} + 2v_1(t) - 2v_2(t) = 0 \qquad (6\text{-}117)$$

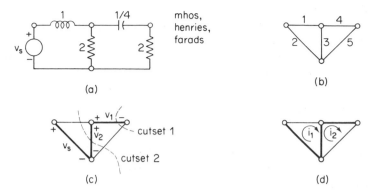

(a)

(b)

(c)

cutset 1

cutset 2

(d)

Figure 6-7.6 Determining the state equations for a second-order *RLC* network.

Now let us assign mesh reference current directions to the two meshes shown in Fig. 6-7.6(d). Applying KVL to the mesh which contains the inductor, we may write

$$\frac{di_1}{dt} + \frac{1}{2} i_1(t) - \frac{1}{2} i_2(t) = v_s(t) \tag{6-118}$$

The two equations derived above must now be transformed so that they contain only the variables $v_1(t)$ and $i_1(t)$ (and their derivatives), and $v_s(t)$. Thus, the variables $i_2(t)$ and $v_2(t)$ must be eliminated. This is easily done by substituting other KCL and KVL relations into the original quations. For example, applying KVL around the right mesh of the circuit, we obtain

$$v_1(t) + \tfrac{1}{2} i_2(t) = \tfrac{1}{2} i_1(t) - \tfrac{1}{2} i_2(t) \tag{6-119}$$

Solving this relation for $i_2(t)$ and substituting it into (6-118), we obtain

$$\frac{di_1}{dt} + \frac{1}{4} i_1(t) + \frac{v_1(t)}{2} = v_s(t) \tag{6-120}$$

This is the first state equation. Similarly, to eliminate the variable $v_2(t)$ in (6-117), we note that the resistor in branch 3 has a terminal relation that may be written in the form

$$v_2(t) = \tfrac{1}{2} i_1(t) - \tfrac{1}{2} i_2(t) \tag{6-121}$$

If we solve (6-119) for $i_2(t)$ and substitute it into (6-121), then we have a relation giving $v_2(t)$ as a function of $i_1(t)$ and $v_1(t)$. Substituting this result in (6-117), we obtain a second state equation, namely,

$$\frac{1}{4} \frac{dv_1}{dt} + v_1(t) - \frac{1}{2} i_1(t) = 0 \tag{6-122}$$

From (6-120) and (6-122), our state equations are easily put in the form

$$\begin{bmatrix} \dfrac{di_1}{dt} \\[2mm] \dfrac{dv_1}{dt} \end{bmatrix} = \begin{bmatrix} -\dfrac{1}{4} & -\dfrac{1}{2} \\[2mm] 2 & -4 \end{bmatrix} \begin{bmatrix} i_1(t) \\[2mm] v_1(t) \end{bmatrix} + \begin{bmatrix} v_s(t) \\[2mm] 0 \end{bmatrix} \tag{6-123}$$

This is obviously of the general form specified in (6-111).

The examples given above illustrating the determination of the state equations for second-order circuits have not included any cases where the circuits contain more than two independent variables, or any cases where the circuits contain dependent sources. Such circuits, however, may be reduced to circuits similar to those discussed in this section by using techniques similar to those presented in Sec. 6-3, in which the resistors and the dependent sources are replaced by equivalent resistors and equivalent Thevenin and Norton sources.

One of the major reasons for the importance of the state-variable form of the equations describing a given circuit is that such equations are readily solved by numerical integration as implemented on a digital computer. A subroutine for performing such an integration is described in Appendix C.

Now let us consider finding the state-variable equations for higher-order circuits. The development of completely general procedures for doing this is a very broad subject which can easily fill a complete book.[12] Here we shall content ourselves with pointing out that, for many networks, choosing the voltages across the capacitors and the currents through the inductors as the state variables, and appropriately applying KCL and KVL, we may obtain the required set of equations. Some examples of this procedure are given in the problems.

6-8 DUAL AND ANALOG CIRCUITS

Frequently the analysis of a given circuit configuration can be simplified by recognizing that the circuit can be described by an equation or by a set of equations that have the same form as another circuit whose solution is already known. For example, consider the series *RLC* circuit containing an ICVS shown in Fig. 6-8.1. Applying KVL to this circuit we may write

$$v_a(t) = v_1(t) + v_2(t) + v_3(t) + v_4(t) \tag{6-124}$$

Figure 6-8.1   Series *RLC* circuit with an ICVS.

After the substitution of the relations for the branch terminal variables the above relation becomes

$$v_a(t) = R_a i_a(t) + L_a \frac{di_a}{dt} + \frac{1}{C_a} \int i_a(t)\, dt + r_a i_a(t) \tag{6-125}$$

Now consider the parallel RLC circuit which includes a VCIS shown in Fig. 6-8.2.

Figure 6-8.2   Parallel *RLC* circuit with a VCIS.

[12]See, for example, R. A. Rohrer, *Circuit Theory: An Introduction to the State Variable Approach*, McGraw-Hill Book Company, New York, 1970.

Applying KCL to this circuit we obtain

$$i_b(t) = i_1(t) + i_2(t) + i_3(t) + i_4(t) \qquad (6\text{-}126)$$

After the substitution of the relations for the branch terminal variables the above equation may be written

$$i_b(t) = G_b v_b(t) + C_b \frac{dv_b}{dt} + \frac{1}{L_b} \int v_b(t) \, dt + g_b v_b(t) \qquad (6\text{-}127)$$

Now let us compare Eqs. (6-124) and (6-125) (for the circuit shown in Fig. 6-8.1) with Eqs. (6-126) and (6-127) (for the circuit shown in Fig. 6-8.2). Clearly, Eqs. (6-124) and (6-126) have the same form and may be directly obtained from each other by interchanging the current and voltage variables (and the subscripts a and b). Similarly, Eqs. (6-125) and (6-127) have the same form and may be directly obtained from each other by interchanging the following quantities: (1) voltage and current, (2) resistance and conductance, and (3) inductance and capacitance; i.e., if we interchange the defining symbols v and i, R and G, and L and C (and the subscripts a and b) we may derive one set of equations from the other. In general, circuits which are described by equations having the same form, but in which the coefficients and variables are interchanged as described above, are said to be *dual* to each other. More specifically, in dual circuits each mesh which provides a KVL equation in one circuit corresponds to a node (and the branches which connect to it) which provides a KCL equation in the dual circuit and vice versa. Thus, elements in series in one circuit correspond with elements in parallel in the dual circuit. A summary of the properties of dual circuits is given in Table 6-8.1.

TABLE 6-8.1 Properties of Dual Circuits

KVL Equations	KCL Equations
Mesh Current	Node Voltage
Voltage Source	Current Source
Series Branch	Parallel Branch
Inductor	Capacitor
Capacitor	Inductor
Resistor (ohms)	Resistor (mhos)
ICVS	VCIS
Controlling Mesh Current for an ICVS	Controlling Node Voltage for a VCIS
Output Voltage for an ICVS	Output Current for a VCIS

Now let us develop a general procedure for finding the dual of a given circuit. To do this we may use the following graphical construction procedure.[13]

1. In the center of each mesh of the given network place a node symbol. Number these symbols sequentially. In addition, place a ground reference node symbol

[13]M. F. Gardner and G. L. Barnes, *Transients in Linear Systems*, John Wiley & Sons, Inc., New York, 1942.

outside of all the network meshes. Finally, draw a duplicate set of node symbols exterior to the network to provide the framework for the construction of the dual network.

As an example of the above step, consider the two-mesh circuit shown in Fig. 6-8.3. The addition of the node symbols in shown in Fig. 6-8.4(a). The node symbols have been redrawn in Fig. 6-8.4(b).

Figure 6-8.3 Two-mesh circuit.

Figure 6-8.4 Finding the dual of a two-mesh circuit.

2. Draw lines between the node symbols in such a way that each line crosses one network element. As each line is drawn add an element to the dual network between corresponding nodes. Use the relations of Table 6-8.1 to determine the type of dual element. The reference polarity of sources in the dual network is readily observed by noting that a voltage source with a reference polarity which produces a positive mesh reference current has as its dual a current source whose reference direction is towards the numbered node symbol and away from the ground node symbol.

This step is illustrated in Fig. 6-8.4(c) and (d). In part (c) the lines drawn between the node symbols are shown. For convenience these have been numbered from 1 to 5. The corresponding elements in the dual network are shown in part (d) of the figure.

Note that since the ICVS has a reference polarity which opposes the mesh current $i_2(t)$ shown in Fig. 6-8.3 the dual element, which is a VCIS, has a reference polarity directed away from node 2. To verify the duality of the two circuits we may write the equations for the circuit shown in Fig. 6-8.3 as

$$v_a(t) = i_1(t) + \frac{1}{2} \int i_1(t)\, dt - \frac{1}{2} \int i_2(t)\, dt$$

$$0 = 4i_1(t) - \frac{1}{2} \int i_1(t)\, dt + \frac{1}{2} \int i_2(t)\, dt + 3\frac{di_2}{dt}$$

(6-128)

The equations for the dual network shown in Fig. 6-8.4(d) are readily found to be

$$i_a(t) = v_1(t) + \frac{1}{2} \int v_1(t)\, dt - \frac{1}{2} \int v_2(t)\, dt$$

$$0 = 4v_1(t) - \frac{1}{2} \int v_1(t)\, dt + \frac{1}{2} \int v_2(t)\, dt + 3\frac{dv_2}{dt}$$

(6-129)

The similarity of the two sets of equations is readily apparent. The method described above may be applied to any network which has a planar graph. Since it is based on a transformation of the graph of the network, it is valid not only for linear time-invariant networks but also for non-linear and time-varying ones.

In the above paragraphs we saw that, given a circuit described by mesh equations we could construct a dual of the circuit in such a way that its node equations had the identical form as the mesh equations of the original circuit. Now let us treat a similar development in which we consider the use of circuits to serve as *analogs* for non-electrical types of systems. Specifically, we will consider mechanical systems. Elementary mechanical systems which have only translation may be considered as being composed of three types of elements, i.e., an element which is characterized only by mass and which is represented by the symbol shown in Fig. 6-8.5(a), an element which is characterized only by friction and which is represented by the symbol shown in Fig. 6-8.5(b), and an element which has a spring coefficient and which is represented

Figure 6-8.5 Elements of a mechanical system.

by the symbol shown in Fig. 6-8.5(c). The defining relations for these three types of elements, using $f(t)$ for force and $v(t)$ for velocity, are

$$f(t) = M\frac{dv}{dt} \qquad \text{Mass (Newton's Law)}$$

$$f(t) = Bv(t) \qquad \text{Friction}$$

$$f(t) = K\int_0^t v(\tau)\, d\tau + f(0) \qquad \text{Spring Coefficient}$$

(6-130)

where $f(0)$ is an initial condition term representing the force on the spring at $t = 0$. Now let us consider a simple mechanical system using these elements as shown in Fig. 6-8.6(a). The equation of motion for this system is found by summing the forces acting on the block. This equation has the form

$$f(t) = Bv(t) + M\frac{dv}{dt} + K\int_0^t v(\tau)\,d\tau \tag{6-131}$$

Figure 6-8.6 Simple mechanical system.

where it is assumed that the spring is not in tension at $t = 0$. This equation, however, is similar to one we have frequently seen in this text. Specifically, if we apply KVL to a series RLC circuit comprised of a resistor of value B ohms, an inductor of value M henries and a capacitor of value $1/K$ farads as shown in Fig. 6-8.6(b), we obtain the relation

$$v(t) = Bi(t) + M\frac{di}{dt} + K\int_0^t i(\tau)\,d\tau \tag{6-132}$$

where $v(t)$ is the voltage of the source. Comparing Eqs. (6-131) and (6-132) we may conclude that the circuit shown in Fig. 6-8.6(b) will function as an analog of the mechanical circuit shown in Fig. 6-8.6(a), where, more specifically, we note that the applied voltage is the analog of the applied force, the mesh current is the analog of the velocity, and the resistance, inductance, and reciprocal capacitance are respectively the analogs of friction, mass, and spring coefficient. The different mechanical and electrical variables for such an analog are summarized in Table 6-8.2.

TABLE 6-8.2 Analogous Electrical and Mechanical Quantities

Mechanical Quantities	Electrical Quantities
Force	Voltage
Velocity	Current
Distance	Charge
Mass	Inductance
Friction	Resistance
Spring Coefficient	Reciprocal Capacitance

Analog electrical quantities are readily derived for a wide variety of types of non-electrical systems such as mechanical systems with rotational displacement, hydraulic systems, etc. In addition, engineering studies are frequently made using

analogs to represent large complex systems such as manufacturing plants, airplanes, watersheds, etc. Such large scale projects are generally made using analog computers which are specifically designed to facilitate the application of basic electrical analogs to a wide range of systems.

6-9 CONCLUSION

In this and the preceding chapter we have presented a set of general methods which may be applied to find the time-domain solution for any of the variables of a second- or higher-order network. The techniques may be applied to any linear time-invariant network.

This chapter marks the end of a fundamental division of the material contained in this book. In this and the preceding chapters we have presented many techniques concerned with the analysis of a wide range of circuits. All the network variables associated with these circuits have been specified as a function of time. Thus, we may collectively refer to these first six chapters as treating the *time-domain* analysis of circuits. In the chapters that follow we shall consider a quite different approach to network analysis. It may be categorized as the *frequency-domain* analysis of circuits. The full implications of the term "frequency-domain" will become apparent as we progress through these chapters.

PROBLEMS

Problem 6-1 (Sec. 6-1)

In the circuit shown in Fig. P6-1, find the roots of the characteristic equation [for the differential equation for $v(t)$] in terms of R, L, and C.

Figure P6-1

Problem 6-2 (Sec. 6-1)

For the circuit shown in Fig. P6-2, $v(0) = 1$ V and $v'(0) = 1$ V/s. Find:
(a) $v(t)$ for $t \geq 0$.
(b) $i_C(t)$ for $t \geq 0$.
(c) $i_L(t)$ for $t \geq 0$.

ohms, henries, farads

Figure P6-2

Problem 6-3 *(Sec. 6-1)*

In the circuit shown in Fig. P6-3, $i_L(0) = \frac{3}{5}$ A and $v_C(0) = 0$. Find $i_2(t)$.

ohms, henries, farads

Figure P6-3

Problem 6-4 *(Sec. 6-1)*

In the circuit shown in Fig. P6-4, find the value of R such that Case II applies, then let $v(0) = 1$ V and $v'(0) = 1$ V/s, and find $v(t)$ for $t \geq 0$, for the value of R that you have calculated.

ohms, henries, farads

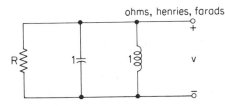

Figure P6-4

Problem 6-5 *(Sec. 6-1)*

For the circuit shown in Fig. P6-5, assume that Case I applies. Find a relation between $i(0)$ and $v_C(0)$ such that only the root with the smaller magnitude appears in the response. [*Hint*: Remember that any fundamental circuit variable will have the form $K_1 e^{s_1 t} + K_2 e^{s_2 t}$. Let $v_C(t)$ have this form and remember that $i(t) = C \, dv_C(t)/dt$.]

Figure P6-5

Problem 6-6 *(Sec. 6-2)*

In the circuit shown in Fig. P6-6, $v(0) = 1$ V and $v'(0) = 0$. Find:
(a) $v(t)$ for $t \geq 0$.
(b) $i_L(t)$ for $t \geq 0$.
(c) $i_C(t)$ for $t \geq 0$.

ohms, henries, farads

Figure P6-6

Problem 6-7 *(Sec. 6-2)*

In the circuit shown in Fig. P6-7, the switch closes at $t = 0$. Prior to this, the capacitor was charged to 3 V. Find $v(t)$ for all $t \geq 0$.

ohms, henries, farads

Figure P6-7

Problem 6-8 (Sec. 6-2)

For the circuit shown in Fig. P6-1:

(a) Plot the roots of the characteristic equation in the complex s plane as R is varied and L and C remain constant.

(b) Plot the roots as C is varied and R and L remain constant.

(c) Plot the roots as L is varied and R and C remain constant.

Problem 6-9 (Sec. 6-2)

For the circuit shown in Fig. P6-5:

(a) Plot the roots of the characteristic equation in the complex s plane as R is varied and L and C remain constant.

(b) Plot the roots as C is varied and R and L remain constant.

(c) Plot the roots as L is varied and R and C remain constant.

Problem 6-10 (Sec. 6-2)

The graph in Fig. P6-10 has the form $f(t) = Ke^{-\alpha t} \cos(\omega t)$. Find values for K, α, and ω.

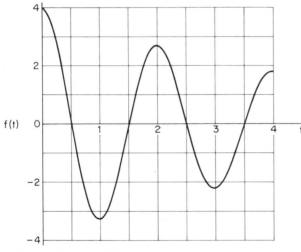

Figure P6-10

Problem 6-11 (Sec. 6-3)

(a) For the circuit shown in Fig. P6-11, find $v(t)$ for $t \geq 0$ for the initial conditions $i_1(0) = 1$ and $i_2(0) = 0$ A by writing mesh equations.

(b) Repeat the problem using the ladder analysis method.

ohms, henries

Figure P6-11

Problem 6-12 (Sec. 6-3)

For the circuit shown in Fig. P6-12, find $i(t)$ for $t \geq 0$ for the initial conditions $v_1(0) = 1$ and $v_2(0) = 1$ V by writing node equations.

Figure P6-12

Problem 6-13 (Sec. 6-3)

For the circuit shown in Fig. P6-12, find $v_2(t)$ for the initial conditions $v_1(0) = 1$ and $v_2(0) = 1$ V by using the ladder analysis method.

Problem 6-14 (Sec. 6-3)

(a) In the circuit shown in Fig. P6-14, find $v(t)$ for $t \geq 0$ for the initial conditions $v_C(0) = 1$ V and $i_L(0) = 1$ A by writing node equations.

(b) Repeat the problem using the ladder analysis method.

ohms, henries, farads

Figure P6-14

Problem 6-15 (Sec. 6-3)

In the circuit shown in Fig. P6-14, let $C = 1$ F, $L = \frac{1}{4}$ H, and $R = \frac{1}{2}\,\Omega$. Find an expression for $v(t)$ for $t \geq 0$ for the initial conditions $v_C(0) = 2$ V and $i_L(0) = 1$ A.

Problem 6-16 (Sec. 6-3)

In the circuit shown in Fig. P6-16, $v(0)$ equals 1 V and $v'(0)$ equals 0. Find $v(t)$ for all $t \geq 0$.

Problem 6-17 (Sec. 6-3)

Find the characteristic equation for the circuit shown in Fig. P6-17.

Figure P6-16   Figure P6-17

Problem 6-18 (Sec. 6-3)

In the circuit shown in Fig. P6-18, find $i_L(t)$ for the initial conditions $i_L(0) = 1$ A, $i'_L(0) = 1$.

ohms, henries

Figure P6-18

Problem 6-19 (Sec. 6-3)

In the circuit shown in Fig. P6-19, find $i_L(t)$ for the initial conditions $i_L(0) = 1$ and $i'_L(0) = 2$.

ohms, henries, farads

Figure P6-19

Problem 6-20 (Sec. 6-3)

For the circuit shown in Fig. P6-20, find the loci of the roots of the characteristic equation as K is varied.

mhos, henries, farads

Figure P6-20

Problem 6-21 (Sec. 6-3)

Without solving the circuit, find $(di_1/dt)(0)$ and $(di_2/dt)(0)$ in the network shown in Fig. P6-21. Assume that $i_1(0) = 1$ and $i_2(0) = 2$ A.

ohms, henries

Figure P6-21

Problem 6-22 (Sec. 6-4)

In the circuit shown in Fig. P6-22, $v(0) = 0$ and $i_L(0) = 0$. Find $v(t)$, $i_C(t)$, and $i_L(t)$ for $t \geq 0$.

ohms, henries, farads

Figure P6-22

Problem 6-23 (Sec. 6-4)

For the circuit shown in Fig. P6-23, $v(0) = 0$ and $i_L(0) = 0$. Find $v(t)$, $i_L(t)$, and $i_C(t)$ for $t \geq 0$.

ohms, henries, farads

Figure P6-23

Problem 6-24 (Sec. 6-4)

In the circuit shown in Fig. P6-24, $v_1(0)$ equals 1 V and $i(0)$ equals 1 A. Find $v_1'(0)$, $v_2(0)$, and $v_2'(0)$.

mhos, henries, farads

Figure P6-24

Problem 6-25 (Sec. 6-4)

Find the complete response for $v_2(t)$ in the circuit shown in Fig. P6-25. Assume that $v_1(0) = 0$ and $v_2(0) = 0$.

mhos, farads

Figure P6-25

Problem 6-26 (Sec. 6-4)

In the circuit shown in Fig. P6-26, find $i_L(t)$ for the initial conditions $i_L(0) = 1$A and $v_C(0) = 1$ V.

mhos, henries, farads

Figure P6-26

Problem 6-27 (Sec. 6-4)

The circuit shown in Fig. P6-27 is valid for $t \geq 0$. Find an expression for $v(t)$. Assume the initial conditions are $v(0) = 1$ V and $v'(0) = 0$.

ohms, henries, farads

Figure P6-27

Second-Order and Higher-Order Circuits   Chap. 6

Problem 6-28 *(Sec. 6-4)*

Find $v(t)$ for $t \geq 0$ in the circuit shown in Fig. P6-28. Assume that $v_C(0) = 0$ and $i_L(0) = 0$.

ohms, henries, farads

Figure P6-28

Problem 6-29 *(Sec. 6-4)*

For the circuit shown in Fig. P6-29, find $v(t)$ for $t \geq 0$ if $i_{L1}(0) = 0$ and $i_{L2}(0) = 0$.

ohms, henries

Figure P6-29

Problem 6-30 *(Sec. 6-4)*

In the circuit shown in Fig. P6-30, $v(0) = 0$ and $v'(0) = 0$. Find $v(t)$, $i_C(t)$, and $i_L(t)$ for $t \geq 0$.

ohms, henries, farads

Figure P6-30

Problem 6-31 *(Sec. 6-4)*

In the circuit shown in Fig. P6-31, $v(0) = 0$ and $v'(0) = 0$. Find $v(t)$, $i_C(t)$, and $i_L(t)$ for $t \geq 0$.

ohms, henries, farads

Figure P6-31

Problem 6-32 *(Sec. 6-4)*

In the circuit shown in Fig. P6-32, starting at t equals zero, the voltage $v(t) = -e^{-3t}$ V is applied. The current that results is $i(t) = e^{-3t} + e^{-t} \cos(2t)$ A. Find the values of the net-

Figure P6-32

work elements R (in ohms), C (in farads), and L (in henries). Note that there is an initial condition on $i(t)$, namely, $i(0) = 2$ A.

Problem 6-33 (Sec. 6-4)

The switch in the circuit shown in Fig. P6-33 moves from position a to position b at $t = 0$. The circuit is in a steady state for $t < 0$. By measurement, it is found that $(dv_{C1}/dt)(0+) = -1$ V/s. Find C_1.

ohms, farads

Figure P6-33

Problem 6-34 (Sec. 6-4)

In the circuit shown in Fig. P6-34, all variables are constant before the switch opens at $t = 0$. Without solving the network equations, find the values of v_1, v_2, v_1', v_2', and v_2'', all evaluated at $t = 0+$.

ohms, henries, farads

Figure P6-34

Problem 6-35 (Sec. 6-4)

In the circuit shown in Fig. P6-35, the switch is moved from a to b at $t = 0$. The circuit is in a steady state for $t < 0$. Find $v(0+)$, $(dv/dt)(0+)$, and $(d^2v/dt^2)(0+)$.

ohms, henries, farads

Figure P6-35

Problem 6-36 (Sec. 6-4)

The switch in the circuit shown in Fig. P6-36 closes at $t = 0$. Assume zero initial conditions and find $(di_L/dt)(0+)$, and $(di_C/dt)(0+)$.

ohms, henries, farads

Figure P6-36

Problem 6-37 (Sec. 6-4)

In the circuit shown in Fig. P6-37, the switch opens at $t = 0$. Find $v_{sw}(0+)$ and $v_{sw}(\infty)$ (where v_{sw} is the voltage across the switch).

ohms, farads

Figure P6-37

Problem 6-38 (Sec. 6-5)

For the active-RC circuit shown in Fig. P6-38, find the differential equation relating $v_1(t)$ and $v_2(t)$.

ohms, farads

Figure P6-38

Problem 6-39 (Sec. 6-5)

In Fig. 6-5.1, assume that the elements have the following values: $R_1 = 75$ kΩ, $C_2 = 0.1$ μF, $R_3 = 3.94$ kΩ, $C_4 = 0.95$ μF, and $K = 10$. Find the output $v_2(t)$ for the case where $v_1(t) = u(t)$. Assume that the initial conditions are zero.

Problem 6-40 (Sec. 6-5)

Find the output voltage $v_2(t)$ for the active-RC circuit shown in Fig. P6-40 for the case where $v_1(t) = tu(t)$. Let all resistors have a value of unity, all capacitors have a value of $\sqrt{2}$, and let $K = 4 - \sqrt{2}$. Assume that all initial conditions are zero.

Figure P6-40

Problem 6-41 (Sec. 6-5)

Repeat Problem 6-40 for the case where all resistors have a value of 1.414 kΩ, all capacitors have a value of 0.1 μF, and K has a value of 3.8586.

Problem 6-42 (Sec. 6-5)

For the active-RC circuit discussed in Example 6-5.2, find expressions for the initial conditions $v_2(0+)$ and $v_2'(0+)$ as functions of the initial conditions $v_{C4}(0+)$ and $v_{C5}(0+)$ on the capacitors.

Problem 6-43 (Sec. 6-5)

For the active-RC circuit discussed in Example 6-5.2, let $v_1(t) = u(t)$ and find $v_2(t)$. Assume that all elements have unity value and that all initial conditions are zero.

Problem 6-44 (Sec. 6-5)

In Fig. 6-5.2, assume that the elements have the following values: $R_1 = 1125\ \Omega$, $R_2 = 11.25$ kΩ, $R_3 = 1020\ \Omega$, $C_4 = 2.2\ \mu F$, and $C_5 = 0.1\ \mu F$. Find the output $v_2(t)$ for the case where $v_1(t) = u(t)$. Assume that the initial conditions are zero.

Problem 6-45 (Sec. 6-5)

Find the output voltage $v_2(t)$ for the active-RC circuit shown in Fig. P6-45 for the case where $v_1(t) = tu(t)$. Let all capacitors have a value of unity, and let $R_1 = R_4 = 1$ and $R_5 = 2$. Assume that all initial conditions are zero.

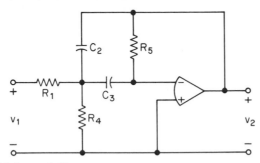

Figure P6-45

Problem 6-46 (Sec. 6-5)

(a) For the automotive ignition system described in Example 6-5.3, find the value of the current reached in Phase 1 if the engine is operated at 5000 rpm and the points have a dwell angle (when the points are closed) of 30°. Assume that the initial conditions are zero.

(b) Use the value of the current found in part (a) to determine the firing time t_f in Phase 2.

Problem 6-47 (Sec. 6-5)

Find an expression for the primary (low-voltage) current in the ignition coil in the automotive ignition system described in Example 6-5.3 in the time period after the spark plug has fired. Assume that the resistance of the secondary (high-voltage) wiring and of the spark gap may be approximated as 160 kΩ. The initial conditions may be taken from the discussion of Phase 2 in the example.

Problem 6-48 (Sec. 6-6)

Find the characteristic equation for the circuit shown in Fig. P6-48.

Figure P6-48

Problem 6-49 (Sec. 6-6)

For the circuit shown in Fig. P6-49, find differential equations relating:

(a) $v_3(t)$ and $i_s(t)$.

(b) $v_2(t)$ and $i_s(t)$.

Second-Order and Higher-Order Circuits   Chap. 6

(c) $v_1(t)$ and $i_s(t)$.
Assume that all elements are unity-valued.

Figure P6-49

Problem 6-50 (Sec. 6-6)

Find $v_3(t)$ for $t \geq 0$ in the circuit shown in Fig. 6-6.1. Assume that $v_3(0) = 1$, $v_3'(0) = 0$, $v_3''(0) = 0$, and $i_s(t) = 0$.

Problem 6-51 (Sec. 6-6)

Find the particular solution for $v_3(t)$ for $t \geq 0$ for the circuit shown in Fig. P6-51 for $i_s(t) = 2e^{-t}$ A.

Figure P6-51

Problem 6-52 (Sec. 6-6)

Find the characteristic equation of the circuit shown in Fig. P6-52. Assume that all elements are unity-valued.

Figure P6-52

Problem 6-53 (Sec. 6-6)

For the circuit shown in Fig. P6-53, find differential equations relating:
(a) $v(t)$ and $i_3(t)$.
(b) $v(t)$ and $i_2(t)$.
(c) $v(t)$ and $i_1(t)$.
Assume that all elements are unity-valued.

Figure P6-53

Problem 6-54 (Sec. 6-6)

In the circuit shown in Fig. P6-54, find $i_{L1}(t)$ for $t \geq 0$. Let $i_{L1}(0) = 1$, $i_{L2}(0) = -1$, and $i_{L3}(0) = 1$ A.

Figure P6-54

Problem 6-55 (Sec. 6-6)

For the circuit shown in Fig. P6-55, find $i(t)$ for $t \geq 0$. Assume that the network is unexcited for $t < 0$.

Figure P6-55

Problem 6-56 (Sec. 6-6)

Solve for $v_2(t)$ in the network shown in Fig. 6-6.4 for the following initial conditions: $v_2(0) = 0$, $v_2'(0) = 1$, $v_2''(0) = 0$. [Assume that $i_s(t) = 0$.]

Problem 6-57 (Sec. 6-6)

Find a solution for $v_2(t)$ for the network shown in Fig. 6-6.4 for the case where $i_s(t) = e^{-2t}$ A, and all the initial conditions are zero.

Problem 6-58 (Sec. 6-7)

Find the state-variable equations for the circuit shown in Fig. P6-58.

Figure P6-58

Problem 6-59 (Sec. 6-7)

Find the state-variable equations for the circuit shown in Fig. P6-59.

Figure P6-59

Problem 6-60 (Sec. 6-7)

Find the state-variable equations for the circuit shown in Fig. P6-60.

Second-Order and Higher-Order Circuits   Chap. 6

Figure P6-60

Problem 6-61 (Sec. 6-7)

Find the state-variable equations for the circuit shown in Fig. P6-61.

Figure P6-61

Problem 6-62 (Sec. 6-7)

Find the state-variable equations for the circuit shown in Fig. P6-62.

Figure P6-62

Problem 6-63 (Sec. 6-7)

Find the state-variable equations for the circuit shown in Fig. P6-63.

Figure P6-63

Problem 6-64 (Sec. 6-7)

Find the state-variable equations for the circuit shown in Fig. P6-64.

Figure P6-64

Problem 6-65 (Sec. 6-7)

Write the state-variable equations for the *RC* network shown in Fig. P6-65. Assume that all the elements have unity value.

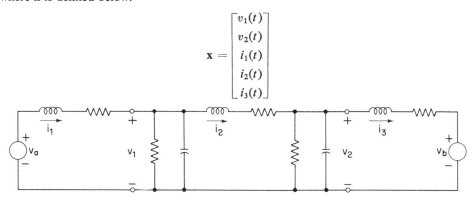

Figure P6-65

Problem 6-66 (Sec. 6-7)

Write the state-variable equations for the *RL* network shown in Fig. P6-66. Assume that all the network elements have unity value.

Figure P6-66

Problem 6-67 (Sec. 6-7)

Write the state-variable equations in matrix form for the network shown in Fig. P6-67. Assume that all elements have unity value. The equations should have the form $\mathbf{x}' = \mathbf{Ax} + \mathbf{u}$, where $\mathbf{x}$ is defined below.

$$\mathbf{x} = \begin{bmatrix} v_1(t) \\ v_2(t) \\ i_1(t) \\ i_2(t) \\ i_3(t) \end{bmatrix}$$

Figure P6-67

Problem 6-68 (Sec. 6-7)

A three-winding transformer may be represented as shown in Fig. P6-68. The winding resistances are considered as lumped external resistances, and the self and mutual inductances are given by the inductance matrix shown in the figure. Find the state-variable equations for the circuit.

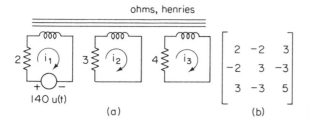

ohms, henries

$$\begin{bmatrix} 2 & -2 & 3 \\ -2 & 3 & -3 \\ 3 & -3 & 5 \end{bmatrix}$$

(a)   (b)   **Figure P6-68**

Problem 6-69 (Sec. 6-7)

A distributed *RC* three-terminal network element is frequently formed in integrated circuit applications by depositing alternate layers of conducting, dielectric, and resistive material as shown in Fig. P6-69(a). A circuit model for such an element is shown in Fig. P6-69(b). Assume that ten sections are used in the model, and that each of the capacitors has a value of 0.15 F and each of the resistors has a value of 1 Ω. Find the state-variable equations for the network.

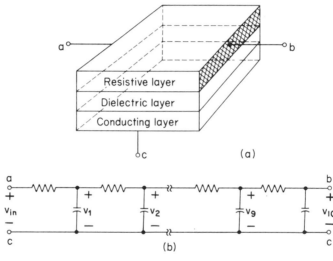

Figure P6-69

Problem 6-70 (Sec. 6-8)

Apply the rules given in Sec. 6-8 for finding a dual network to determine the dual of the network in Fig. 6-8.4(d). Show that the resulting dual is actually the network shown in Fig. 6-8.3. That is, show that, in general, taking the dual of a dual produces the original network.

Problem 6-71 (Sec. 6-8)

Construct a table similar to Table 6-8.2, giving electrical and mechanical quantities which are analogous, by using a parallel *RLC* analog circuit for the mechanical system shown in Fig. 6-8.6(a).

Problem 6-72 (Sec. 6-8)

(a) Find the differential equation relating the input voltage $v_1(t)$ and the output voltage $v_2(t)$ in the network shown in Fig. P6-72.

(b) Find a network which is the dual of the one shown in Fig. P6-72. Use a source conversion

ohms, henries, farads

Figure P6-72

to define an input voltage $v_1(t)$ for the dual network. Find the differential equation relating the input and output voltages in the dual network, and compare it with the one obtained in part (a).

Problem 6-73 (Sec. 6-8)

Repeat Problem 6-72 for the network shown in Fig. P6-73.

ohms, henries, farads

Figure P6-73

7

Sinusoidal

Steady-State Analysis

In the preceding chapters we have concerned ourselves with finding the response of network variables for various excitations produced by initial conditions and sources. We have seen that the complete solution to the differential equation (or equations) describing the circuit could always be divided into two parts, namely, the complementary solution and the particular solution. In general, since the complementary solution decays exponentially as time increases, *for large values of time, only the particular solution is of interest*. This part of the complete solution always has a form similar to the form of the excitation provided by the source or sources. There are two excitation forms that occur more frequently than any others in actual practice. These are the constant excitation and the sinusoidal excitation. Methods for treating the constant excitation case were discussed in Sec. 5-4. In this chapter we shall discuss the sinusoidal case. Specifically, we shall be concerned with developing methods for finding the particular solution for network variables under the conditions that the network is excited only by sinusoids. In this case, the particular solution is frequently referred to as the *sinusoidal steady-state response*. The methods to be discussed in this chapter are of considerable importance to the electrical engineer, since sinusoids appear in many engineering applications, from the generation and transmission of electric power to satellite communication and control.

7-1 PROPERTIES OF SINUSOIDAL FUNCTIONS

Since we are planning to spend an entire chapter discussing sinusoidal steady-state response, it is important that we first explore some properties of sinusoidal functions. There are several ways in which such functions may be expressed. As a first method, consider the expression given in Table 5-2.1 for the general form that the particular solution of a network variable will have when the excitation function is a sinusoid. If we let $f(t)$ be the particular solution for the network variable, then, from the table, we may write

$$f(t) = A \cos \omega t + B \sin \omega t \qquad (7\text{-}1)$$

where A and B are constants, and ω is the angular frequency in radians per second. This means of characterizing a sinusoid is called the *quadrature* or *rectangular* form. It is particularly convenient when the method of undetermined coefficients is to be used to find the particular solution. This was the method introduced in Sec. 5-2. In this chapter, however, we shall find a different method of expressing sinusoids to be more convenient. This second method expresses the particular solution in an *amplitude-phase* or *polar* form,

$$f(t) = F_m \cos (\omega t + \alpha) \qquad (7\text{-}2)$$

where F_m is a positive quantity referred to as the *amplitude* or *peak magnitude* of the sinusoid, α is called the *argument* or *phase angle* (in radians), and ω is the *angular frequency* in radians per second. A plot of $f(t)$ is shown in Fig. 7-1.1. From this figure,

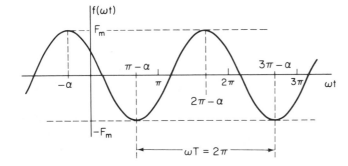

Figure 7-1.1 Sinusoidal waveform.

we see that when $\omega t = -\alpha \pm k\pi$ ($k = 0, 1, 2, 3, \ldots$), i.e., when the argument of the cosine function equals $0, \pm\pi, \pm 2\pi, \pm 3\pi, \pm 4\pi$, etc., the function $f(t)$ reaches its peak magnitude, i.e., $\pm F_m$. The time between identical points on the waveform, such as the time between successive positive peak values, the time between successive negative peak values, the time between successive zero-crossings of the abscissa (with the same slope), etc., is called the *period T* of the sinusoidal function. It is readily seen that the period $T = 2\pi/\omega$.

To see how the constants A and B of Eq. (7-1) are related to the constants F_m and α of Eq. (7-2), we may use the trigonometric identity

$$\cos (x + y) = \cos x \cos y - \sin x \sin y \qquad (7\text{-}3)$$

Applying this to (7-2), we obtain

$$f(t) = F_m \cos \alpha \cos \omega t - F_m \sin \alpha \sin \omega t \qquad (7\text{-}4)$$

Comparing this with the expression given in (7-1), we obtain the following relations expressing A and B in terms of F_m and α.

$$A = F_m \cos \alpha \qquad B = -F_m \sin \alpha \qquad (7\text{-}5)$$

Sinusoidal Steady-State Analysis   Chap. 7

The inverse relations which specify F_m and α in terms of A and B will now be found. We begin by squaring both members of each equation given in (7-5). Thus, we obtain

$$A^2 = F_m^2 \cos^2 \alpha \qquad B^2 = F_m^2 \sin^2 \alpha \qquad (7\text{-}6)$$

Adding the two equations gives

$$A^2 + B^2 = F_m^2(\cos^2 \alpha + \sin^2 \alpha) = F_m^2 \qquad (7\text{-}7)$$

From this equation and from the ratio of the two equations in (7-5) we obtain the following inverse relations:[1]

$$F_m = \sqrt{A^2 + B^2} \qquad \alpha = \tan^{-1}\left(\frac{-B}{A}\right) \qquad (7\text{-}8)$$

By using a simple graphical representation, the relations given above are easy to visualize, apply, and remember. To do this we need simply note that if we define an x-y plane as shown in Fig. 7-1.2(a), then a vector z as shown in the figure will

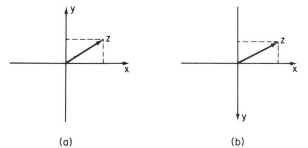

(a)   (b)   **Figure 7-1.2**   Vector in an x-y plane.

be defined by the relations

$$|z| = \sqrt{x^2 + y^2} \qquad \text{Arg } z = \tan^{-1}\left(\frac{y}{x}\right) \qquad (7\text{-}9)$$

If we now reverse the positive direction of the y-axis as shown in Fig. 7-1.2(b), then these relations become

$$|z| = \sqrt{x^2 + y^2} \qquad \text{Arg } z = \tan^{-1}\left(\frac{-y}{x}\right) \qquad (7\text{-}10)$$

These relations have the same form as those given in (7-8). Thus we may define an A-B plane as shown in Fig. 7-1.3, in which the value of the coefficient A of (7-1) is

A
(coefficient of cos ωt)

B
(coefficient of sin ωt)

Figure 7-1.3 A-B plane for representing sinusoidal functions.

[1] In taking the arctangent function as indicated in (7-8) and in other places in this chapter, due note must be taken of the polarity of the quantities $-B$ and A in determining the quadrant in which α lies. For example, if $-B = -1$, and $A = -1$, then $\alpha = \tan^{-1}(-1/-1) = -3\pi/4$. However, if $-B = 1$, and $A = 1$, then $\alpha = \tan^{-1}(1/1) = \pi/4$. In both cases, the argument of the arctangent function is plus one, but the angle is different.

plotted along the "x" axis, and the value of the coefficient B of (7-1) is plotted along the negative "y" axis. Any given values for A and B then define a vector. The magnitude and argument of the vector in turn define the quantities F_m and α, respectively, in (7-2). As examples, the following equalities are illustrated in Fig. 7-1.4.[2]

$$\cos \omega t + \sin \omega t = \sqrt{2} \cos \left(\omega t - \frac{\pi}{4}\right) \qquad \text{[Fig. 7-1.4(a)]} \quad (7\text{-}11)$$

$$-3 \cos \omega t + 4 \sin \omega t = 5 \cos (\omega t - 126.87°) \qquad \text{[Fig. 7-1.4(b)]} \quad (7\text{-}12)$$

$$3 \cos \omega t - 4 \sin \omega t = 5 \cos (\omega t + 53.13°) \qquad \text{[Fig. 7-1.4(c)]} \quad (7\text{-}13)$$

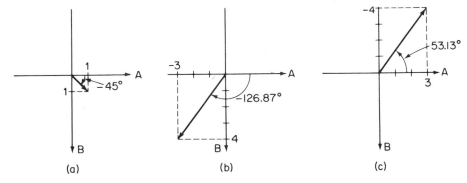

Figure 7-1.4 Examples of the use of the A-B plane.

In addition, it is readily verified that the A-B plane may be used for computing the phase angles associated with any other type of sinusoidal expression than the positive cosine function. For example, it is readily shown that

$$\cos \left(\omega t + \frac{\pi}{6}\right) = -\sin \left(\omega t - \frac{\pi}{3}\right) = \sin \left(\omega t + \frac{2\pi}{3}\right)$$

$$= -\cos \left(\omega t - \frac{5\pi}{6}\right) \qquad (7\text{-}14)$$

The various angles are illustrated in Fig. 7-1.5.

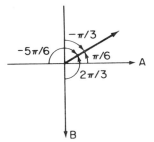

Figure 7-1.5 Some equivalent arguments.

[2]To be dimensionally correct, the phase angle which is included as part of the argument of the sinusoid must be expressed in units of radians. Practically, however, for angles which are not easily expressible as simple fractions of π, we will use angles measured in degrees and identified by the superscript ° to more clearly indicate the actual value of these angles.

There are several properties of sinusoids that will be useful in succeeding sections of this chapter. One of the most important of these has to do with the addition of sinusoidal quantities. Let us consider two functions $h(t)$ and $g(t)$ defined as follows.

$$h(t) = H_m \cos (\omega t + \beta)$$
$$g(t) = G_m \cos (\omega t + \gamma) \tag{7-15}$$

From the discussion given above we know that each of these functions may be expressed in a form similar to that given in (7-1). Let these expressions be

$$h(t) = A_h \cos \omega t + B_h \sin \omega t \qquad g(t) = A_g \cos \omega t + B_g \sin \omega t \tag{7-16}$$

Now let us add the two functions, using the relations of (7-16). We obtain

$$h(t) + g(t) = (A_h + A_g) \cos \omega t + (B_h + B_g) \sin \omega t \tag{7-17}$$

If we define $A = A_h + A_g$ and $B = B_h + B_g$, then the sum of the two sinusoids may be written in the form

$$h(t) + g(t) = F_m \cos (\omega t + \alpha) \tag{7-18}$$

where F_m and α are defined by Eq. (7-8). Thus, the result of the addition of the two sinusoids is another sinusoid *of the same frequency*. The A-B plane shown in Fig. 7-1.3 can be used to advantage when it is desired to form such a sum, since the addition of the functions is then simply reduced to the addition of the vectors representing the magnitude and phase angle of the related functions. For example, suppose that it is desired to find a single sinusoid $f(t)$ comprised of the sum of $h(t)$ and $g(t)$, where

$$h(t) = 5 \cos (\omega t + 36.87°)$$
$$g(t) = 3\sqrt{2} \sin (\omega t - 45°) \tag{7-19}$$

The vector representations of the magnitude and phase of $h(t)$ and $g(t)$ are shown in Fig. 7-1.6. The addition of these vectors yields the magnitude and phase of the

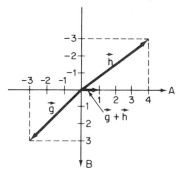

Figure 7-1.6 Adding sinusoidal functions on the A-B plane.

sinusoid (of the same frequency) $f(t)$. Thus, we see that

$$f(t) = h(t) + g(t) = \cos \omega t \tag{7-20}$$

We shall find frequent application of these techniques in the remainder of this chapter.

Since sinusoidal waveforms occur with such frequency in engineering situations, it is useful to be able to characterize them according to their effects. One of the most important of such effects has to do with the power dissipated by a sinusoidal voltage or current when it is one of the variables associated with a resistive two-terminal element. To be more specific, let us see if we can make a comparison between the power dissipated in a resistor (of value $R\ \Omega$) in two different situations. The first of these will be when a sinusoidal current $i(t)$ is passed through the resistor, and the second will be when a constant current of value I_0 is passed through the same resistor. For the sinusoidal case, let $i(t)$ have the form

$$i(t) = I_m \cos(\omega t + \phi) \tag{7-21}$$

The power is given by the expression

$$p(t) = Ri^2(t) = RI_m^2 \cos^2(\omega t + \phi) = RI_m^2 \frac{\cos(2\omega t + 2\phi) + 1}{2} \tag{7-22}$$

Correspondingly, in the constant current case, the power is given by the expression

$$p(t) = RI_0^2 \tag{7-23}$$

The expressions given in (7-22) and (7-23) cannot be directly compared, since the first fluctuates sinusoidally between the limits of 0 and RI_m^2 with a frequency of 2ω rad/s, while the second is constant. Thus, to permit a comparison between the two quantities, let us calculate the *average* value of the power over some time interval. For the sinusoidal case, if we compute the average power over a single period, then this will be the same as the average power for any integer number of periods. The period is given by $T = 2\pi/\omega$. Since the computation is to be made over one period, the phase angle is unimportant, and for convenience of integration, we may assume that it is zero. Thus, our computation becomes

$$P_{\text{avg}} = \frac{1}{T} \int_0^T RI_m^2 \cos^2(\omega t)\, dt = \frac{RI_m^2}{T} \int_0^T \cos^2(\omega t)\, dt$$

$$= \frac{RI_m^2}{T} \int_0^T \frac{\cos 2\omega t + 1}{2}\, dt = \frac{RI_m^2}{2} \tag{7-24}$$

The average power for the constant current case is obviously equal to the instantaneous power, namely, RI_0^2. If the values of the average power for the sinusoidal current case and the constant current case are to be the same, then we must have $RI_0^2 = RI_m^2/2$. This requires that

$$I_0 = \frac{I_m}{\sqrt{2}} \tag{7-25}$$

Thus, a sinusoidal waveform of current with a peak magnitude of I_m will dissipate the same average power in a resistor as a constant current of value $I_m/\sqrt{2}$ A. A similar development is easily made to show that a sinusoidal waveform of voltage with a peak magnitude of V_m applied to a resistor will produce the same average power dissipation in a resistor as a constant potential of value $V_m/\sqrt{2}$ V. The details are left as an exercise to the reader.

Let us consider some of the individual steps which were made in the above development, and their effect on the function $i(t)$ of Eq. (7-21). First of all, since power considerations were involved, we *squared* the function. Second, in Eq. (7-24), we found the *mean* (or average) value of the squared function. Finally, we took the square *root* of the result. Thus, in summary, we may refer to the result as the *root-*

mean-square or *rms* value of $i(t)$. This process may be summarized by the relation

$$I_{rms} = \sqrt{\frac{1}{T} \int_0^T i^2(t)\, dt} \qquad (7\text{-}26)$$

where T is the length of one period of the sinusoidal variation of $i(t)$. Although our primary consideration here is with sinusoidal functions, it should be noted that the relation given in (7-26) applies equally well to the determination of the rms value of any periodic function, i.e., any function which repeats its waveform with a constant period of T s. Such a periodic function satisfies the relation

$$f(t) = f(t + T) \qquad (7\text{-}27)$$

for all t.

The results given above may be summarized by pointing out that *the rms value of a periodic current or voltage is equal to the value of an equivalent constant current or voltage which will produce the same average power dissipation in a given resistor.*

As an example of the use of rms values, note that the familiar distribution of household electricity is predominantly done using sinusoidal variables, and the value of the voltage is specified in terms of rms units. Thus, a "120-V outlet" provides a sinusoidal voltage which varies between the (peak) limits of $\pm\, 120 \times \sqrt{2}$ or about ± 169.7 V. The frequency of such a sinusoid is usually 60 Hz (cycles per second), and this is $60 \times 2\pi$ or about 377 rad/s. Thus, the sinusoidal voltage which is provided by most electric power companies may be characterized as $169.7 \cos (377t + \phi)$ V, where ϕ is the phase angle.

7-2 THE EXPONENTIAL FUNCTION

Let us begin our study of sinusoidal steady-state response by considering the circuit shown in Fig. 7-2.1. Applying KCL to this circuit, if $i(t)$ is a sinusoid, we may write

$$C\frac{dv}{dt} + Gv(t) = i(t) = I_m \cos (\omega t + \phi) \qquad (7\text{-}28)$$

Figure 7-2.1 Simple *RC* network.

where I_m is the positive peak magnitude of the excitation sinusoid and ϕ is the phase angle. From Table 5-2.1 we know that the particular solution for $v(t)$ will also be a sinusoid. From the discussion of the preceding section, such a particular solution may be written in the form

$$v(t) = V_m \cos (\omega t + \theta) \qquad (7\text{-}29)$$

Instead of directly substituting this expression into Eq. (7-28) to determine the value of the constants V_m and θ, let us first convert the expression given in (7-29) to the exponential form for the cosine function. Thus, we may write

$$v(t) = \frac{V_m}{2}(e^{j(\omega t+\theta)} + e^{-j(\omega t+\theta)}) = \frac{V_m e^{j\theta}}{2}e^{j\omega t} + \frac{V_m e^{-j\theta}}{2}e^{-j\omega t} \qquad (7\text{-}30)$$

The coefficients of the exponential functions in the right member of (7-30) have equal magnitudes but opposite angles, thus they are complex conjugates. To simplify our notation we may use a script letter $\mathcal{V}$ to represent the numerator of these complex quantities. Thus, we may write

$$\mathcal{V} = V_m e^{j\theta} \qquad \mathcal{V}^* = V_m e^{-j\theta} \qquad (7\text{-}31)$$

where the superscript asterisk is used to indicate the complex conjugate. The node voltage $v(t)$ may now be written

$$v(t) = \frac{\mathcal{V}}{2}e^{j\omega t} + \frac{\mathcal{V}^*}{2}e^{-j\omega t} \qquad (7\text{-}32)$$

In a similar fashion, we may express the excitation current $i(t)$ as

$$i(t) = \frac{I_m}{2}[e^{j(\omega t+\phi)} + e^{-j(\omega t+\phi)}] = \frac{I_m e^{j\phi}}{2}e^{j\omega t} + \frac{I_m e^{-j\phi}}{2}e^{-j\omega t} \qquad (7\text{-}33)$$

If we define

$$\mathcal{I} = I_m e^{j\phi} \qquad \mathcal{I}^* = I_m e^{-j\phi} \qquad (7\text{-}34)$$

then (7-33) may be put in the form

$$i(t) = \frac{\mathcal{I}}{2}e^{j\omega t} + \frac{\mathcal{I}^*}{2}e^{-j\omega t} \qquad (7\text{-}35)$$

Now let us substitute the expressions for $v(t)$ and $i(t)$ given in (7-32) and (7-35) in the original differential equation of (7-28). After rearranging terms, we obtain

$$[(G + j\omega C)\mathcal{V} - \mathcal{I}]\frac{e^{j\omega t}}{2} + [(G - j\omega C)\mathcal{V}^* - \mathcal{I}^*]\frac{e^{-j\omega t}}{2} = 0 \qquad (7\text{-}36)$$

Since the exponentials are functions of t, the only way in which the equation given above will be satisfied *for all values of t* is for the terms multiplying the exponentials to be zero. Thus, we see that

$$\mathcal{V} = \frac{\mathcal{I}}{G + j\omega C} \qquad \mathcal{V}^* = \frac{\mathcal{I}^*}{G - j\omega C} \qquad (7\text{-}37)$$

These equations provide the information that we need to determine the particular solution of the network. For the moment, let us only use the first equation of (7-37). Thus, we may write

$$\mathcal{V} = V_m e^{j\theta} = \frac{\mathcal{I}}{G + j\omega C} = \frac{I_m e^{j\phi}}{|G + j\omega C|e^{j\,\tan^{-1}(\omega C/G)}}$$

$$= \frac{I_m}{\sqrt{G^2 + (\omega C)^2}}e^{j[\phi - \tan^{-1}(\omega C/G)]} \qquad (7\text{-}38)$$

Thus we conclude that the separate relations for the magnitude and phase of $v(t)$ of Eq. (7-29) are

$$V_m = \frac{I_m}{\sqrt{G^2 + (\omega C)^2}} \qquad \theta = \phi - \tan^{-1}\left(\frac{\omega C}{G}\right) \qquad (7\text{-}39)$$

When these terms are substituted in Eq. (7-29), the particular solution for $v(t)$ is completely defined. An example follows.

Example 7-2.1

The elements shown in the circuit diagram in Fig. 7-2.1 are specified as having the values $G = 2$ mhos and $C = 1$ F. It is desired to find the sinusoidal steady-state response for $v(t)$ under the conditions that $i(t) = 5 \cos [2t - (\pi/4)]$. From (7-34), $\mathcal{I} = 5e^{-j\pi/4}$ and, from Eq. (7-37), we obtain

$$\mathcal{V} = \frac{\mathcal{I}}{G + j\omega C} = \frac{5e^{-j\pi/4}}{2 + j2} = \frac{5e^{-j\pi/4}}{2\sqrt{2}\, e^{j\pi/4}} = \frac{5}{2\sqrt{2}} e^{-j\pi/2}$$

Thus, our solution for $v(t)$ is

$$v(t) = \frac{5}{2\sqrt{2}} \cos \left(2t - \frac{\pi}{2} \right) = \frac{5}{2\sqrt{2}} \sin 2t$$

It should be noted that the most laborious part of the procedure given above is the determination of the equation of the form of (7-37) relating the excitation to the response. Thus, even though the procedure for finding $v(t)$ as illustrated in the numerical example is simple, the overall process may appear more complicated than the method of undetermined coefficients presented in Chaps. 5 and 6. We will shortly show, however, that this approach may be applied very simply to arbitrary network configurations, and thus it is of considerable value.

Now let us return to consider again the two equations of (7-37). It is easily seen that if we take the conjugate of both sides of the first equation, we obtain the second. Thus, the second equation is always satisfied if the first one is satisfied. Therefore, we see that once the first equation is known, finding the second equation does not give us any new information concerning the network. The first equation, however, is the equation associated with the exponential term $e^{j\omega t}$ (the *positive* exponential term) in the expression for the excitation current $i(t)$ and the expression for the particular solution for $v(t)$. Thus, this equation can be generated if we adopt the mathematical artifice of assuming that the excitation has the form[3]

$$i(t) = \frac{\mathcal{I}}{2} e^{j\omega t} \tag{7-40}$$

where $\mathcal{I}$ is defined in Eq. (7-34). Similarly, we may assume that the particular solution for the node voltage has the form

$$v(t) = \frac{\mathcal{V}}{2} e^{j\omega t} \tag{7-41}$$

where $\mathcal{V}$ is defined in (7-31). Finally, substituting Eqs. (7-40) and (7-41) into the basic differential equation defined by the left members of (7-28), we obtain the identical relation given as the first equation of (7-37). The effort involved in obtaining this result, however, is exactly half of that required for the original determination of Eq. (7-37), a considerable computational saving.

---

[3]Note that the quantities given for $i(t)$ and $v(t)$ in Eqs. (7-40) and (7-41), respectively, cannot actually be equal to physical variables, since they are *complex* quantities, while the physical variables of current and voltage must, of course, be *real*. However, there is no reason why, in an equation such as that shown in (7-28), we cannot mathematically make a substitution of the form indicated in Eqs. (7-40) and (7-41). This is what we imply by the words "mathematical artifice." In the next section we shall use a more sophisticated approach which is acceptable both from a mathematical viewpoint and a physical viewpoint.

It is readily shown that the simplified procedure outlined above may be applied to a differential equation of any order. As an example of this, consider the circuit shown in Fig. 7-2.2. The differential equation relating the excitation $i_s(t)$ and the node

ohms, farads

Figure 7-2.2 Third-order RC network.

voltage $v_3(t)$ for this network may be shown to be

$$i_s(t) = 3v_3''' + 9v_3'' + 6v_3' \tag{7-42}$$

Let us now assume that $i_s(t)$ is a sinusoid. Using the method presented above, we may use the notation

$$i_s(t) = I_s \cos(\omega t + \phi) = \frac{\mathcal{I}_s}{2}e^{j\omega t} + \frac{\mathcal{I}_s^*}{2}e^{-j\omega t} \tag{7-43}$$

where $\mathcal{I}_s = I_s e^{j\phi}$. The particular solution for $v_3(t)$, i.e., the sinusoidal steady-state response, will also have the form of a sinusoid with the same frequency. Thus, we define

$$v_3(t) = V_3 \cos(\omega t + \theta) = \frac{\mathcal{V}_3}{2}e^{j\omega t} + \frac{\mathcal{V}_3^*}{2}e^{-j\omega t} \tag{7-44}$$

where $\mathcal{V}_3 = V_3 e^{j\theta}$. Substituting the positive exponential terms of Eqs. (7-43) and (7-44) into the differential equation of (7-42), we obtain

$$\{[3(j\omega)^3 + 9(j\omega)^2 + 6(j\omega)]\mathcal{V}_3 - \mathcal{I}_s\}e^{j\omega t} = 0 \tag{7-45}$$

Thus we see that

$$\mathcal{V}_3 = \frac{\mathcal{I}_s}{3(j\omega)^3 + 9(j\omega)^2 + 6(j\omega)} \tag{7-46}$$

Using Eq. (7-44), we may now find the steady-state solution for $v_3(t)$.

It should be noted that the procedure given is also easily applied if the network is described by an integral equation or an integro-differential equation, since in such a case we need merely differentiate the given expression to obtain a purely differential equation. For example, if we apply KVL to a circuit consisting of a series-connected voltage source, a resistor, and a capacitor, we obtain

$$v(t) = Ri(t) + \frac{1}{C}\int_{-\infty}^{t} i(\tau)\,d\tau \tag{7-47}$$

Differentiating this equation, we obtain

$$\frac{dv}{dt} = R\frac{di}{dt} + \frac{1}{C}i(t) \tag{7-48}$$

From this point, substitutions similar to those described above may be used to find the sinusoidal steady-state response for $i(t)$ for a given sinusoidal excitation $v(t)$. The details are left to the reader.

7-3 PHASORS AND SINUSOIDAL STEADY-STATE RESPONSE

In the preceding section we formulated a method for finding the sinusoidal steady-state response of a given network by assuming that the excitation and response variables could be expressed as quantities involving a positive exponential term and having the form

$$\mathfrak{F}e^{j\omega t} \tag{7-49}$$

where $\mathfrak{F} = F_m e^{j\phi}$ and the quantities F_m and ϕ are, respectively, the amplitude and phase of the sinusoidally varying quantity $f(t)$, which has the form

$$f(t) = F_m \cos(\omega t + \phi) \tag{7-50}$$

A more rigorous justification of the method may be presented by noting that, from (7-49) we may write

$$\mathfrak{F}e^{j\omega t} = F_m e^{j\phi}e^{j\omega t} = F_m e^{j(\omega t + \phi)}$$
$$= F_m[\cos(\omega t + \phi) + j\sin(\omega t + \phi)] \tag{7-51}$$

Comparing Eqs. (7-50) and (7-51), we now may write

$$f(t) = \text{Re}\,(\mathfrak{F}e^{j\omega t}) \tag{7-52}$$

where "Re" stands for "the real part of." Thus, we see that, using the formal relation given in Eq. (7-52), we may use the complex quantity $\mathfrak{F}e^{j\omega t}$ as defined by Eq. (7-51) to represent the sinusoidal function $F_m \cos(\omega t + \phi)$. Such a complex quantity is called a *phasor*.[4] A phasor has a definite representation on the complex plane, i.e., the plane on which the value of the real part of a complex quantity is plotted in the "x" direction and on which the value of the imaginary part of a complex quantity is plotted in the "y" direction. Such a representation is shown in Fig. 7-3.1 for various values of ωt. For example, at $t = 0$, the real part of $\mathfrak{F}e^{j\omega t}$ is $F_m \cos \phi$ and the imaginary part is $F_m \sin \phi$. These two values determine a vector of magnitude F_m and of angle ϕ as shown in Fig. 7-3.1(a). Since, from Eq. (7-52), $f(t)$ is the real part of $\mathfrak{F}e^{j\omega t}$, it is equal to the component of $\mathfrak{F}e^{j\omega t}$ that lies along the real axis, i.e., it is equal to the projection of $\mathfrak{F}e^{j\omega t}$ on the real axis. Thus, when t equals zero, $f(0)$ is equal to Re $\mathfrak{F}$ or $F_m \cos \phi$, as shown in Fig. 7-3.1(a). Now let us consider the situation where $\omega t = (\pi/2) - \phi$, i.e., the case where $t = [(\pi/2) - \phi)]/\omega$. In this case, the real part of $\mathfrak{F}e^{j\omega t}$ is zero, and the imaginary part is F_m. Thus, for this value of t the phasor $\mathfrak{F}e^{j\omega t}$ appears as shown in Fig. 7-3.1(b). The corresponding projection on the real axis tells us that $f(t)$ evaluated for $t = [(\pi/2) - \phi]/\omega$ equals zero, as is also apparent from Eq. (7-50). Continuing in the same fashion, we see that when $\omega t = \pi/2$, i.e., when $t = \pi/2\omega$, the phasor $\mathfrak{F}e^{j\omega t}$ appears as shown in Fig. 7-3.1(c), namely, it has a magnitude of F_m and an angle of $(\pi/2) + \phi$. Thus, $f(\pi/2\omega) = -F_m \sin \phi$. Similarly, when $t = \pi/\omega$, the phasor will have the value shown in Fig. 7-3.1(d), etc. Considering all the above,

[4]It is customary to use the same units for the phasor as were used for the original sinusoidally varying quantity. Thus, a phasor used to represent a sinusoidally varying voltage will be treated as having units of volts. A phasor used to represent a sinusoidally varying current will be treated as having units of amperes, etc.

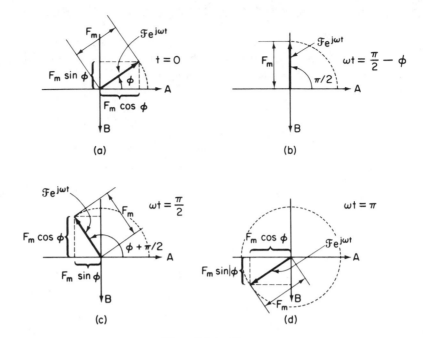

Figure 7-3.1 Phasor.

we note that the locus of $\mathcal{F}e^{j\omega t}$ is a circle of radius F_m centered at the origin of the complex plane.[5] The preceding discussion may be summarized as follows:

SUMMARY 7-3.1

Definition of a Phasor. A phasor $\mathcal{F} = F_m e^{j\phi}$ is a complex quantity which is used to represent a sinusoidally varying function of time. The locus of $\mathcal{F}e^{j\omega t}$ is a circle of radius F_m, centered at the origin of a complex plane, and generated by a rotating vector with an angular velocity of ω rad/s. The projection of this vector on the real axis generates the sinusoidal function[6]

$$f(t) = F_m \cos(\omega t + \phi) \tag{7-53}$$

∎

In order to discuss the use of functions such as the one given in Eq. (7-52) in solving differential equations, we need some information about what happens to such a function when it is differentiated. Suppose that we begin by defining a complex

---

[5]Note that the "rotation" of the phasor for increasing values of time is what distinguishes it from a stationary *vector*, i.e. a representation of a complex number.

[6]Since a phasor represents a sinusoidally varying function of time, a quantity such as $\mathcal{F}$ may be considered as a representation of a sinusoid evaluated at t equals 0, while a quantity $\mathcal{F}e^{j\omega t}$ may be considered as a representation of a sinusoid at all values of time. For convenience we shall refer to both types of expressions as phasors.

function $g(t)$, with a real part $x(t)$ and an imaginary part $y(t)$. Thus, we may write

$$g(t) = x(t) + jy(t) \qquad (7\text{-}54)$$

Now let us consider a function $f(t)$ defined as

$$f(t) = \operatorname{Re}[g(t)] = \operatorname{Re}[x(t) + jy(t)] = x(t) \qquad (7\text{-}55)$$

If we desire to differentiate $f(t)$, then clearly we may write

$$\frac{df}{dt} = \frac{d}{dt}\operatorname{Re}[g(t)] = \frac{dx}{dt} \qquad (7\text{-}56)$$

Let us now investigate what happens if we perform the operations in the opposite order, i.e., if we first differentiate $g(t)$ and then take the real part of the result. If we do this, we may write

$$\operatorname{Re}\left[\frac{dg}{dt}\right] = \operatorname{Re}\left[\frac{dx}{dt} + j\frac{dy}{dt}\right] = \frac{dx}{dt} \qquad (7\text{-}57)$$

Comparing Eqs. (7-56) and (7-57), we see that

$$\frac{d}{dt}\operatorname{Re}[g(t)] = \operatorname{Re}\left[\frac{dg}{dt}\right] \qquad (7\text{-}58)$$

Thus we may conclude that if a complex function of t is differentiated, and then its real part is extracted, the result is the same as if the order of the two operations had been reversed, i.e., as if first the real part had been taken and then the result differentiated. A similar result may be shown to hold with respect to integration.

A second item of information that we shall require before we can discuss the use of relations of the type given in (7-52) is the relative implications of an equality between two complex functions having the form given in Eq. (7-49), and an equality between the real parts of such functions. Let the two complex functions be $g_1(t)$ and $g_2(t)$, where

$$\begin{aligned}
g_1(t) &= \mathcal{G}_1 e^{j\omega t} = (G_{1r} + jG_{1i})e^{j\omega t} \\
g_2(t) &= \mathcal{G}_2 e^{j\omega t} = (G_{2r} + jG_{2i})e^{j\omega t}
\end{aligned} \qquad (7\text{-}59)$$

As indicated, G_{1r} and G_{1i} are the real and imaginary parts of the complex quantity $\mathcal{G}_1$. Similarly, G_{2r} and G_{2i} are the real and imaginary parts of $\mathcal{G}_2$. It should be apparent that, if we specify $g_1(t) = g_2(t)$ for all t, then, by definition of an equality for complex functions, it must hold true that $\operatorname{Re}[g_1(t)] = \operatorname{Re}[g_2(t)]$. However, here we are more interested in the reverse situation. That is, for functions of the type given in Eq. (7-59), if the real parts are specified as being equal, what can we conclude about the functions themselves? To find the answer to this question, let us first evaluate the functions of Eq. (7-59) for t equals 0. We obtain

$$g_1(0) = G_{1r} + jG_{1i} \qquad g_2(0) = G_{2r} + jG_{2i} \qquad (7\text{-}60)$$

If we require that $\operatorname{Re}[g_1(t)] = \operatorname{Re}[g_2(t)]$ for all t, then, for the particular value $t = 0$, we conclude from (7-60), that $G_{1r} = G_{2r}$. Now let us evaluate the relations of (7-59) for $t = \pi/2\omega$. For this value of t, $e^{j\omega t} = j$; thus, from (7-59) we see that

$$g_1\left(\frac{\pi}{2\omega}\right) = -G_{1i} + jG_{1r} \qquad g_2\left(\frac{\pi}{2\omega}\right) = -G_{2i} + jG_{2r} \qquad (7\text{-}61)$$

From Eq. (7-61) we see that requiring $\operatorname{Re}[g_1(t)] = \operatorname{Re}[g_2(t)]$ for this value of t implies that $G_{1i} = G_{2i}$. Since we have shown that the real and the imaginary parts of G_1 and G_2 are separately equal, the complex quantities themselves are obviously equal, thus, $g_1(t)$ and $g_2(t)$ must also be equal.

We have now developed sufficient background to present a more rigorous justification of the method for the determination of the sinusoidal steady-state response presented in the previous section. Let us again consider the network shown in Fig. 7-2.1. The differential equation describing this circuit is

$$C\frac{dv}{dt} + Gv(t) = i(t) \tag{7-62}$$

If the excitation is a sinusoid of magnitude I_m, phase ϕ, and frequency ω, then the sinusoid may be represented by the phasor defined as $\mathcal{I}e^{j\omega t}$, where $\mathcal{I} = I_m e^{j\phi}$. Thus, following the form of (7-52), we may write

$$i(t) = \text{Re}\,(\mathcal{I}e^{j\omega t}) \tag{7-63}$$

Similarly, since we know that the response for the voltage will be a sinusoid, let us assume that it will have an amplitude V_m and a phase θ. Then we may represent the voltage sinusoid by a phasor $\mathcal{V}e^{j\omega t}$, where $\mathcal{V} = V_m e^{j\theta}$. Thus, the expression for $v(t)$ becomes

$$v(t) = \text{Re}\,(\mathcal{V}e^{j\omega t}) \tag{7-64}$$

Now let us substitute Eqs. (7-63) and (7-64) into the differential equation of (7-62). We obtain

$$C\frac{d}{dt}\,\text{Re}\,(\mathcal{V}e^{j\omega t}) + G\,\text{Re}\,(\mathcal{V}e^{j\omega t}) = \text{Re}\,(\mathcal{I}e^{j\omega t}) \tag{7-65}$$

From the result given above, we may interchange the operations of differentiation and real part extraction in the first term of the left member of Eq. (7-65). Doing this and performing the differentiation, we obtain

$$C\,\text{Re}\,(j\omega\mathcal{V}e^{j\omega t}) + G\,\text{Re}\,(\mathcal{V}e^{j\omega t}) = \text{Re}\,(\mathcal{I}e^{j\omega t}) \tag{7-66}$$

In the left member of Eq. (7-66), the constants may be moved inside the Re operation. Also, since the sum of two real parts is equal to the real part of the sum of the two corresponding complex quantities, the preceding equation may be rewritten in the form

$$\text{Re}\,[(j\omega C\mathcal{V} + G\mathcal{V})e^{j\omega t}] = \text{Re}\,(\mathcal{I}e^{j\omega t}) \tag{7-67}$$

From the preceding discussion, we know that the equality of the real parts given above implies the equality of the entire functions. Thus, we may write

$$(j\omega C\mathcal{V} + G\mathcal{V})e^{j\omega t} = \mathcal{I}e^{j\omega t} \tag{7-68}$$

If we multiply both sides of the above by $e^{-j\omega t}$, we find that the solution for the phasor $\mathcal{V}$ is

$$\mathcal{V} = \frac{\mathcal{I}}{G + j\omega C} \tag{7-69}$$

which is the same result as was obtained in the last section. We may obviously use the value of $\mathcal{V}$ found in Eq. (7-69) to specify the value of $v(t)$ using Eq. (7-64). The procedure given above is readily generalized. An example follows.

Example 7-3.1

The excitation $i_1(t)$ and the node voltage $v_2(t)$ of the network shown in Fig. 7-3.2 may be shown to be related by the differential equation

$$\frac{5}{4}v_2(t) + \frac{9}{4}\frac{dv_2}{dt} + 2\frac{d^2v_2}{dt^2} + \frac{d^3v_2}{dt^3} = \frac{5}{8}i_1(t)$$

If $i_1(t)$ is a sinusoid, then we may use a phasor $\mathcal{I}_1 e^{j\omega t}$ to represent it. Similarly, the sinusoidal steady-state component of $v_2(t)$ may be represented by the phasor $\mathcal{V}_2 e^{j\omega t}$.

ohms,
henries,
farads

Figure 7-3.2 *RLC* network.

From the development given above, we see that the relation between $\mathcal{V}_2$ and $\mathcal{I}_1$ may be directly written as

$$\frac{\mathcal{V}_2}{\mathcal{I}_1} = \frac{\frac{5}{8}}{\frac{5}{4} + \frac{9}{4}(j\omega) + 2(j\omega)^2 + (j\omega)^3}$$

This is more conveniently written by grouping the real and the imaginary terms together. Thus, we may write

$$\frac{\mathcal{V}_2}{\mathcal{I}_1} = \frac{\frac{5}{8}}{\frac{5}{4} - 2\omega^2 + j(\frac{9}{4}\omega - \omega^3)}$$

The above equation relates $\mathcal{V}_2$ and $\mathcal{I}_1$ at any value of ω. For example, if $i_1(t) = 10\cos[t + (\pi/6)]$, then $\omega = 1$ rad/s, and $\mathcal{I}_1 = 10e^{j\pi/6}$. From the above equation, we see that

$$\mathcal{V}_2 = \frac{5}{-6 + j10}10e^{j\pi/6} = 4.287e^{-j90.97°}$$

Therefore, the sinusoidal steady-state value of $v_2(t)$ is

$$v_2(t) = 4.287\cos(t - 90.97°)$$

A procedure of the type illustrated in the preceding example is referred to as the *phasor method* of determining the sinusoidal steady-state response. We may summarize it as follows:

SUMMARY 7-3.2

Use of Phasors in Solving for the Sinusoidal Steady-State Response of a Circuit. If a given circuit is described by a differential equation having the form

$$a_1 f(t) + a_2 \frac{df}{dt} + a_3 \frac{d^2f}{dt^2} + \cdots + a_{n+1}\frac{d^nf}{dt^n}$$

$$= b_1 g(t) + b_2 \frac{dg}{dt} + b_3 \frac{d^2g}{dt^2} + \cdots + b_{m+1}\frac{d^mg}{dt^m} \qquad (7\text{-}70)$$

which relates some response variable $f(t)$ to a sinusoidal excitation variable $g(t)$ with magnitude G_m and phase ϕ, we may find the sinusoidal steady-state component of $f(t)$ by substituting $\mathcal{G} = G_m e^{j\phi}$ in the following:

$$\mathcal{F} = \frac{b_1 + b_2(j\omega) + b_2(j\omega)^2 + \cdots + b_{m+1}(j\omega)^m}{a_1 + a_2(j\omega) + a_3(j\omega)^2 + \cdots + a_{n+1}(j\omega)^n} \mathcal{G} \qquad (7\text{-}71)$$

The magnitude and phase of the phasor $\mathcal{F}$ are the amplitude and phase of the sinusoidal steady-state component of $f(t)$, respectively. In comparing Eqs. (7-70) and (7-71), note that we may directly write the relation between the excitation phasor $\mathcal{G}$ and the response phasor $\mathcal{F}$ by simply making the following substitutions in the differential equation:

$$\frac{d^i f}{dt^i} \longrightarrow (j\omega)^i \mathcal{F} \qquad \frac{d^i g}{dt^i} \longrightarrow (j\omega)^i \mathcal{G} \qquad (7\text{-}72)$$

■

7-4 PHASOR SOLUTIONS OF SIMULTANEOUS SETS OF INTEGRO-DIFFERENTIAL EQUATIONS

The procedure described in the preceding section for using phasors to determine the sinusoidal steady-state solution for a response variable in a *single* differential equation is readily applied to a simultaneous set of differential equations and/or integro-differential equations. For example, consider the network shown in Fig. 7-4.1. Applying KCL to the two nodes of this network, we obtain the equations

mhos,
henries,
farads

Figure 7-4.1 *RLC* network.

$$\frac{dv_1}{dt} + 2v_1(t) - 2v_2(t) = i_1(t)$$

$$4 \int_{-\infty}^{t} v_2(\tau)\, d\tau + 4v_2(t) - 2v_1(t) = i_2(t) \qquad (7\text{-}73)$$

If we desire to find the sinusoidal steady-state response for $v_1(t)$ and $v_2(t)$, we may begin by differentiating the second of the equations given in Eq. (7-73) to convert it to a differential equation rather than an integro-differential equation. Thus, the set of equations becomes

$$\frac{dv_1}{dt} + 2v_1(t) - 2v_2(t) = i_1(t)$$

$$4v_2(t) + 4\frac{dv_2}{dt} - 2\frac{dv_1}{dt} = \frac{di_2}{dt} \qquad (7\text{-}74)$$

Now let us consider the case where the excitation to the network is purely sinusoidal, i.e., where the source currents $i_1(t)$ and $i_2(t)$ have the form

$$i_1(t) = I_1 \cos(\omega t + \phi_1) \qquad i_2(t) = I_2 \cos(\omega t + \phi_2) \qquad (7\text{-}75)$$

It is assumed that the magnitudes I_1 and I_2 and the phase angles ϕ_1 and ϕ_2 are known. The two magnitudes may, of course, be different, as may the two phase angles. The frequency ω, however, is assumed to be the same for the two excitations. The particular solution for the node voltages, i.e., the steady-state solution for these network variables, will also have a sinusoidal form. Thus, we may write

$$v_1(t) = V_1 \cos(\omega t + \theta_1) \qquad v_2(t) = V_2 \cos(\omega t + \theta_2) \tag{7-76}$$

where the values of V_1, V_2, θ_1, and θ_2 are to be found. To determine these quantities, we need simply use the phasors $\mathcal{I}_1 e^{j\omega t}$ and $\mathcal{I}_2 e^{j\omega t}$ to represent the two excitation variables $i_1(t)$ and $i_2(t)$, and the phasors $\mathcal{U}_1 e^{j\omega t}$ and $\mathcal{U}_2 e^{j\omega t}$ to represent the two response variables $v_1(t)$ and $v_2(t)$. Thus, we may write

$$\begin{aligned} i_1(t) = \operatorname{Re}(\mathcal{I}_1 e^{j\omega t}) \qquad i_2(t) = \operatorname{Re}(\mathcal{I}_2 e^{j\omega t}) \\ v_1(t) = \operatorname{Re}(\mathcal{U}_1 e^{j\omega t}) \qquad v_2(t) = \operatorname{Re}(\mathcal{U}_2 e^{j\omega t}) \end{aligned} \tag{7-77}$$

where the complex constants $\mathcal{I}_1$, $\mathcal{I}_2$, $\mathcal{U}_1$, and $\mathcal{U}_2$ are defined as

$$\begin{aligned} \mathcal{I}_1 = I_1 e^{j\phi_1} \qquad \mathcal{I}_2 = I_2 e^{j\phi_2} \\ \mathcal{U}_1 = V_1 e^{j\theta_1} \qquad \mathcal{U}_2 = V_2 e^{j\theta_2} \end{aligned} \tag{7-78}$$

Substituting the relations of Eq. (7-77) into the differential equations given in (7-74) and interchanging the order of differentiation and real part extraction, we obtain

$$\operatorname{Re}\left\{ \begin{bmatrix} 2+j\omega & -2 \\ -j2\omega & 4+j4\omega \end{bmatrix} \begin{bmatrix} \mathcal{U}_1 e^{j\omega t} \\ \mathcal{U}_2 e^{j\omega t} \end{bmatrix} \right\} = \operatorname{Re}\left\{ \begin{bmatrix} \mathcal{I}_1 e^{j\omega t} \\ j\omega \mathcal{I}_2 e^{j\omega t} \end{bmatrix} \right\} \tag{7-79}$$

Equating the entire complex matrices of (7-79) and multiplying both sides by $e^{-j\omega t}$, we obtain

$$\begin{bmatrix} 2+j\omega & -2 \\ -j2\omega & 4+j4\omega \end{bmatrix} \begin{bmatrix} \mathcal{U}_1 \\ \mathcal{U}_2 \end{bmatrix} = \begin{bmatrix} \mathcal{I}_1 \\ j\omega \mathcal{I}_2 \end{bmatrix} \tag{7-80}$$

Note that in the right member of (7-80), the term $\mathcal{I}_2$ is multiplied by $j\omega$, since this term represents the derivative of $i_2(t)$ with respect to time. If we divide the entire second equation of (7-80) by $j\omega$, the set of equations becomes

$$\begin{bmatrix} 2+j\omega & -2 \\ -2 & 4+\dfrac{4}{j\omega} \end{bmatrix} \begin{bmatrix} \mathcal{U}_1 \\ \mathcal{U}_2 \end{bmatrix} = \begin{bmatrix} \mathcal{I}_1 \\ \mathcal{I}_2 \end{bmatrix} \tag{7-81}$$

If we now compare the left members of (7-81) with the original equations of (7-73), we see that the integral operation in the original integro-differential equations produces a factor $1/j\omega$ in the equations relating the phasors, and the derivative operation produces a factor of $j\omega$. This is easily shown to be always true. Thus, in general, phasor methods may be directly applied to integro-differential equations, without the necessity of differentiating these equations so as to convert them to the form of purely differential equations.

The relations given in (7-81) are a simultaneous set of equations in which all the elements are complex and are functions of $j\omega$. Solving such a set of equations follows exactly the procedures which were given in Chap. 3, since the methods developed there apply irrespective of whether the matrices have elements which are real or complex, or whether the elements are constant or are functions of some variable. Thus, using the relations given in Sec. 3-1 for the solution of a set of two simultaneous equations, the relations of (7-81) become

$$\begin{bmatrix} \mathcal{U}_1 \\ \mathcal{U}_2 \end{bmatrix} = \frac{j\omega}{4(j\omega)^2 + 8(j\omega) + 8} \begin{bmatrix} 4 + \dfrac{4}{j\omega} & 2 \\ 2 & 2 + j\omega \end{bmatrix} \begin{bmatrix} \mathcal{I}_1 \\ \mathcal{I}_2 \end{bmatrix} \tag{7-82}$$

This set of equations provides the solution for the network. That is, given any specified values for $\mathcal{I}_1$ and $\mathcal{I}_2$, we may determine the corresponding values of $\mathcal{U}_1$ and $\mathcal{U}_2$, and, using Eqs. (7-76) and (7-78), we obtain the expressions for the sinusoidal steady-state components of $v_1(t)$ and $v_2(t)$. The procedure is illustrated by the following example.

Example 7-4.1

The network shown in Fig. 7-4.1 is excited as follows.

$$i_1(t) = \cos t \qquad i_2(t) = 2 \cos \left(t + \frac{\pi}{2} \right)$$

It is desired to find the sinusoidal steady-state response for $v_1(t)$ and $v_2(t)$. To do this, we first note from (7-75) and (7-78) that $\mathcal{I}_1 = 1e^{j0}$, $\mathcal{I}_2 = 2e^{j\pi/2}$, and $\omega = 1$. Substituting these values in (7-82), we obtain

$$\begin{bmatrix} \mathcal{U}_1 \\ \mathcal{U}_2 \end{bmatrix} = \frac{j1}{4 + j8} \begin{bmatrix} 4 - j4 & 2 \\ 2 & 2 + j1 \end{bmatrix} \begin{bmatrix} 1e^{j0} \\ 2e^{j\pi/2} \end{bmatrix}$$

Performing the indicated multiplication, we find that

$$\mathcal{U}_1 = \frac{j1}{1 + j2} = \frac{1}{\sqrt{5}} e^{j26.57°} \qquad \mathcal{U}_2 = \frac{-1}{1 + j2} = \frac{1}{\sqrt{5}} e^{j116.57°}$$

The corresponding solutions for the quantities $v_1(t)$ and $v_2(t)$ are

$$v_1(t) = \frac{1}{\sqrt{5}} \cos (t + 26.57°) \qquad v_2(t) = \frac{1}{\sqrt{5}} \cos (t + 116.57°)$$

These are the sinusoidal steady-state values of the nodal voltages, i.e., the particular solutions for these network variables.

The procedure outlined above is easily generalized to more complex networks. An example follows.

Example 7-4.2

The network shown in Fig. 7-4.2 is excited by three voltage sources having the values

$$v_1(t) = 4 \cos 2t$$

$$v_2(t) = -6 \cos 2t$$

$$v_3(t) = 2\sqrt{2} \cos \left(2t - \frac{\pi}{4} \right)$$

ohms, henries, farads

Figure 7-4.2 Network for Example 7-4.2.

Sinusoidal Steady-State Analysis   Chap. 7

It is desired to find the sinusoidal steady-state components of the mesh currents $i_1(t)$, $i_2(t)$, and $i_3(t)$. To do this, we may first apply KVL to the three meshes to find the integro-differential equations describing the network. These are

$$\frac{1}{2}\frac{di_1}{dt} + i_1(t) - i_2(t) = v_1(t)$$

$$\frac{1}{2}\frac{di_2}{dt} + 2i_2(t) - i_1(t) - i_3(t) = v_2(t)$$

$$2\int i_3(t)\,dt + i_3(t) - i_2(t) = v_3(t)$$

The voltage phasors are readily found to be[7]

$$\mathcal{V}_1 = 4\underline{/0°} = 4 + j0$$
$$\mathcal{V}_2 = 6\underline{/180°} = -6 + j0$$
$$\mathcal{V}_3 = 2\sqrt{2}\underline{/-45°} = 2 - j2.$$

Applying the method described in this section, we may directly write the equations relating the voltage phasors $\mathcal{V}_1$, $\mathcal{V}_2$, and $\mathcal{V}_3$ to the current phasors $\mathcal{I}_1$, $\mathcal{I}_2$, and $\mathcal{I}_3$. Thus, we obtain

$$\begin{bmatrix} \dfrac{j\omega}{2}+1 & -1 & 0 \\[2mm] -1 & \dfrac{j\omega}{2}+1 & -1 \\[2mm] 0 & -1 & \dfrac{2}{j\omega}+1 \end{bmatrix} \begin{bmatrix} \mathcal{I}_1 \\[2mm] \mathcal{I}_2 \\[2mm] \mathcal{I}_3 \end{bmatrix} = \begin{bmatrix} \mathcal{V}_1 \\[2mm] \mathcal{V}_2 \\[2mm] \mathcal{V}_3 \end{bmatrix}$$

Inserting the values for the voltage phasors given above and the value of ω ($= 2$ rad/s), we find

$$\begin{bmatrix} 1+j1 & -1 & 0 \\ -1 & 2+j1 & -1 \\ 0 & -1 & 1-j1 \end{bmatrix} \begin{bmatrix} \mathcal{I}_1 \\ \mathcal{I}_2 \\ \mathcal{I}_3 \end{bmatrix} = \begin{bmatrix} 4+j0 \\ -6+j0 \\ 2-j2 \end{bmatrix}$$

To solve this matrix equation, we first find the determinant (see Appendix A). Calling the 3×3 square matrix $\mathbf{Z}$ we obtain

$$\det \mathbf{Z} = (1+j1)(2+j1)(1-j1) - (1+j1) - (1-j1) = 2+j2$$

The cofactors Z_{ij} (see Sec. 3-3) are found as follows:

$$Z_{11} = \begin{vmatrix} 2+j1 & -1 \\ -1 & 1-j1 \end{vmatrix} = 2 - j1$$

$$Z_{12} = -\begin{vmatrix} -1 & -1 \\ 0 & 1-j1 \end{vmatrix} = 1 - j1 = Z_{21}$$

$$Z_{13} = \begin{vmatrix} -1 & 2+j1 \\ 0 & -1 \end{vmatrix} = 1 = Z_{31}$$

[7]Here we use an alternative representation for the phase of a complex quantity which is frequently more convenient than the exponential format and which is often found in the literature. It is defined by the relation

$$Ae^{j\alpha} = A\underline{/\alpha}$$

$$Z_{22} = \begin{vmatrix} 1 + j1 & 0 \\ 0 & 1 - j1 \end{vmatrix} = 2$$

$$Z_{23} = - \begin{vmatrix} 1 + j1 & -1 \\ 0 & -1 \end{vmatrix} = 1 + j1 = Z_{32}$$

$$Z_{33} = \begin{vmatrix} 1 + j1 & -1 \\ -1 & 2 + j1 \end{vmatrix} = j3$$

Forming the adjoint matrix of cofactors we obtain

$$\begin{bmatrix} \mathscr{I}_1 \\ \mathscr{I}_2 \\ \mathscr{I}_3 \end{bmatrix} = \frac{1}{2 + j2} \begin{bmatrix} 2 - j1 & 1 - j1 & 1 \\ 1 - j1 & 2 & 1 + j1 \\ 1 & 1 + j1 & j3 \end{bmatrix} \begin{bmatrix} 4 + j0 \\ -6 + j0 \\ 2 - j2 \end{bmatrix}$$

The result of such a solution is

$$\mathscr{I}_1 = 1 - j1 = \sqrt{2}\underline{/-45°}$$
$$\mathscr{I}_2 = -2 = 2\underline{/180°}$$
$$\mathscr{I}_3 = 1 - j1 = \sqrt{2}\underline{/-45°}$$

The corresponding values for the sinusoidal steady-state components of the mesh currents are

$$i_1(t) = \sqrt{2} \cos(2t - 45°)$$
$$i_2(t) = -2 \cos 2t$$
$$i_3(t) = \sqrt{2} \cos(2t - 45°)$$

Thus, we have found the solution for the network. The validity of this solution is readily demonstrated by substituting the steady-state components of the mesh currents into the original integro-differential equations and verifying that they do indeed satisfy these equations. This is left as an exercise for the reader.

It is readily apparent that the procedure outlined above requires considerable computational effort, even for the case where only three simultaneous equations are to be solved. As a result of this conclusion, we will find that it is advantageous to employ digital computational methods for circuits with considerably fewer numbers of independent variables than would be the case for a resistance network. A digital computer subroutine which may be used to solve such circuits is described in Appendix C.

7-5 IMPEDANCE AND ADMITTANCE

All the techniques of network analysis which we have developed to this point in this text are based on the use of KVL and KCL to determine a set of network equations. In this section we shall begin an investigation of the way in which these techniques may be extended to include the sinusoidal steady-state case. For example, consider the simple one-mesh circuit shown in Fig. 7-5.1. Applying KVL, we may write

$$v_L(t) + v_C(t) + v_R(t) = v_S(t) \tag{7-83}$$

Now let us consider what happens to (7-83) when only the sinusoidal steady-state

Figure 7-5.1 Simple one-mesh *RLC* circuit.

response is desired. We shall use the phasors $\mathcal{V}_L$, $\mathcal{V}_C$, $\mathcal{V}_R$, and $\mathcal{V}_S$ to represent the sinusoidally varying variables in the circuit. Thus, following the procedure given in the preceding section, we may write

$$v_L(t) = \text{Re}\,(\mathcal{V}_L e^{j\omega t}) \qquad v_C(t) = \text{Re}\,(\mathcal{V}_C e^{j\omega t})$$
$$v_R(t) = \text{Re}\,(\mathcal{V}_R e^{j\omega t}) \qquad v_S(t) = \text{Re}\,(\mathcal{V}_S e^{j\omega t}) \tag{7-84}$$

If we substitute the relations of (7-84) into (7-83), we obtain

$$\text{Re}\,(\mathcal{V}_L e^{j\omega t}) + \text{Re}\,(\mathcal{V}_C e^{j\omega t}) + \text{Re}\,(\mathcal{V}_R e^{j\omega t}) = \text{Re}\,(\mathcal{V}_S e^{j\omega t}) \tag{7-85}$$

Since the sum of a group of real parts of complex quantities is equal to the real part of the sum of the complex quantities, we may write Eq. (7-85) in the form

$$\text{Re}\,[(\mathcal{V}_L + \mathcal{V}_C + \mathcal{V}_R)e^{j\omega t}] = \text{Re}\,(\mathcal{V}_S e^{j\omega t}) \tag{7-86}$$

As was shown in the last section, this implies the equality of the complex members on both sides of the equation. Thus we may write

$$\mathcal{V}_L + \mathcal{V}_C + \mathcal{V}_R = \mathcal{V}_S \tag{7-87}$$

Comparing Eqs. (7-87) and (7-83), we see that the phasors $\mathcal{V}_L$, $\mathcal{V}_C$, $\mathcal{V}_R$, and $\mathcal{V}_S$ satisfy the same KVL relation that the sinusoidally varying quantities $v_L(t)$, $v_C(t)$, $v_R(t)$, and $v_S(t)$ satisfy. This conclusion is readily generalized. Thus we may say that when a set of voltage variables are related by KVL, the phasors which represent these voltage variables also satisfy an identical relation. The same conclusion holds true when KCL is applied. For example, consider the circuit shown in Fig. 7-5.2. Applying KCL, we may write

$$i_L(t) + i_C(t) + i_R(t) = i_S(t) \tag{7-88}$$

Figure 7-5.2 Simple one-node *RLC* circuit.

If we now assume that the current variables are all sinusoidal functions, then, under steady-state conditions, they may be represented by a set of phasors $\mathcal{I}_L$, $\mathcal{I}_C$, $\mathcal{I}_R$, and $\mathcal{I}_S$. The relations between the phasors and the original currents are

$$i_L(t) = \text{Re}\,(\mathcal{I}_L e^{j\omega t}) \qquad i_C(t) = \text{Re}\,(\mathcal{I}_C e^{j\omega t})$$
$$i_R(t) = \text{Re}\,(\mathcal{I}_R e^{j\omega t}) \qquad i_S(t) = \text{Re}\,(\mathcal{I}_S e^{j\omega t}) \tag{7-89}$$

Substituting the relations of Eq. (7-89) into (7-88) and making manipulations identical with those which were made for the KVL example, we obtain

$$\mathcal{I}_L + \mathcal{I}_C + \mathcal{I}_R = \mathcal{I}_S \tag{7-90}$$

Thus, we see that phasors will also satisfy relations identical with the KCL equations that relate the variables that the phasors represent. These conclusions are easily generalized, and may be summarized as follows:

SUMMARY 7-5.1

Phasors and Kirchhoff's Laws. If a set of sinusoidally varying quantities are related by a group of KVL and KCL equations, the set of phasors which represent these sinusoidally varying quantities are related by an identical group of equations. Thus, we may say that Kirchhoff's laws may be directly applied to phasors.[8]   ∎

Now let us consider again the circuit shown in Fig. 7-5.1. The equations relating the terminal variables of each of the passive elements shown in this circuit are

$$v_L(t) = L\frac{di}{dt} \qquad v_C(t) = \frac{1}{C}\int i(t)\,dt \qquad v_R(t) = Ri(t) \tag{7-91}$$

Now let us define a current phasor $\mathcal{I}$ to represent the variable $i(t)$ of (7-91). The defining relation is

$$i(t) = \mathrm{Re}\,(\mathcal{I}e^{j\omega t}) \tag{7-92}$$

If we substitute the relations of (7-84) and (7-92) in (7-91), we obtain the expressions

$$\mathrm{Re}\,(\mathcal{V}_L e^{j\omega t}) = L\frac{d}{dt}\,\mathrm{Re}\,(\mathcal{I}e^{j\omega t})$$

$$\mathrm{Re}\,(\mathcal{V}_C e^{j\omega t}) = \frac{1}{C}\int \mathrm{Re}\,(\mathcal{I}e^{j\omega t})\,dt \tag{7-93}$$

$$\mathrm{Re}\,(\mathcal{V}_R e^{j\omega t}) = R\,\mathrm{Re}\,(\mathcal{I}e^{j\omega t})$$

Since the operations of differentiation and taking the real part, and the operations of integration and taking the real part may be interchanged, if we perform the indicated differentiation and integration, we obtain

$$\mathrm{Re}\,(\mathcal{V}_L e^{j\omega t}) = \mathrm{Re}\,(j\omega L\mathcal{I}e^{j\omega t})$$

$$\mathrm{Re}\,(\mathcal{V}_C e^{j\omega t}) = \mathrm{Re}\left(\frac{1}{j\omega C}\mathcal{I}e^{j\omega t}\right) \tag{7-94}$$

$$\mathrm{Re}\,(\mathcal{V}_R e^{j\omega t}) = \mathrm{Re}\,(R\mathcal{I}e^{j\omega t})$$

Equating the complex members of the above and multiplying both sides of each relation by $e^{-j\omega t}$, we obtain

$$\mathcal{V}_L = j\omega L\mathcal{I} \qquad \mathcal{V}_C = \frac{1}{j\omega C}\mathcal{I} \qquad \mathcal{V}_R = R\mathcal{I} \tag{7-95}$$

The expressions given in (7-95) are the phasor equivalents of the terminal relations

---

[8]Note that due account must be taken of the reference polarity of phasors when applying KVL and KCL. These reference polarities are identical with the reference polarities of the variables that the phasors represent.

given in (7-91). Comparing the two sets of equations, we see that the operation of differentiation has been replaced by multiplication by the factor $j\omega$, and the operation of integration has been replaced by multiplication by the factor $1/j\omega$. Thus, the *integral and differential relations* of Eq. (7-91) are replaced by the *algebraic relations* of Eq. (7-95) when phasors are used to represent the sinusoidal network variables. Let us now express the relations of (7-95) in the form

$$\frac{\mathcal{U}_L}{\mathcal{I}} = j\omega L \qquad \frac{\mathcal{U}_C}{\mathcal{I}} = \frac{1}{j\omega C} \qquad \frac{\mathcal{U}_R}{\mathcal{I}} = R \qquad (7\text{-}96)$$

The left members of the expressions given in Eq. (7-96) each consist of the ratio of a voltage phasor to a current phasor. Thus, the right members of these expressions, namely, the expressions $j\omega L$, $1/j\omega C$, and R must have the same dimensional units. The name given to such quantities is *impedance*, and the units are *ohms*. When an impedance is purely real, it is referred to as a *resistive* impedance. When it is purely imaginary, it is referred to as a *reactive* impedance. An impedance which is neither purely real nor purely imaginary is simply referred to as a *complex* impedance. In general, we may use the letter Z to refer to an impedance and the letters R and X to refer to its *resistive part* and its *reactive part*, respectively. Thus, in general, we may write

$$Z = R + jX \qquad (7\text{-}97)$$

For example, considering the first relation given in (7-96), we see that an inductor has an impedance which is purely imaginary and has the value $j\omega L$. Thus, the reactance of an inductor is ωL. Obviously, this reactance will always be positive (for a positive-valued L). From the second relation given in (7-96), we see that a capacitor has an impedance which is purely imaginary and which has the value $-j/\omega C$. Thus, the reactance of a capacitor equals $-1/\omega C$, and it is always negative. Finally, a resistor has an impedance which is purely real and has a value of R.

Impedance, since it is a complex quantity, specifies not only the relation between the magnitudes of the voltage and current phasors, but also the relation between their phase angles. For example, from the first relation of Eq. (7-95), we see that the phase angle of $\mathcal{I}$ plus the phase angle of the impedance $j\omega L$ (which is 90°) must equal the phase angle of $\mathcal{U}_L$. We conclude that the phase angle of $\mathcal{U}_L$ will always be 90° greater than the phase angle of $\mathcal{I}$. Thus, we may say that *for an inductor, under conditions of sinusoidal steady-state excitation, the current sinusoid lags the voltage sinusoid by 90°*. It is frequently useful to express such relations by drawing a phasor diagram. Thus, for the variables $\mathcal{U}_L$ and $\mathcal{I}$ of the first relation of Eq. (7-95), the phasor diagram would appear as shown in Fig. 7-5.3(a). The relative position of the two vectors, i.e., the 90° angle, will be the same for all values of frequency ω and all (positive) values of L, although the relative lengths of the $\mathcal{U}_L$ and $\mathcal{I}$ vectors will be different. A phasor diagram of the type drawn in Fig. 7-5.3(a) only shows the *relative phase between phasors*. The actual phase angle is of little importance. Thus the phasor diagrams of Fig. 7-5.3(b) and (c) have exactly the same significance as the one shown in Fig. 7-5.3(a), namely, they show that the voltage across an inductor leads the current through it by 90° under conditions of sinusoidal steady-state excitation.

Figure 7-5.3 Phasor diagrams for an inductor.

A similar situation, as defined by the second relation given in (7-95), holds for a capacitor. Since the impedance of a capacitor is $-j/\omega C$, its phase angle will always be $-90°$. Thus, *for a capacitor under conditions of sinusoidal steady-state excitation, the current sinusoid leads the voltage sinusoid by 90°.* Some phasor diagrams for the voltage and current phasors of a capacitor are given in Fig. 7-5.4. Finally, for a resistor, since

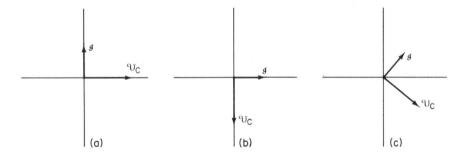

Figure 7-5.4 Phasor diagrams for a capacitor

the impedance of a resistor has a phase angle of zero degrees, we may say that *the voltage and current sinusoids for a resistor are in phase*, i.e., there is no phase lead or lag, under conditions of sinusoidal steady-state excitation. We may summarize the results given above as follows:

SUMMARY 7-5.2

Definition of Impedance. If voltage and current phasors are defined for the sinusoidal voltage and current variables of a two-terminal network element, then the expression relating the voltage phasor to the current phasor is defined as the impedance of the element. The real part of the impedance is called the resistance of the element, and the imaginary part is called the reactance. An inductor has a positive reactance which is proportional to frequency, and zero resistance. A capacitor has a negative reactance which is inversely proportional to frequency, and zero resistance. Under sinusoidal steady-state conditions, the current variable of an inductor lags the

Sinusoidal Steady-State Analysis   Chap. 7

voltage variable by 90°, while the current variable of a capacitor leads the voltage variable by 90°. For a resistor, the voltage and current variables are in phase.[9] ■

We may develop a set of conclusions similar to those given above by starting with the circuit shown in Fig. 7-5.2. The equations relating the terminal variables are

$$i_L(t) = \frac{1}{L} \int v(t)\, dt \qquad i_C(t) = C\frac{dv}{dt} \qquad i_R(t) = Gv(t) \qquad (7\text{-}98)$$

Now let us define a voltage phasor $\mathcal{V}$ to represent the variable $v(t)$ under conditions of sinusoidal steady-state. The defining relation is

$$v(t) = \text{Re}\,(\mathcal{V}e^{j\omega t}) \qquad (7\text{-}99)$$

If we substitute the relations of (7-89) and (7-99) in (7-98) and reduce the equations in a manner identical with that used in (7-93) through (7-95), we obtain the relations

$$\frac{\mathcal{I}_L}{\mathcal{V}} = \frac{1}{j\omega L} \qquad \frac{\mathcal{I}_C}{\mathcal{V}} = j\omega C \qquad \frac{\mathcal{I}_R}{\mathcal{V}} = G \qquad (7\text{-}100)$$

In these expressions, the left members all consist of the ratio of a current phasor to a voltage phasor. Thus, the expressions on the right, namely, $1/j\omega L$, $j\omega C$, and G, must have the same dimensional units. The name given to these quantities is *admittance*, and they have the units of mhos. When an admittance is purely real, it is referred to as a *conductive admittance*. When it is purely imaginary, it is referred to as a *susceptive admittance*. In general, we use the latter Y to refer to an admittance. The letters G and B are frequently used to refer to the *conductive part* and the *susceptive part* of an admittance, respectively. Thus, in general, an admittance may be written as follows:

$$Y = G + jB \qquad (7\text{-}101)$$

where G and B are real and Y is complex. For example, considering the first relation given in Eq. (7-100), we see that an inductor has an admittance which is purely imaginary and equals $-j/\omega L$. The susceptance of an inductor is $-1/\omega L$ and it is always negative. Similarly, a capacitor has an admittance of $j\omega C$. Its susceptance is ωC, and it is always positive. Finally, a resistor has an admittance which is purely real and which is equal to the value G. Comparing these results with those determined for the impedance of the inductor, the resistor, and the capacitor, we see that the admittance of these elements is the reciprocal of their impedance, and vice versa.

The conclusions given above may be summarized as follows:

SUMMARY 7-5.3

Definition of Admittance. If voltage and current phasors are defined for the sinusoidal voltage and current variables of a two-terminal network element, then the expression relating the current phasor to the voltage phasor is defined as the admit-

[9]The conclusions given here are based on the requirement that the relative reference polarities for the voltage and current phasors are the associated ones normally used for a two-terminal element, i.e., the reference arrow for the current phasor enters the terminal marked by the positive reference sign of the voltage phasor.

tance of the element. The real part of the admittance is referred to as the conductance of the element, and the imaginary part is called the susceptance. The admittance is the reciprocal of the impedance for a two-terminal network element. ■

A summary of the properties of the impedance and admittance of the resistor, the inductor, and the capacitor are given in Table 7-5.1.

TABLE 7-5.1 Impedance and Admittance of Passive Two-Terminal Network Elements

Element	Impedance	Resistance	Reactance	Admittance	Conductance	Susceptance
R	$Z_R = R$	$R_R = R$	$X_R = 0$	$Y_R = \dfrac{1}{R}$	$G_R = \dfrac{1}{R}$	$B_R = 0$
L	$Z_L = j\omega L$	$R_L = 0$	$X_L = \omega L$	$Y_L = \dfrac{1}{j\omega L}$	$G_L = 0$	$B_L = \dfrac{-1}{\omega L}$
C	$Z_C = \dfrac{1}{j\omega C}$	$R_C = 0$	$X_C = \dfrac{-1}{\omega C}$	$Y_C = j\omega C$	$G_C = 0$	$B_C = \omega C$

In the preceding paragraphs of this section we have defined the quantities of impedance and admittance, and we have pointed out that, for a given two-terminal element, these quantities are reciprocally valued. Thus, if we know the impedance, we obviously know the admittance and vice versa. Frequently, it will be convenient in our future discussions to refer to the concept implied by impedance and admittance without specifically choosing one or the other of these quantities. To do this we shall use the word *immittance*, which is formed from the words *imp*edance and ad*mittance*. Thus, the immittance of a two-terminal network element may be either the ratio of a voltage phasor to a current phasor or the (reciprocal) ratio of a current phasor to a voltage phasor. The connections between the terms immittance, impedance, admittance, resistance, reactance, conductance, and susceptance, as used in sinusoidal steady-state phasor analysis, are summarized in Table 7-5.2.

TABLE 7-5.2 Terminology for Phasor Analysis: Immittance-Impedance or Admittance

Impedance—Ratio of $\dfrac{\mathcal{V}}{\mathcal{I}}$		Admitance—Ratio of $\dfrac{\mathcal{I}}{\mathcal{V}}$	
Resistance $\mathrm{Re}\left(\dfrac{\mathcal{V}}{\mathcal{I}}\right)$	Reactance $\mathrm{Im}\left(\dfrac{\mathcal{V}}{\mathcal{I}}\right)$	Conductance $\mathrm{Re}\left(\dfrac{\mathcal{I}}{\mathcal{V}}\right)$	Susceptance $\mathrm{Im}\left(\dfrac{\mathcal{I}}{\mathcal{V}}\right)$

Now let us investigate the immittance of two terminal networks formed by interconnecting resistors, capacitors, and inductors. As a first example, consider the circuit shown in Fig. 7-5.1. If we substitute the relations of Eq. (7-95) in the phasor KVL equation given in (7-87), we obtain

$$j\omega L \mathcal{I} + \frac{1}{j\omega C} \mathcal{I} + R \mathcal{I} = \mathcal{V}_S \qquad (7\text{-}102)$$

This may be written in the form

$$\frac{\mathcal{V}_S}{\mathcal{I}} = R + j\omega L + \frac{1}{j\omega C} \qquad (7\text{-}103)$$

Thus, we see that if we consider the entire circuit (as seen from the terminals of the voltage source) as a two-terminal network element, the impedance of this circuit is equal to the sum of the separate impedances of each of the individual network elements. To see that this is true in general, consider the circuit consisting of a series connection of impedances Z_i ($i = 1, 2, \ldots, n$) as shown in Fig. 7-5.5. Voltage and

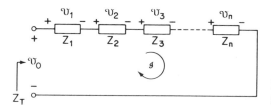

Figure 7-5.5 Series connection of impedances.

current phasors are defined as indicated in the figure. Applying KVL to the voltage phasors (see Summary 7-5.1), we may write

$$\mathcal{V}_1 + \mathcal{V}_2 + \mathcal{V}_3 + \cdots + \mathcal{V}_n = \mathcal{V}_0 \qquad (7\text{-}104)$$

If we now substitute the branch relations for each of the individual impedances, we may rewrite (7-104) as follows

$$Z_1 \mathcal{I} + Z_2 \mathcal{I} + Z_3 \mathcal{I} + \cdots + Z_n \mathcal{I} = (Z_1 + Z_2 + Z_3 + \cdots + Z_n) \mathcal{I} = \mathcal{V}_0 \qquad (7\text{-}105)$$

The impedance of the series connection of the individual impedances is simply equal to the ratio of the phasors $\mathcal{V}_0$ and $\mathcal{I}$. If we define this total impedance as Z_T, then, from (7-105), we may write

$$\frac{\mathcal{V}_0}{\mathcal{I}} = Z_T = Z_1 + Z_2 + Z_3 + \cdots + Z_n \qquad (7\text{-}106)$$

Thus, we see that the total impedance is the sum of the individual impedances when the elements are connected in series.

A similar development is possible with admittances. To see this, let us substitute the relations of (7-100) in the phasor KCL equation for Fig. 7-5.2 given in (7-90). We obtain

$$\frac{1}{j\omega L} \mathcal{V} + j\omega C \mathcal{V} + G \mathcal{V} = \mathcal{I}_S \qquad (7\text{-}107)$$

This may be rewritten in the form

$$\frac{\mathcal{I}_S}{\mathcal{V}} = \frac{1}{j\omega L} + j\omega C + G \qquad (7\text{-}108)$$

Thus, if we consider the entire circuit as a single two-terminal element with phasor variables $\mathcal{I}_S$ and $\mathcal{V}$, we see that the admittance of the circuit is given as the sum of the individual admittances which are connected in parallel. This conclusion is easily generalized. To see this, consider the circuit consisting of a parallel connection of admittances Y_i $(i = 1, 2, \ldots, n)$ as shown in Fig. 7-5.6. Applying KCL to the current phasors indicated in the figure, we obtain

$$\mathcal{I}_1 + \mathcal{I}_2 + \mathcal{I}_3 + \cdots + \mathcal{I}_n = \mathcal{I}_0 \tag{7-109}$$

Figure 7-5.6 Parallel connection of admittances.

If we now substitute the branch relations for each of the admittances in (7-109), we obtain

$$Y_1\mathcal{V} + Y_2\mathcal{V} + Y_3\mathcal{V} + \cdots + Y_n\mathcal{V}$$
$$= (Y_1 + Y_2 + Y_3 + \cdots + Y_n)\mathcal{V} = \mathcal{I}_0 \tag{7-110}$$

The total admittance Y_T of the entire network is simply the ratio of the phasors $\mathcal{I}_0$ and $\mathcal{V}$. Thus, we see that

$$\frac{\mathcal{I}_0}{\mathcal{V}} = Y_T = Y_1 + Y_2 + Y_3 + \cdots + Y_n \tag{7-111}$$

We conclude that the total admittance of a number of two-terminal elements connected in parallel is simply the sum of the admittances of the individual elements.

The developments given above are readily extended to other network configurations. For example, consider the voltage divider shown in Fig. 7-5.7. For this circuit we readily find that

$$\frac{\mathcal{V}_2}{\mathcal{V}_1} = \frac{Z_2}{Z_1 + Z_2} \tag{7-112}$$

Figure 7-5.7 Voltage divider.

As another example, consider the ladder network shown in Fig. 7-5.8. Following the development given in Sec. 2-4, we find that the input admittance Y_{in} of the ladder is given by the continued fraction

$$Y_{\text{in}} = Y_5 + \cfrac{1}{Z_4 + \cfrac{1}{Y_3 + \cfrac{1}{Z_2 + \cfrac{1}{Y_1}}}} \tag{7-113}$$

Sinusoidal Steady-State Analysis   Chap. 7

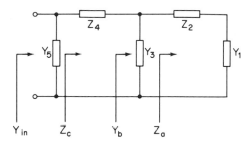

Figure 7-5.8 Ladder network of immittances.

Rationalizing this expression we obtain

$$Y_{in} = \frac{\begin{array}{c}Y_1 + Y_3 + Y_5 + Y_1 Z_4 Y_5 + Y_3 Z_4 Y_5 + Y_1 Z_2 Y_5 \\ + Y_1 Z_2 Y_3 + Y_1 Z_2 Y_3 Z_4 Y_5\end{array}}{1 + Y_1 Z_4 + Y_3 Z_4 + Y_1 Z_2 + Y_1 Z_2 Y_3 Z_4} \qquad (7\text{-}114)$$

Example 7-5.1

It is desired to find the impedance Z_0 of the ladder network shown in Fig. 7-5.9 at the terminals *a-b*. This is easily done by using a continued fraction similar to that given in Eq. (7-113). Thus, we may write

$$Z_0 = 1 + \cfrac{1}{j\dfrac{\omega}{2} + \cfrac{1}{j\omega + \cfrac{1}{1}}} = \frac{2 - (\omega^2/2) + j(3\omega/2)}{1 - (\omega^2/2) + j(\omega/2)}$$

ohms, henries, farads

Figure 7-5.9 Network for Example 7-5.1.

To separate the real part and the imaginary part of the above expression, we need merely multiply the numerator and the dominator by the conjugate of the dominator. Doing this and rearranging the terms, we obtain

$$Z_0 = \frac{2 - (3\omega^2/4) + (\omega^4/4) + j[(\omega/2) - (\omega^3/2)]}{1 - (3\omega^2/4) + (\omega^4/4)}$$

Thus, the resistive component of the impedance Z_0 is $[(2 - (3\omega^2/4) + (\omega^4/4)]/[1 - (3\omega^2/4) + (\omega^4/4)]$ and the reactive component is $[(\omega/2) - (\omega^3/2)]/[1 - (3\omega^2/4) + (\omega^4/4)]$.

We may make some interesting observations about the basic concept introduced in this section by studying the results of the example given above. First of all, we may note that the expression for the real or resistive part of the impedance is affected not only by the values of the resistive elements of the circuit, *but also by the values of the inductor and the capacitor.* As such, the resistive component of the impedance is a function of ω. Thus, in general, we must be careful to differentiate between the *resistance of the resistors* which are included in a given circuit (and which are clearly not functions of ω), and the *resistive portion of an impedance* which, except for the simplest of circuits, will be a function of frequency. Thus, if we let R stand for the real part of an impedance, we may write it more explicitly as $R(\omega)$, indicating its functional dependence.

In a similar manner, from the example given above, we may note that the reactive portion of an impedance is also a function of frequency. Thus, if we let X stand for such a reactive part, we may indicate the functional dependence by writing $X(\omega)$. Now let us consider the impedance itself. It is obviously comprised of algebraic combinations of terms which have three possible forms: (1) terms which are constant; (2) terms which are proportional to $j\omega$; and (3) terms which are proportional to $1/j\omega$. Thus, if we let Z stand for impedance, we may more explicitly write $Z(j\omega)$ to indicate the dependence of Z on the variable $j\omega$. Considering all the above, we may rewrite our general expression for impedance given in Eq. (7-97) as

$$Z(j\omega) = R(\omega) + jX(\omega) \tag{7-115}$$

A similar development is easily applied to the general expression for admittance given in Eq. (7-101). Thus, in general, we may write

$$Y(j\omega) = G(\omega) + jB(\omega) \tag{7-116}$$

Now let us consider what happens when we evaluate an expression of the type given in Eq. (7-115) at a particular value of ω. In such a case, we simply obtain a complex number for the impedance, since the dependence on ω has been eliminated by substituting a particular value of this variable. Such an impedance is equivalent (at that specific frequency) to a series connection of a single resistor and a single inductor or capacitor, As an example of this, consider the circuit shown in Fig. 7-5.9. If we evaluate the expression for Z_0 given in Example 7-5.1 with $\omega = 2$ rad/s, we obtain $Z_0 = \frac{3}{2} - j\frac{3}{2}$. We may construct an equivalent circuit which has the same impedance (at $\omega = 2$ rad/s) by using a single resistor and capacitor as shown in Fig. 7-5.10(a). Similarly, if we evaluate the impedance of the circuit of Fig. 7-5.9 at $\omega = \frac{1}{2}$ rad/s, we obtain $Z_0 = \frac{117}{53} + j\frac{12}{53}$. Thus, the given circuit has the same impedance as the equivalent circuit shown in Fig. 7-5.10(b) at the specified frequency. It should be noted that, depending on whether the reactive component is positive or negative at a given frequency, the equivalent circuit may include either an inductor or a capacitor. It is also possible for the reactive component to be zero. For example, if we evaluate the impedance Z_0 at $\omega = 1$ rad/s, we obtain $Z_0 = 3 + j0$. Thus, at this frequency, the circuit shown in Fig. 7-5.9 has exactly the same properties as an equivalent circuit

ohms,
henries,
farads

Figure 7-5.10 Equivalent circuits for the network of Fig. 7-5.9.

consisting of a single resistor of value 3 Ω, as shown in Fig. 7-5.10(c). The circuits shown in Fig. 7-5.10 are, of course, only equivalent with respect to sinusoidal steady-state conditions. More specifically, they are not equivalent if we are also considering transient behavior. Conclusions similar to those given above may be made on an admittance basis. In this case, we may derive an equivalent circuit consisting of a resistor in parallel with a single inductor or capacitor to represent any two-terminal admittance at a given frequency. As an example of this, since the impedance of the circuit shown in Fig. 7-5.9 has been shown to be $\frac{3}{2} - j\frac{3}{2}$ at $\omega = 2$ rad/s, its admittance at that same frequency is obviously equal to $1/(\frac{3}{2} - j\frac{3}{2}) = \frac{1}{3} + j\frac{1}{3}$. At the specified frequency, this is the same admittance as that of the equivalent circuit shown in Fig. 7-5.10(d). Thus, at $\omega = 2$ rad/s the circuits shown in Figs. 7-5.9 and 7-5.10(a) and (d) are all equivalent.

We may summarize the discussion given above as follows:

SUMMARY 7-5.4

Equivalence of Circuits Under Conditions of Sinusoidal Steady-State Excitation. Any two-terminal immittance formed by an arbitrary number of resistors, capacitors, and inductors may be replaced at a specified value of frequency by an equivalent two-terminal immittance consisting either of a series connection of a resistor with a capacitor or inductor, or a parallel connection of a resistor with a capacitor or inductor.

■

7-6 MESH AND NODE ANALYSIS WITH IMPEDANCES AND ADMITTANCES

In this section we will extend the concepts which have been presented in the preceding sections so as to develop general methods of phasor analysis for networks which are under conditions of sinusoidal steady-state excitation. The methods are very similar to those for resistance networks which were presented in Chap. 3. A review of that chapter will prove very worthwhile in enabling the reader to understand the material which follows.

We may begin our study of general phasor analysis methods by considering the circuit shown in Fig. 7-6.1. Let us assume that the circuit is to be analyzed under sinusoidal steady-state conditions, and that the values given for the various elements in the figure are impedances. Three mesh-current phasors $\mathcal{I}_1$, $\mathcal{I}_2$, and $\mathcal{I}_3$ are defined

Figure 7-6.1 *RLC* network analyzed by applying KVL.

as indicated. If we apply KVL directly to the phasor quantities and the impedances indicated in the figure, we obtain the following (phasor) mesh equations:

$$\mho_1 = j\frac{\omega}{2}\mathcal{I}_1 + (\mathcal{I}_1 - \mathcal{I}_2)$$

$$\mho_2 = j\frac{\omega}{2}\mathcal{I}_2 + (\mathcal{I}_2 - \mathcal{I}_1) + (\mathcal{I}_2 - \mathcal{I}_3) \qquad (7\text{-}117)$$

$$\mho_3 = -j\frac{2}{\omega}\mathcal{I}_3 + (\mathcal{I}_3 - \mathcal{I}_2)$$

These equations may be more conveniently written in matrix form as

$$
\begin{bmatrix} \mho_1 \\ \mho_2 \\ \mho_3 \end{bmatrix} =
\begin{bmatrix}
1 + j\dfrac{\omega}{2} & -1 & 0 \\
-1 & 2 + j\dfrac{\omega}{2} & -1 \\
0 & -1 & 1 - j\dfrac{2}{\omega}
\end{bmatrix}
\begin{bmatrix} \mathcal{I}_1 \\ \mathcal{I}_2 \\ \mathcal{I}_3 \end{bmatrix}
\qquad (7\text{-}118)
$$

Note that the elements of the square matrix given in Eq. (7-118) each relate a voltage phasor to a current phasor, thus they have the dimensions of impedance. We will refer to such a matrix as an *impedance matrix*. If we use the symbol $\mathbf{Z}(j\omega)$ for this matrix, then we may consider (7-118) as an example of a general matrix equation having the form

$$\mathbf{V} = \mathbf{Z}(j\omega)\mathbf{I} \qquad (7\text{-}119)$$

where $\mathbf{V}$ is a column matrix, each element of which is a phasor $\mho_i$, and $\mathbf{I}$ is a column matrix, each element of which is a phasor $\mathcal{I}_i$. A set of equations having the form of Eq. (7-119) will be referred to as a set of *mesh impedance equations*. It should be noted that the above determination of the mesh impedance equations by the use of the phasor version of KVL yields exactly the same set of equations as would be obtained by the substitution of expressions of the form $v_i(t) = \mathrm{Re}\,(\mho_i e^{j\omega t})$ and $i_i(t) = \mathrm{Re}\,(\mathcal{I}_i e^{j\omega t})$ into the original integro-differential equations describing the network. As an example of this, note that if we let $\omega = 2$ rad/s in (7-118) the equations become

$$
\mathbf{V} = \begin{bmatrix} \mho_1 \\ \mho_2 \\ \mho_3 \end{bmatrix} =
\begin{bmatrix}
1 + j1 & -1 & 0 \\
-1 & 2 + j1 & -1 \\
0 & -1 & 1 - j1
\end{bmatrix}
\begin{bmatrix} \mathcal{I}_1 \\ \mathcal{I}_2 \\ \mathcal{I}_3 \end{bmatrix}
= \mathbf{Z}(j2)\mathbf{I}
\qquad (7\text{-}120)
$$

The impedance matrix $\mathbf{Z}(j2)$ of Eq. (7-120) is readily seen to be identical with the square matrix obtained in Example 7-4.2 (the network shown in Fig. 7-4.2 is identical with the one shown in Fig. 7-6.1).

As another example of the determination of the mesh impedance equations, consider the network shown in Fig. 7-6.2. This network consists of a set of six impedances $Z_i(j\omega)$. Applying KVL to the three meshes defined by the phasor mesh current variables $\mathcal{I}_i$ we obtain

$$\mho_1 - \mho_3 = Z_1\mathcal{I}_1 + Z_3(\mathcal{I}_1 - \mathcal{I}_2) + Z_2(\mathcal{I}_1 - \mathcal{I}_3)$$

$$-\mho_2 = Z_5\mathcal{I}_2 + Z_3(\mathcal{I}_2 - \mathcal{I}_1) + Z_4(\mathcal{I}_2 - \mathcal{I}_3) \qquad (7\text{-}121)$$

$$\mho_3 = Z_6\mathcal{I}_3 + Z_2(\mathcal{I}_3 - \mathcal{I}_1) + Z_4(\mathcal{I}_3 - \mathcal{I}_2)$$

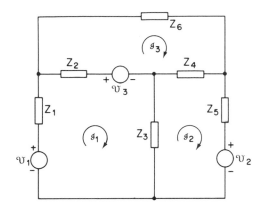

Figure 7-6.2 Network analyzed using KVL.

Putting these equations in matrix format, we may write

$$
\begin{bmatrix} \mathcal{V}_1 - \mathcal{V}_3 \\ -\mathcal{V}_1 \\ \mathcal{V}_3 \end{bmatrix} = \begin{bmatrix} Z_1 + Z_2 + Z_3 & -Z_3 & -Z_2 \\ -Z_3 & Z_3 + Z_4 + Z_5 & -Z_4 \\ -Z_2 & -Z_4 & Z_2 + Z_4 + Z_6 \end{bmatrix} \begin{bmatrix} \mathcal{I}_1 \\ \mathcal{I}_2 \\ \mathcal{I}_3 \end{bmatrix} \qquad (7\text{-}122)
$$

If we examine the elements of the matrices of the mesh impedance equations derived for the above examples, we readily see that we may formulate a set of rules for directly determining the expressions for these elements. These rules exactly parallel the rules given in Sec. 3-2 for determining the elements of the matrix equation $\mathbf{v}(t) = \mathbf{R}\mathbf{i}(t)$ for a resistance network. They may be summarized as follows:

SUMMARY 7-6.1

Rules for Forming the Mesh Impedance Equations.   The elements of the matrices of the mesh impedance equation given in (7-119) may be found as follows:

$\mathcal{V}_i$ the summation of the phasors defining all the voltage sources in the ith mesh, with sources whose reference polarity constitutes a voltage rise (in the positive reference direction of the ith mesh current) treated as positive, and with sources whose reference polarity constitutes a voltage drop treated as negative

$Z_{ii}(j\omega)$ the summation of the values of all the impedances which are located in the ith mesh

$Z_{ij}(j\omega)$ $(i \neq j)$ the negative of the sum of the values of any impedances which are common to the ith and the jth meshes.[10]

$\mathcal{I}_j$ the mesh current phasors which are the unknown variables for the network   ∎

The negative polarity associated with the elements $Z_{ij}(j\omega)$ $(i \neq j)$ only occurs,

[10]The quantity $Z_{ii}(j\omega)$ is sometimes called the *self impedance* of the ith mesh, while the quantity $Z_{ij}(j\omega)$ $(i \neq j)$ is called the impedance *mutual* to the ith and jth meshes.

of course, if all the mesh current phasors are given the same positive reference direction. An example of the application of these rules follows.

Example 7-6.1

It is desired to obtain the mesh impedance equations for the circuit shown in Fig. 7-6.3. Applying the rules given in Summary 7-6.1, we obtain

$$
\begin{bmatrix} \mathcal{U}_1 \\ -\mathcal{U}_2 \\ 0 \end{bmatrix} =
\begin{bmatrix}
R_1 + j\omega L_1 - \dfrac{j}{\omega C_1} & -R_1 + \dfrac{j}{\omega C_1} & -j\omega L_1 \\
-R_1 + \dfrac{j}{\omega C_1} & R_1 + R_2 + R_4 - \dfrac{j}{\omega C_1} & -R_2 \\
-j\omega L_1 & -R_2 & R_2 + R_3 + j\omega L_1 - \dfrac{j}{\omega C_2}
\end{bmatrix}
\begin{bmatrix} \mathcal{I}_1 \\ \mathcal{I}_2 \\ \mathcal{I}_3 \end{bmatrix}
$$

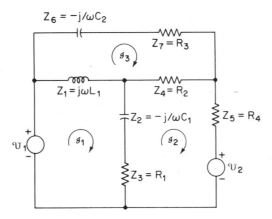

Figure 7-6.3 Circuit for Example 7-6.1.

Once the mesh impedance equations have been formulated, their solution is accomplished by the same techniques used in Chap. 3 for resistance networks.

Example 7-6.2

As an example of the solution of a set of mesh impedance equations, consider the network shown in Fig. 7-6.4. Its impedance matrix is readily found to be

$$
\mathbf{Z}(j\omega) =
\begin{bmatrix}
1 + j\omega & -1 & 0 \\
-1 & 2 + j\omega & -1 \\
0 & -1 & 1 + j\omega
\end{bmatrix}
$$

Figure 7-6.4 Three-mesh *RL* circuit.

To find the inverse matrix, we first find the determinant. Following the rules given in Appendix A, we obtain

$$
\det \mathbf{Z}(j\omega) = (1 + j\omega)(2 + j\omega)(1 + j\omega) - (1 + j\omega) - (1 + j\omega)
$$
$$
= -4\omega^2 + j(3\omega - \omega^3)
$$

If we let the quantities Z_{ij} be the cofactors of $\mathbf{Z}(j\omega)$, then these cofactors are readily found as follows:

$$Z_{11} = \begin{vmatrix} 2+j\omega & -1 \\ -1 & 1+j\omega \end{vmatrix} = 1 - \omega^2 + j3\omega$$

$$Z_{12} = -\begin{vmatrix} -1 & -1 \\ 0 & 1+j\omega \end{vmatrix} = 1 + j\omega = Z_{21}$$

$$Z_{13} = \begin{vmatrix} -1 & 2+j\omega \\ 0 & -1 \end{vmatrix} = 1 = Z_{31}$$

$$Z_{22} = \begin{vmatrix} 1+j\omega & 0 \\ 0 & 1+j\omega \end{vmatrix} = 1 - \omega^2 + j2\omega$$

$$Z_{23} = -\begin{vmatrix} 1+j\omega & -1 \\ 0 & -1 \end{vmatrix} = 1 + j\omega = Z_{32}$$

$$Z_{33} = \begin{vmatrix} 1+j\omega & -1 \\ -1 & 2+j\omega \end{vmatrix} = 1 - \omega^2 + j3\omega$$

Substituting the above values in the general relation for the inverse of a 3×3 matrix,

$$\mathbf{Z}(j\omega)^{-1} = \frac{1}{\det \mathbf{Z}(j\omega)} \begin{bmatrix} Z_{11} & Z_{21} & Z_{31} \\ Z_{12} & Z_{22} & Z_{32} \\ Z_{13} & Z_{23} & Z_{33} \end{bmatrix}$$

we obtain

$$\mathbf{Z}(j\omega)^{-1} = \frac{1}{-4\omega^2 + j(3\omega - \omega^3)} \begin{bmatrix} 1 - \omega^2 + j3\omega & 1 + j\omega & 1 \\ 1 + j\omega & 1 - \omega^2 + j2\omega & 1 + j\omega \\ 1 & 1 + j\omega & 1 - \omega^2 + j3\omega \end{bmatrix}$$

Thus, using the relation $\mathbf{I} = \mathbf{Z}(j\omega)^{-1}\mathbf{V}$, we can find the mesh currents $\mathcal{I}_i$ for any specified set of excitations $\mathcal{V}_i$, and for any value of frequency ω. It is readily verified that the product $\mathbf{Z}(j\omega)\mathbf{Z}(j\omega)^{-1}$ is equal to the identity matrix.

It should be apparent that the general procedure given above becomes quite involved for networks having three or more independent variables, i.e., networks having three or more simultaneous equations in their impedance or admittance formulation. As a matter of fact, the network used in the example given above for a 3×3 impedance matrix was purposely selected so that the elements of the impedance matrix were polynomials rather than rational functions (ratios of polynomials). In the latter (and more usual) case, the computations would be considerably more complex.

A development similar to that given in the preceding paragraphs is readily made if it is desired to write a set of equations relating the dependent and independent variables of a network using quantities of admittance rather than impedance. The procedure is quite similar to that given for the determination of the node equations for a resistance network in Sec. 7-5. As an example of the procedure, consider the network shown in Fig. 7-6.5. If we apply KCL to the two nodes of this network, then, specifying the branch currents in terms of the branch admittances and the phasor node voltages, we may write

Figure 7-6.5 *RLC* network analyzed using KCL.

$$\mathscr{I}_1 = G_1 \mathcal{V}_1 - \frac{j}{\omega L_1}(\mathcal{V}_1 - \mathcal{V}_2) + G_2(\mathcal{V}_1 - \mathcal{V}_2)$$

$$\mathscr{I}_2 = \frac{-j}{\omega L_2}\mathcal{V}_2 + j\omega C_1 \mathcal{V}_2 - \frac{j}{\omega L_1}(\mathcal{V}_2 - \mathcal{V}_1) + G_2(\mathcal{V}_2 - \mathcal{V}_1)$$

(7-123)

Rearranging the terms and putting the equations in matrix format, we obtain

$$
\begin{bmatrix} \mathscr{I}_1 \\ \mathscr{I}_2 \end{bmatrix} =
\begin{bmatrix}
G_1 + G_2 - \dfrac{j}{\omega L_1} & -G_2 + \dfrac{j}{\omega L_1} \\[2ex]
-G_2 + \dfrac{j}{\omega L_1} & G_2 - \dfrac{j}{\omega L_1} - \dfrac{j}{\omega L_2} + j\omega C_1
\end{bmatrix}
\begin{bmatrix} \mathcal{V}_1 \\ \mathcal{V}_2 \end{bmatrix}
$$

(7-124)

Note that the elements of the square matrix given in (7-124) each relate a voltage phasor to a current phasor; thus, these elements each have the dimensions of admittance. It is therefore logical to refer to the square matrix as an *admittance matrix*. If we use the designation $\mathbf{Y}(j\omega)$ for this matrix, then Eq. (7-124) may be written in the general form

$$\mathbf{I} = \mathbf{Y}(j\omega)\mathbf{V}$$

(7-125)

where $\mathbf{I}$ and $\mathbf{V}$ are column matrices whose elements are the phasors $\mathscr{I}_i$ and $\mathcal{V}_i$, respectively. A set of equations having the form of (7-125) will be referred to as a set of *node admittance equations*.

As another example of the determination of the node admittance equations, consider the network shown in Fig. 7-6.6. Applying KCL to each of the nodes of this figure, we obtain the equations

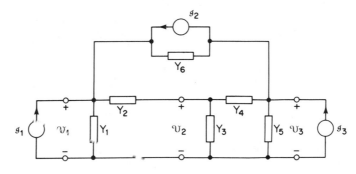

Figure 7-6.6 Network analyzed using KCL.

$$\mathscr{I}_1 + \mathscr{I}_2 = Y_1 \mathcal{V}_1 + Y_2(\mathcal{V}_1 - \mathcal{V}_2) + Y_6(\mathcal{V}_1 - \mathcal{V}_3)$$

$$0 = Y_3 \mathcal{V}_2 + Y_2(\mathcal{V}_2 - \mathcal{V}_1) + Y_4(\mathcal{V}_2 - \mathcal{V}_3) \qquad (7\text{-}126)$$

$$\mathscr{I}_3 - \mathscr{I}_2 = Y_5 \mathcal{V}_3 + Y_4(\mathcal{V}_3 - \mathcal{V}_2) + Y_6(\mathcal{V}_3 - \mathcal{V}_1)$$

Putting these in the form of (7-125), we obtain

$$
\begin{bmatrix} \mathscr{I}_1 + \mathscr{I}_2 \\ 0 \\ \mathscr{I}_3 - \mathscr{I}_2 \end{bmatrix}
=
\begin{bmatrix}
Y_1 + Y_2 + Y_6 & -Y_2 & -Y_6 \\
-Y_2 & Y_2 + Y_3 + Y_4 & -Y_4 \\
-Y_6 & -Y_4 & Y_4 + Y_5 + Y_6
\end{bmatrix}
\begin{bmatrix} \mathcal{V}_1 \\ \mathcal{V}_2 \\ \mathcal{V}_3 \end{bmatrix}
\qquad (7\text{-}127)
$$

It we examine the elements of the matrices of the node admittance equations for the above examples, we readily note that we may formulate a set of rules for directly determining the expressions for these elements. These rules exactly parallel the rules given in Sec. 3-5 for determining the elements of the matrix equation $\mathbf{i}(t) = \mathbf{G}\mathbf{v}(t)$ for a resistance network. They may be summarized as follows:

SUMMARY 7-6.2

Rules for Forming the Node Admittance Equations. The elements of the matrices of the node admittance equation given in (7-125) may be found as follows:

$\mathscr{I}_i$ the summation of the phasors representing the current sources connected to the ith node; sources whose positive reference direction is toward the node are treated as positive, and those whose reference direction is away from the node are treated as negative

$Y_{ii}(j\omega)$ the summation of the values of all the admittances connected to the ith node

$Y_{ij}(j\omega)\ (i \neq j)$ the negative of the sum of the values of any admittances which are connected between the ith and jth nodes[11]

$\mathcal{V}_i$ the node voltage phasors which are the unknown variables for the network ∎

An example of the application of these rules follows.

Example 7-6.3

It is desired to obtain the node admittance equations for the circuit shown in Fig. 7-6.7. Applying the rules given in Summary 7-6.2, we obtain

$$
\begin{bmatrix} \mathscr{I}_1 \\ \mathscr{I}_2 \\ -\mathscr{I}_2 \\ \mathscr{I}_3 \end{bmatrix}
=
\begin{bmatrix}
j4 & j2 & -j4 & 0 \\
j2 & 14 + j10 & -j7 & -8 \\
-j4 & -j7 & 19 & -10 + j11 \\
0 & -8 & -10 + j11 & 30 + j2
\end{bmatrix}
\begin{bmatrix} \mathcal{V}_1 \\ \mathcal{V}_2 \\ \mathcal{V}_3 \\ \mathcal{V}_4 \end{bmatrix}
$$

Practically any of the other analysis techniques for resistance networks which

---

[11]The quantity $Y_{ii}(j\omega)$ is sometimes called the *self-admittance* of the ith node, while the quantity $Y_{ij}(j\omega)\ (i \neq j)$ is called the *mutual admittance* between the ith and jth nodes.

Figure 7-6.7 Network for Example 7-6.3.

were introduced in Chap. 3 and summarized in the flow chart of Fig. 3-7.4 may be applied to the phasor analysis of *RLC* networks. For example, a non-ideal source consisting of the series connection of a voltage source and an impedance may be converted to an equivalent non-ideal source consisting of the parallel connection of a current source and an admittance. The general form of the two sources is shown in Fig. 7-6.8. It is easily shown that the sources are equivalent in terms of their behavior

(a)   (b)   **Figure 7-6.8** Non-ideal sources.

at the external terminals *a-b* under the conditions that

$$Y(j\omega) = \frac{1}{Z(j\omega)} \qquad \mathcal{V} = \mathcal{I}Z(j\omega) \qquad (7\text{-}128)$$

As an example of such an equivalence, the two non-ideal sources shown in Fig. 7-6.9(a) and (b) are equivalent. In addition, since the series impedance $1 + j1 \ \Omega$ is

Figure 7-6.9 Example of non-ideal sources.

Sinusoidal Steady-State Analysis   Chap. 7

equal to an admittance of $1/(1 + j1)$ or $\frac{1}{2} - j\frac{1}{2}$ mho, the immittance itself may also be realized by a parallel connection of elements as shown in Fig. 7-6.9(c) and (d). Such an equivalence is only valid, of course, at the specified frequency.

Example 7-6.4

As an example of the use of equivalent non-ideal sources, consider the shunt-peaked transistor stage shown in Fig. 7-6.10(a). In this circuit an inductor is used to extend the frequency response of the common-emitter transistor amplifier. For convenience, the dc biasing components have been omitted. Under small-signal operation, the circuit may be modeled as shown in Fig. 7-6.10(b). To analyze this circuit, it is desired to

Figure 7-6.10 Shunt-peaked transistor amplifier.

find the relation between the output current phasor $\mathcal{I}_{\text{out}}$ and the input current phasor $\mathcal{I}_{\text{in}}$. As an intermediate step, we may find the relation between $\mathcal{U}$ and $\mathcal{I}_{\text{in}}$. To do this, we may first transform the Norton type non-ideal source shown at the left of Fig. 7-6.10(b) into a Thevenin type as shown in Fig. 7-6.10(c). Applying the voltage divider relation to this circuit and using the relation $\mathcal{I}_{\text{out}} = -g_m \mathcal{U}$, we obtain

$$\frac{\mathcal{I}_{\text{out}}}{\mathcal{I}_{\text{in}}} = \frac{(-g_m/C_i)(j\omega + R_1/L)}{(j\omega)^2 + j\omega[(r_b + R_1)/L + 1/R_iC_i] + (r_b + R_1 + R_i)/LR_iC_i} \qquad (7\text{-}129)$$

This is of the general form

$$\frac{\mathcal{I}_{\text{out}}}{\mathcal{I}_{\text{in}}} = H\frac{j\omega + z}{(j\omega)^2 + aj\omega + b} \qquad (7\text{-}130)$$

A condition which is frequently used to extend the frequency range of a circuit is called the *maximally flat magnitude* condition (see the problems and Sec. 7-9). Applying this to the above equation we require that $a^2 - 2b = (b/z)^2$. If we use the following typical circuit values

$$r_b = 100\ \Omega \qquad R_i = 250\ \Omega \qquad R_L = 500\ \Omega$$
$$R_1 = 87.5\ \Omega \qquad g_m = 0.2\ \text{mho} \qquad C_i = 290 \times 10^{-9}\ \text{F}$$

then we find that, to satisfy the maximally flat magnitude condition, $L = 3.4\ \text{mH}$.

Another parallel between sinusoidal steady-state networks and resistance networks lies in the fact that any collection of sources and two-terminal elements may be replaced, at a given pair of terminals *and at a given frequency*, by a Thevenin or Norton equivalent circuit consisting of a single source plus an immittance, the latter in turn consisting of a single resistor and an inductor or capacitor. The techniques for determining the value of the source and the value of the immittance are similar to those given in Sec. 3-8 for a resistance network. Thus, if a Thevenin equivalent circuit is to be found, it is necessary to determine the open-circuit voltage at the specified pair of terminals, and the impedance seen at those terminals when all *independent* sources are set to zero. As an example of this, consider the circuit shown in Fig. 7-6.11(a). To determine the open-circuit voltage at terminals *a-b* it is con-

Figure 7-6.11   Thevenin and Norton equivalent circuits.

venient to first convert the non-ideal source consisting of a voltage source and a series inductor to a current source and a parallel inductor as shown in Fig. 7-6.11(b). Combining the two parallel current sources and adding the admittances of the paralleled passive elements, we obtain the circuit shown in Fig. 7-6.11(c). From this figure we see that the open-circuit voltage phasor is $-j1$. The shunt admittance is readily converted to a series impedance of value $\frac{1}{2} - j\frac{1}{2}\,\Omega$. Thus, the Thevenin equivalent circuit has the form shown in Fig. 7-6.11(d). A Norton equivalent for the circuit shown in Fig. 7-6.11(a) is readily found by determining the short-circuit current at the terminals *a-b*, and the admittance at these terminals when the independent sources in the network are set to zero. Alternatively, the Norton equivalent may be found directly from the Thevenin equivalent circuit given in Fig. 7-6.11(d), using the rules for conversion of non-ideal sources described above.

It should be apparent from the above discussion that, using the techniques of source conversion, adding voltage sources in parallel, and adding current sources in series as discussed in connection with resistance networks in Sec. 3-6, it is possible as a preliminary step to finding the solution of the network (i.e., of determining the value of some variable or variables) to write either a set of mesh impedance equations *or* a set of node admittance equations for a given network. The choice of which formulation to use is determined by two main factors, namely, computational efficiency and convenience. Computational efficiency is mainly concerned with the number of simultaneous equations which are to be solved. The number of equations for the two types of formulation can be found as follows: If n is the number of nodes, and b the number of branches, there will be $b - (n - 1)$ mesh impedance equations and $n - 1$ node admittance equations. The "convenience" factor is somewhat harder to evaluate. For example, for a network with only voltage sources, with all elements specified in terms of their impedances, and in which it is desired to find the value of some network current, it may be more convenient to use a mesh impedance formulation than a node admittance formulation. This avoids the necessity of transforming sources, taking the reciprocal of the impedance values to find the admittances, etc., even though the number of mesh impedance equations is greater than the corresponding number of node equations. This conclusion would be even more applicable if digital computer methods are to be used to obtain the solution. Other examples where the convenience factor dominates are easily found.

7-7 NETWORK FUNCTIONS

In Sec. 7-5 we introduced the concept of immittance, as embodied in impedance and admittance. These latter quantities were defined, respectively, as the ratio of a voltage phasor to a current phasor, and the ratio of a current phasor to a voltage phasor, with the added conditions that the voltage and current phasors were both defined at the same pair of terminals, and that their relative reference polarities followed the usual associated convention for two-terminal elements. It was also pointed out that impedance was reciprocal to admittance, i.e., once one of these quantities was known, the other could be easily found by simple inversion. The general form of an immittance was shown to be that of a rational function, i.e., a ratio of polynomials which were functions of the variable $j\omega$. In this section we shall show that impedance and admittance are two examples of a more general class of rational functions which are referred to as *network functions*. More specifically, we may define a network function as the ratio of a response phasor to an excitation phasor. In general, it is not necessary that these phasors be located at the same pair of terminals (as they must be to satisfy the definitions of immittances which we have used previously). As an example of a network function, consider the circuit shown in Fig. 7-7.1. Let us assume that we desire to find the network function defined as the ratio of the response phasor $\mathcal{I}_2$ to the excitation

Figure 7-7.1   *RLC* network.

phasor $\mathcal{V}_1$. We may begin by writing the phasor KVL equations around the two meshes. Using matrix notation, we obtain

$$\begin{bmatrix} \mathcal{V}_1 \\ 0 \end{bmatrix} = \begin{bmatrix} 1 + j\omega & -j\omega \\ -j\omega & 1 + j\left(\omega - \dfrac{1}{\omega}\right) \end{bmatrix} \begin{bmatrix} \mathcal{I}_1 \\ \mathcal{I}_2 \end{bmatrix} \tag{7-131}$$

To determine $\mathcal{I}_2$ as a function of $\mathcal{V}_1$, the inverse set of equations is required. Using the techniques of Sec. 3-1, we obtain

$$\begin{bmatrix} \mathcal{I}_1 \\ \mathcal{I}_2 \end{bmatrix} = \frac{\omega}{2\omega + j(2\omega^2 - 1)} \begin{bmatrix} 1 + j\left(\omega - \dfrac{1}{\omega}\right) & j\omega \\ j\omega & 1 + j\omega \end{bmatrix} \begin{bmatrix} \mathcal{V}_1 \\ 0 \end{bmatrix}$$

The desired network function is easily seen to be

$$\frac{\mathcal{I}_2}{\mathcal{V}_1} = \frac{j(\omega)^2}{2\omega + j(2\omega^2 - 1)} \tag{7-132}$$

Since it represents the ratio of a current phasor to a voltage phasor, this network function has the dimensions of admittance (units of mhos). The function, however, does not satisfy the definition previously given for an admittance, since the excitation and response phasors are located at different points in the network. We shall use the term *transfer admittance* to define such a network function. The type of admittance defined earlier in this chapter (referring to two phasors located at the *same pair* of network terminals) may be called a *driving-point admittance* to emphasize its character. In a similar manner, we may define a *transfer impedance* as the ratio of a voltage response phasor to a current excitation phasor, where the phasors are *not* located at the same terminal pair. A network function relating a voltage phasor to a current phasor at the same pair of terminals is called a *driving-point impedance*. The term *transfer immittance* is used to collectively refer to transfer admittances and transfer impedances. Similarly, the term *driving-point immittance* is used to collectively refer to driving-point impedances and admittances.

A second class of transfer functions consists of *dimensionless transfer functions*. The first of these is defined as the ratio of a voltage response phasor to a voltage excitation phasor. Such a network function is called a *voltage transfer function*. Similarly, we may define a *current transfer function* as the ratio of a current response phasor to a current excitation phasor. An example is given in Eq. (7-129) of Example 7-6.4. The various network functions described above are summarized in Table 7-7.1. Note that they are all defined as the ratio of response to excitation, and *not* as the ratio of excitation to response. It is possible to also define network functions relating charge and/or flux linkages to voltage, current, charge, etc. In practice, however, these are rarely used.

In analyzing the properties of a given network function, it is frequency advantageous to make a plot of its characteristics as the frequency variable ω is varied over some range. The range usually starts from 0 rad/s and goes to some value of frequency which is large enough so that all the significant variations of the network function have been included. Such a plot shows the *frequency response* of the network function. Since the value of the network function is complex, two loci are generally required to

TABLE 7-7.1 Types of Network Functions

Driving-Point Functions		Transfer Functions			
Immittance Functions		Immittance Functions		Dimensionless Functions	
Driving-Point Impedance	Driving-Point Admittance	Transfer Impedance	Transfer Admittance	Transfer Voltage Ratio	Transfer Current Ratio
$\dfrac{\mathcal{V}}{\mathcal{I}}$	$\dfrac{\mathcal{I}}{\mathcal{V}}$	$\dfrac{\mathcal{V}}{\mathcal{I}}$	$\dfrac{\mathcal{I}}{\mathcal{V}}$	$\dfrac{\mathcal{V}_2}{\mathcal{V}_1}$	$\dfrac{\mathcal{I}_2}{\mathcal{I}_1}$

display its frequency response. In this text we shall usually use the magnitude and the phase as the quantities plotted against frequency.[12]

As an introduction to the study of the frequency response of network functions, let us consider the *RC* network shown in Fig. 7-7.2(a). The network function that relates the response voltage phasor $\mathcal{V}_2$ to the excitation voltage phasor $\mathcal{V}_1$ is

$$\frac{\mathcal{V}_2}{\mathcal{V}_1} = \frac{1}{1 + j\omega RC} \tag{7-133}$$

Sketches of the magnitude and phase of this network function as ω is varied are shown in Fig. 7-7.2(b) and (c). Note that in the magnitude curve shown in Fig. 7-7.2(b), the magnitude is unity at $\omega = 0$, but that it approaches zero as ω approaches ∞. Such a function is referred to as a *low-pass function*, since the region where its magnitude is large is lower in frequency than the region where its magnitude is small and, thus, more of any low-frequency sinusoidal excitations which are applied are "passed" than are high-frequency ones. Since the function is only of first-order in $j\omega$, it may also be referred to as a *first-order low-pass function*. Note that the phase of the network function as shown in Fig. 7-7.2(c) varies from $0°$ (at $\omega = 0$) to $-90°$ (when ω becomes very large). Thus, the response sinusoid will lag behind the excitation sinusoid by a phase angle of 0 to 90°, the exact angle depending on the value of ω. Considering this, it is customary to refer to a network of this type as a *lag network*.

[12]In studying some properties of networks, it is also useful to plot the real part and the imaginary part of the network function.

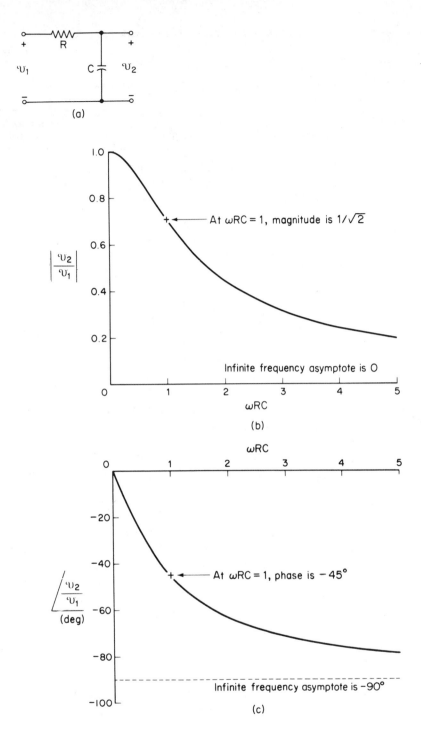

Figure 7-7.2 Low-pass *RC* network and its magnitude and phase plots.

In general, for a network with a low-pass characteristic, each reactive element such as an inductor or a capacitor will contribute 90° to the total phase lag provided by the network.

One of the important characteristics of low-pass functions of the type given in Eq. (7-133) is the frequency at which the magnitude characteristic decreases to 0.707 of its zero frequency value. The frequency at which this occurs is called the *3-dB frequency*, where dB is an abbreviation for *decibel*, a logarithmic measure of the magnitude of the network function which is defined by the relation

$$\text{dB} = 20 \log |N(j\omega)| \tag{7-134}$$

where $N(j\omega)$ is the network function and log is the logarithm to the base 10. For a transfer function, when the magnitude drops to 0.707 of its zero frequency value, the power being transmitted drops to the square of this, i.e., one-half. Thus the 3-dB frequency is also referred to as the *half-power frequency*. The 3-dB frequency for the function given in Eq. (7-133) is readily found by noting that the magnitude at $\omega = 0$ rad/s is unity. Thus, at the 3-dB frequency $\omega_{3\,\text{dB}}$, we have

$$\left| \frac{1}{1 + j\omega_{3\,\text{dB}}RC} \right| = 0.707 = \frac{1}{\sqrt{2}} \tag{7-135}$$

If we multiply both sides of the above equation by their complex conjugates to convert the expression to one involving only magnitudes, we obtain

$$\frac{1}{1 + (\omega_{3\,\text{dB}}RC)^2} = \frac{1}{2} \tag{7-136}$$

Solving this equation, we find that $\omega_{3\,\text{dB}} = 1/RC$.

As a second example of a network function, let us consider the voltage transfer function for the network shown in Fig. 7-7.3(a). The function is easily found to be

$$\frac{\mathcal{V}_2}{\mathcal{V}_1} + \frac{j\omega RC}{1 + j\omega RC} \tag{7-137}$$

Sketches of the magnitude and phase of this network function as ω is varied are shown in Fig. 7-7.3(b) and (c). It is easily verified that the magnitude characteristic has a shape which is inverse to that shown for a low-pass network function in Fig. 7-7.2(b). Since this network function provides most of its transmission in the high-frequency range, and since it has one reactive element, it is appropriately referred to as a *first-order high-pass network function*. Note that the phase of the network function as shown in Fig. 7-7.3(c) varies from 90° (at $\omega = 0$) to 0° (for the range where ω becomes very large). Therefore, the response sinusoid will lead the excitation sinusoid by a phase angle of 90° to 0°, depending on the frequency. Thus, the network is appropriately called a *lead network*. In general, each reactive element, such as an inductor or a capacitor, in a high-pass network will contribute 90° to the total phase shift of the network function. It should be noted that *causality*, i.e., the principle that an output cannot precede the input that produces it, naturally precludes the output sinusoid from "leading" the input sinusoid. Thus, a lead network such as the one shown in Fig. 7-7.3(a) actually produces a negative phase which ranges from −270° to −360° rather than a positive phase going from +90° to 0°. However, since individual cycles of the sinusoidal excitation lose their identity during sinusoidal steady-

(a)

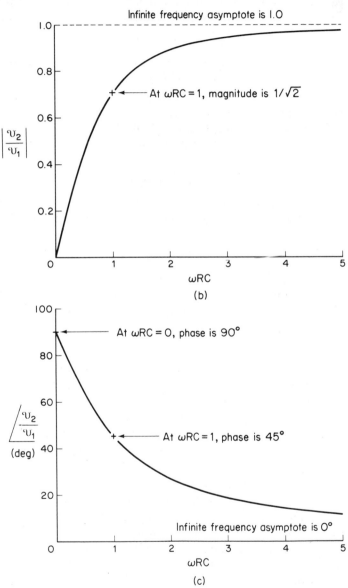

Infinite frequency asymptote is 1.0

At $\omega RC = 1$, magnitude is $1/\sqrt{2}$

$\left|\dfrac{\upsilon_2}{\upsilon_1}\right|$

ωRC

(b)

At $\omega RC = 0$, phase is 90°

$\underline{/\dfrac{\upsilon_2}{\upsilon_1}}$
(deg)

At $\omega RC = 1$, phase is 45°

Infinite frequency asymptote is 0°

ωRC

(c)

Figure 7-7.3 High-pass RC network and its magnitude and phase plots.

state conditions, it is customary to refer to such a situation by using the term lead rather than lag.

In a manner similar to that used for low-pass functions, high-pass functions are frequently described in terms of the frequency at which the magnitude characteristic decreases to 0.707 of its infinite frequency value. The frequency at which this occurs is also referred to as the 3-dB frequency. For the high-pass network shown in Fig. 7-7.3(a), let us write the network function in the form

$$\frac{\mathcal{V}_2}{\mathcal{V}_1} = \frac{1}{1 + (1/j\omega RC)} \tag{7-138}$$

A development similar to that made for the low-pass network readily shows that the 3-dB frequency for this function is $1/RC$.

As a slightly more complicated example of a low-pass network, let us consider the network shown in Fig. 7-7.4(a). A second-order voltage transfer function may be defined by using the voltage phasor $\mathcal{V}_C$ as the response quantity, and the input voltage phasor $\mathcal{V}_1$ as the excitation. The network function is

$$\frac{\mathcal{V}_C}{\mathcal{V}_1} = \frac{1}{1 - \omega^2 LC + j\omega RC} \tag{7-139}$$

For convenience in emphasizing the properties of such a network function, let us choose unity values for L and C. Thus the network function may be written[13]

$$\frac{\mathcal{V}_C}{\mathcal{V}_1} = \frac{1}{1 - \omega^2 + j\omega R} \tag{7-140}$$

The magnitude and phase of the function defined in Eq. (7-140) are shown for several values of R in Fig. 7-7.4(b) and (c). There are several points of interest that may be observed in these frequency response curves. First it should be noted that the magnitude is non-zero at $\omega = 0$, but that it approaches zero as ω approaches ∞. This behavior is readily verified from Eq. (7-140), since the function equals unity when $\omega = 0$, and its magnitude is inversely proportional to ω^2 for large values of ω. Thus the function is a low-pass function. Next, note that the phase or argument of the denominator polynomial goes from $0°$ at $\omega = 0$, to $90°$ when $\omega \approx 1$ or, more generally, when $\omega \approx 1/\sqrt{LC}$, and to $-180°$ when ω becomes very large. Thus, the phase of the entire network function goes from $0°$ to $-180°$. This result is consistent with the statement made for the first-order low-pass network, namely, that each reactive element in a low-pass network contributes $90°$ to the total phase shift of the function. A further examination of the various curves shown in Fig. 7-7.4(b) and (c) shows that as the value R of the resistor is decreased, the magnitude curve develops a peak at $\omega = \sqrt{1 - (R^2/2)}$ rad/s. As R approaches 0, the frequency approaches 1 rad/s. In the more general case, where L and C are not set to unity, this peak appears at

$$\omega = \frac{1}{\sqrt{LC}}\left[1 - \frac{R^2}{2L^2}\right]^{1/2}$$

In this case, as R approaches 0, the frequency approaches $1/\sqrt{LC}$ rad/s. It should be

---

[13]We will show in Sec. 7-9 that such a choice of element values, in effect, constitutes a *frequency and impedance normalization* of the network's frequency characteristics, and thus it does not affect the general properties of the network function.

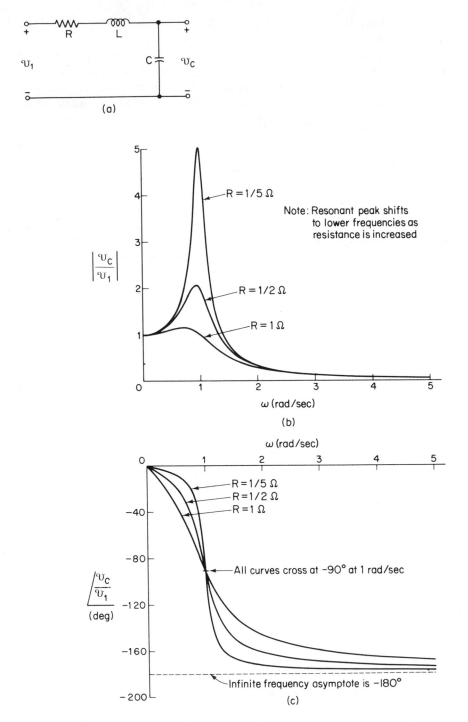

Figure 7-7.4 Low-pass *RLC* network and its magnitude and phase plots.

noted that the phase curve passes through 90° at the same frequency where the peak of the magnitude curve would occur if R was set to 0.

As a second example of a high-pass network function, let us use the phasor $\mathcal{V}_L$ of the network shown in Fig. 7-7.5(a) as a response phasor. The ratio of this phasor to the input excitation phasor $\mathcal{V}_1$ is readily shown to be

$$\frac{\mathcal{V}_L}{\mathcal{V}_1} = \frac{-\omega^2 LC}{1 - \omega^2 LC + j\omega RC} \qquad (7\text{-}141)$$

As was done for the low-pass case, if we set the values of L and C to unity, the network function takes the form

$$\frac{\mathcal{V}_L}{\mathcal{V}_1} = \frac{-\omega^2}{1 - \omega^2 + j\omega R} \qquad (7\text{-}142)$$

The magnitude and phase of the network function defined by (7-142) are plotted for several values of R in Fig. 7-7.5(b) and (c). An examination of these curves shows that the two reactive elements produce a total of 180° of phase shift in the network, and that the peak of the magnitude curve which appears for low values of resistance occurs when the phase shift is 90°.

As an example of another class of network functions, we will use the phasor $\mathcal{V}_R$ of the network shown in Fig. 7-7.6(a) as the response phasor. The ratio of this to the excitation phasor $\mathcal{V}_1$ is readily shown to be

$$\frac{\mathcal{V}_R}{\mathcal{V}_1} = \frac{j\omega RC}{1 - \omega^2 LC + j\omega RC} \qquad (7\text{-}143)$$

If we set the values of L and C to unity, the network function becomes

$$\frac{\mathcal{V}_R}{\mathcal{V}_1} = \frac{j\omega R}{1 - \omega^2 + j\omega R} \qquad (7\text{-}144)$$

The magnitude and phase of the network function defined by (7-144) are plotted for several values of R in Fig. 7-7.6(b) and (c). The magnitude characteristic in these plots approaches zero for very low and very high values of ω and reaches a maximum when $\omega = 1$ rad/s. Thus, most of the response for this network function occurs for the band of frequencies which are near 1 rad/s (more generally for ω near $1/\sqrt{LC}$). A network function which has a magnitude curve of this type is referred to as a *band-pass network function*. The network is called a *band-pass network*. In this case, since the network contains two reactive elements, it may also be referred to as a *second-order* band-pass network.

The peaking of the magnitude characteristic of Fig. 7-7.6(b) is also referred to as a *resonant* effect. Thus, the circuit of Fig. 7-7.6(a) is frequently referred to as a *resonant circuit*. Since all the elements are connected in series, and since there are two reactive elements, this circuit may also be called a *second-order series-resonant* circuit. The frequency at which the magnitude characteristic of Fig. 7-7.6(b) reaches its peak value is called the *center frequency* or the *resonant frequency* of the band-pass network. On each side of this center frequency, there is a frequency at which the magnitude characteristic has a value which is 0.707 of its value at the center frequency. These frequencies are referred to as the *upper and lower 3-dB frequencies*. The difference between these frequencies is called the *bandwidth* of the function. If we let B be the

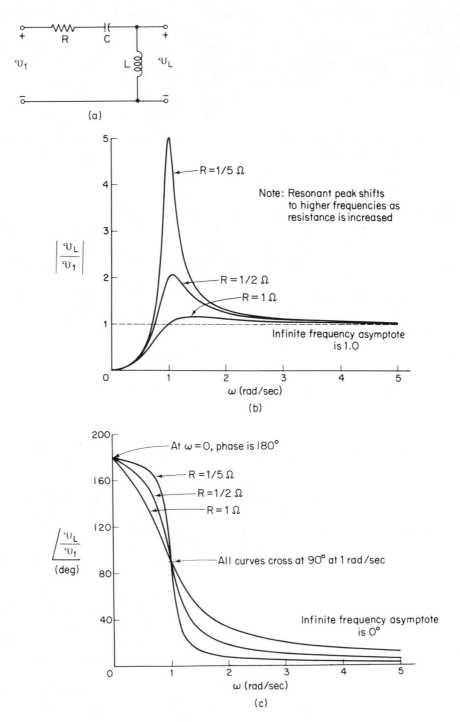

Figure 7-7.5 High-pass *RLC* network and its magnitude and phase plots.

(a)

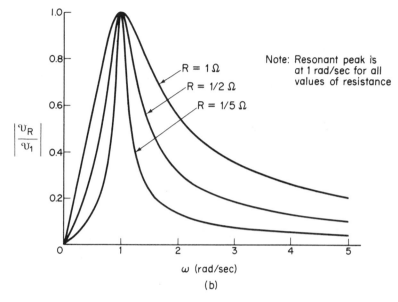

Note: Resonant peak is at 1 rad/sec for all values of resistance

(b)

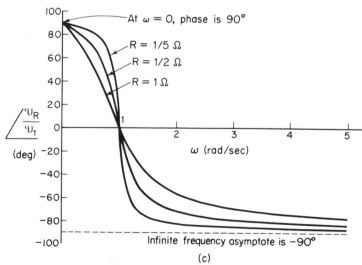

(c)

Figure 7-7.6 Second-order series-resonant *RLC* network and its magnitude and phase plots.

bandwidth, then[14]

$$B = \omega_{\text{upper 3 dB}} - \omega_{\text{lower 3 dB}} \qquad (7\text{-}145)$$

The resonant effect in a second-order network is brought about by a cancellation of the imaginary components in the denominator of the network function. This is the result of the fact that reactances may be positive or negative in value. To see how such a cancellation comes about, let us write the band-pass network function given in (7-143) in the following form.

$$\frac{\mathcal{V}_R}{\mathcal{V}_1} = \frac{R}{R + j[\omega L - (1/\omega C)]} \qquad (7\text{-}146)$$

The magnitude of this function has its maximum value when the imaginary term in the denominator becomes zero. If we let ω_0 be the frequency at which this occurs, then

$$\omega_0 L - \frac{1}{\omega_0 C} = 0 \qquad \omega_0 = \frac{1}{\sqrt{LC}} \qquad (7\text{-}147)$$

Thus, ω_0 is the resonant frequency for this circuit. When ω has this value, the circuit is said to be in *resonance*.

One measure of the sharpness of the resonant peak in a band-pass circuit is the *quality factor*, commonly designated as Q. This may be defined as

$$Q = \frac{\text{Resonant frequency}}{\text{Bandwidth}} = \frac{\omega_0}{B} \qquad (7\text{-}148)$$

From the magnitude curves shown in Fig. 7-7.6(b) we may note that the Q of a series-resonant circuit is inversely proportional to the series resistance in the circuit. The relation for Q given in (7-148) may also be related directly to the values of network elements. To see this, we note from (7-146) that at the 3-dB frequency

$$\frac{R}{\sqrt{R^2 + [\omega L - (1/\omega C)]^2}} = \frac{1}{\sqrt{2}} \qquad (7\text{-}149)$$

This may be put in the form

$$\omega L - \frac{1}{\omega C} = \pm R \qquad (7\text{-}150)$$

Expressing this result as a quadratic equation in ω and solving, we obtain

$$\omega = \pm \frac{R}{2L} \pm \sqrt{\left(\frac{R}{2L}\right)^2 + \frac{1}{LC}} \qquad (7\text{-}151)$$

Since ω must be positive, we may identify the upper and lower 3-dB frequencies as

$$\omega_{\text{upper 3 dB}} = \frac{R}{2L} + \sqrt{\left(\frac{R}{2L}\right)^2 + \frac{1}{LC}} \qquad (7\text{-}152)$$

$$\omega_{\text{lower 3 dB}} = -\frac{R}{2L} + \sqrt{\left(\frac{R}{2L}\right)^2 + \frac{1}{LC}} \qquad (7\text{-}153)$$

Thus, the bandwidth may be expressed

$$B = \omega_{\text{upper 3 dB}} - \omega_{\text{lower 3 dB}} = \frac{R}{L} \qquad (7\text{-}154)$$

[14]For a *low-pass* network the bandwidth is defined as being equal to the frequency at which the magnitude is down 3 dB from its zero frequency value.

Substituting this result in (7-148), we obtain the following expression for the Q of this network function.

$$Q = \frac{\omega_0 L}{R} \qquad (7\text{-}155)$$

The process of resonance may be illustrated in more detail by the use of phasor diagrams. To make the example more concrete, let us use the numerical values $L = 1$, $C = 1$, and $R = 0.5$ for the circuit shown in Fig. 7-7.6(a). Since only the relative position of the phasors will be of interest to us, let us choose the reference angle for $\mathcal{V}_R$ as zero degrees. Similarly, since we are primarily concerned with the ratio of $\mathcal{V}_R$ to $\mathcal{V}_1$ rather than the actual values of these phasors, let us set the magnitude of $\mathcal{V}_R$ to unity. Thus, we may draw the phasor for $\mathcal{V}_R$ as shown in the phasor diagram in Fig. 7-7.7(a). The phasor $\mathcal{I}$ is easily found from $\mathcal{V}_R$, using the relation $\mathcal{I} = \mathcal{V}_R/R$.

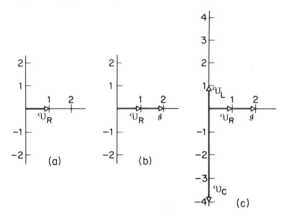

Figure 7-7.7 Phasor diagrams for the network shown in Fig. 7-7.6(a).

Thus, $\mathcal{I} = 2\underline{/0^\circ}$ and it may be added to the phasor diagram as shown in Fig. 7-7.7(b). Next, we may compute the phasors $\mathcal{V}_C = \mathcal{I}/j\omega C$, and $\mathcal{V}_L = j\omega L \mathcal{I}$. Let us make our first determination of these quantities at $\omega = \frac{1}{2}$ rad/s. We find that $\mathcal{V}_C = 4\underline{/-90^\circ}$ and $\mathcal{V}_L = 1\underline{/90^\circ}$. If these quantities are added to the phasor diagram, the result is shown in Fig. 7-7.7(c). We may now reposition the phasors on the phasor diagram, in exactly the same way that vectors are positioned on a vector diagram, to indicate the sum $\mathcal{V}_R + \mathcal{V}_C + \mathcal{V}_L$ of the phasors. From KVL, this quantity is equal to $\mathcal{V}_1$. The construction is shown in Fig. 7-7.8(a). From the figure it is easily seen that $\mathcal{V}_R/\mathcal{V}_1 = 1/\sqrt{10}\underline{/-71.6^\circ}$. This conclusion is readily verified if the network function given in (7-144) is evaluated at $\omega = 0.5$ rad/s, and $R = 0.5$. Now let us construct a phasor diagram to show the relation of the phasors when $\omega = 1$ rad/s. In this case, again assuming that $\mathcal{V}_R = 1\underline{/0^\circ}$, we find that $\mathcal{I} = 2\underline{/0^\circ}$, $\mathcal{V}_C = 2\underline{/-90^\circ}$, and $\mathcal{V}_L = 2\underline{/90^\circ}$. The resulting phasor diagram is shown in Fig. 7-7.8(b). Obviously $\mathcal{V}_R/\mathcal{V}_1 = 1\underline{/0^\circ}$, which may be verified from (7-144). As a final example, consider the phasor diagram for $\omega = 2$ rad/s, still assuming that $\mathcal{V}_R = 1\underline{/0^\circ}$. In this case, $\mathcal{V}_C = 1\underline{/-90^\circ}$, and $\mathcal{V}_L = 4\underline{/90^\circ}$. The phasor diagram appears as shown in Fig. 7-7.8(c). From the figure we see that $\mathcal{V}_R/\mathcal{V}_1 = 1/\sqrt{10}\underline{/71.6^\circ}$. Thus, this is the value of the network function at $\omega = 2$ rad/s, as may readily be verified from Eq. (7-144). The three phasor diagrams shown in Fig. 7-7.8 illustrate typical conditions in a series-resonant circuit

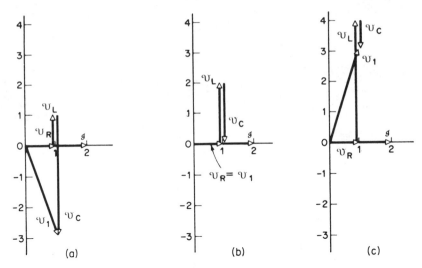

Figure 7-7.8 Phasor diagrams for the network shown in Fig. 7-7.6(a).

at frequencies below resonance, at resonance, and above resonance. Since the response phasor is maintained at unity magnitude, the magnitude of the network function is inversely proportional to the length of the phasor $\mathcal{V}_1$. From the diagrams, we readily see that the minimum magnitude of this phasor occurs at the resonant frequency. In addition, we note that at frequencies below resonance, the circuit acts as an RC circuit, in that the current $\mathcal{I}$ leads the applied voltage $\mathcal{V}_1$. Similarly, at frequencies above resonance, the current $\mathcal{I}$ lags the applied voltage $\mathcal{V}_1$ and thus the circuit behaves like an RL circuit, exhibiting the phase lag characteristic typical of an inductor. Finally, at resonance, the circuit behaves like a pure resistance, since the applied voltage $\mathcal{V}_1$ is in phase with the current $\mathcal{I}$ flowing through the circuit.

The above discussion may be summarized as follows:

SUMMARY 7-7.1

Properties of a Second-Order Series-Resonant Circuit. A circuit consisting of a series connection of a resistor, a capacitor, and an inductor is called a second-order series-resonant circuit. The resonant frequency is $\omega_0 = 1/\sqrt{LC}$ rad/s. Below resonance, the circuit acts like an RC network. Above resonance, the circuit acts like an RL circuit. At resonance, the circuit acts like a resistor. If the resistance is zero, at resonance the circuit acts like a short circuit. ∎

It should be noted that at resonance, the voltages across the capacitor and the inductor are equal and opposite. In high-Q circuits, i.e., ones in which the resistance is low, these voltages can become very large, and can actually be far larger than the applied voltage. As an example of this, suppose that a series-resonant circuit in which $R = 100 \ \Omega$, $L = 1$ H, and $C = 7.04 \ \mu$F (7.04×10^{-6} F) is connected to an electrical wall outlet which provides 120 V (rms) at 60 Hz (377 rad/s). The magnitude of the

the voltage phasor is 169.7 V (see Sec. 7-1). Substitution of the values for L and C in Eq. (7-147) shows that the circuit is resonant at the applied frequency, namely, 377 rad/s. The magnitude of the current phasor is thus determined only by the resistance. It equals 169.7/100 or 1.697 A. The magnitude of the phasor for the voltage appearing across the inductor is $\mathcal{I}\omega L$, i.e., $1.697 \times 377 \times 1$ or 640 V! The magnitude of the phasor voltage across the capacitor is also 640 V. Note that these voltages are so much higher than the applied voltage that they may present a dangerous shock hazard. A number of people who failed to realize that, in a series-resonant circuit, the voltages across component network elements may be many times larger than the applied voltage are now learning circuit theory in "heavenly" universities. It is hoped that the readers of this text will not join them prematurely.

It is also possible to define a second-order *parallel-resonant circuit*. As an example of a network function for such a circuit, consider the driving-point impedance of the circuit shown in Fig. 7-7.9. In terms of the (phasor) variables defined in this circuit, we may write

Figure 7-7.9 Second-order parallel resonant circuit.

$$\frac{\mathcal{V}}{\mathcal{I}_1} = \frac{1}{G + j[\omega C - (1/\omega L)]} \qquad (7\text{-}156)$$

Resonance occurs in this circuit when the imaginary part of the denominator (the susceptance) goes to zero. Thus, we see that the resonant frequency ω_0 is defined by

$$\omega_0 C - \frac{1}{\omega_0 L} = 0 \qquad \omega_0 = \frac{1}{\sqrt{LC}} \qquad (7\text{-}157)$$

This is the same relation as was determined for the series-resonant circuit. Phasor diagrams for the parallel-resonant circuit for frequencies below resonance, at resonance, and above resonance are shown in Fig. 7-7.10. For convenience in drawing

(a)   (b)   (c)

Figure 7-7.10 Phasor diagrams for the circuit shown in Fig. 7-7.9.

these phasor diagrams, it has been assumed that $\mathcal{I}_G$, the current through the resistor, has the same magnitude in each of the three cases. From the diagram we may conclude that below resonance, a parallel-resonant circuit is inductive in behavior, since the applied current lags the voltage across the circuit. Similarly, for frequencies above the resonant frequency, the circuit acts in a capacitive manner, in the sense that the current leads the voltage. Finally, at resonance, the circuit acts as a pure conductance, i.e., the applied current is in phase with the voltage across the three parallel circuit elements.

The discussion given above may be summarized as follows:

SUMMARY 7-7.2

Properties of a Second-Order Parallel-Resonant Circuit. A circuit consisting of a parallel connection of a resistor, a capacitor, and an inductor is called a second-order parallel-resonant circuit. The resonant frequency is $\omega_0 = 1/\sqrt{LC}$ rad/s. Below resonance, the circuit acts like an *RL* circuit. Above resonance, the circuit acts like an *RC* circuit. At resonance, the circuit acts like a resistor. If the conductance is zero (the resistance is infinite), at resonance the circuit acts like an open circuit. ∎

In a parallel-resonant circuit at resonance, conditions dual to those described for the series-resonant circuit exist, i.e., the magnitude of the phasor currents $\mathcal{I}_C$ and $\mathcal{I}_L$ shown in Fig. 7-7.10(b) may become considerably larger than the magnitude of the input current $\mathcal{I}_1$. These large currents are sometimes referred to as *circulating currents* since they circulate in the loop formed by the inductor and capacitor. Unless proper consideration is given to the magnitude of these currents, they may become large enough to destroy circuit elements.

7-8 FREQUENCY RESPONSE PLOTS

In the preceding section we introduced the concept of a network function and showed how it could be used to relate a pair of input and output phasors. To illustrate the properties of such functions, magnitude and phase plots were used to graphically display the manner in which a given network function was affected by its sinusoidal frequency variable ω. More specifically, if a network function was defined as $N(j\omega)$, we used a magnitude plot to display $|N(j\omega)|$ as a function of ω and a phase plot to display Arg $N(j\omega)$ vs. ω. In this section we shall consider another useful graph which may be used for the display of magnitude and phase information. This is one in which logarithmic scales are used for $|N(j\omega)|$ and for the variable ω. Such graphs are called Bode plots (or Bode diagrams). Obviously, such graphs contain the same information which is present in the (non-logarithmic) plots previously discussed. We shall see, however, that the logarithmic plots are considerably easier to construct. As a matter of fact, we shall find that they can be approximated to a surprising degree of accuracy by using only straight-line segments.

To begin our discussion of Bode plots, let us consider a network function $N(j\omega)$ which is characterized in terms of its magnitude and phase components. Thus, we may write

$$N(j\omega) = |N(j\omega)| \, e^{j \, \text{Arg} \, N(j\omega)} \tag{7-158}$$

If we now take the natural logarithm of both sides of Eq. (7-158) and note that the logarithm of a product of terms is the sum of the logarithms of the individual terms, we obtain

$$\ln[N(j\omega)] = \ln|N(j\omega)| + \ln[e^{j \, \text{Arg} \, N(j\omega)}] \qquad (7\text{-}159)$$

Since $\ln(e^a) = e^{\ln(a)} = a$, the above expression may be put in the form

$$\ln[N(j\omega)] = \ln|N(j\omega)| + j \, \text{Arg} \, N(j\omega) \qquad (7\text{-}160)$$

The significance of this result is that by taking the natural logarithm of the network function $N(j\omega)$, we convert the function to a sum of two quantities, namely, a real part which is a function only of the magnitude of $N(j\omega)$, and an imaginary part which is a function only of its phase. In the relation given in Eq. (7-160), the unit for $\ln|N(j\omega)|$ is the *neper* and the unit for the Arg $N(j\omega)$ is the *radian*. To convert these quantities to the more usual *dB* and *degree* units we may use the following relations:

Magnitude in decibels (dB) $= 8.6859 \times$ magnitude in nepers

Argument in degrees $= 57.2958 \times$ argument in radians

To see how the use of logarithmic quantities simplifies the graphical representation of network functions, let us consider a simple network function $N_z(j\omega)$ having the form

$$N_z(j\omega) = 1 + \frac{j\omega}{z} \qquad (7\text{-}161)$$

Taking the natural logarithm of both sides of this, we obtain

$$\ln[N_z(j\omega)] = \ln\left|1 + \frac{j\omega}{z}\right| + j \tan^{-1} \frac{\omega}{z} \qquad (7\text{-}162)$$

The first term on the right-hand side of this relation is concerned only with the magnitude of the network function. Thus, it may equally well be expressed in dB. If we define such a magnitude as $M_z(\omega)$, we may write

$$M_z(\omega) = 8.6859 \ln\left|1 + \frac{j\omega}{z}\right| = 20 \log\left|1 + \frac{j\omega}{z}\right| \qquad (7\text{-}163)$$

This term is obviously a function of the variable ω. However, for values of ω such that $\omega/z \ll 1$, i.e., such that $\omega \ll z$, this term becomes

$$M_z(\omega)\big|_{\omega \ll z} \approx 20 \log 1 = 0 \text{ dB} \qquad (7\text{-}164)$$

Thus, for such a range of ω the term $M_z(\omega)$ may be plotted as a straight line at 0 dB on a dB vs. log ω plot as shown in Fig. 7-8.1(a). For values of ω such that $\omega \gg z$, the term $M_z(\omega)$ given in (7-163) may be approximated as

$$M_z(\omega)\big|_{\omega \gg z} \approx 20 \log \frac{\omega}{z} = 20(\log \omega - \log z) \text{ dB} \qquad (7\text{-}165)$$

On a dB vs. log ω plot this is simply a straight line with a slope of 20. It intersects the 0-dB line (which represents $M_z(\omega)$ for small values of ω) when log $\omega = $ log z, i.e., when $\omega = z$. Thus, it may be characterized as shown in Fig. 7-8.1(b). Since the slope of this line is 20, in any region for which this characterization is valid, a unit change in log ω such as log ω going from 1 to 2 or from 9 to 10, or from 9.7 to 10.7, etc., must produce a resulting change in $M_z(\omega)$ of 20 dB. The corresponding change in ω

Figure 7-8.1 Bode magnitude plots for a first-order numerator term.

for such a unit change in $\log \omega$ can be determined by letting ω_1 and ω_2 be any two frequencies the difference of whose logarithms is unity. Thus, we may write (assuming $\omega_2 > \omega_1$)

$$\log \omega_2 - \log \omega_1 = \log \frac{\omega_2}{\omega_1} = 1 \qquad (7\text{-}166)$$

Thus, the ratio ω_2/ω_1 must equal 10. A pair of frequencies ω_2 and ω_1 such that their ratio is 10 is said to encompass a *decade* of frequency. Thus the solid line shown in Fig. 7-8.1(b) is frequently referred to as having a slope of 20 dB/decade of frequency or, more concisely, simply as a *20 dB/decade* slope. Another frequently used expression for the slope of the line shown in Fig. 7-8.1(b) is *6 dB/octave* where an *octave* is defined as the frequency range encompassed by two frequencies, one of which is twice the value of the other. If we extend the two straight-line characteristics shown in Fig. 7-8.1(a) and (b) until they intersect as shown in Fig. 7-8.1(c), we obtain an asymptotic straight-line approximation for the quantity $M_z(\omega)$ defined in (7-163). The actual plot of this function is also shown in the figure. It is readily observed that the maximum error in the asymptotic approximation is 3 dB and that this occurs at a frequency $\omega = z$. In Fig. 7-8.1(c) it is important to note that values of ω itself rather than values of $\log \omega$ are given on the abscissa scale. From the values of ω, however, we see that the scale is obviously logarithmic. Such a convention is usually considerably more convenient than using values of $\log \omega$ for the scale.

Now let us consider the imaginary part of ln $[N_z(j\omega)]$, i.e., the part which gives the phase of $N_z(j\omega)$. From (7-160) and (7-162) we have

$$\text{Arg } N_z(j\omega) = \tan^{-1}\frac{\omega}{z} \tag{7-167}$$

A straight-line asymptotic approximation which is frequently used to plot this function on a degree vs. log ω plot consists of a horizontal straight line at zero degrees (for $\omega \ll z$) and a second straight line at 90° (for $\omega \gg z$). The two horizontal lines are joined by a straight line with a slope of 45° per decade passing through the point (log z, 45°). Such a straight-line phase approximation is shown in Fig. 7-8.2(a). A portion of the actual phase curve as defined in Eq. (7-167) is shown in Fig. 7-8.2(b). The maximum error in the straight-line approximation is about 6°. An example of the use of the technique described above follows.

(a)

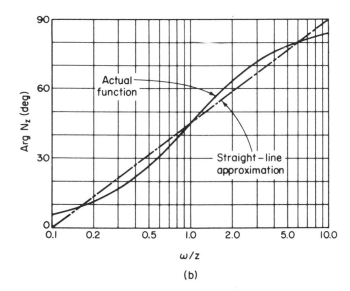

(b)

Figure 7-8.2 Bode phase plots for a first-order numerator term.

Example 7-8.1

As an example of the use of Bode plots to approximate the magnitude and phase characteristics of a simple network function, consider the driving-point admittance $Y(j\omega)$ for the network shown in Fig. 7-8.3. For this network

$$Y(j\omega) = 3(j\omega + 10)$$

mhos, farads

$Y(j\omega) \longrightarrow$ $\doteq 3$ $\gtrless 30$

Figure 7-8.3 Circuit for Example 7-8.1.

Putting the term in parentheses in a form similar to that given in Eq. (7-161), we obtain

$$Y(j\omega) = 30\left(1 + \frac{j\omega}{10}\right)$$

The magnitude in dB may be expressed as

$$20\,|\log Y(j\omega)| = 20\log 30 + 20\log\left|1 + \frac{j\omega}{10}\right| = 29.54 + 20\log\left|1 + \frac{j\omega}{10}\right|$$

Using the method described above to obtain the asymptotic characteristics for the second term of the right member of the equation and adding the constant 29.54, we obtain the Bode magnitude plot shown in Fig. 7-8.4(a). Since the phase of the multi-

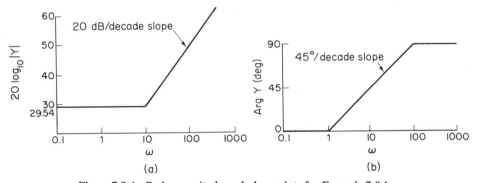

Figure 7-8.4 Bode magnitude and phase plots for Example 7-8.1.

plicative constant is zero, the phase of the network function $Y(j\omega)$ is simply that of the term $[1 + (j\omega/10)]$. Thus, applying the techniques described above, we obtain the Bode phase plot shown in Fig. 7-8.4(b).

An analysis similar to that given in the preceding paragraphs may be used to construct the Bode diagrams for a network function $N_p(j\omega)$ having the form

$$N_p(j\omega) = \frac{1}{1 + (j\omega/p)} \tag{7-168}$$

To see this, let us again take the natural logarithm of both members of the equation. We obtain

$$\ln\,[N_p(j\omega)] = \ln\left|\frac{1}{1 + (j\omega/p)}\right| - j\tan^{-1}\frac{\omega}{p} \tag{7-169}$$

Since the logarithm of the reciprocal of a quantity is equal to the negative of the logarithm of the quantity, the above relation may be put in the form

$$\ln\,[N_p(j\omega)] = -\ln\left|1 + \frac{j\omega}{p}\right| - j\tan^{-1}\frac{\omega}{p} \tag{7-170}$$

It is readily observed that this is simply the negative of the expression given in Eq. (7-162) (with the substitution of p for z). Thus, the asymptotic magnitude and phase Bode plots will be as shown in Fig. 7-8.5. These are simply the negative of those shown in Figs. 7-8.1(c) and 7-8.2(b). The actual plots are also shown. It is readily seen that the magnitude characteristic again has a maximum error of 3 dB and the phase characteristic has a maximum error of 6°. Thus, we see that an asymptotic straight-line approximation for the dB vs. $\log \omega$ plot for $|N_p(j\omega)|$ may be constructed by drawing a horizontal 0-dB line from the ordinate to the value $\log p$ and continuing this with a straight line with a *minus* 20 dB/decade slope. An asymptotic straight-line plot of degrees vs. $\log \omega$ for Arg $N_p(j\omega)$ may be constructed by drawing a straight line with a slope of $-45°$/decade through the point $(\log p, -45°)$, and extending this line with 0° and $-90°$ horizontal asymptotes.

(a)

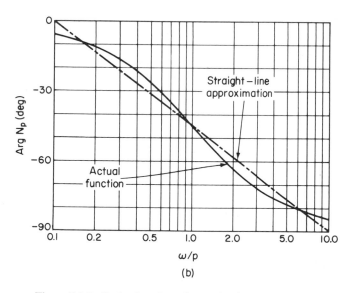

(b)

Figure 7-8.5 Bode plots for a first-order denominator term.

In the preceding paragraphs we have discussed the construction of Bode plots for two types of simple network functions, i.e., those with a first-order factor in either the numerator or the denominator. There is one other type of first-order network function that also occurs frequently. This is one having the form $j\omega$ (or $1/j\omega$). The magnitude Bode plot for such a function simply consists of a straight line which crosses the 0-dB abscissa at $\log \omega = 0$, i.e., at $\omega = 1$, and which has a slope of $+20$ dB/decade (-20 dB/decade for the function $1/j\omega$). The Bode phase plot simply consists of a horizontal line at $+90°$ ($-90°$ for the function $1/j\omega$).

The results developed in this section to this point are readily extended to provide a means of constructing Bode plots for any function consisting of a product of numerator factors having the form shown in Eq. (7-161) and denominator factors of the form shown in Eq. (7-168). To see this, consider the following network function:

$$N(j\omega) = \frac{A(j\omega)}{B(j\omega)} = K\frac{[1 + (j\omega/z_1)][1 + (j\omega/z_2)] \cdots [1 + (j\omega/z_m)]}{[1 + (j\omega/p_1)][1 + (j\omega/p_2)] \cdots [1 + (j\omega/p_n)]} \qquad (7\text{-}171)$$

If we take the logarithm of both members of this equation, we obtain

$$\ln [N(j\omega)] = \ln K + \ln\left|1 + \frac{j\omega}{z_1}\right| + \ln\left|1 + \frac{j\omega}{z_2}\right|$$

$$+ \cdots + \ln\left|1 + \frac{j\omega}{z_m}\right| + \ln\left|\frac{1}{1 + (j\omega/p_1)}\right|$$

$$+ \ln\left|\frac{1}{1 + (j\omega/p_2)}\right| + \cdots + \ln\left|\frac{1}{1 + (j\omega/p_n)}\right|$$

$$+ j\left[\tan^{-1}\frac{\omega}{z_1} + \tan^{-1}\frac{\omega}{z_2} + \cdots + \tan^{-1}\frac{\omega}{z_m}\right.$$

$$\left. - \tan^{-1}\frac{\omega}{p_1} - \tan^{-1}\frac{\omega}{p_2} - \cdots - \tan^{-1}\frac{\omega}{p_n}\right] \qquad (7\text{-}172)$$

If we now define $|N(j\omega)|$ in dB as $M(\omega)$, we may write

$$M(\omega) = 20 \log|N(j\omega)| = 20 \log K + 20 \log\left|1 + \frac{j\omega}{z_1}\right|$$

$$+ 20 \log\left|1 + \frac{j\omega}{z_2}\right| + \cdots + 20 \log\left|1 + \frac{j\omega}{z_m}\right|$$

$$+ 20 \log\left|\frac{1}{1 + (j\omega/p_1)}\right| + 20 \log\left|\frac{1}{1 + (j\omega/p_2)}\right|$$

$$+ \cdots + 20 \log\left|\frac{1}{1 + (j\omega/p_n)}\right| \qquad (7\text{-}173)$$

Thus we see that an asymptotic plot of $M(\omega)$ can be made by summing the separate asymptotic plots of factors having the form $20 \log|1 + (j\omega/z_i)|$ and $20 \log|1/[1 + (j\omega/p_i)]|$. Since these latter plots simply consist of straight lines, such a summation is readily done graphically. In a similar manner, if we investigate the behavior of Arg $N(j\omega)$, we may write

$$\text{Arg } N(j\omega) = \tan^{-1}\frac{\omega}{z_1} + \tan^{-1}\frac{\omega}{z_2} + \cdots + \tan^{-1}\frac{\omega}{z_m}$$

$$- \tan^{-1}\frac{\omega}{p_1} - \tan^{-1}\frac{\omega}{p_2} - \cdots - \tan^{-1}\frac{\omega}{p_n} \qquad (7\text{-}174)$$

From this result we again see that the addition of separate Bode phase plots for the numerator and denominator factors of $N(j\omega)$ gives us the resulting overall phase plot for the network function. An example follows.

Example 7-8.2

It is desired to construct Bode plots for the network function

$$N(j\omega) = 10\frac{1 + (j\omega/10)}{[1 + (j\omega/1)][1 + (j\omega/100)]}$$

The separate magnitude and phase plots for the factors $1 + (j\omega/10)$, $1/[1 + (j\omega/1)]$, and $1/[1 + (j\omega/100)]$ are shown in Figs. 7-8.6(a) and 7-8.7(a). The addition of these plots and the inclusion of the constant term $20 \log 10 = 20$ for the constant multiplier of $N(j\omega)$ gives the Bode magnitude and phase plots shown in Figs. 7-8.6(b) and 7-8.7(b). The actual magnitude and phase curves are also shown in these plots for comparison.

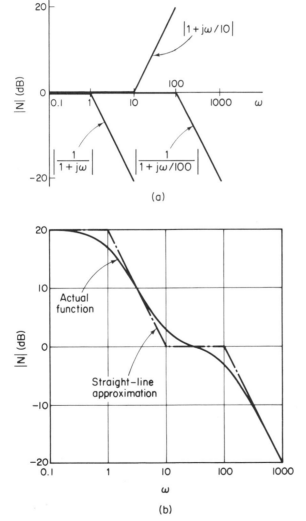

Figure 7-8.6 Bode magnitude plots for Example 7-8.2.

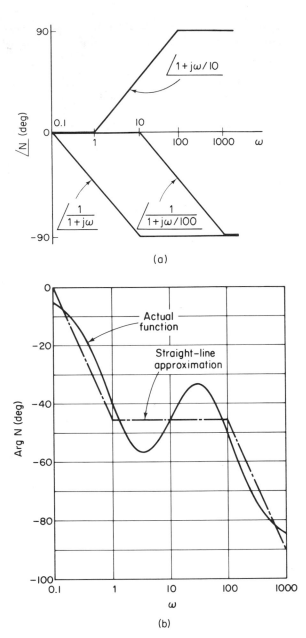

(a)

(b)

Figure 7-8.7 Bode phase plots for Example 7-8.2.

In the preceding paragraphs we showed how to construct Bode plots for network functions consisting of first-order factors in the numerator and denominator. Now let us consider the effect of second-order factors which have complex conjugate zeros. For simplicity, we will consider a network function $N_c(j\omega)$ consisting of a second-order denominator factor. Such a network function may be written in the form

$$N_c(j\omega) = \frac{1}{1 + 2\zeta(j\omega/\omega_0) + (j\omega/\omega_0)^2} \qquad (7\text{-}175)$$

where ζ is called the damping factor and ω_0 is called the undamped natural frequency. Taking the natural log of both sides of (7-175), we obtain

$$\ln\left[N_c(j\omega)\right] = \ln\left|\frac{1}{1 + 2\zeta(j\omega/\omega_0) + (j\omega/\omega_0)^2}\right| - j\tan^{-1}\left(\frac{2\zeta\omega/\omega_0}{1 - (\omega/\omega_0)^2}\right) \qquad (7\text{-}176)$$

If we investigate the real and imaginary parts of the above function in a manner similar to that done for the first-order factors, we see that a plot of $20\log|N_c(j\omega)|$ has a low-frequency asymptote at 0 dB, and a high-frequency asymptotic locus with a slope of -40 dB/decade (-12 dB/octave). The break point is at $\omega = \omega_0$, and is independent of the value of ζ. At the break point, however, we obtain from the actual equation

$$20\log|N_c(j\omega)| = -20\log 2\zeta \qquad (7\text{-}177)$$

Since the asymptotic curve always has a value of 0 dB at this point, we see that the error of the approximation for the magnitude increases as the value of ζ deviates from 0.5. These results are readily observed in the plots of $20\log|N_c(j\omega)|$ vs. $\log\omega$ for various values of ζ shown in Fig. 7-8.8(a). Similar considerations are readily established for the phase. From (7-176), we see that the low-frequency phase approaches $0°$ while the high-frequency phase equals $-180°$. The phase at $\omega = \omega_0$ is $-90°$ and it is independent of ζ. A series of plots of the actual phase for various values of ζ is shown in Fig. 7-8.8(b). These magnitude and phase plots may also be applied to network func-

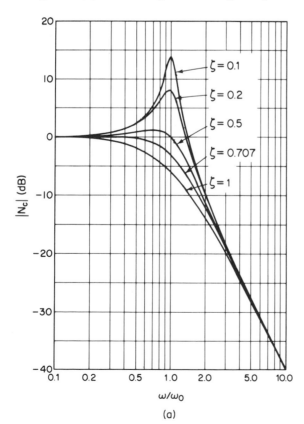

ω/ω_0

(a)

Figure 7-8.8 Bode magnitude and phase plots for a second-order denominator factor.

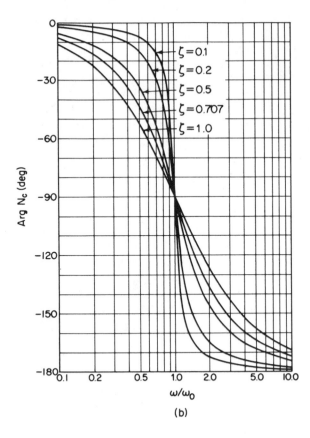

(b)

Figure 7-8.8 Continued

tions that have second-order numerator factors by simply reversing the sign of the ordinate scales.

The damping factor and undamped natural frequency ω_0 introduced in the preceding paragraph may be directly related to the network Q which was defined in Sec. 7-7. To see this, let us solve for the 3-dB frequencies of the network function $N_c(j\omega)$ given in Eq. (7-175). Selecting only the positive solutions, we obtain

$$\omega_{\text{upper 3 dB}} = \omega_0(\sqrt{\zeta^2 + 1} + \zeta)$$
$$\omega_{\text{lower 3 dB}} = \omega_0(\sqrt{\zeta^2 + 1} - \zeta) \tag{7-178}$$

Thus, the bandwidth, which is the difference of the two 3-dB frequencies, is

$$B = \omega_{\text{upper 3 dB}} - \omega_{\text{lower 3 dB}} = 2\zeta\omega_0 \tag{7-179}$$

The expression for the Q is now readily found as

$$Q = \frac{\omega_0}{B} = \frac{1}{2\zeta} \tag{7-180}$$

Frequently, for small values of the damping factor, i.e., high values of Q, simplified versions of the expressions given in Eq. (7-178) may be used. These are

$$\omega_{\text{upper 3 dB}} = \omega_0(1 + \zeta)$$
$$\omega_{\text{lower 3 dB}} = \omega_0(1 - \zeta) \qquad (7\text{-}181)$$

Since the numerator and denominator of any network function may be factored into a product of first- and second-order factors, the techniques introduced above may be used to construct the Bode plots for such a factored function. The plots shown in Fig. 7-8.8 may readily be used to apply corrections to the asymptotic plots made by assuming simple straight-line asymptotes for second-degree factors. Such a correction is illustrated in the following example.

Example 7-8.3

It is desired to find the Bode magnitude plot for the network function

$$N(j\omega) = \frac{j\omega}{(1 + j\omega)[1 + (0.4j\omega/100) + (j\omega/100)^2]} \qquad (7\text{-}182)$$

The separate asymptotic Bode plots for the numerator and for the two denominator factors are shown in Fig. 7-8.9(a). It should be noted that the second-order denominator factor has an asymptote of -40 dB/decade for high frequencies, while the first-order factor has an asymptotic slope of -20 dB/decade. The sum of these loci is given in Fig. 7-8.9(b). A correction to the asymptotic straight-line approximation for the second-order denominator factor may be obtained from the curves of Fig. 7-8.8(a) for the case where $\zeta = 0.2$. This is readily apparent in the locus giving the actual magnitude of $N(j\omega)$, which is also shown in Fig. 7-8.9(b).

There is one other type of graphical representation which is frequently used to display the dependence of network functions on the sinusoidal frequency variable ω. This representation consists of a two-dimensional plot in which the real part of the network function is displayed along the abscissa or x axis and the imaginary part is displayed along the ordinate or y axis. The locus which represents a given network function is generated by evaluating the function for a range of frequency starting from zero and going to infinity. The result is called a *Nyquist* plot. As an example of such a plot, consider the network shown in Fig. 7-8.10. If we define $N(j\omega)$ as the open-circuit voltage transfer function, then $N(j\omega) = 1/[1 + (j\omega/p)]$. This may be written in the form

$$N(j\omega) = \frac{1}{1 + (\omega/p)^2} - j\frac{\omega/p}{1 + (\omega/p)^2} \qquad (7\text{-}183)$$

A plot of this function for a range of values of ω from zero to infinity is shown by the solid line in Fig. 7-8.11. Also shown in this figure is a dashed line which represents the locus for a range of negative values of ω from $-\infty$ to 0. Nyquist diagrams have considerable use in studies of the stability of active networks and control systems. Such applications are covered in detail in books on modern control theory.

The use of the digital computer is especially attractive when it is desired to make frequency response plots, since the multitude of computations required by such displays are, in general, very laborious. A set of subroutines which may be used in performing such computations is given in Appendix C.

(a)

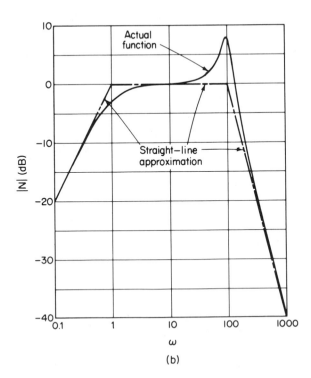

(b)

Figure 7-8.9 Bode magnitude plots for Example 7-8.3.

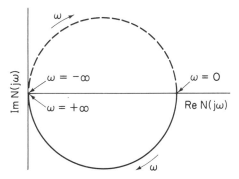

Figure 7-8.10 *RL* network.

Figure 7-8.11 Nyquist plot for the network shown in Fig. 7-8.10.

7-9 FILTERS

One of the most important applications of the sinusoidal steady-state analysis techniques presented in this chapter is in defining the characteristics of filters. Some examples of simple filters were given in Sec. 7-7. In this section we shall extend our treatment to cover a representative variety of passive and active filters which are actually used in practice.

As an introduction, let us consider *low-pass* filters. The network function $N(j\omega)$ for a low-pass filter will have the form

$$N(j\omega) = \frac{H}{(j\omega)^n + a_{n-1}(j\omega)^{n-1} + \cdots + a_1 j\omega + a_0} \tag{7-184}$$

where H is some arbitrary constant. The properties of such a filter, in general, are given in terms of specifications on either the magnitude characteristic or the phase characteristic. Here we shall restrict ourselves to the most common case and consider only the magnitude characteristic. One of the most frequently encountered magnitude specifications is the *maximally flat magnitude* or *Butterworth* one. It is achieved by specifying that, at zero frequency, as many derivatives of the magnitude function be zero as is possible. Thus, the maximally flat magnitude criteria requires that

$$\left. \frac{d^i |N(j\omega)|}{d\omega^i} \right|_{\omega=0} = 0 \qquad i = 1, 2, \ldots, n-1 \tag{7-185}$$

In general, as indicated in the equation, for an nth-order network function, $n-1$ derivatives may be set to zero. The coefficients of the denominator of Eq. (7-184) required to achieve this for several different values of n are given in Table 7-9.1. For the maximally flat magnitude function, the magnitude characteristic is given by

$$|N(j\omega)| = \frac{H}{(1 + \omega^{2n})^{1/2}} \tag{7-186}$$

TABLE 7-9.1 Denominator Coefficients of Maximally Flat Magnitude Network Functions of the Form of Eq. (7-184) (with $a_0 = 1$)

n	a_1	a_2	a_3	a_4
2	1.4142			
3	2.0000	2.0000		
4	2.6131	3.4142	2.6131	
5	3.2361	5.2361	5.2361	3.2361

Note that, independent of the value of n, this magnitude has a 3-dB frequency (see Sec. 7-7) of 1 rad/s, since, at this frequency we have $|N(j1)| = H/\sqrt{2}$. The frequencies below this 3-dB frequency are said to constitute the *passband*. Thus the passband is from $0 - 1$ rad/s. In a low-pass filter, the passband is also said to define the *bandwidth* of the filter. Thus, for the function of Eq. (7-186), the bandwidth is 1 rad/s. The frequencies above the 3-dB frequency, namely from $1 - \infty$ rad/s, constitute the *stopband*. Some typical magnitude plots for various values of n are given in Fig. 7-9.1. In the figure note that, as n increases, the loci approach the ideal (non-attainable) low-pass characteristic indicated by the dashed line.

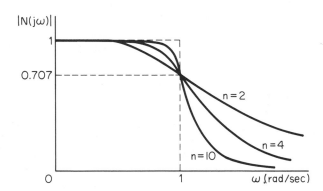

Figure 7-9.1 Butterworth magnitude characteristics.

For a Butterworth filter, the order is usually determined by giving a stopband requirement. This typically has the form, "At least K_s dB attenuation for all frequencies greater than or equal to ω_s rad/s." The attenuation specified by the constant K_s is taken with respect to the magnitude of the network function at zero frequency. We may use Eq. (7-186) to determine the required order n to meet such a specification. Since the function of Eq. (7-186) decreases monotonically as ω increases, satisfying the stopband requirement at $\omega = \omega_s$, also satisfies it for all $\omega > \omega_s$. Thus, from (7-186) we obtain

$$-20 \log |N(j\omega_s)| + 20 \log |N(j0)| = -20 \log \frac{H}{(1 + \omega_s^{2n})^{1/2}} + 20 \log H \geq K_s$$

$$(7-187)$$

Note that the minus sign in the left member is required since K_s is given as an attenuation. The right members of this equation may be written in the form

$$20 \log (1 + \omega_s^{2n})^{1/2} \geq K_s \tag{7-188}$$

Solving this, we obtain

$$n \geq \log_{\omega_s} (10^{K_s/10} - 1)^{1/2} \tag{7-189}$$

where $\log_{\omega_s}$ is the logarithm to the base ω_s. As an example of the application of this result, for $K_s = 15$ dB and $\omega_s = 2$ rad/s, we require $n \geq \log_2 5.534 = \log_{10} 5.534/ \log_{10} 2 = 2.467$. The required order is given by the next highest integer, namely 3.

Synthesis techniques (as opposed to the *analysis* techniques presented in this text) may be applied, using the coefficients given in Table 7-9.1, to develop passive filter configurations which realize network functions having a maximally flat magnitude frequency response. The development of such techniques, however, is beyond the scope of this text. Some results from these techniques are given in Table 7-9.2.

TABLE 7-9.2 Element Values for Butterworth Filters of the Form of Fig. 7-9.2

n	C_1	L_2	C_3	L_4	C_5
2	1.4142	1.4142			
3	1.0000	2.0000	1.0000		
4	0.7654	1.8478	1.8478	0.7654	
5	0.6180	1.6180	2.0000	1.6180	0.6180

The elements designated in the table relate to the network configuration shown in Fig. 7-9.2. The network function realized is a voltage transfer function in which the dc

Figure 7-9.2 Passive low-pass filter realization.

value is 0.5. As an example of the use of this table, a fourth-order filter would have the form shown in Fig. 7-9.3(a). Using Table 7-9.1, the voltage transfer function for this filter would be

$$\frac{\upsilon_2}{\upsilon_1} = \frac{0.5}{(j\omega)^4 + 2.6131(j\omega)^3 + 3.4142(j\omega)^2 + 2.6131 j\omega + 1} \tag{7-190}$$

The filters realized using the element values listed in Table 7-9.2 all have a bandwidth of 1 rad/s. Such a filter is called a *frequency normalized* one. Obviously, most filter applications will require other bandwidths. These may be readily found by using a process called *frequency denormalization*. If we let ω_n be the normalized

ohms, henries, farads

(a)

ohms, mH, mF

(b)

ohms, mH, mF

(c)

kilohms, henries, μF

(d)

Figure 7-9.3 Fourth-order low-pass Butterworth filter.

(original) frequency variable, and ω be the denormalized (new) one, then these variables are related by

$$\omega = \Omega_n \omega_n \qquad (7\text{-}191)$$

where Ω_n is the *frequency-denormalization constant*. Let us now see how such a denormalization changes the values of the inductors in the filter. Let L_{fn} be the normalized value of an inductor, and $Z_{fn}(j\omega_n)$ be its impedance. Then, using (7-191),

$$Z_{fn}(j\omega_n) = j\omega_n L_{fn} = j\left(\frac{\omega}{\Omega_n}\right)L_{fn} = j\omega\left(\frac{L_{fn}}{\Omega_n}\right) = Z(j\omega) \qquad (7\text{-}192)$$

where $Z(j\omega)$ is the denormalized impedance. Comparing the last two members of the above, we see that the value of the denormalized inductance is $L = L_{fn}/\Omega_n$, i.e., the value of the normalized inductance divided by the frequency denormalization constant. A similar development may be made for a capacitor. Let C_{fn} be the normalized capacitor and let $Y_{fn}(j\omega_n)$ be its admittance. Using (7-191), we obtain

Sinusoidal Steady-State Analysis   Chap. 7

$$Y_{fn}(j\omega_n) = j\omega_n C_{fn} = j\left(\frac{\omega}{\Omega_n}\right)C_{fn} = j\omega\left(\frac{C_{fn}}{\Omega_n}\right) = Y(j\omega) \tag{7-193}$$

where $Y(j\omega)$ is the denormalized admittance. Thus, the denormalized capacitance is $C = C_{fn}/\Omega_n$. As an example of these frequency denormalization relations, the filter shown in Fig. 7-9.3(a) with a normalized bandwidth of 1 rad/s, can be used to find a filter with a bandwidth of 10 krad/s by specifying a frequency denormalization constant $\Omega_n = 10^4$. The result is shown in Fig. 7-9.3(b). The L and C element values have simple been divided by 10^4. As another example, since 10^4 rad/s $= 10^4/2\pi$ Hz, a filter with a bandwidth of 10^4 Hz is readily found by dividing the elements shown in Fig. 7-9.3(b) by 2π. The result is shown in Fig. 7-9.3(c).

In addition to being frequency normalized for a bandwidth of 1 rad/s, the filter realizations specified by the entries of Table 7-9.2 have also been *impedance normalized*. This is evidenced by the unity-valued resistors at each end of the realization shown in Fig. 7-9.2. In general, practical filter realizations will require an *impedance denormalization* (in addition to a frequency denormalization). Such a denormalization is specified by the relation

$$Z(j\omega) = z_n Z_{zn}(j\omega) \tag{7-194}$$

where $Z_{zn}(j\omega)$ is the normalized impedance, $Z(j\omega)$ is the denormalized one, and z_n is the *impedance denormalization constant*. To see how such a denormalization affects an inductor, let L_{zn} be its impedance normalized value and L be its denormalized value. Then, using (7-194),

$$Z(j\omega) = j\omega L = z_z Z_{zn}(j\omega) = z_n j\omega L_{zn} = j\omega(z_n L_{zn}) \tag{7-195}$$

From the second and fifth members of the above equation we see that $L = z_n L_n$. A similar development shows that a denormalized resistor $R = z_n R_n$, where R_n is the the normalized value. Now let us consider a normalized capacitor C_{zn} which is to be denormalized to a value C. From (7-194)

$$Z(j\omega) = \frac{-j}{\omega C} = z_n Z_{zn}(j\omega) = \frac{-z_n j}{\omega C_{zn}} = \frac{-j}{\omega(C_{zn}/z_n)} \tag{7-196}$$

Thus we see that $C = C_{zn}/z_n$. As an example of an impedance denormalization, let $z_n = 1000$ in the network shown in Fig. 7-9.3(c). The result is shown in Fig. 7-9.3(d). Note that the element values are given in kilohms, henries, and micro-farads. In comparing the voltage transfer functions for the networks shown in Fig. 7-9.3(c) and (d), note that, since a voltage transfer function is dimensionless, it is not affected by an impedance transformation. Thus, the two filters will have the same voltage transfer function, and thus the same frequency behavior. The values of the elements shown in Fig. 7-9.3(d), however, are far more representative of actual practice. A summary of the various frequency- and impedance-denormalization effects is given in Table 7-9.3. It should be noted that the two types of denormalizations can be performed in any order.

Now let us consider *high-pass* filters. The network function $N(j\omega)$ for such a filter will have the form

$$N(j\omega) = \frac{H(j\omega)^n}{(j\omega)^n + a_{n-1}(j\omega)^{n-1} + \cdots + a_1 j\omega + a_0} \tag{7-197}$$

TABLE 7-9.3 Effect of Frequency and Impedance Denormalizations on Network Elements

Denormalization	R	C	L
$\omega = \Omega_n \omega_n$ (frequency)	R	C/Ω_n	L/Ω_n
$Z(j\omega) = z_n Z_{zn}(j\omega)$ (impedance)	$z_n R$	C/z_n	$z_n L$

where H is an arbitrary constant. A *normalized* high-pass filter with a maximally flat magnitude characteristic can be obtained from a normalized low-pass one by using a procedure called the *low-pass to high-pass transformation*. Its steps are:

1. Replace each inductor in the low-pass filter by a capacitor whose value is the reciprocal of the inductor's.
2. Replace each capacitor in the low-pass filter by an inductor whose value is the reciprocal of the capacitor's.

This procedure effectively substitutes $1/j\omega$ for $j\omega$ in the low-pass network function. As such, it inverts the frequency axis about the 1 rad/s frequency, interchanging the 0 and ∞ frequency points. The high-pass filter will have the following properties:

1. Its passband will include the frequencies from 1 to ∞ rad/s (the same as the stopband of the low-pass filter).
2. Its stopband will include the frequencies from 0 to 1 rad/s (the same as the passband of the low-pass filter).
3. Its magnitude characteristic will be maximally flat at infinite frequency.
4. If its magnitude at infinite frequency is H, at 1 rad/s it will have a magnitude of $H/\sqrt{2}$, independent of the order, n, of the filter.
5. If the low-pass filter function has a magnitude K_s at any frequency ω_s, the high-pass filter function will have a magnitude K_s at the frequency $1/\omega_s$.
6. The denominator polynomial of the high-pass network function will be the same as that of the low-pass one.[15]

As an example of the use of the low-pass to high-pass transformation, consider the third-order normalized Butterworth low-pass filter and its magnitude plot shown in Fig. 7-9.4(a). The high-pass filter produced by the transformation is shown in Fig. 7-9.4(b). This filter can, of course, be frequency- and impedance-denormalized to obtain any desired passband and impedance level.

Now let us consider *band-pass* filters. The network function $N(j\omega)$ for such a

---

[15]This is only true for normalized Butterworth functions, and not for other types of magnitude characteristics.

ohms, henries, farads

(a)

(b)

Figure 7-9.4 Low-pass to high-pass transformation.

filter will have the form

$$N(j\omega) = \frac{H(j\omega)^{n/2}}{(j\omega)^n + a_{n-1}(j\omega)^{n-1} + \cdots + a_1 j\omega + a_0} \qquad (7\text{-}198)$$

where n must be even, and where H is an arbitrary constant. A *normalized* band-pass filter can be obtained from a low-pass one by a procedure called the *low-pass to band-pass transformation*. A *normalized band-pass filter* is defined as one which has a resonant (or center) frequency $\omega_0 = 1$ rad/s. This is the frequency where the magnitude of the filter's network function has its maximum (peak) magnitude. The bandwidth B of the band-pass filter is defined by the two frequencies ω_1 and ω_2 at which the magnitude characteristic is 3 dB down from its maximum value. Assuming that $\omega_2 > \omega_1$, the following relations hold

$$\omega_2 - \omega_1 = B \qquad \sqrt{\omega_1 \omega_2} = \omega_0 = 1 \qquad (7\text{-}199)$$

To realize a normalized band-pass filter of order n, the procedure is:

1. Frequency-denormalize a low-pass filter of order $n/2$ so that its bandwidth is equal to the bandwidth B desired in the normalized band-pass filter.

2. In series with each inductor in the denormalized low-pass filter connect a capacitor whose value is the reciprocal of the inductor's.

3. In parallel with each capacitor in the denormalized low-pass filter connect an inductor whose value is the reciprocal of the capacitor's.

The resulting band-pass filter will have the following properties:

1. Its magnitude characteristic will be maximally flat at the normalized center frequency $\omega_0 = 1$ rad/s.

2. If the magnitude of the low-pass filter at $\omega = 0$ is H, the magnitude of the band-pass one at $\omega = \omega_0 = 1$ rad/s will also be H.

3. The magnitude of the band-pass filter function at the two 3-dB frequencies will be $H/\sqrt{2}$. These frequencies may be found using the relations of Eq. (7-199).

4. If the low-pass filter function has a magnitude K_s at a frequency ω_s, the band-pass one will have a magnitude K_s at two frequencies ω_{1s} and ω_{2s}. These frequencies can be found by solving the equations $\omega_{2s} - \omega_{1s} = \omega_s$ and $\omega_{1s}\omega_{2s} = 1$ (assuming that $\omega_{2s} > \omega_{1s}$).

5. The network function for the band-pass filter can be found by substituting $j\omega + 1/j\omega$ for $j\omega$ in the network function for the *de*normalized low-pass filter.

Frequency and impedance denormalizations may, of course, be applied to the normalized band-pass filter. It should be noted that a frequency denormalization will affect both the resonant frequency and the bandwidth. As a result, the Q defined in Eq. (7-148) will remain invariant under such a denormalization. As an example of the application of the low-pass to band-pass transformation, consider the realization of a sixth-order normalized maximally flat magnitude band-pass filter having a bandwidth of 0.5 rad/s. From Eq. (7-199) we obtain $\omega_2 - (1/\omega_2) = 0.5$. Solving we find $\omega_2 = 1.281$ rad/s, and $\omega_1 = 1/\omega_2 = 0.781$ rad/s. A third-order low-pass filter will be required as the starting point. Using Table 7-9.2 and denormalizing to obtain a bandwidth of 0.5 rad/s we obtain the network and magnitude characteristic shown in Fig. 7-9.5(a). Applying the low-pass to band-pass transformation, we obtain the band-pass network and the magnitude characteristic shown in Fig. 7-9.5(b).

Figure 7-9.5 Low-pass to band-pass transformation.

Frequently, passive filter realizations of the type described in the preceding paragraphs are undesirable because they require the use of inductors. There are many disadvantages to such a use. For example, if a filter is being designed for an airborne or satellite application, the relatively large weight and size of inductors as compared to other network components is very undesirable. Another disadvantage is that inductors, in general, have a smaller signal range over which they exhibit a linear

Sinusoidal Steady-State Analysis   Chap. 7

behavior than do resistors or capacitors. Finally, it is not possible to construct inductors by the fabrication techniques used in making integrated circuits. Thus, network realizations which require inductors, in general, cannot be realized in integrated form. From the above discussion we may conclude that it is highly desirable in many applications to replace *RLC* circuits with some other class of circuits in which inductors are not present. This is readily done by using resistors and capacitors and some type of active element. The resulting circuits are called *active-RC* filters. Not only are they able to provide many more types of network functions than their passive *RLC* counterparts, but, in addition, they can be used to provide gain. One of the most convenient ways of realizing such filters is to use separate active-*RC* filter sections for each second-order factor of the desired network function, and a passive-*RC* first-order section (see Fig. 7-7.2) for any first-order factor. The cascade connection of the individual sections then realizes the desired filter function. To apply this method, the network function must be factored. For the maximally flat magnitude functions given in Table 7-9.1, this has been done in Table 7-9.4. The coefficient of the first-degree term in each of the

TABLE 7-9.4 Factors of the Denominator of Maximally Flat Magnitude Network Functions

n	Denominator
2	$(j\omega)^2 + 1.4142j\omega + 1$
3	$[j\omega + 1][(j\omega)^2 + j\omega + 1]$
4	$[(j\omega)^2 + 0.7654j\omega + 1][(j\omega)^2 + 1.8478j\omega + 1]$
5	$[j\omega + 1][(j\omega)^2 + 0.6180j\omega + 1][(j\omega)^2 + 1.6180j\omega + 1]$

second-order factors shown in this table is the reciprocal of the Q (see Sec. 7-7) of that second-order factor. For example, a fourth-degree network function is comprised of two second-order factors, one with $Q = 1/0.7654 = 1.3066$ and the other with $Q = 1/1.8478 = 0.5412$.

A commonly used active-*RC* filter, which realizes a second-order low-pass voltage transfer function is shown in Fig. 7-9.6. A noninverting VCVS of gain K is

Figure 7-9.6 Active-*RC* filter using a VCVS.

used as the active element and is represented by the triangle. This element also provides the necessary isolation when the second-order sections are cascaded. The voltage transfer function for this circuit is

$$\frac{\mathcal{V}_2}{\mathcal{V}_1} = \frac{K/R_1R_3C_2C_4}{(j\omega)^2 + j\omega(1/R_3C_4 + 1/R_1C_2 + 1/R_3C_2 - K/R_3C_4) + 1/R_1R_3C_2C_4}$$

(7-200)

A frequency and impedance normalized design (in which the zero-degree coefficient of the denominator polynomial is unity) is obtained by letting $R_1 = R_3 = C_2 = C_4 = 1$, and using K to match the first-degree coefficient in the factors shown in Table 7-9.4. From (7-200) we see that for this normalization, $3 - K$ is the value of the first-degree coefficient. As an example of the use of this circuit, consider the realization of a normalized fourth-order maximally flat magnitude low-pass network function. The resulting circuit is shown in Fig. 7-9.7. The overall voltage transfer function realized is

Figure 7-9.7 Normalized fourth-order maximally flat magnitude low-pass filter.

$$\frac{\mathcal{V}_2}{\mathcal{V}_1} = \frac{2.2346 \times 1.1522}{[(j\omega)^2 + 0.7654j\omega + 1][(j\omega)^2 + 1.8478j\omega + 1]}$$

(7-201)

Note that the overall dc gain is $2.2346 \times 1.1522 = 2.5747$, i.e., the product of the dc gains of the individual stages. If the two denominator factors of Eq. (7-201) are multiplied, the resulting denominator will be identical with the one given in Eq. (7-190). Frequency and impedance denormalizations may, of course, be applied to the filter of Fig. 7-9.7. The gain of the VCVS, since it is dimensionless, and since it is not a function of frequency, is not affected by such denormalizations.

In the active-RC filter circuit described above, the gain K of the VCVS has two functions, namely, to determine the value of the Q of the second-order factor, and also to set the dc gain. Frequently, it is desirable to be able to control these two parameters independently. Another commonly used active-RC circuit which has this capability is shown in Fig. 7-9.8. An operational amplifier is used as the active element. For this circuit the voltage transfer function is

$$\frac{\mathcal{V}_2}{\mathcal{V}_1} = \frac{-1/R_1R_3C_4C_5}{(j\omega)^2 + j\omega(1/R_1 + 1/R_2 + 1/R_3)/C_4 + 1/R_2R_3C_4C_5}$$

(7-202)

A normalized design (in which the zero-degree coefficient of the denominator poly-

Figure 7-9.8 Active-RC filter using an operational amplifier.

nomial is unity) is obtained for specified values of H (the dc gain) and Q as follows:

1. Let $C_4 = 1$.
2. Choose C_5 as a convenient value subject to the restriction $C_5 < 1/[4Q^2(1 + |H|)]$
3. $R_2 = [1 \pm \sqrt{1 - 4C_5Q^2(1 + |H|)}]/2C_5Q$
4. $R_1 = R_2/|H|$
5. $R_3 = 1/R_2C_5$

As an example, let us realize the normalized fourth-order function with an overall dc gain of 10. To do this, we may let each of the two stages have a dc gain magnitude of $|H| = \sqrt{10} = 3.162$ (since the filter is an inverting one, the actual gain realized by each stage will be -3.162). For stage 1, with $Q = 1.3066$, we find $C_5 < 0.03518$. Choosing $C_5 = 0.03$, and applying the relations given above, we obtain the element values shown in the first stage in Fig. 7-9.9. For the second stage, with $Q = 0.5412$, we find $C_5 < 0.2051$. Choosing $C_5 = 0.2$ we obtain the element values shown in the second stage of Fig. 7-9.9. The resulting overall voltage transfer function is the same as the ones given in Eqs. (7-190) and (7-201) except that the numerator constant is 10.

Figure 7-9.9 Normalized fourth-order maximally flat magnitude low-pass filter.

A more complex network which uses operational amplifiers as active elements is shown in Fig. 7-9.10. This configuration has the advantage that it may be used to realize very high-Q network functions. In addition, the resulting characteristic is extremely stable, i.e., the magnitude and phase characteristics produced by the network are relatively unaffected by changes in the values of the elements or changes in the gain of the operational amplifiers. A further advantage of this circuit is that it may be used to produce low-pass, high-pass, or band-pass transfer functions depending on the point from which the output is taken. In terms of the variables defined in the figure, the voltage transfer functions for this circuit may be shown to be

$$\frac{\mathcal{V}_{LP}}{\mathcal{V}_{IN}} = \frac{2}{1 - \omega^2 + j2\omega G}$$

$$\frac{\mathcal{V}_{HP}}{\mathcal{V}_{IN}} = \frac{-2\omega^2}{1 - \omega^2 + j2\omega G} \qquad (7\text{-}203)$$

$$\frac{\mathcal{V}_{BP}}{\mathcal{V}_{IN}} = \frac{-j2\omega}{1 - \omega^2 + j2\omega G}$$

ohms, farads

Figure 7-9.10 Active *RC* network using operational amplifiers.

These relations only hold for the case where $G \ll 1$. High-pass and band-pass versions of the active-*RC* filters shown in Figs. 7-9.6 and 7-9.8 are also frequently used. Some configurations for such filters may be found in the problems.

7-10 SINUSOIDAL STEADY-STATE POWER

In Sec. 7-1 we introduced the use of rms values to characterize sinusoidally varying quantities. Specifically, we showed that a sinusoidal function $f(t)$ having the form

$$f(t) = F_m \cos(\omega t + \phi) \tag{7-204}$$

where F_m is the peak value of $f(t)$, has an rms value $F_{\text{rms}} = F_m / \sqrt{2}$. The definition of an rms value was shown to be such that a sinusoidal steady-state current with an rms value I_{rms} would dissipate the same average power in a given resistor as a constant current with a magnitude equal to I_{rms}. In this section we shall delve more deeply into the use of rms values for voltage and current and show how these may be used to determine the manner in which power is dissipated and stored in circuits under sinusoidal steady-state conditions.

To begin our discussion, consider a two-terminal network with voltage and current variables defined using the usual associated reference directions as shown in Fig. 7-10.1. We shall also assume that there are no independent sources in the network. Let us now define a network function giving the driving-point impedance of the network as

$$Z(j\omega) = R(\omega) + jX(\omega) = |Z| e^{j\theta} = |Z| \cos\theta + j|Z| \sin\theta \tag{7-205}$$

Figure 7-10.1 Associated reference polarity directions.

where $R(\omega)$ is the real part of $Z(j\omega)$ and $X(\omega)$ is the imaginary part. The relation between $R(\omega)$, $X(\omega)$, $|Z|$, and θ may be represented by a vector diagram as shown in Fig. 7-10.2. Such a diagram is sometimes called an *impedance triangle*. Now let us

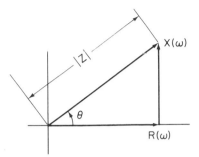

Figure 7-10.2 Impedance triangle.

assume that a sinusoidal voltage is applied as an excitation to the network shown in Fig. 7-10.1. If we let this voltage be

$$v(t) = V_m \cos(\omega t + \alpha) = \sqrt{2}\, V_{rms} \cos(\omega t + \alpha) \qquad (7\text{-}206)$$

then the sinusoidal steady-state current that results may be found by applying the phasor methods derived in this chapter. Thus, we may write

$$i(t) = I_m \cos(\omega t + \alpha - \theta) = \sqrt{2}\, I_{rms} \cos(\omega t + \alpha - \theta) \qquad (7\text{-}207)$$

where

$$I_m = \frac{V_m}{|Z|} \qquad I_{rms} = \frac{V_{rms}}{|Z|} \qquad (7\text{-}208)$$

The instantaneous power supplied to this network is found by multiplying (7-206) and (7-207). Thus, since $\cos a \cos b = \tfrac{1}{2}\cos(a-b) + \tfrac{1}{2}\cos(a+b)$, we may write

$$p(t) = v(t)i(t) = V_{rms}I_{rms}\,[\cos\theta + \cos(2\omega t + 2\alpha - \theta)] \qquad (7\text{-}209)$$

Now let us find the average value of the instantaneous power $p(t)$. This is done by integrating $p(t)$ in Eq. (7-209) over one period of the sinusoidal excitation and dividing the result by the value of the period. The average value of the second term inside the bracket in the right member of Eq. (7-209) is zero. Thus, we find that

$$P_{avg} = \frac{1}{T}\int_0^T p(t)\,dt = V_{rms}I_{rms}\cos\theta \qquad (7\text{-}210)$$

where T is the period. From this expression, we see that the average power is always less than or equal to the product of the rms values of the voltage and the current. To indicate the ratio of these quantities, we may define the *power factor* (abbreviated PF) as follows:

$$PF = \cos\theta = \frac{P_{avg}}{V_{rms}I_{rms}} \qquad (7\text{-}211)$$

Using (7-208) to eliminate V_{rms} in (7-210) and applying (7-205), we may also write

$$P_{avg} = I_{rms}^2 |Z|\cos\theta = I_{rms}^2 R(\omega) \qquad (7\text{-}212)$$

In a similar manner, if we define the network function specifying the driving-point admittance of the network shown in Fig. 7-10.1 as

$$Y(j\omega) = G(\omega) + jB(\omega) = |Y| e^{j\phi} = |Y| \cos\phi + j|Y| \sin\phi \tag{7-213}$$

then since $\phi = -\theta$ and $|Y| = 1/|Z|$, using (7-208) to eliminate I_{rms} in (7-210) we obtain

$$P_{avg} = \frac{V_{rms}^2}{|Z|} \cos\theta = V_{rms}^2 G(\omega) \tag{7-214}$$

It should be noted that $G(\omega) = 1/R(\omega)$ only if $X(\omega)$, and thus $B(\omega)$, is zero.

To more clearly indicate the relation between instantaneous power and average power, let us consider three cases in detail. The first of these is the case for which the imaginary part of $Z(j\omega)$ is zero, i.e., the phase angle θ is zero and the power factor is unity. Typical waveforms of the voltage, current, and power variables for this case are shown in Fig. 7-10.3(a). It is readily observed that the instantaneous power is always positive, similarly, from Eq. (7-210) we see that the average power P_{avg} is simply the product $V_{rms} I_{rms}$. Now let us consider a second case defined by letting $R(\omega) = X(\omega)$. Thus, $\theta = 45°$ and the power factor is 0.707. Typical plots of the waveforms of the voltage, current, and power are shown for this case in Fig. 7-10.3(b). In this case, although the instantaneous power $p(t)$ goes positive and negative, the average power P_{avg} is clearly positive and, from Eq. (7-210), it has the value $V_{rms} I_{rms} \cos 45°$. As a third case, let us see what happens if the real part of the driving-

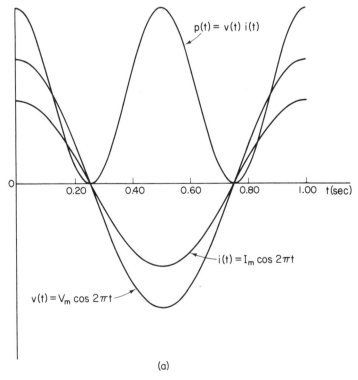

(a)

Figure 7-10.3 Voltage, current, and power waveforms: (a) $\theta = 0°$; (b) $\theta = 45°$; (c) $\theta = 90°$.

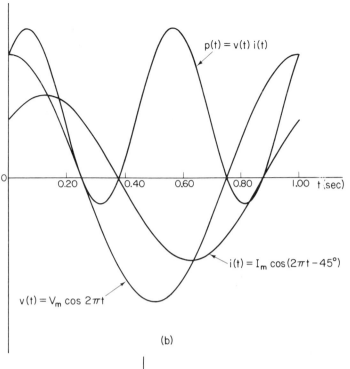

$p(t) = v(t)\, i(t)$

$i(t) = I_m \cos(2\pi t - 45°)$

$v(t) = V_m \cos 2\pi t$

(b)

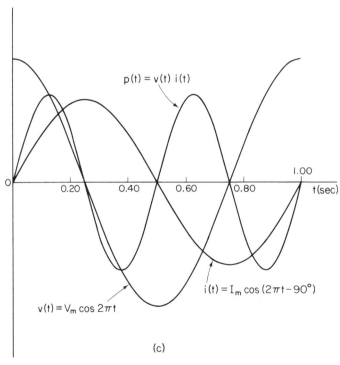

$p(t) = v(t)\, i(t)$

$i(t) = I_m \cos(2\pi t - 90°)$

$v(t) = V_m \cos 2\pi t$

(c)

Figure 7-10.3   Continued

point impedance $Z(j\omega)$ is zero. In this case, $\theta = 90°$ and, from Eq. (7-210), the average power is zero. Typical waveforms of voltage, current, and power for this case are shown in Fig. 7-10.3(c). From the waveforms in this figure, we may note that the *instantaneous* power has quite large values for some instants of time. The *average* value of the power, however, is zero. The positive and negative values of the instantaneous power represent the transfer of energy back and forth between the source which excites the circuit and the elements of the circuit itself. In other words, during the portions of the cycle where $p(t)$ is positive, energy is stored in the electric and magnetic fields associated with the capacitive and inductive elements, respectively. During portions of the applied cycle where $p(t)$ is negative, this energy is delivered back to the source. The power which oscillates back and forth between source and network in this manner is frequently spoken of as being *reactive power*. Although it is not dissipated in the network, i.e., it is not turned into heat, it does represent an actual flow of current into and out of the network. Thus, in designing power systems it is important to provide sufficient wiring capacity to permit the flow of such reactive currents. This is especially important since all "real-world" power lines will have some resistance present in them. Thus, even though the reactive power does not dissipate energy in the load, in practice it will actually produce energy losses in the power lines themselves. To more clearly indicate the component of the instantaneous power which is reactive, we may use the relation $\cos(\alpha + \beta) = \cos\alpha \cos\beta + \sin\alpha \sin\beta$ to write Eq. (7-209) in the following form:

$$p(t) = V_{\text{rms}} I_{\text{rms}} \cos\theta[1 + \cos(2\omega t + 2\alpha)] + V_{\text{rms}} I_{\text{rms}} \sin\theta[\sin(2\omega t + 2\alpha)] \qquad (7\text{-}215)$$

When $\theta = 90°$, the instantaneous power is obviously determined only by the second term in the right member of the above. On the other hand, for values of θ in the range $0 < \theta < 90°$, the magnitude of this second term is $V_{\text{rms}} I_{\text{rms}} \sin\theta$. Thus, we may define a reactive average power represented by this term. Specifically, using the symbol Q for average *reactive power*, we may write[16]

$$Q = V_{\text{rms}} I_{\text{rms}} \sin\theta \qquad (7\text{-}216)$$

The units of Q are VARS (*V*olt *A*mperes *R*eactive). It should be noted that for an inductive circuit, θ is greater than 0, and therefore Q is positive. This condition is referred to as having a *lagging power factor* (the current *lags* the voltage). For a capacitive circuit, θ is less than zero, and Q will be negative. This condition is referred to as having a *leading power factor* (the current leads the voltage). Thus, in general, reactive power can be positive or negative. The relation between P, the average (or real) power, and Q, the reactive power, is simply that of a right triangle. To illustrate this, we may define a *complex power W* by the relation

$$W = V_{\text{rms}} I_{\text{rms}} e^{j\theta} = V_{\text{rms}} I_{\text{rms}}(\cos\theta + j\sin\theta) = P + jQ \qquad (7\text{-}217)$$

The units for W are VA (*V*olt *A*mperes). The right-triangle relation may be emphasized by drawing a power triangle as shown in Fig. 7-10.4.

Complex power may be conveniently expressed in terms of the phasors used to represent the voltage and current variables. If we let $\mathcal{U}$ and $\mathcal{I}$ be the phasors for these

---

[16]Note that the symbol Q as used here is different than the one used for quality factor in Sec. 7-7.

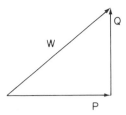

Figure 7-10.4 Power triangle.

variables, where[17]

$$\mathcal{V} = V_{rms}e^{j\alpha} \qquad \mathcal{I} = I_{rms}e^{j\beta} \tag{7-218}$$

Then, if these phasors are the terminal variables associated with the driving-point impedance $Z(j\omega) = |Z|e^{j\theta}$ we may write

$$\mathcal{V} = V_{rms}e^{j\alpha} = |Z|e^{j\theta}I_{rms}e^{j\beta} = |Z|I_{rms}e^{j(\beta+\theta)} \tag{7-219}$$

Now let us multiply both sides of Eq. (7-219) by $\mathcal{I}^* = I_{rms}e^{-j\beta}$. Thus, we obtain

$$\mathcal{V}\mathcal{I}^* = |Z|I_{rms}^2e^{j\theta} \tag{7-220}$$

Substituting for one of the I_{rms} quantities from (7-208), we obtain

$$\mathcal{V}\mathcal{I}^* = V_{rms}I_{rms}e^{j\theta} = W \tag{7-221}$$

The results given above may be summarized as follows:

SUMMARY 7-10.1

Sinusoidal Steady-State Power. If $\mathcal{V}$ and $\mathcal{I}$ are the phasors representing the rms values of sinusoidal steady-state voltage and current variables with associated reference directions at a pair of terminals of a network, the complex power W (in volt amperes) delivered to the pair of terminals is defined as $\mathcal{V}\mathcal{I}^*$. The real part of W is the average power P (in watts) transmitted to the network and the imaginary part of W is the reactive power Q (in vars). ∎

Now let us investigate the power supplied to two two-terminal impedances connected in parallel and excited by a voltage source. Such a connection is shown in Fig. 7-10.5(a). In terms of the phasors defined in the figure, we may write

$$W = \mathcal{V}\mathcal{I}^* = \mathcal{V}(\mathcal{I}_1^* + \mathcal{I}_2^*) = \mathcal{V}\mathcal{I}_1^* + \mathcal{V}\mathcal{I}_2^* \tag{7-222}$$

where W is the total complex power supplied to both networks. If we define the component power to network 1 as $P_1 + jQ_1$, and that to network 2 as $P_2 + jQ_2$, we may write, from (7-222),

$$W = (P_1 + jQ_1) + (P_2 + jQ_2) = P_1 + P_2 + j(Q_1 + Q_2) \tag{7-223}$$

Thus, we conclude that the total real power supplied to the parallel connection of the two networks is the sum of the powers consumed by the individual networks. Similarly the total reactive power is the sum of the individual reactive powers. Now

[17]For convenience, here we define the phasors in terms of the rms value of the sinusoidal functions which they represent rather than the peak magnitude. Such a convention is usually followed when considering sinusoidal steady-state power.

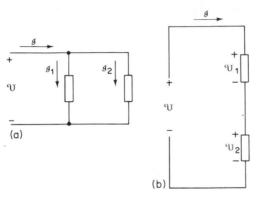

(a)

(b)

Figure 7-10.5 Series and parallel connections of two two-terminal impedances.

let us consider a similar situation in which two networks are connected in series as shown in Fig. 7-10.5(b). For this case, the total complex power W is

$$W = \mathcal{V}\mathcal{I}^* = \mathcal{V}_1\mathcal{I}^* + \mathcal{V}_2\mathcal{I}^* \tag{7-224}$$

Again we see that the total real and reactive powers are simply given as the sum of the real and reactive powers of the individual networks. The conclusion given above for parallel and series connected networks may be readily shown to apply to any arbitrary interconnection of two-terminal networks. As such, we may refer to these results as the *conservation of complex power*. These results may be summarized as follows:

SUMMARY 7-10.2

Conservation of Complex Power. If, in a set of two-terminal networks N_i, the real and reactive powers are P_i and Q_i respectively, the total real and reactive power for *any* interconnection of the N_i is the sum of the P_i and the sum of the Q_i, respectively. Thus, the total complex power for such an interconnection of networks is the sum of the complex powers of the component networks. ■

It should be noted that the units of voltamperes, watts, and vars used for complex power, real power, and reactive power, respectively, are too small for many practical power computations. Units which are more frequently used in practice are kVA (kilovoltamperes), kW (kilowatts), and kVAR (kilovoltamperes reactive), respectively. As an example of a typical power computation, consider the following.

Example 7-10.1

Two loads represented by the two two-terminal networks labeled A and B in Fig. 7-10.6 are fed by a power line with an impedance $Z = 0.1 + j0.1$ Ω as shown. Load A

Figure 7-10.6 Network for Example 7-10.1.

is inductive and consumes 10 kVA of power at 0.6 PF and load B is capacitive and consumes 5 kVA of power at 0.8 PF. The voltage $\mathcal{V}$ at the loads is 100 V rms. It is desired to find the rms value of the generator voltage $\mathcal{V}_g$. To solve this problem, we may first determine the complex power for each of the loads. For load A the real power is $10 \times 0.6 = 6$ kW. Using a power triangle similar to the one shown in Fig. 7-10.4, we find that the reactive power is $+8$ kVAR. Thus the complex power is $6 + j8$ kVA. Similarly, for network B, the complex power is $4 - j3$ kVA. Summing these powers, we find that the total complex load power is $10 + j5$ kVA, thus $|W| = \sqrt{125}$ kVA. Since the total load is inductive, the load voltage leads the total load current by $\tan^{-1}\left(\frac{5}{10}\right)$. If we assume a reference angle of $0°$ for the load voltage phasor, then, using rms quantities, the load voltage phasor is $100\underline{/0°}$ V or $0.1\ \underline{/0°}$ kV. Thus, from (7-221), the total current phasor $\mathcal{I}$ may be computed as follows:

$$\mathcal{I}^* = \frac{W}{\mathcal{V}} = \frac{10 + j5}{0.1} = 100 + j50 \qquad \mathcal{I} = 100 - j50 \text{ A}$$

It should be noted that the current lags the voltage, as must be the case for an inductive load. The phasor voltage drop caused by the power line impedance Z is now found as $\mathcal{I}Z = (100 - j50) \times (0.1 + j0.1) = 15 + j5$ V. Thus the generator voltage $\mathcal{V}_g = \mathcal{I}Z + \mathcal{V} = (15 + j5) + 100 = 115 + j5$ V, and the magnitude of $\mathcal{V}_g$ is very nearly equal to 115 V rms. We see that the generator voltage in this case is 15 percent greater than the load voltage.

In the example given above we noted that the generator voltage was approximately 15 percent higher than the load voltage. Such a situation is usually very undesirable since, if the loads A and B shown in Fig. 7-10.6 are disconnected, the voltage at the load terminals will rise from 100 V to 115 V. If some relatively low-wattage electronic apparatus should remain connected to these load terminals, and if the rated voltage for this equipment was 100 V, such a rise in voltage might seriously affect the operation of the equipment (or even cause it to smoke). Thus, it is desirable to minimize the change in voltage that occurs when a large load is connected to or removed from a given power line. One way of doing this for loads which are highly reactive is to raise the power factor of the load. Such an approach is treated in the following example.

Example 7-10.2

As an example of power factor correction, consider the addition of a third load to the two loads used in the preceding example. We will choose this load so that the resulting power factor of all the loads is unity. From the computations made in the previous example, we see that the power factor correcting load must have a complex power $W = 0 - j5$ kVA, i.e., it must consume zero real power and have a reactive power of -5 kVAR. Physically, such a load simply consists of a capacitor. With the addition of this compensating load, the total complex load power W is now $10 + j0$. The total phasor current to the loads is thus $\mathcal{I} = 100 + j0$ A and the generator voltage is $\mathcal{V}_g = \mathcal{I}Z + \mathcal{V} = (10 + j10) + 100 = 110 + j10$ V. Thus, the magnitude of the generator voltage is now very nearly equal to 110 V rms and the regulation has been reduced from 15 percent to 10 percent.

Load factor correction techniques similar to the one illustrated in the preceding example are greatly encouraged by power companies to the extent that for large industrial users, low power factor rates, which are usually considerably higher than high power factor rates, are levied. As a result, it is frequently economically attractive for large companies to install devices whose sole function is power factor correction. Such corrective techniques may range in size from fluorescent light ballasts specially designed for high power factor all the way to large rotating machines called synchronous condensers where sole purpose is the simulation of a capacitive load.

It is frequently desirable to maximize the power transferred to a given load by adjusting the load itself or by inserting some network between the generator and the load. Such a maximization is important in a broad range of applications that are encountered in communication systems and electronic circuits. One approach to the general problem is illustrated in Fig. 7-10.7. In this figure, a voltage generator with

Figure 7-10.7 Non-ideal source and a load.

phasor output voltage $\mathcal{U}_g$ and series impedance $Z_g = R_g + jX_g$ drives a load impedance $Z_L = R_L + jX_L$. Such a generator circuit may be considered as the Thevenin equivalent for a broad range of network configurations. Now let us assume that R_L and X_L are free to be varied such that the power delivered by the generator to the load is maximized. The phasor current $\mathcal{I}_g$ is readily found to be

$$\mathcal{I}_g = \frac{\mathcal{U}_g}{Z_g + Z_L} = \frac{\mathcal{U}_g}{(R_g + R_L) + j(X_g + X_L)} \tag{7-225}$$

The average power delivered to the load is given as

$$P_L = |\mathcal{I}_g|^2 R_L = \frac{|\mathcal{U}_g|^2 R_L}{(R_g + R_L)^2 + (X_g + X_L)^2} \tag{7-226}$$

Let us first consider the case where R_L is a constant and find a value of X_L to maximize the power P_L. Taking the partial derivative of the expression in Eq. (7-226) with respect to X_L we obtain

$$\frac{\partial P_L}{\partial X_L} = |\mathcal{U}_g|^2 R_L \left[\frac{-2(X_g + X_L)}{[(R_g + R_L)^2 + (X_g + X_L)^2]^2} \right] \tag{7-227}$$

Setting this derivative to zero, we obtain the relation $X_g = -X_L$. Substituting this value in Eq. (7-226) we obtain

$$P_L |_{X_L = -X_g} = \frac{|\mathcal{U}_g|^2 R_L}{(R_g + R_L)^2} \tag{7-228}$$

To determine the effect R_L has on power maximization, we may now differentiate this result again, this time with respect to R_L, and set the result to zero. Doing this, we find the maximum occurs when $R_G = R_L$. Considering both of the above results, we see that the maximum power transfer is obtained when

$$Z_L = Z_g^* \tag{7-229}$$

Sinusoidal Steady-State Analysis   Chap. 7

Thus, maximum power transfer occurs *when the source impedance and the load impedance are conjugates.*

Another example of the maximization of power transfer occurs when both the generator impedance and the load impedance are fixed. In this situation a method which is frequently used to maximize the transfer of power is the insertion of a transformer between the generator and the load. The turns ratio of the transformer is chosen such that the desired maximum is obtained. As an example of such a situation, consider the network shown in Fig. 7-10.8 in which an ideal transformer of

Figure 7-10.8 Use of a transformer to maximize power transfer.

turns ratio $1:n$ is inserted between a source impedance Z_g and a load impedance Z_L. The driving-point impedance seen at the left end of the transformer $Z_0 = Z_L/n^2$. If the phase of Z_L is θ, then

$$Z_0 = |Z_0| e^{j\theta} = |Z_0| \cos\theta + j |Z_0| \sin\theta \qquad (7\text{-}230)$$

Since the transformer is a lossless device, the power dissipated in Z_0 is the same as that dissipated in Z_L. Thus, we may write

$$P_L = |\mathcal{I}_g|^2 |Z_0| \cos\theta = \frac{|\mathcal{V}_g|^2 |Z_0| \cos\theta}{(R_g + |Z_0| \cos\theta)^2 + (X_g + |Z_0| \sin\theta)^2} \qquad (7\text{-}231)$$

Since $|Z_0|$ is adjusted by changing the value of n, the turns ratio, let us now take the derivative of P_L with respect to $|Z_0|$. Doing this and setting the result to zero, we obtain $|Z_0| = |Z_g|$ Thus, since $|Z_0| = |Z_L|/n^2$, we obtain

$$n = \sqrt{\frac{|Z_L|}{|Z_g|}} \qquad (7\text{-}232)$$

as the value of n which provides maximum power transfer. An example follows.

Example 7-10.3

As an example of the use of a transformer to maximize power transfer, let us consider an audio amplifier which has an output impedance of 500 Ω. It is desired to connect the amplifier to a loudspeaker with a coil impedance of 8 Ω. To achieve the desired maximum power transfer, we may insert a transformer between the amplifier and the loudspeaker with the turns ratio $n = \sqrt{8/500}$, i.e., approximately 7.9:1 as shown in Fig. 7-10.9.

Amplifier   Loudspeaker

Figure 7-10.9 Network for Example 7-10.3.

7-11 THREE-PHASE CIRCUITS

In this section we shall show how the phasor methods developed to analyze circuits operating under sinusoidal steady-state conditions can be applied to the study of three-phase ac circuits. Such circuits have wide use in the generation, transmission, and distribution of electric power since they provide constant power rather than sinusoidally varying power. Thus, three-phase circuits provide an important example of the application of circuit-analysis techniques to an everyday engineering situation. An ideal three-phase power generator may be visualized as consisting of three independent voltage sources, each generating a sinusoidal voltage displaced in phase by $120°$ from the other two. For example, consider the three sources with outputs $v_a(t)$, $v_b(t)$, and $v_c(t)$ shown in Fig. 7-11.1(a). For clarity we have indicated expressions for

Figure 7-11.1 Three sources and their phasor diagram.

both the time-domain and the phasor representations for these sources. It should be noted that the magnitude and frequency of the output of each source is the same, namely, $\sqrt{2} V_o$ V and ω rad/s. In addition, it should be noted that the phasors are defined using rms quantities for their magnitudes. A phasor diagram for the three phasor voltages $\mathcal{V}_a$, $\mathcal{V}_b$, and $\mathcal{V}_c$ is shown in Fig. 7-11.1(b). Now let us consider the case where the negative reference terminals of the three sources shown in Fig. 7-11.1(a) are connected to a common point as shown in Fig. 7-11.2(a). The common point

Figure 7-11.2 Wye-connected three-phase source and its phasor diagrams.

is identified by the letter *n*. It is called a *neutral*. Such a network is called a *wye-con-nected* three-phase generator. The relative orientation of the three voltage sources with respect to terminals *a*, *b*, and *c* which is shown in Fig. 7-11.2(a) is called *abc* rotation. If the outputs of voltage sources $\mathcal{V}_b$ and $\mathcal{V}_c$ were interchanged, the result would be referred to as *acb* rotation. Since the terminals labeled *a*, *b*, and *c* are usually connected to a set of three transmission lines which carry power to some load remotely located from the generator, it is convenient to refer to the source voltages shown in Fig. 7-11.2(a) as *line-to-neutral* voltages. We may also define a set of *line-to-line* voltages $\mathcal{V}_{ab}$, $\mathcal{V}_{bc}$, and $\mathcal{V}_{ca}$ as shown in Fig. 7-11.2(a). The phasor representations for these voltages are readily found in terms of those for the line-to-netural voltages as follows:

$$\mathcal{V}_{ab} = \mathcal{V}_a - \mathcal{V}_b$$
$$\mathcal{V}_{bc} = \mathcal{V}_b - \mathcal{V}_c \qquad (7\text{-}233)$$
$$\mathcal{V}_{ca} = \mathcal{V}_c - \mathcal{V}_a$$

The graphical construction used to determine the line-to-line phasor voltages is shown in 7-11.2(b). From this figure, we see that the magnitude of the line-to-line voltages may be computed as $2\cos 30° V_o = \sqrt{3} V_o$ and that the phase of the line-to-line voltage phasors is displaced 30° from that of the line-to-neutral phasors. Thus, we may write

$$\mathcal{V}_{ab} = \sqrt{3} V_o / 30° \qquad v_{ab}(t) = \sqrt{6} V_o \cos (\omega t + 30°)$$
$$\mathcal{V}_{bc} = \sqrt{3} V_o / -90° \qquad v_{bc}(t) = \sqrt{6} V_o \cos (\omega t - 90°) \qquad (7\text{-}234)$$
$$\mathcal{V}_{ca} = \sqrt{3} V_o / 150° \qquad v_{ca}(t) = \sqrt{6} V_o \cos (\omega t + 150°)$$

For brevity, the line-to-neutral voltages are frequently referred to simply as *phase* voltages and the line-to-line voltages simply as *line* voltages. It is somewhat easier to visualize the relation between the phase and the line voltages if the phasors for the latter are removed from the origin and drawn as shown in Fig. 7-11.2(c). From this latter phasor diagram, we see that if KVL is applied to the line voltages we obtain

$$\mathcal{V}_{ab} + \mathcal{V}_{bc} + \mathcal{V}_{ca} = 0 \qquad (7\text{-}235)$$

Frequently, three-phase voltages are generated by connecting voltage sources directly to the lines rather than to a neutral point. Such a connection is shown in Fig. 7-11.3(a). It is called a *delta connection*. A three-phase transformer bank is one means of providing such a connection which is frequently encountered. The phasor diagram for this connection is shown in Fig. 7-11.3(b). For such a connection no neutral point is available. It should be noted that if any of the sources shown in Fig. 7-11.3(a) are reversed, the results could be disastrous. In this case, the sum of the phasor voltages no longer equals zero. Thus there will be a net non-zero voltage around the loop and, since the impedance of the loop is that of the voltage sources, namely, zero, in theory infinite current will flow. In practice, if there is no circuit breaker to protect the circuit, the heat generated by the resulting large currents will rapidly produce smoke and may even cause the transformer to explode.

As a first example of the type of computations which are normally encountered for three-phase circuits, let us consider the case of a *balanced* wye-connected load,

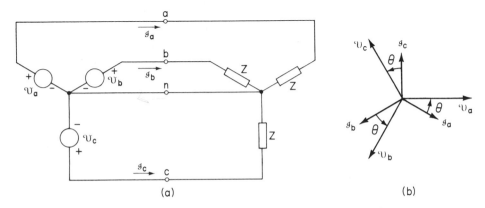

Figure 7-11.3 Delta-connected three-phase source and its phasor diagram.

i.e., a load consisting of three equal impedances $Z(j\omega)$, all of which have a common neutral terminal. Such a load, driven by a wye-connected three-phase generator, is shown in Fig. 7-11.4(a). If we assume that the phase angle of $Z(j\omega)$ is θ, then the

Figure 7-11.4 Balanced wye-connected load and its phasor diagram.

phasor diagram (using $\mathcal{U}_a$ as the reference) for the circuit is as shown in Fig. 7-11.4(b). From this phasor diagram we see that the sum of the line currents $\mathcal{I}_a$, $\mathcal{I}_b$, and $\mathcal{I}_c$ is zero. Thus, applying KCL to the circuit shown in Fig. 7-11.4(a), we see that the current in the neutral wire must be zero. Thus, for a balanced wye-connected load the neutral connection may be omitted without any change occurring in the circuit variables. If we let V_P be the rms magnitude of the phase voltage, and V_L be the rms magnitude of the line voltage, then, as we have previously shown, $V_L = \sqrt{3}\,V_P$. The average real and reactive power delivered to each phase of the wye-connected load may be designated as P_P and Q_P, respectively. Thus, applying the results of the preceding section, we may write

$$P_P = V_P I \cos\theta = \left(\frac{V_L}{\sqrt{3}}\right) I \cos\theta \tag{7-236a}$$

$$Q_P = V_P I \sin\theta = \left(\frac{V_L}{\sqrt{3}}\right) I \sin\theta \tag{7-236b}$$

Where I is the rms value of the line current. Since there are three phases, we may make a total power analysis of the circuit as follows:

$$P_{\text{total}} = 3P_P = 3V_P I \cos \theta = \sqrt{3} \, V_L I \cos \theta \qquad (7\text{-}237\text{a})$$

$$Q_{\text{total}} = 3Q_P = 3V_P I \sin \theta = \sqrt{3} \, V_L I \sin \theta \qquad (7\text{-}237\text{b})$$

$$|W_{\text{total}}| = (P_{\text{total}}^2 + Q_{\text{total}}^2)^{1/2} = \sqrt{3} \, V_L I \qquad (7\text{-}237\text{c})$$

$$\text{PF} = \frac{P_{\text{total}}}{|W_{\text{total}}|} = \cos \theta \qquad (7\text{-}237\text{d})$$

As an example of a typical problem involving a balanced wye-connected load, consider the following.

Example 7-11.1

A three-phase voltage generator with line voltages of 120 V rms is used to drive a balanced wye-connected load consisting of three impedances $Z_L = 4 + j3\ \Omega$ through a set of power lines, each of which has an impedance $Z_G = 0.1 + j0.1\ \Omega$ as shown in Fig. 7-11.5. It is desired to find the line voltages at the load and the power delivered to

Figure 7-11.5 Network for Example 7-11.1.

the load. To do this we first find I, the rms magnitude of the line currents. Since the load is balanced, such a determination may be made on the basis of phase (line-to-neutral) quantities even though no neutral connection is present. Thus, we see that

$$I = \frac{V_P}{|Z_G + Z_L|} = \frac{120/\sqrt{3}}{|0.1 + j0.1 + 4 + j3|} = 13.48 \text{ A rms}$$

The line voltage at the load is $\sqrt{3}\, I |Z_L| = \sqrt{3} \times 13.48 \times 5 = 116.73$ V rms. The total power delivered to the load is $3I^2 \operatorname{Re} Z_L = 3 \times (13.48)^2 \times 4 = 2180$ W.

A second example of a three-phase power system frequently encountered in practice is one in which a balanced delta-connected load is used. Such a system is shown in Fig. 7-11.6(a). Since the system is balanced, the phasor load currents $\mathcal{I}_1$, $\mathcal{I}_2$, and $\mathcal{I}_3$ must be separated 120° in phase from each other. Thus, arbitrarily using $\mathcal{I}_1$ as reference and assuming abc rotation, the load currents may be represented as shown in Fig. 7-11.6(b). The individual phasor line currents $\mathcal{I}_a$, $\mathcal{I}_b$, and $\mathcal{I}_c$ defined in the figure are now related to the phasor load currents as follows:

$$\mathcal{I}_a = \mathcal{I}_1 - \mathcal{I}_3$$

$$\mathcal{I}_b = \mathcal{I}_2 - \mathcal{I}_1 \qquad (7\text{-}238)$$

$$\mathcal{I}_c = \mathcal{I}_3 - \mathcal{I}_2$$

These relations are also indicated in the phasor diagram shown in Fig. 7-11.6(b). It

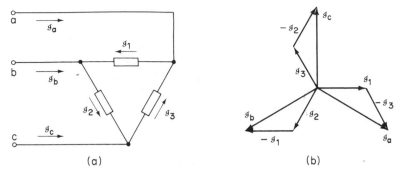

Figure 7-11.6 Balanced delta-connected load and its phasor diagram.

is readily seen that the magnitude of the line currents is $\sqrt{3}$ times the magnitude of the load currents. If we let I_P be the rms value of the phase (or line) currents and I_L be the rms value of the load currents, the real and reactive power delivered to each of the individual loads of the delta connection is

$$P_L = VI_L \cos \theta = V\left(\frac{I_P}{\sqrt{3}}\right) \cos \theta \qquad (7\text{-}239\text{a})$$

$$Q_L = VI_L \sin \theta = V\left(\frac{I_P}{\sqrt{3}}\right) \sin \theta \qquad (7\text{-}239\text{b})$$

where V is the rms value of the line voltage. Summing these quantities to determine the total power delivered to the load, we find

$$P_{\text{total}} = 3P_L = \sqrt{3}\,VI_L \cos \theta \qquad (7\text{-}240\text{a})$$

$$Q_{\text{total}} = 3Q_L = \sqrt{3}\,VI_P \sin \theta \qquad (7\text{-}240\text{b})$$

$$|W_{\text{total}}| = (P_{\text{total}}^2 + Q_{\text{total}}^2)^{1/2} = \sqrt{3}\,VI_P \qquad (7\text{-}240\text{c})$$

$$\text{PF} = \frac{P_{\text{total}}}{|W_{\text{total}}|} = \cos \theta \qquad (7\text{-}240\text{d})$$

It should be noted that these relations are identical with those given for the balanced wye-connected load.

In solving problems in which the power lines feeding power to a delta-connected load have some non-zero impedance, it is usually convenient to transform the delta-connected load to a wye-connected load. This can be accomplished using relations similar to those given for making delta-wye transformations of resistance networks in Sec. 3-6. Extending these relations, for the balanced case, we obtain

$$Z_\Delta = 3Z_Y \qquad (7\text{-}241)$$

where Z_Δ and Z_Y are the load impedances in the delta- and wye-connected cases, respectively. An example of the use of such a transformation follows.

Example 7-11.2

It is desired to find a balanced delta-connected load which will create the same circuit conditions in the power lines as those found in connection with the preceding example in which the individual impedances of the wye-connected load had the value $Z_Y = 4 + j3 = 5\underline{/36.9°}$. Applying (7-241), we see that $Z_\Delta = 12 + j9$. Thus, for the circuit shown in Fig. 7-11.7, using the results of the preceding example, we see that the line currents are 13.48 A rms and that the line voltages at the load are 116.73 V rms. The power delivered to the total load consisting of the three impedances Z_Δ is still 2180 W. The currents in the individual loads are $13.48/\sqrt{3} = 7.78$ A rms.

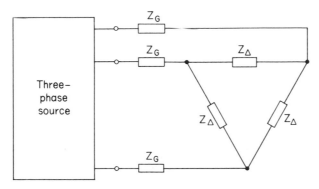

Figure 7-11.7 Network for Example 7-11.2.

The one remaining situation illustrating three-phase power circuits which will be treated in this section is the case of an unbalanced load, either wye- or delta-connected. Frequently, delta-wye or wye-delta transformations may be used to simplify the resulting analysis. For the unbalanced case, the delta-to-wye transformations are readily shown to be

$$Z_a = \frac{Z_1 Z_3}{Z_1 + Z_2 + Z_3} \qquad Z_b = \frac{Z_1 Z_2}{Z_1 + Z_2 + Z_3} \qquad Z_c = \frac{Z_2 Z_3}{Z_1 + Z_2 + Z_3} \qquad (7\text{-}242)$$

where the component impedances are indicated in Fig. 7-11.8. Similarly, the wye-to-

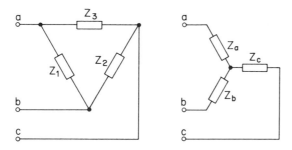

Figure 7-11.8 Wye and delta networks.

delta relations are

$$Z_1 = \frac{Z_a Z_b + Z_a Z_c + Z_b Z_c}{Z_c} \qquad Z_2 = \frac{Z_a Z_b + Z_a Z_c + Z_b Z_c}{Z_a}$$

$$Z_3 = \frac{Z_a Z_b + Z_a Z_c + Z_b Z_c}{Z_b} \qquad (7\text{-}243)$$

In using these relations to realize a wye-connected load from a delta-connected load, it should be pointed out that the potential of the netural point in the resulting wye-connected load will, in general, be such that it is not located at the center of the line voltage phasor triangle, i.e., that such a neutral will not be at the same potential as the neutral point of the source voltages (assuming that the source voltages are wye-connected). This being the case, a connection cannot be made between the source and neutral points without changing the circuit variables. An example of the computations associated with an unbalanced load follows.

Example 7-11.3

The delta-connected load shown in Fig. 7-11.9(a) is excited by a three-phase set of 200-V rms line voltages with *abc* rotation (generated by three line-to-neutral sources). The line voltage phasor triangle is shown in Fig. 7-11.9(b). To analyze this circuit, let us first convert the delta of load impedances to a wye. Applying the relations given in Eq. (7-242), we find that

$$Z_a = 10 - j10$$

$$Z_b = 10$$

$$Z_c = 10 + j10$$

The resulting wye of impedances is shown in Fig. 7-11.9(c). As indicated in this figure, this circuit can be analyzed as a two-mesh circuit with the indicated phasor voltage sources. Defining phasor mesh currents $\mathcal{I}_1$ and $\mathcal{I}_2$ as shown in the figure, the mesh equations in matrix form are

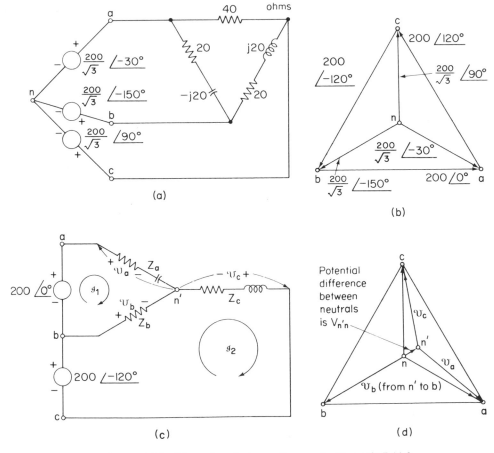

Figure 7-11.9 Network and phasor diagram for Example 7-11.3.

$$\begin{bmatrix} 200 \\ -100 - j173.2 \end{bmatrix} = \begin{bmatrix} 20 - j10 & -10 \\ -10 & 20 + j10 \end{bmatrix} \begin{bmatrix} \mathscr{I}_1 \\ \mathscr{I}_2 \end{bmatrix}$$

The determinant of the square matrix is $(20 - j10)(20 + j10) - 100 = 400$. Using the techniques of Sec. 3-1, the solution to the set of simultaneous equations is

$$\begin{bmatrix} \mathscr{I}_1 \\ \mathscr{I}_2 \end{bmatrix} = \frac{1}{400} \begin{bmatrix} 20 + j10 & 10 \\ 10 & 20 - j10 \end{bmatrix} \begin{bmatrix} 200 \\ -100 - j173.2 \end{bmatrix}$$

Solving for the phasor mesh currents, we obtain $\mathscr{I}_1 = 7.53\underline{/5.1°}$ and $\mathscr{I}_2 = 7.53\underline{/-125°}$. The line-to-neutral or phase voltages for the load may thus be computed as

$$\mathscr{V}_a = \mathscr{I}_1 Z_a = 106.5\underline{/-39.9°}$$
$$\mathscr{V}_b = (\mathscr{I}_2 - \mathscr{I}_1)Z_b = 136.6\underline{/-150°}$$
$$\mathscr{V}_c = -\mathscr{I}_2 Z_c = 106.5\underline{/99.9°}$$

Drawing these phasors inside the line-to-line phasor voltage triangle given in Fig. 7-11.9(b), we obtain the phasor diagram shown in Fig. 7-11.9(d). Thus, we see that the magnitude of the potential between the neutral of the load and the neutral of the generator (assuming a wye-connected generator) is 21.13 V, i.e., that the two points are at different potentials.

7-12 CONCLUSION

This chapter represents a significant departure from the topics and the techniques presented in the first six chapters of this book. The earlier chapters were primarily concerned with the analysis of circuits in the time domain. In this chapter, however, we have been concerned with the behavior of networks in the frequency domain, i.e., under conditions of sinusoidal steady-state excitation. The solution techniques presented in this chapter relate directly to the time-domain material presented in Chaps. 1-6 since they provide a systematic and practical method for determining the particular solution, i.e., the forced response, for the case where the network excitation is sinusoidal. A review of the examples in Chaps. 5 and 6 will readily show that the mathematics connected with the determination of particular solutions is considerably simplified by using the phasor methods introduced in this chapter. In addition to its relation to time-domain network analysis, however, the sinusoidal steady-state analysis techniques presented here have considerable additional significance since they directly relate to the experimental determination of the properties of a given circuit. This is, of course, due to the fact that sinusoidal quantities are so easily generated and measured in practice. Thus, in many cases such quantities provide a definition of network performance which is more closely related to actual usage than are corresponding time-domain results.

The frequency-domain analysis of networks, however, is not limited only to the sinusoidal steady-state studies presented in this chapter. In the following chapter we shall see that the concept of a frequency domain may be extended to include all the analysis results which were obtained in the time-domain studies made in Chaps. 1-6. In addition, we shall find that these general frequency-domain solution techniques are frequently easier and more convenient to apply than the time-domain ones. Thus, collectively, they represent an analysis approach which has major importance.

PROBLEMS

Problem 7-1 (Sec. 7-1)

Put the following expressions in the form $F_m \cos(\omega t + \phi)$ and in the form $F_m \sin(\omega t + \theta)$.

(a) $\sqrt{3} \cos \omega t + \sin \omega t$ (b) $-2 \cos \omega t + 4 \sin \omega t$

(c) $2 \cos \omega t - 5 \sin \omega t$ (d) $-\cos \omega t - \sin \omega t$

Problem 7-2 (Sec. 7-1)

Put the following expressions into the form $A \cos \omega t + B \sin \omega t$.

(a) $(2/\sqrt{3}) \cos(\omega t + 30°)$ (b) $5 \sin(\omega t - 53.13°)$

(c) $-2 \cos(\omega t + 45°)$ (d) $7 \sin(\omega t + 160°)$

Problem 7-3 (Sec. 7-1)

If $\cos[(\pi/2) - \omega t] + \cos(\omega t) = F_m \cos(\omega t - \alpha)$, find α and F_m.

Problem 7-4 (Sec. 7-1)

Find the rms values of the periodic functions shown in Fig. P7-4 (let $T = 2$ s).

(a)

(b)

Figure P7-4

Problem 7-5 (Sec. 7-1)

Find the rms values of the periodic functions shown in Fig. P7-5 (let $T = 4$ s).

(a)

(b)

Figure P7-5

Problem 7-6 (Sec. 7-1)

Find the average value and the rms value of the voltage waveform shown in Fig. P7-6 (let $T = 4$ s).

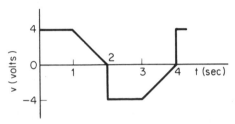

Figure P7-6

Sinusoidal Steady-State Analysis   Chap. 7

Problem 7-7 (Sec. 7-1)

Find the average value and the rms value of the waveform shown in Fig. P7-7 (let $T = 4$ s).

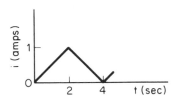

Figure P7-7

Problem 7-8 (Sec. 7-1)

The characteristics of ammeters which are used to measure current fall into four categories as summarized in the following table.

Category	Type of Meter	Characteristics
1	Thermocouple, iron-vane, dynamometer	Reads the rms value of the current.
2	D'Arsonval (dc)	Reads the average value of the current.
3	Full-wave rectifier	Reads 1.11 times the average value of the magnitude of the current.
4	Half-wave rectifier	Reads 2.22 times the average value of the positive portion of the current waveform.

Assume that the waveshapes shown in the following figures represent current and find the values that each type of meter will read for each of the waveshapes:
(a) Fig. P7-4.   (b) Fig. P7-5.
(c) Fig. P7-6.   (d) Fig. P7-7.

Problem 7-9 (Sec. 7-2)

Use the technique described in Sec. 7-2 to find the particular solution for the circuit given in Problem 5-12.

Problem 7-10 (Sec. 7-2)

Use the technique described in Sec. 7-2 to find the particular solution for the circuit given in Problem 5-13.

Problem 7-11 (Sec. 7-2)

Use the technique described in Sec. 7-2 to find the particular solution for $v(t)$ for the circuit given in Problem 6-31.

Problem 7-12 (Sec. 7-3)

What is the locus of a phasor $\mathcal{F}e^{-j\omega t}$ where $\mathcal{F} = 3e^{j(\pi/4)}$?

Problem 7-13 (Sec. 7-3)

Prove that the solution given for the sinusoidal steady-state component of $v_2(t)$ in Example 7-3.1 is indeed the particular solution of the original differential equation given in the example by substituting it into that equation.

Problem 7-14 (Sec. 7-3)

In the circuit shown in Fig. P7-14, find the steady-state component of $i(t)$ if $v(t) = 2 \sin t$ V.

Figure P7-14

Problem 7-15 (Sec. 7-3)

In the circuit shown in Fig. P7-15 the input voltage $v_1(t) = \cos t$ V. Find the sinusoidal steady-state component of the output voltage $v_2(t)$.

Figure P7-15

Problem 7-16 (Sec. 7-3)

The circuit shown in Fig. P7-16 is in a steady-state condition at 1 rad/s. Draw a phasor diagram for the phasor quantities $\mathcal{I}_0$, $\mathcal{I}_G$, $\mathcal{I}_L$, $\mathcal{I}_C$, and $\mathcal{U}$.

Figure P7-16

Problem 7-17 (Sec. 7-3)

When a current $i(t) = 8 \cos t - 11 \sin t$ A is applied to the circuit shown in Fig. P7-17, it is found that $v(t) = \sin t + 2 \cos t$ V for steady-state conditions. Find the values of L, $i_R(t)$, $i_L(t)$, and $i_C(t)$.

Figure P7-17

Problem 7-18 (Sec. 7-4)

Find the inverse of the following matrix.

$$\mathbf{A} = \begin{bmatrix} 1+j1 & -1 & 0 \\ -1 & 2+j1 & -1 \\ 0 & -1 & 1-j1 \end{bmatrix}$$

Problem 7-19 (Sec. 7-4)

Find the particular solutions for $f_1(t)$ and $f_2(t)$ if $g_1(t) = \cos t$ and $g_2(t) = u(t)$ in the equations that follow:

$$3f_1' - 2f_2' + 2f_1 - 2f_2 = g_1$$
$$f_1' + 2f_1 + 2f_2' - f_2 = g_2$$

Problem 7-20 (Sec. 7-4)

In the network shown in Fig. P7-20, the two voltage sources have the values $v_1(t) = \cos 2t$ and $v_2(t) = \sqrt{2} \cos(2t - 45°)$ V. Find the sinusoidal steady-state components of $i_1(t)$ and $i_2(t)$.

ohms, henries, farads

Figure P7-20

Problem 7-21 (Sec. 7-4)

If all variables in the network shown in Fig. P7-21 have reached the steady state, find $v_2(t)$.

ohms, henries, farads

Figure P7-21

Problem 7-22 (Sec. 7-4)

Assuming that all variables in the network shown in Fig. P7-22 have reached the steady state, find $v(t)$.

ohms, henries, farads

Figure P7-22

Problem 7-23 (Sec. 7-4)

Find an expression for the sinusoidal steady-state current $i_C(t)$ in the circuit shown in Fig. P7-23. Note that the value of the capacitor C has *not* been specified.

ohms, henries

Figure P7-23

Problem 7-24 (Sec. 7-4)

In the circuit shown in Fig. P7-24, find the sinusoidal steady-state component of $v_1(t)$.

ohms, henries, farads

Figure P7-24

Problem 7-25 (Sec. 7-4)

In the circuit shown in Fig. P7-25, find an expression for the sinusoidal steady-state component of the current $i_1(t)$.

ohms, henries, farads

Figure P7-25

Problem 7-26 (Sec. 7-4)

Verify the values given for the mesh currents in Example 7-4.2.

Problem 7-27 (Sec. 7-4)

Assume that all the voltages and currents in the network shown in Fig. P7-27 have reached the steady state. Find $i(t)$.

henries, farads

Figure P7-27

Problem 7-28 (Sec. 7-4)

In the circuit shown in Fig. P7-28, find $v_1(t)$ and $v_2(t)$ for steady-state conditions.

ohms, henries, farads

Figure P7-28

Problem 7-29 (Sec. 7-4)

Find the sinusoidal steady-state value for $i(t)$ in the circuit shown in Fig. P7-29.

ohms, henries, farads

Figure P7-29

Problem 7-30 (Sec. 7-5)

Find expressions for the input impedance $Z(j\omega)$ for each of the circuits shown in Fig. P7-30.

(a) (b) **Figure P7-30**

Problem 7-31 (Sec. 7-5)

Find expressions for the input admittance $Y(j\omega)$ for each of the circuits shown in Fig. P7-31.

(a) (b) **Figure P7-31**

Problem 7-32 (Sec. 7-5)

For each of the circuits shown in Fig. P7-30, find two two-element circuits which have the same driving-point impedance at $\omega = 2$ rad/s.

Problem 7-33 (Sec. 7-5)

For each of the circuits shown in Fig. P7-31, find two two-element circuits which have the same driving-point admittance at $\omega = 2$ rad/s.

Problem 7-34 (Sec. 7-5)

Find the driving-point impedance $Z(j\omega)$ for the network containing an ideal transformer shown in Fig. P7-34.

Figure P7-34

Problem 7-35 (Sec. 7-5)

In the circuit shown in Fig. P7-35, three phasor voltages are identified. The following conditions apply: (1) $|\mathcal{V}_1|$ is twice $|\mathcal{V}_2|$, and (2) $\mathcal{V}_g$ leads $\mathcal{V}_2$ by 30°. If, under sinusoidal steady-state conditions, $v_g(t) = 10 \cos 3t$, find $v_2(t)$. (*Hint:* The rule of sines may be helpful here.)

Figure P7-35

Problem 7-36 (Sec. 7-5)

Find the input admittance $Y(j1)$ for the network shown in Fig. P7-36.

Figure P7-36 Figure P7-37

Problem 7-37 (Sec. 7-5)

Draw a phasor diagram showing the magnitude and phase relationships between the variables $i(t)$, $v_1(t)$, and $v_2(t)$ in the network shown in Fig. P7-37.

Problem 7-38 (Sec. 7-5)

Voltage and current phasors $\mathcal{V}_0$ and $\mathcal{I}_0$ are defined at the terminals of a network as shown in Fig. P7-38(a). Their values are shown in the phasor diagram of Fig. P7-38(b). Inside the network are three circuit elements Z_1, Z_2, and Z_3, with phasor currents $\mathcal{I}_1$, $\mathcal{I}_2$, and $\mathcal{I}_3$, respectively. These phasors are also shown in Fig. P7-38(b). Draw the circuit configuration inside of the network. Specify the numerical value and type of each element. Assume that $\omega = 2$ rad/s.

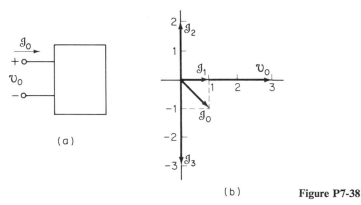

(a)

(b) Figure P7-38

Problem 7-39 (Sec. 7-6)

Write, but do not solve, the mesh impedance equations in matrix form for the network shown in Fig. P7-39 when
(a) $\omega = 1$ rad/s
(b) $\omega = 2$ rad/s

Figure P7-39

Problem 7-40 (Sec. 7-6)

Write, but do not solve, the node admittance equations in matrix form for the network shown in Fig. P7-40 when

(a) $\omega = 1$ rad/s

(b) $\omega = 2$ rad/s

Treat the network as a three-node network.

Figure P7-40

Problem 7-41 (Sec. 7-6)

Write, but do not solve, the mesh impedance equations in matrix form for the network shown in Fig. P7-41.

Figure P7-41

Problem 7-42 (Sec. 7-6)

Find, but do not solve, the node admittance matrix for the network shown in Fig. P7-42. Treat the network as having three nodes.

Figure P7-42

Problem 7-43 (Sec. 7-6)

Write a set of matrix equations describing the network shown in Fig. P7-43.

Figure P7-43

Problem 7-44 (Sec. 7-6)

In Example 7-6.2, show that the product $\mathbf{Z}(j\omega)\mathbf{Z}(j\omega)^{-1}$ is equal to the identity matrix.

Problem 7-45 (Sec. 7-6)

For the transistor amplifier circuit described in Example 7-6.4, show that the choice for R_1 of 87.5Ω provides a dc gain of -10.

Problem 7-46 (Sec. 7-6)

Verify the value found for the inductor L in Example 7-6.4.

Problem 7-47 (Sec. 7-6)

Show that the maximally flat magnitude criteria $a^2 - 2b = (b/z)^2$ used in Example 7-6.4 may be derived by obtaining an expression for $|\mathcal{I}_{\text{out}}/\mathcal{I}_{\text{in}}|^2$ from Eq. (7-130), putting the numerator and denominator polynomials in the form $1 + \alpha\omega^2 + \beta\omega^4$, and equating corresponding second-degree coefficients in the numerator and denominator polynomials.

Problem 7-48 (Sec. 7-6)

Assume that, in Eq. (7-130), the coefficients a, b, and z satisfy the maximally flat magnitude criteria $a^2 - 2b = (b/z)^2$ given in the example. Find an expression for the frequency in rad/s at which the $|\mathcal{I}_{\text{out}}/\mathcal{I}_{\text{in}}|^2$ has one-half the value that it does at dc.

Problem 7-49 (Sec. 7-6)

Find a Thevenin equivalent circuit for the network shown in Fig. P7-49.

Figure P7-49

Problem 7-50 (Sec. 7-6)

Find Thevenin and Norton equivalent circuits for the network shown in Fig. P7-50 at the terminals *a-b* when $\omega = 2$ rad/s.

Figure P7-50

Problem 7-51 (Sec. 7-6)

For the circuit shown in Fig. P7-51, find the Thevenin equivalent circuit at the terminals *a-b*.

Figure P7-51

Problem 7-52 (Sec. 7-7)

Find the network function $\mathcal{I}/\mathcal{V}$, where $\mathcal{I}$ and $\mathcal{V}$ are the phasors defined in Fig. P7-52.

Figure P7-52

Problem 7-53 (Sec. 7-7)

Find the network function $\mathcal{V}_2/\mathcal{I}_0$, where $\mathcal{V}_2$ and $\mathcal{I}_0$ are the phasors defined in Fig. P7-53.

Figure P7-53

Problem 7-54 (Sec. 7-7)

Find the network function $\mathcal{V}_2/\mathcal{V}_1$, where $\mathcal{V}_2$ and $\mathcal{V}_1$ are the phasors defined in Fig. P7-54.

Figure P7-54

Problem 7-55 (Sec. 7-7)

Find the network function $\mathcal{V}_2/\mathcal{V}_1$, where $\mathcal{V}_2$ and $\mathcal{V}_1$ are the phasors defined in the network shown in Fig. P7-55.

Figure P7-55

Problem 7-56 (Sec. 7-7)

Find the network function $\mathcal{V}_2/\mathcal{V}_1$ for the circuit shown in Fig. P7-56.

Figure P7-56

Problem 7-57 (Sec. 7-7)

Find the half-power (3-dB) frequency for the network function $\mathcal{V}_2/\mathcal{V}_1$ in terms of the quantities R, G, and C, for the circuit shown in Fig. P7-57.

Figure P7-57

Problem 7-58 (Sec. 7-7)

For the circuit shown in Fig. P7-58, find the network function $Z(j\omega)$.

mhos, henries, farads

Figure P7-58

Problem 7-59 (Sec. 7-7)

For the circuit shown in Fig. P7-59:

(a) Find the voltage transfer function $\mathcal{V}_2/\mathcal{V}_1$.

(b) If $v_1(t) = 3\cos 4t$, find the sinusoidal steady-state expression for $v_2(t)$.

Figure P7-59

Problem 7-60 (Sec. 7-7)

Find expressions for the bandwidth and the Q of the network function $\mathcal{V}/\mathcal{I}$ for the network shown in Fig. P7-60.

Figure P7-60

Problem 7-61 (Sec. 7-7)

Plot the magnitude and phase as a function of ω for the driving-point impedance of the network shown in Fig. P7-61.

ohms, henries, farads

Figure P7-61

Problem 7-62 (Sec. 7-7)

Plot the magnitude and phase as a function of ωRC for the voltage transfer function of the network shown in Fig. P7-62.

Figure P7-62

Problem 7-63 (Sec. 7-7)

Plot the magnitude and phase as function of ωRC for the voltage transfer function of the network shown in Fig. P7-63.

378 *p. loaded w/.1C*

Sinusoidal Steady-State Analysis   Chap. 7

Figure P7-63

Problem 7-64 (Sec. 7-7)

    (a) Find the magnitude of the network function $\mathcal{V}_2/\mathcal{V}_1$ for the network shown in Fig. P7-64 at $\omega = 1$ rad/s, assuming that $L = \frac{1}{3}$ H.

    (b) Repeat the problem for $L = 1$ H.

ohms, henries, farads

Figure P7-64

Problem 7-65 (Sec. 7-8)

    Construct Bode magnitude and phase plots for the following network functions:

    (a) $N_a(j\omega) = \dfrac{10}{(1 + j\omega)[1 + (j\omega/10)]}$

    (b) $N_b(j\omega) = \dfrac{10j\omega}{(1 + j\omega)[1 + (j\omega/10)]}$

    (c) $N_c(j\omega) = \dfrac{10(j\omega)^2}{(1 + j\omega)[1 + (j\omega/10)]}$

Problem 7-66 (Sec. 7-8)

    Construct Bode magnitude and phase plots for the following network function:

$$N(j\omega) = \frac{10[1 + j\omega + (j\omega)^2]}{[1 + (j\omega/0.1)][1 + (j\omega/10)]}$$

Problem 7-67 (Sec. 7-8)

    Construct Bode magnitude and phase plots for the following network function:

$$N(j\omega) = \frac{100j\omega(10 + j\omega)}{(1 + j\omega)[10^4 + 10^2 j\omega + (j\omega)^2]}$$

Problem 7-68 (Sec. 7-8)

    Construct Bode magnitude and phase plots for the following network function:

$$N(j\omega) = \frac{10^4[(j\omega)^2 + 20j\omega + 100]}{(j\omega)^3 + 1100(j\omega)^2 + 10^5 j\omega}$$

Problem 7-69 (Sec. 7-8)

    Construct Nyquist plots for the networks shown in Figs. P7-61, P7-62, and P7-63.

Problem 7-70 (Sec. 7-8)

    A Bode plot of the magnitude of a network function $N(j\omega)$ is shown in Fig. P7-70. Find the network function.

Figure P7-70

Problem 7-71 (Sec. 7-8)

A Bode plot of the phase of a network function $N(j\omega)$ is shown in Fig. P7-71. Find the network function.

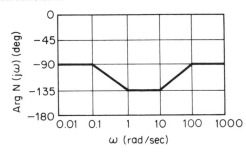

Figure P7-71

Problem 7-72 (Sec. 7-8)

The transistor circuit (without biasing elements) shown in Fig. P7-72(a) is modeled for small-signal analysis as shown in Fig. P7-72(b). Typical element values in the model are $r_\pi = 5226\ \Omega$, $r_x = 100\ \Omega$, $C_\pi = 60$ pF, $R_L = 10$ kΩ, and $g_m = \frac{1}{26}$ mho.

(a) For a simplified analysis, assume that $C_\mu = 0$, and find the Bode plots for the voltage transfer function $\mathcal{V}_2/\mathcal{V}_1$.

(b) For a more accurate analysis, let $C_\mu = 1.25$ pF, and find the Bode plots.

(a)

(b)

Figure P7-72

Sinusoidal Steady-State Analysis   Chap. 7

Problem 7-73 (Sec. 7-9)

Find the order required for a normalized Butterworth low-pass filter with a stopband attenuation of at least 20 dB at each of the following frequencies.
(a) 1.6 rad/s
(b) 1.8 rad/s
(c) 2.2 rad/s

Problem 7-74 (Sec. 7-9)

Find the order required for a normalized Butterworth low-pass filter with each of the following stopband attenuations, all specified at 1.8 rad/s.
(a) 15 dB
(b) 20 dB
(c) 25 dB

Problem 7-75 (Sec. 7-9)

Show that, by using duality (see Sec. 6-8), the element values given in Table 7-9.2 may be used to define a different network configuration from that shown in Fig. 7-9.2. Use this result to realize a fifth-order low-pass Butterworth filter with three inductors, two capacitors, and two resistors. Give the voltage transfer function for this filter.

Problem 7-76 (Sec. 7-9)

Design a third-order passive Butterworth low-pass filter with terminating resistors of 500 Ω and a bandwidth of 1 kHz.

Problem 7-77 (Sec. 7-9)

Derive the voltage transfer function for a passive third-order Butterworth filter with a bandwidth of 2 rad/s.

Problem 7-78 (Sec. 7-9)

Draw an approximate magnitude characteristic for each of the networks shown in Fig. 7-9.3. Give numerical values for the magnitude of the network function at zero frequency, the 3-dB frequency, and the magnitude of the network function at two times the 3-dB frequency.

Problem 7-79 (Sec. 7-9)

Design a passive third-order Butterworth high-pass filter with terminating resistors of 1000 Ω and a 3-dB frequency of 1 kHz.

Problem 7-80 (Sec. 7-9)

Derive the voltage transfer function for a third-order Butterworth high-pass filter with a 3-dB frequency of 2 rad/s.

Problem 7-81 (Sec. 7-9)

Derive the voltage transfer function for a fourth-order band-pass normalized filter with a bandwidth of 1 rad/s. Draw a sketch of the magnitude of the function giving numerical values for the magnitude at the resonant frequency, the values of the 3-dB frequencies, and the magnitude at two times the upper 3-dB frequency and one-half the lower 3-dB frequency.

Problem 7-82 (Sec. 7-9)

Design a passive sixth-order band-pass filter with 1000-Ω terminating resistors, a 10-kHz center frequency, and a Q of 10. Draw a sketch of the magnitude characteristic for the filter.

Problem 7-83 (Sec. 7-9)

Design a fifth-order active-*RC* filter in which all the resistors have a value of 1000 Ω. The filter should be low-pass and have a bandwidth of 1 kHz. Use the circuit shown in Fig. 7-9.6 for the second-order functions.

Problem 7-84 (Sec. 7-9)

A normalized high-pass active-RC filter is shown in Fig. P7-84.

(a) Find the voltage transfer function.

(b) Use this filter to realize a normalized fourth-order Butterworth high-pass voltage transfer function.

ohms, farads

Figure P7-84

Problem 7-85 (Sec. 7-9)

A band-pass active-RC filter is shown in Fig. P6-40. Let $R_1 = R_2 = R_4 = R$ and $C_3 = C_5 = C$.

(a) Find the voltage transfer function.

(b) Develop design equations relating R, C, and K to the coefficients of a standardized band-pass second-order function having the form $Hj\omega(\omega_0/Q)/[(j\omega)^2 + j\omega(\omega_0/Q) + \omega_0^2]$.

(c) Use this circuit to realize a normalized fourth-order Butterworth bandpass voltage transfer function with a bandwidth of 0.5 rad/s. [*Hint:* For this function, the quadratic factors of the denominator are $(j\omega)^2 + 0.4160j\omega + 1.492$ and $(j\omega)^2 + 0.2911j\omega + 0.6997$.]

Problem 7-86 (Sec. 7-9)

A bandpass active-RC filter is shown in Fig. P6-45. Let $C_2 = C_3 = C$. The design equations are given in terms of the standardized function given in Problem 7-85. They are

$$R_1 = \frac{Q}{\omega_0 C |H|} \qquad R_4 = \frac{Q}{(2Q^2 - |H|)\omega_0 C} \qquad R_5 = \frac{2Q}{\omega_0 C}$$

Use this circuit to realize the filter specifications given in Problem 7-85. Assume that each stage has $|H| = 3.162$.

Problem 7-87 (Sec. 7-10)

For the circuit shown in Fig. P7-87:

(a) Find the rms value of the current $\mathcal{I}$.

(b) Find the real power (watts) dissipated in the load Z_L.

(c) Find the reactive power (vars) consumed by the load Z_L.

(d) Find the rms voltage $\mathcal{V}_L$ across the load Z_L.

1 + j1   ohms

100 V (rms)

$\mathcal{I}$

Z_L
$3 - j4$

$\mathcal{V}_L$

Figure P7-87

Problem 7-88 (Sec. 7-10)

In the circuit shown in Fig. P7-88, the load consumes 4 kW and -3 kVAR. Find the rms value of the generator voltage.

Figure P7-88

Problem 7-89 (Sec. 7-10)

In the circuit shown in Fig. P7-89, the load impedance $Z_L = 100 - j50$ is required to have an rms voltage of at least 100 V across it. The line impedance which supplies this load is $1 + j1$ Ω/mi. What is the maximum number of miles of line which can be used if the generator has an output voltage of 110 V (rms)?

Figure P7-89

Problem 7-90 (Sec. 7-10)

The square symbol (with four terminals) shown in Fig. P7-90 represents a wattmeter, a device used to measure average power in circuits under sinusoidal steady-state conditions. As shown, two of the terminals are connected in series with a load Z_L (to measure the current), and two are connected across the load (to measure the voltage). Determine the readings that the wattmeter, ammeter, and voltmeter will indicate.

Figure P7-90

Problem 7-91 (Sec. 7-10)

In the circuit shown in Fig. P7-91, a load impedance Z_L is connected to a generator by an ideal transformer with a turns ratio of 1:2. The circuit variables are measured by two (rms reading) ammeters and two wattmeters. Find the reading of each of the four meters. (See Problem 7-90 for a discussion of the convention used in designating a wattmeter.)

Figure P7-91

Problem 7-92 (Sec. 7-10)

In the circuit shown in Fig. P7-92, a set of rms meter readings for the voltage and current variables are given. Use these values to find the impedance Z_1. Assume that the overall power factor is lagging and that Z_2 is purely resistive.

Figure P7-92   Figure P7-93

Problem 7-93 (Sec. 7-10)

In the circuit shown in Fig. P7-93, a set of rms values for the voltage and current variables are given. Assume that the overall power factor is leading, and find the magnitude and phase of the admittance Y.

Problem 7-94 (Sec. 7-10)

Prove that the conservation of complex power applies in the circuits shown in the following figures.
(a) Fig. P7-90.
(b) Fig. P7-92.

Problem 7-95 (Sec. 7-10)

The real and reactive powers consumed by the load shown in Fig. P7-95 are 6 W and +8 VAR, respectively (without considering the capacitor). Find the value of capacitance C which will make the source see a unity power factor load.

Problem 7-96 (Sec. 7-10)

In the network shown in Fig. P7-96, load A consumes 10 kW of power at 0.6 PF lagging, and load B consumes 8 kW with a power factor of 0.8 lagging. Find a value for the resistor such that the overall power factor will be 0.9. Is this an effective way of achieving power factor correction? Why, or why not?

Figure P7-95   Figure P7-96

Problem 7-97 (Sec. 7-10)

For the network shown in Fig. P7-97, find a value for the load impedance Z_L such that maximum power is delivered to it.

Figure P7-97

384   Sinusoidal Steady-State Analysis   Chap. 7

Problem 7-98 (Sec. 7-10)

For the circuit shown in Fig. P7-98, find the value of the turns ratio n of the ideal transformer such that maximum power is delivered to the load resistor R_L.

Figure P7-98

Problem 7-99 (Sec. 7-11)

Draw the phasor diagram for the line and load voltages and the line currents for the three-phase circuit shown in Fig. P7-99. What is the total power delivered to the load?

$$V_a = \frac{120}{\sqrt{3}} \underline{/0°}$$

$$V_b = \frac{120}{\sqrt{3}} \underline{/-120°}$$

$$V_c = \frac{120}{\sqrt{3}} \underline{/120°}$$

$$Z_L = 10 - j5$$

Figure P7-99

Problem 7-100 (Sec. 7-11)

Draw the phasor diagram for the line and phase voltages and the line and load currents for the three-phase circuit shown in Fig. P7-100. What is the total power delivered to the load?

$$\mathcal{V}_a = \frac{120}{\sqrt{3}} \underline{/0°}$$

$$\mathcal{V}_b = \frac{120}{\sqrt{3}} \underline{/-120°}$$

$$\mathcal{V}_c = \frac{120}{\sqrt{3}} \underline{/120°}$$

$$Z_L = 5 + j3$$

Figure P7-100

Problem 7-101 (Sec. 7-11)

Solve for the line currents for the three-phase circuit shown in Fig. P7-101. Draw the phasor diagram. Find the phasor $V_{nn'}$ which represents the voltage between the neutral n of the three-phase generator and the common point n' of the wye of impedances.

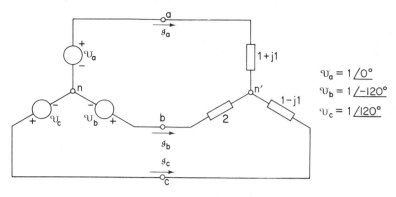

Figure P7-101

$$\mathcal{V}_a = 1\underline{/0°}$$
$$\mathcal{V}_b = 1\underline{/-120°}$$
$$\mathcal{V}_c = 1\underline{/120°}$$

Problem 7-102 (Sec. 7-11)

For the three-terminal circuits shown in Fig. P7-102, find delta equivalents for the wye circuits, and wye equivalents for the delta circuits.

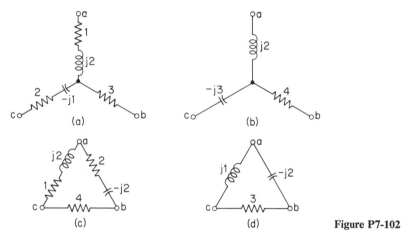

Figure P7-102

Problem 7-103 (Sec. 7-11)

Solve for the line and load currents for the three-phase circuit containing three impedances in a delta connection shown in Fig. P7-103.

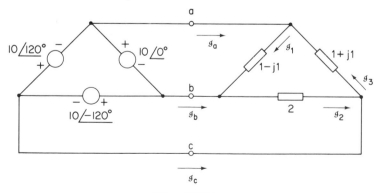

Figure P7-103

Problem 7-104 (Sec. 7-11)

A balanced three-phase circuit is shown in Fig. P7-104. In this circuit three single-phase transformers with a turns ratio of 1 : 2 are used in a wye-delta transformer bank (the windings connected in the wye portion have half the turns of the windings connected in the delta portion). Find the rms values of the line currents from the three-phase 100-V (line-to-line) source.

Figure P7-104

Problem 7-105 (Sec. 7-11)

Three transformer secondary windings are shown in Fig. P7-105 together with the rms voltages measured across them and across various connections of them. Use the data to construct: (a) a balanced three-phase wye-connected source, and (b) a balanced three-phase delta-connected one, by correctly connecting the windings together. Assume that the voltages of each winding are 120° out of phase.

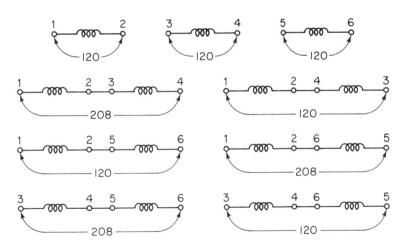

Figure P7-105

8

The Laplace

Transformation

In the preceding chapters of this book we have been concerned with the formulation and solution of sets of network equations relating variables which are functions of time. If the network is comprised only of resistors, such sets of equations are algebraic. More generally, however, when inductors and capacitors are present, sets of integro-differential equations result. It has been shown that the solution of such sets of equations requires the development of many specialized techniques and skills. There exists another quite different approach to finding the solution of sets of integro-differential equations than the ones we have studied so far. This is the use of transformed network variables, i.e., network variables which have been submitted to some mathematical process so that the equations which involve these variables are more amenable to solution. There are many such mathematical transformations which are useful in engineering applications. In this chapter we shall study the simplest, most basic, and most used of such transformations, the *Laplace transformation*. We shall see that it has the property of transforming integro-differential equations into algebraic equations, thus providing a valuable tool for the analysis of networks.

8-1 DEVELOPING THE LAPLACE TRANSFORMATION

Let us see if we can develop the Laplace transformation by simple reasoning. Our goal will be to "invent" some mathematical relationship that transforms a function $f(t)$ of a real variable t into a function of some other variable. Let us use F for the transformed function and s for the new variable. The process may be indicated symbolically as[1]

$$f(t) \longrightarrow F(s) \tag{8-1}$$

where the arrow represents the process of transformation. In describing such a process, it is common to refer to functions of t, where t is a real variable, as being in the t (or time) domain, and the resulting transformed function as being in the s domain. Since at this point we do not know anything about the mathematical nature of the variable s, we must be alert for the possibility that it will be complex rather than real. This will actually be shown to be the case. Now let us assume that the purpose of our transformation is to convert the operation of differentiation in the time domain to multiplication by the transformed variable in the s domain. This result may be indicated symbolically as

$$\frac{d}{dt} f(t) \longrightarrow s F(s) \tag{8-2}$$

where $F(s)$ is defined by (8-1). This result is preliminary, since it is possible that the transformation of the differentiated function will produce terms other than the term $sF(s)$; however, this point we may defer until later. Now let us specify the form that the transformation must have. Let us assume that we are interested only in positive values of t, i.e., that all the functions $f(t)$ that we shall consider are zero for $t < 0$. For the transformation process to accurately represent a function $f(t)$ for all positive values of t, the transformation must take into account all values that $f(t)$ takes on. One way of doing this is by integrating $f(t)$ over the range from zero to infinity. To ensure that the effect of any singularity functions (such as impulses) which are applied at $t = 0$ are included, we will specify the lower limit of integration as $0-$ rather than 0.[2] In evaluating such a definite integral, the variable of integration, namely, t, is eliminated from the resulting expression. Therefore, to have such an integration produce a function $F(s)$, a term containing the variable s must be added to the integrand. In general, such a term must be a function of both t and s. Thus, if we use the letter h to represent the function, we may write it in the form $h(t, s)$. The actual form of $h(t, s)$ is yet to be determined. However, as a preliminary step, we have now specified the general form that our transformation will have. Thus, we may write

$$F(s) = \int_{0-}^{\infty} f(t) h(t, s) \, dt \tag{8-3}$$

To determine $h(t, s)$ (this is sometimes referred to as the *kernel* of the transformation integral), we must satisfy the specified condition that differentiation is transformed to multiplication. Thus, from (8-2) and (8-3), we may write

$$s F(s) = \int_{0-}^{\infty} f'(t) h(t, s) \, dt \tag{8-4}$$

where $f'(t) = df/dt$. Let us integrate the right member of (8-4) using integration by parts. The general rule for integration by parts for two functions $w(t)$ and $u(t)$ is

---

[1]The symbol p is also widely used as the variable for the transformed function.

[2]The impulse function was briefly introduced in Sec. 4-2. A more detailed treatment of this function will be given later in this chapter. The quantity $0-$ is defined in Sec. 4-3.

$$\int_a^b w'(t)u(t)\,dt = w(t)u(t)\Big|_a^b - \int_a^b w(t)u'(t)\,dt \qquad (8\text{-}5)$$

where the prime indicates differentiation with respect to the independent variable t. Thus, if we let $w'(t) = f'(t)$, and let $u(t) = h(t, s)$, we may write

$$\int_{0-}^{\infty} f'(t)h(t, s)\,dt = f(t)h(t, s)\Big|_{0-}^{\infty} - \int_{0-}^{\infty} f(t)h'(t, s)\,dt \qquad (8\text{-}6)$$

The first term in the right member of (8-6) only involves the initial and final values of t, namely, zero and infinity. Thus, this term is unlikely to affect the transformation, which is necessarily based on the entire range of values of $f(t)$. For the moment, therefore, let us ignore this term. Thus, (8-6) may be written as

$$s\int_{0-}^{\infty} f(t)h(t, s)\,dt = -\int_{0-}^{\infty} f(t)h'(t, s)\,dt \qquad (8\text{-}7)$$

where we have used (8-3) and (8-4) to modify the form of the left member of (8-6). For this equation to be satisfied, we require

$$sh(t, s) = -h'(t, s) = -\frac{dh(t, s)}{dt} \qquad (8\text{-}8)$$

This equation is easily solved by separating variables. Thus, we may write

$$\frac{dh(t, s)}{h(t, s)} = -s\,dt \qquad (8\text{-}9)$$

Integrating both sides of the above, we obtain

$$\ln h(t, s) = -st \qquad (8\text{-}10)$$

If we now raise both sides of (8-10) to the exponential power, we obtain

$$h(t, s) = e^{-st} \qquad (8\text{-}11)$$

This is our desired expression for $h(t, s)$. Thus, from (8-3) we may define the Laplace transformation by the relation

$$\mathcal{L}[f(t)] = F(s) = \int_{0-}^{\infty} f(t)e^{-st}\,dt \qquad (8\text{-}12)$$

where the term $\mathcal{L}[f(t)]$ may be read as "the Laplace transformation of $f(t)$," and $F(s)$ is called the *Laplace transform* of $f(t)$ or, more simply, the transform of $f(t)$. In general, s is a complex variable which may take on any value. In actually performing the integration specified in (8-12), however, Re (s) (the real part of s) must be somewhat restricted. As an example of such a restriction, consider the determination of the Laplace transform of the unit step function $u(t)$. If we define $s = \sigma + j\omega$, then from (8-12) we may write

$$\mathcal{L}[u(t)] = F(s) = \int_{0-}^{\infty} u(t)e^{-st}\,dt = \int_{0-}^{\infty} e^{-\sigma t}(\cos \omega t - j \sin \omega t)\,dt \qquad (8\text{-}13)$$

If σ (the real part of s) is negative, then, since the since and cosine terms do not approach a given value as t approaches infinity, the definite integral is not convergent and cannot be evaluated. The same is true if $\sigma = 0$. On the other hand, if σ is positive, then the integrand approaches zero as t goes to infinity and the integral exists. Specifically, we may write

The Laplace Transformation   Chap. 8

$$\mathcal{L}[u(t)] = \int_{0-}^{\infty} e^{-st}\, dt = -\frac{1}{s}\, e^{-st}\Big|_{0-}^{\infty} = \frac{1}{s} \qquad \text{Re } s > 0 \qquad (8\text{-}14)$$

Restrictions on Re s similar to that used in determining the transform of the unit step function in (8-14) are readily defined for any of the functions that one normally encounters in network theory applications. Specifically, for a function $f(t)$ to be transformable, it is sufficient to require that

$$\int_{0-}^{\infty} |f(t)|\, e^{-\sigma_0 t}\, dt < \infty \qquad (8\text{-}15)$$

for some real positive value of σ_0. It should be noted that even if a function does not satisfy (8-15), it may still be transformed if we suitably restrict the range of our interest in the function. For example, the function e^{t^2} is not transformable; however, a function $f(t)$ defined as

$$f(t) = \begin{cases} e^{t^2} & t < t_0 \\ 0 & t > t_0 \end{cases} \qquad (8\text{-}16)$$

is transformable, and may be used in place of e^{t^2} by simply restricting the range of time over which the function is considered, to be $0 \le t \le t_0$ where t_0 may be selected to have any positive finite value.

At this point, it should be noted that there are differences, both conceptual and practical, between the *Laplace transformation*, i.e., the integral defined by (8-12), and the resulting *Laplace transform*, i.e., the function $F(s)$. For example, consider the unit step function $u(t)$ which was found in (8-14) to have the transform $1/s$. The defining integral transformation has been shown to exist only for values of s satisfying the relation Re $s > 0$. The resulting transform, however, obviously has a value for all values of s (including those in the left half-plane) except $s = 0$. Mathematically, we may justify the use of the transform for ranges of values of s for which the original transformation integral was not defined by a process referred to as *analytic continuation*. A treatment of the details of this procedure may be found in any standard text on complex variable theory.

Now let us determine whether the transformation defined by (8-12) actually accomplishes the desired goal, i.e., whether it transforms the operation of differentiation into the operation of multiplication. To find this out we may proceed by applying the transformation to the derivative of $f(t)$ with respect to t. Using (8-12) and integrating by parts, we obtain

$$\mathcal{L}[f'(t)] = \int_{0-}^{\infty} f'(t) e^{-st}\, dt = f(t) e^{-st}\Big|_{0-}^{\infty} - \int_{0-}^{\infty} - sf(t) e^{-st}\, dt \qquad (8\text{-}17)$$

In evaluating the first term in the right member of the above equation at its upper (infinite) limit, if we assume that $f(t)$ satisfies (8-15), then[3]

$$\lim_{t \to \infty} f(t) e^{-st} = 0 \qquad (8\text{-}18)$$

Almost all the functions with which we shall deal may be shown to have this property.

---

[3] Some restrictions must also be placed on Re s to correctly define the convergence of the function in the limit. In general, these are the same as those required to ensure that the Laplace transformation of $f(t)$ exists.

Evaluating this same term at its lower limit, we simply obtain $-f(0-)$. Now consider the last term of the right member of (8-17). Since the integration is with respect to t, the term $-s$ may be moved outside the integral. From (8-12), the remaining integral is simply equal to $F(s)$. From the above, we see that we may simplify (8-17) to the form

$$\mathcal{L}[f'(t)] = sF(s) - f(0-) \qquad (8\text{-}19)$$

Thus, we see that the Laplace transformation as defined by (8-12) has the property that the operation of differentiation (in the t domain) is transformed into the operation of multiplication by s (in the s domain), and that the initial condition is introduced explicitly by the transformation.

Now let us consider what properties the variable s must have. It is always true that the coefficient of an exponential term must be dimensionless, therefore, from (8-12) we see that the quantity $-st$ must be dimensionless. Since t has the dimensions of time, s must have the dimensions of *reciprocal* time, i.e., *frequency*. In later developments in this chapter, we shall show that, in general, s will be a complex quantity. Thus, we may appropriately refer to s as the *complex frequency variable*. The relationship between the complex frequency variable s introduced here and the frequency variable used in the last chapter will be made clearer in a later section of this chapter.

The development given above may be summarized as follows:

SUMMARY 8-1.1

The Laplace Transformation. The Laplace transformation is defined by the relation

$$F(s) = \mathcal{L}[f(t)] = \int_{0-}^{\infty} f(t)e^{-st}\, dt$$

where $f(t)$ is a time-domain function of the real variable t and $F(s)$ is a function of the complex frequency variable s. The transformation has the property that

$$\mathcal{L}\left[\frac{df}{dt}\right] = sF(s) - f(0-)$$

where $F(s)$ is defined above. ■

In the following section we will investigate some basic properties of the Laplace transformation. It should be noted that the transformed quantities that result from the Laplace transformation process are not "physical" network variables, i.e., if $v(t)$ is some voltage (which may be physically measured by connecting a voltmeter across the terminals at which the voltage is defined), then the transformed quantity $V(s) = \mathcal{L}[v(t)]$ is *not* a physical quantity. That is, there is no voltmeter yet built which will measure such a term. Despite this fact, it is sometimes found convenient to use the same units in referring to $V(s)$ as were used for $v(t)$. It must be strongly emphasized that such use *is only a convenient way of "remembering" what units were originally associated with the time-domain quantity*, and that such a use does not imply any physical properties of the transformed quantity.

8-2 PROPERTIES OF THE LAPLACE TRANSFORMATION

In the last section we defined the Laplace transformation which transformed a function $f(t)$ of a real variable t into a function $F(s)$ of a complex variable s. The transformation was shown to have the property that a derivative operation on an untransformed function was changed into an algebraic operation on a transformed one. In this section we shall develop some additional properties of the Laplace transformation. These properties will be of considerable use when we begin applying the transformation to the solution of network problems later in this chapter.

One of the most important properties of the Laplace transformation is that it is a *linear* transformation. Thus, it satisfies the property of *homogeneity*, i.e., the Laplace transformation of $K f(t)$ is K times the Laplace transformation of $f(t)$. To see this, we may write

$$\mathcal{L}[Kf(t)] = \int_{0-}^{\infty} Kf(t)e^{-st}\, dt = K \int_{0-}^{\infty} f(t)e^{-st}\, dt = KF(s) \tag{8-20}$$

For this result to hold, of course, the quantity K cannot be a function of either t or $f(t)$. Since the Laplace transformation is linear, the property of *superposition* also applies, namely, the Laplace transformation of the function $f_1(t) + f_2(t)$ is the sum of the separate transformations of $f_1(t)$ and $f_2(t)$. To see this, we may write

$$\mathcal{L}[f_1(t) + f_2(t)] = \int_{0-}^{\infty} [f_1(t) + f_2(t)]e^{-st}\, dt$$

$$= \int_{0-}^{\infty} f_1(t)e^{-st}\, dt + \int_{0-}^{\infty} f_2(t)e^{-st}\, dt = \mathcal{L}[f_1(t)] + \mathcal{L}[f_2(t)] \tag{8-21}$$

As a consequence of the linearity of the Laplace transformation, it should be noted that in a given network situation each variable is transformed separately. As an example of this, consider the equation relating the terminal variables of a capacitor. This has the form

$$i(t) = C\frac{dv}{dt} \tag{8-22}$$

Now let us define transforms of the time-domain functions $i(t)$ and $v(t)$ by the relations

$$\mathcal{L}[i(t)] = \int_{0-}^{\infty} i(t)e^{-st}\, dt = I(s)$$

$$\mathcal{L}[v(t)] = \int_{0-}^{\infty} v(t)e^{-st}\, dt = V(s) \tag{8-23}$$

For the general case, where specific expressions are not given for $i(t)$ and $v(t)$, we may use the symbolic forms $I(s)$ and $V(s)$ to represent the transformed variables. Thus, from (8-12) and (8-19), the transform of the equation given in (8-22) may be written

$$I(s) = sCV(s) - Cv(0-) \tag{8-24}$$

Now let us consider a second example of the application of the linearity properties of the Laplace transformation. In Fig. 8-2.1 a parallel RC circuit is shown. The time-domain equation for this circuit is

$$i(t) = C\frac{dv}{dt} + Gv(t) \tag{8-25}$$

Figure 8-2.1 Parallel *RC* circuit.

A problem usually associated with such a circuit is to find $v(t)$ when $i(t)$ is known. To do this let us transform the equation. Following the rules given above, we obtain

$$I(s) = sCV(s) - Cv(0-) + GV(s)$$
$$= (sC + G)V(s) - Cv(0-)$$

(8-26)

The original differential equation given in (8-25) may be thought of as having two unknown quantities, namely, dv/dt and $v(t)$ [it is assumed that $i(t)$ is known]. In the transformed equation given in (8-26), however, there is only one unknown quantity, namely, $V(s)$ since the initial condition term $v(0-)$ is assumed to be known. Thus, by transformation we have reduced the number of unknown quantitites.

As a second property of the Laplace transformation, let us investigate what happens when higher-order derivative functions are transformed. We will start with the second derivative. To do this, we define $g(t) = df/dt$. Then we may write the second derivative of $f(t)$ in the form

$$\frac{d^2f}{dt^2} = \frac{d}{dt}\frac{df}{dt} = \frac{dg}{dt}$$

(8-27)

Now let us take the Laplace transformation of the right member of the above equation. From (8-19) we may write

$$\mathcal{L}\left[\frac{dg}{dt}\right] = s\mathcal{L}[g(t)] - g(0-)$$

(8-28)

Now let us substitute the relation $g(t) = df/dt$ in (8-28). We obtain

$$\mathcal{L}\left[\frac{d^2f}{dt^2}\right] = s\mathcal{L}\left[\frac{df}{dt}\right] - \frac{df}{dt}(0-)$$

(8-29)

where $df(0-)/dt$ is the value of the derivative of $f(t)$ evaluated at $t = 0-$. We may again use (8-19) to evaluate the term $\mathcal{L}[df/dt]$ in the above equation. Thus, we find

$$\mathcal{L}\left[\frac{d^2f}{dt^2}\right] = s[sF(s) - f(0-)] - \frac{df}{dt}(0-)$$

(8-30)

This may be put in the form

$$\mathcal{L}\left[\frac{d^2f}{dt^2}\right] = s^2F(s) - sf(0-) - \frac{df}{dt}(0-)$$

(8-31)

The development given above is easily extended to treat a derivative of any order. The details are left as an exercise for the reader. The results obtained may be summarized as follows:

SUMMARY 8-2.1

The Laplace Transform of Derivatives. If the Laplace transform of a function $f(t)$ of a real variable t is a function $F(s)$ of a complex variable s, then the transform

The Laplace Transformation   Chap. 8

of the nth derivative of $f(t)$ is given as

$$\mathcal{L}\left[\frac{d^n f}{dt^n}\right] = s^n F(s) - s^{n-1} f(0-) - s^{n-2} \frac{df}{dt}(0-) - \cdots - \frac{d^{n-1}}{dt^{n-1}} f(0-)$$ ■

The development given above makes it possible for us to obtain the Laplace transform of differential equations of any order. An example follows.

Example 8-2.1

The circuit shown in Fig. 8-2.2 may be analyzed using the techniques given in Chap. 6 to find a differential equation relating the input voltage $v(t)$ and the current $i(t)$. The differential equation is

ohms, henries

Figure 8-2.2 Network for Example 8-2.1.

$$v(t) = 6\frac{d^2 i}{dt^2} + 7\frac{di}{dt} + i(t) \tag{8-32}$$

Applying the results of Summary 8-2.1 and using the transformed variables $V(s)$ and $I(s)$, we may write

$$V(s) = 6\left[s^2 I(s) - si(0-) - \frac{d}{dt}i(0-)\right] + 7[sI(s) - i(0-)] + I(s)$$

Rearranging the terms of the equation, we may write

$$V(s) = (6s^2 + 7s + 1)I(s) - (6s + 7)i(0-) - 6\frac{d}{dt}i(0-) \tag{8-33}$$

This is the Laplace transform of the original differential equation of (8-32).

The Laplace transform of the differential equation which was found in the example given above illustrates some important points concerning the use of the Laplace transformation to transform differential equations. In general, the transformed equation will consist of two parts. The first part, as illustrated by the first term in the right member of Eq. (8-33), consists of a polynomial in s which multiplies the transformed dependent variable $I(s)$. Comparing Eqs. (8-32) and (8-33) we see that the coefficients of this polynomial are the same as those of the original differential equation, and that the power of s which multiplies each term corresponds with the order of the individual derivatives. Thus, this polynomial is identical with the one that is obtained when the trial solution $f(t) = Ke^{st}$ is substituted in the original differential equation and the common term Ke^{st} is factored out. Because this is so, this polynomial is identical to the *characteristic polynomial* which was introduced in Sec. 6-1. The second part of the transformed equation, i.e., the remaining terms in the right member of Eq. (8-33), is a function of the coefficients, the complex frequency variable s, and the initial conditions. Thus, the entire transformed equation includes the differential equation and the initial conditions. These conclusions are easily shown to be true in general, and may be summarized as follows:

The Laplace Transform of a Differential Equation. If a general differential equation which relates some excitation function $g(t)$ and some response function $f(t)$ has the form

$$a_1 f(t) + a_2 \frac{df}{dt} + a_3 \frac{d^2 f}{dt^2} + \cdots + a_{n+1} \frac{d^n f}{dt^n}$$

$$= b_1 g(t) + b_2 \frac{dg}{dt} + b_3 \frac{d^2 g}{dt^2} + \cdots + b_{m+1} \frac{d^m g}{dt^m}$$

then the Laplace transform of this equation will have the form

$$(a_1 + a_2 s + a_3 s^2 + \cdots + a_{n+1} s^n) F(s) + F_{\text{i.c.}}$$

$$= (b_1 + b_2 s + b_3 s^2 + \cdots + b_{m+1} s^m) G(s)$$

where $F(s)$ and $G(s)$ are the Laplace transforms of $f(t)$ and $g(t)$, respectively, and where $F_{\text{i.c.}}$ is a function of the initial conditions $f(0-)$, $df(0-)/dt$, ..., the coefficients a_i ($i = 1, 2, \ldots$), and s. The polynomial which multiplies $F(s)$ is the characteristic polynomial. ■

It should be noted that, in the above development, the results apply for the case where the quantities a_i and b_i are not functions of t, i.e., *the equation (and thus) the network represented by it) must be time-invariant*. In addition, no terms of the type $f^2(t)$, $(df/dt)^2$, etc. can be present, i.e., *the equation must be linear*. Thus the Laplace transformation is only applicable to linear time-invariant networks.

In a manner similar to that used in the last section for finding the transform of a derivative function, we may find the transform of an integral function. From (8-12) we may write

$$\mathcal{L}\left[\int_{0-}^{t} f(\tau)\, d\tau \right] = \int_{0-}^{\infty} \left[\int_{0-}^{t} f(\tau)\, d\tau \right] e^{-st}\, dt \qquad (8\text{-}34)$$

where, for clarity, we have used τ as the variable of integration in the integrand. The integration is easily carried out by parts, using (8-5) and letting

$$u(t) = \int_{0-}^{t} f(\tau)\, d\tau \qquad u'(t) = f(t)$$

$$w'(t) = e^{-st}\, dt \qquad w(t) = -\frac{e^{-st}}{s} \qquad (8\text{-}35)$$

Thus, we obtain

$$\mathcal{L}\left[\int_{0-}^{t} f(\tau)\, d\tau \right] = -\frac{e^{-st}}{s} \int_{0-}^{t} f(\tau)\, d\tau \Big|_{0-}^{\infty} + \frac{1}{s} \int_{0-}^{\infty} f(t) e^{-st}\, dt \qquad (8\text{-}36)$$

Now let us consider the first term of the right member of this equation. Evaluating the term at the upper limit of infinity, we obtain zero. The reasoning is similar to that used in connection with Eq. (8-14). Evaluating the same term at its lower limit, we again obtain zero, since the integral vanishes when t is set to zero and thus the entire first term is zero. Now let us consider the second term of the right member of Eq. (8-36).

The integral term here is the same as the one given in Eq. (8-12), thus, it simply defines the transformed function $F(s)$. Summarizing the above discussion, we may conclude

$$\mathcal{L}\left[\int_{0-}^{t} f(\tau)\,d\tau\right] = \frac{F(s)}{s} \tag{8-37}$$

As an example of the application of this result, let us consider the integral relation for the voltage and current variables of a capacitor. This is[4]

$$v(t) = \frac{1}{C}\int_{0-}^{t} i(\tau)\,d\tau + v(0-) \tag{8-38}$$

The right member of this equation contains a constant, namely, $v(0-)$. Therefore, before we can take the Laplace transformation of this equation, we must determine how to take the transformation of a constant. Let K be such a constant. Since, in applying the Laplace transformation, it is assumed that all functions are zero for $t < 0$, such a constant must be defined in connection with the unit step operator $u(t)$, i.e., we must find the transform of $Ku(t)$. From (8-14) we see that this is simply K/s. Now let us return to the problem of transforming the equation given in (8-38). From (8-37) we may write

$$V(s) = \frac{1}{sC} I(s) + \frac{v(0-)}{s} \tag{8-39}$$

where $V(s)$ is the transform of $v(t)$, and $I(s)$ is the transform of $i(t)$. This conclusion may be combined with the preceding results obtained for differential equations to obtain the solution of integro-differential equations. As an example of such a situation, consider the network shown in Fig. 8-2.3. If the switch S is closed at $t = 0$, then the excitation applied to the circuit is $V_0 u(t)$. Thus, from KVL, we may write

$$V_0 u(t) = Ri(t) + L\frac{di}{dt} + \frac{1}{C}\int_{0-}^{t} i(\tau)\,d\tau + v_C(0-) \tag{8-40}$$

Transforming the equation and rearranging terms, we obtain

$$\frac{V_0}{s} = \left(R + sL + \frac{1}{sC}\right) I(s) + \frac{v_C(0-)}{s} - Li(0-) \tag{8-41}$$

It is readily noted that the first term in the right member of this equation is the transformed equivalent of the right member of the original integro-differential equation with multiplication by the variable s replacing differentiation, and with

Figure 8-2.3 Series RLC network.

[4]In this relation we have written $v(0-)$ to be consistent with our notation throughout this treatment of the Laplace transformation. Actually, since the continuity condition applies to the voltage across the capacitor, we might as well have written $v(0)$, except for the fact that frequently in Laplace transform analysis of networks, impulses may be used to establish initial conditions, thus, integral functions may be discontinuous. More will be said about this in a later section.

division by s replacing integration. The second and third terms in the right member are functions of the initial conditions on the various elements of the network. The results given in this example are readily shown to be true in general. Thus we may summarize the above discussion as follows:

SUMMARY 8-2.3

The Laplace Transform of an Integro-Differential Equation. If the Laplace transformation is applied to an integro-differential equation, the result is a transformed equation, each member of which consists of two terms. One of these terms will be identical in form with the original equation, except that differentiation of any order is replaced by multiplication of the transformed quantity by the complex frequency variable s raised to the same order, and that integration is similarly replaced by division by s. The second term of each member will be a function of the initial conditions in the network (and also of s). ∎

In this section we have developed the transformations for the elementary operations of integration and differentiation, and showed how these may be used to transform integro-differential equations describing networks. Later in this chapter we shall see how to use these transformed relations to actually solve for the unknown network variable, i.e., to retransform the equation in the complex frequency variable s back to an equation in the time domain in such a manner as to find an expression in t for some unknown network variable. Before doing this, however, we must develop some additional transformations for various functions of time. This will be done in the next section.

8-3 LAPLACE TRANSFORMS OF FUNCTIONS OF TIME

In the first section of this chapter we derived the Laplace transform of one simple function of time, namely, the step function or constant-valued (for $t > 0$) function. Specifically, we showed that

$$\mathcal{L}[u(t)] = \frac{1}{s} \tag{8-42}$$

A pair of terms such as those given in (8-42), namely, the terms $u(t)$ and $1/s$, are referred to as a *transform pair*. In this section we shall derive several such transform pairs, corresponding with the time-domain functions most frequently encountered in network theory. We shall assume that all such functions are zero for $t < 0$. Thus, the functions will all be shown as being multiplied by $u(t)$.

As a first example of such a transform pair, let us consider the Laplace transformation of an impulse function. Just as we did in Sec. 4-2, let us first define an approximation $\delta_\epsilon(t)$ to the impulse function by the plot shown in Fig. 8-3.1. This waveform has the property

$$\int_{-\infty}^{\infty} \delta_\epsilon(t)\, dt = 1 \tag{8-43}$$

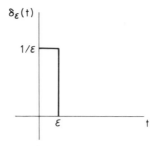

Figure 8-3.1 Approximation to the impulse function.

In the limit, as the quantity ϵ shown in Fig. 8-3.1 approaches zero, the resulting function may be defined as the impulse function $\delta(t)$. Thus, we may write for the impulse

$$\int_{-\infty}^{\infty} \delta(t)\, dt = 1 \qquad (8\text{-}44)$$

Now let us investigate an integral over the range of $-\infty$ to ∞ of the *product* of the approximation to the impulse $\delta_\epsilon(t)$ and some other function $f(t)$. Using the definition of $\delta_\epsilon(t)$ given in Fig. 8-3.1, we obtain

$$\int_{-\infty}^{\infty} \delta_\epsilon(t) f(t)\, dt = \int_{0}^{\epsilon} \frac{1}{\epsilon} f(t)\, dt = \frac{1}{\epsilon} \int_{0}^{\epsilon} f(t)\, dt \qquad (8\text{-}45)$$

As ϵ becomes very small, the value of $f(t)$ over the range $0 \le t \le \epsilon$ becomes essentially constant, and may be approximated as $f(0)$. Thus, we write

$$\int_{-\infty}^{\infty} \delta_\epsilon(t) f(t)\, dt = \frac{1}{\epsilon} f(0) \int_{0}^{\epsilon} dt = f(0) \qquad (8\text{-}46)$$

In the limit as ϵ approaches zero, the above relation may be written in terms of $\delta(t)$ rather than $\delta_\epsilon(t)$. Thus, we obtain

$$\int_{-\infty}^{\infty} \delta(t) f(t)\, dt = f(0) \qquad (8\text{-}47)$$

This result is sometimes referred to as the *sampling property* of a unit impulse, since it samples or selects a single value of the function $f(t)$, namely $f(0)$, from the continuum of values which constitute such a function.[5] The result given above is readily extended to impulses occurring at other values of time. Thus, in general, we may write

$$\int_{-\infty}^{\infty} \delta(t - a) f(t)\, dt = f(a) \qquad (8\text{-}48)$$

Now let us apply the result given in Eq. (8-47) to the determination of the Laplace transform of an impulse. From the general definition of the transformation, we may write

$$\mathcal{L}[\delta(t)] = \int_{0-}^{\infty} \delta(t) e^{-st}\, dt = e^{-st}\Big|_{t=0-} = 1 \qquad (8\text{-}49)$$

---

[5]It is possible to rigorously justify the result given in Eq. (8-47). Such a development, however, requires the use of an area of mathematics called the theory of distributions, which is beyond the scope of this text.

Thus, we see that the transformation of a unit impulse occurring at $t = 0$ is simply the constant unity.

As another example of the transformation of a frequently encountered time function, consider the simple exponential function $e^{-at}u(t)$ for $a \geq 0$. The Laplace transformation of this function may be found as follows:

$$\mathcal{L}[e^{-at}u(t)] = \int_{0-}^{\infty} e^{-at}e^{-st}\,dt = \int_{0-}^{\infty} e^{-(s+a)t}\,dt = \frac{-1}{s+a}e^{-(s+a)t}\bigg|_{0-}^{\infty} \qquad (8\text{-}50)$$

In evaluating the right member of (8-50), we note that the exponential term evaluated at $t = \infty$ is zero.[6] At the lower limit, namely $t = 0-$, the exponential term is 1. Thus, we may define a transform pair relating the function $e^{-at}u(t)$ for $a \geq 0$, and its transform, by

$$\mathcal{L}[e^{-at}u(t)] = \frac{1}{s+a} \qquad (8\text{-}51)$$

It is readily shown that this transform pair also exists for $a < 0$. It may also be shown that, when $a = 0$, this transform pair is the same as the one given in Eq. (8-42).

Now let us consider the transformation of the imaginary exponential function $e^{j\omega t}u(t)$. The Laplace transform of this function may be found as follows:

$$\mathcal{L}[e^{j\omega t}u(t)] = \int_{0-}^{\infty} e^{j\omega t}e^{-st}\,dt = \int_{0-}^{\infty} e^{-(s-j\omega)t}\,dt = \frac{-1}{s-j\omega}e^{-(s-j\omega)t}\bigg|_{0-}^{\infty} \qquad (8\text{-}52)$$

In the evaluation of the right member of the above equation, we again find that the term disappears when evaluated at the upper limit of $t = \infty$. Thus, our transform pair becomes

$$\mathcal{L}[e^{j\omega t}u(t)] = \frac{1}{s-j\omega} \qquad (8\text{-}53)$$

By a similar process, we may show that the transformation of the exponential $e^{-j\omega t}u(t)$ is defined as

$$\mathcal{L}[e^{-j\omega t}u(t)] = \frac{1}{s+j\omega} \qquad (8\text{-}54)$$

We may use the transform pairs defined in (8-53) and (8-54) to determine the transformation of the trigonometric functions $\sin \omega t$ and $\cos \omega t$. From Euler's identity (see Appendix B) we know that $\cos \omega t = (e^{j\omega t} + e^{-j\omega t})/2$. Thus, from (8-53) and (8-54), we may write

$$\mathcal{L}[\cos \omega t\, u(t)] = \frac{1}{2}\left[\frac{1}{s-j\omega} + \frac{1}{s+j\omega}\right] = \frac{s}{s^2 + \omega^2} \qquad (8\text{-}55)$$

Similarly, since $\sin \omega t = (e^{j\omega t} - e^{-j\omega t})/2j$, we may define the transform pair for the sine function as

$$\mathcal{L}[\sin \omega t\, u(t)] = \frac{1}{2j}\left[\frac{1}{s-j\omega} - \frac{1}{s+j\omega}\right] = \frac{\omega}{s^2 + \omega^2} \qquad (8\text{-}56)$$

It has already been shown that the Laplace transform of the derivative of a function of time is

[6]Convergence of the integral also requires that Re $s > -a$.

$$\mathcal{L}\left[\frac{df}{dt}\right] = sF(s) - f(0-) \tag{8-57}$$

where $F(s) = \mathcal{L}[f(t)]$. The importance of choosing $f(0-)$ rather than $f(0)$ for the point at which the initial condition term is evaluated can be appreciated in terms of the two transforms derived above. To see this, first let us consider the transform of the derivative of the function $\cos \omega t \, u(t)$. Applying the formula for differentiation, we obtain

$$\mathcal{L}\left[\frac{d}{dt}\cos \omega t \, u(t)\right] = s\mathcal{L}[\cos \omega t \, u(t)] - \cos \omega t \, u(t)\Big|_{t=0-}$$

$$= s\frac{s}{s^2 + \omega^2} - 0 = \frac{s^2}{s^2 + \omega^2} \tag{8-58}$$

Note that if we had used $0+$ as the value of t to be used in evaluating the initial condition term, we would obtain $-\omega^2/(s^2 + \omega^2)$ for the transform of the derivative of $\cos \omega t \, u(t)$. To show that this would be incorrect, let us actually differentiate the time-domain function $\cos \omega t \, u(t)$. We obtain

$$\frac{d}{dt}[\cos \omega t \, u(t)] = -\omega \sin \omega t \, u(t) + \cos \omega t \, \delta(t) \tag{8-59}$$

Since the impulse function is non-zero only at $t = 0$, and since the cosine function evaluated at $t = 0$ is unity, the term $\cos \omega t \, \delta(t)$ at the right of the above equation simply equals $\delta(t)$. Using this value, and taking the transform of the right member of (8-59), we obtain

$$\mathcal{L}[-\omega \sin \omega t \, u(t) + \delta(t)] = \frac{-\omega^2}{s^2 + \omega^2} + 1 = \frac{s^2}{s^2 + \omega^2} \tag{8-60}$$

This is the same result as was obtained from (8-58), thus it was necessary that $0-$ rather than 0 or $0+$ be used in evaluating the initial condition term in Eq. (8-58).

Now let us consider the *ramp function* $Ktu(t)$ (when K equals unity, this function may be called a *unit* ramp function). In the preceding section we showed that the transform of the integral of a function of time was equal to the transform of the function divided by s. It is easy to see that the ramp function $Ktu(t)$ is simply the integral of the step function $Ku(t)$. Since the transform of the step function is K/s, the transform of the ramp is simply K/s^2. Thus, we may write directly

$$\mathcal{L}[Ktu(t)] = \frac{K}{s^2} \tag{8-61}$$

This procedure is easily applied to determine the transform of the general class of functions $Kt^n u(t)$. The details are left as an exercise for the reader.

A very useful transformation is one which treats the *shifted unit step function* $u(t - a)$. This represents a unit step applied at $t = a$. We need only concern ourselves with values of $a > 0$, since the Laplace transformation is only defined for positive values of t. Thus, a unit step which begins at negative values of time is no different (when transformed) than a unit step starting at $t = 0$. The Laplace transformation of the shifted step function is defined by

$$\mathcal{L}[u(t - a)] = \int_{0-}^{\infty} u(t - a)e^{-st}\,dt = \int_{a-}^{\infty} e^{-st}\,dt = \frac{-e^{-st}}{s}\Big|_{a-}^{\infty} \tag{8-62}$$

The right member of the above, evaluated at $t = \infty$, is zero. Therefore, we may write

$$\mathcal{L}[u(t - a)] = e^{-as}\frac{1}{s} \tag{8-63}$$

If we examine the transform pair given in Eq. (8-63), we find that the transformed expression is the same as that given in (8-42) for a unit step applied in $t = 0$, namely, $1/s$, multiplied by the additional factor e^{-as}. Thus, the effect of shifting the unit step from $t = 0$ to $t = a$ is simply to multiply the transform of the unit step applied at $t = 0$ by e^{-as}. Many different functions may be constructed from combinations of shifted step functions. An example follows.

Example 8-3.1

We may use a step function and a shifted step function to define a pulse of magnitude K starting at $t = 0$ and terminating at $t = a$. Such a pulse has been designated $f(t)$ in Fig. 8-3.2(a). The pulse may be considered as the summation of two step functions: the first, of strength K, starting at $t = 0$ as shown in Fig. 8-3.2(b), and the second, of strength $-K$, starting at $t = a$ as shown in Fig. 8-3.2(c). Thus, we may write

$$f(t) = K[u(t) - u(t - a)]$$

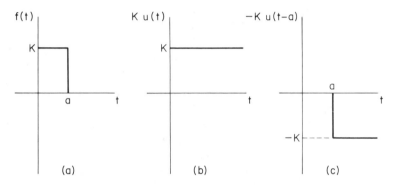

Figure 8-3.2   Waveforms for Example 8-3.1.

From the discussion given above, the Laplace transforms of the two step functions are readily found to be

$$\mathcal{L}[Ku(t)] = \frac{K}{s} \qquad \mathcal{L}[-Ku(t - a)] = \frac{-Ke^{-as}}{s}$$

Thus, since the transform of a sum of terms is equal to the sum of the transforms, we may write

$$\mathcal{L}[f(t)] = \frac{K(1 - e^{-as})}{s}$$

where $f(t)$ is the pulse defined in the time domain in Fig. 8-3.2(a). The term $u(t) - u(t - a)$ used in this example is sometimes referred to as a *gating function*, since, when used to multiply a given function of time, the result is a function which is zero for all values of time less than 0 and greater than a. Thus, this term provides a "gate" through which only those values of the function which exist for the range $0 \le t \le a$ are permitted to pass.

The Laplace Transformation   Chap. 8

Now let us show that the results obtained above, namely, that shifting the time scale by a quantity a for a unit step corresponds with multiplication of the transform of the unit step by the quantity e^{-as}, applies to time functions in general. To see this, consider an arbitrary function $f(t)u(t)$. Now let us assume that we desire to find the transform of a function which has a waveform identical with that of $f(t)u(t)$, but which has been shifted to the right along the real t axis by the distance a, where $a \geq 0$. Such a function may be described as $f(t - a)u(t - a)$. To begin, let us write the general integral expression for the Laplace transformation of a function $f(t)u(t)$ using a "dummy" variable of integration τ. The expression is

$$\mathcal{L}[f(t)u(t)] = F(s) = \int_{0-}^{\infty} f(\tau)e^{-s\tau}\, d\tau \qquad (8\text{-}64)$$

Since the variable of integration is arbitrary, let us define a new variable by the relation $\tau = t - a$. The upper limit of integration remains ∞, but the lower limit of integration becomes $a-$. Finally, $d\tau$ becomes dt. Thus, Eq. (8-64) may be written in the form

$$\mathcal{L}[f(t)u(t)] = F(s) = \int_{a-}^{\infty} f(t - a)e^{-s(t-a)}\, dt = \int_{a-}^{\infty} f(t - a)e^{-st}e^{as}\, dt \qquad (8\text{-}65)$$

The lower limit of integration may be changed to $0-$ by including the term $u(t - a)$ inside the integral. Also, since e^{as} is not a function of t, it may be moved outside the integral. Making these changes, we obtain

$$\mathcal{L}[f(t)u(t)] = e^{as} \int_{0-}^{\infty} f(t - a)u(t - a)e^{-st}\, dt \qquad (8\text{-}66)$$

The integral term in the right member of the above equation is simply the Laplace transformation of $f(t - a)u(t - a)$. Thus, multiplying both sides of the above equation by e^{-as}, we obtain

$$\mathcal{L}[f(t - a)u(t - a)] = e^{-as}\mathcal{L}[f(t)u(t)] \qquad (8\text{-}67)$$

Thus, we see that the transformation of a function $f(t)u(t)$ which is delayed to begin at $t = a$ is simply the transformation of the original function multiplied by e^{-as}. Thus, we have generalized the result originally obtained for the shifted unit step function in Eq. (8-63).

The result derived above is called the *shifting theorem* or, since it provides a translation of the real variable t, the *real translation theorem*. It is of considerable application in determining the Laplace transform of functions for cases where the direct application of the defining integral for the transformation might be difficult. An example follows.

Example 8-3.2

It is desired to find the Laplace transform for a single sinusoidal pulse defined as $f(t)u(t)$ in Fig. 8-3.3(a). Such a pulse is readily seen to be the result of adding two sinusoidal waves of period $2a$, one starting at $t = 0$, as shown in Fig. 8-3.3(b), the other starting at $t = a$, as shown in Fig. 8-3.3(c). Thus, we may write

$$f(t)u(t) = \sin \frac{\pi t}{a} u(t) + \sin \frac{\pi(t - a)}{a} u(t - a)$$

Figure 8-3.3 Waveforms for Example 8-3.2.

From (8-56) we see that the transform of $\sin(\pi t/a)u(t)$ is

$$\mathcal{L}\left[\sin\frac{\pi t}{a}u(t)\right] = \frac{\pi/a}{s^2 + (\pi/a)^2}$$

Thus, using (8-67), we may write

$$\mathcal{L}[f(t)u(t)] = \frac{(1 + e^{-as})\pi/a}{s^2 + (\pi/a)^2}$$

The shifting theorem discussed above may be summarized as follows:

SUMMARY 8-3.1

The Shifting Theorem. If a function $f(t)u(t)$ which is zero for $t < 0$ has a Laplace transform $F(s)$, then for a positive real value of a, a function $f(t-a)u(t-a)$ which is zero for $t < a$ and which corresponds to the original function shifted in time by a seconds, has a Laplace transform $e^{-as}F(s)$. ∎

An extension of the shifting theorem is readily made to functions which are periodic for $t > 0$. If we let $f(t)$ be such a function, and $f_1(t)$ be a function which describes *only the first cycle* of the periodic function and is zero for all other values of t, then, if the period of $f(t)$ is T, we may write $f(t)$ in the form

$$f(t) = f_1(t) + f_1(t-T)u(t-T) + f_1(t-2T)u(t-2T) + \cdots \qquad (8\text{-}68)$$

Here $f_1(t-T)u(t-T)$ is the first cycle of $f(t)$ shifted one period, thus it is the second cycle of $f(t)$; similarly, $f_1(t-2T)u(t-2T)$ is the third cycle, etc. If we now take the Laplace transform of Eq. (8-68), simultaneously applying the shifting theorem given in Summary 8-3.1, we obtain

$$\mathcal{L}[f(t)] = F(s) = F_1(s)(1 + e^{-Ts} + e^{-2Ts} + \cdots) \qquad (8\text{-}69)$$

where $F_1(s)$ is the Laplace transform of $f_1(t)$. Equation (8-69) may be written in the form

$$\mathcal{L}[f(t)] = F(s) = \frac{F_1(s)}{1 - e^{-Ts}} \qquad (8\text{-}70)$$

where the form of the denominator term in the right member of (8-70) is easily related to the term in parentheses in the right member of Eq. (8-69) by simple division. An example of the application of this result follows.

Example 8-3.3

It is desired to obtain the Laplace transform of the periodic pulse train with period b shown as $f(t)$ in Fig. 8-3.4. From the result of Example 8-3.1, we know that the transform of the first cycle of this pulse train is simply the transform of the pulse defined

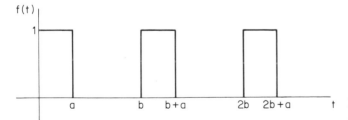

Figure 8-3.4 Waveform for Example 8-3.3.

over $0 \le t \le a$. Thus, following the terminology given preceding Eq. (8-68), we may write

$$F_1(s) = \frac{1 - e^{-as}}{s}$$

Applying Eq. (8-70), we see that the transform of $f(t)$ is given as

$$\mathcal{L}[f(t)] = F(s) = \frac{1 - e^{-as}}{s(1 - e^{-bs})}$$

Note that when $b = a$, then $F(s) = 1/s$, i.e., the function becomes a unit step.

Another important property of the Laplace transformation follows from the transform of the term $e^{-at}f(t)u(t)$. For this, we may write

$$\mathcal{L}[e^{-at}f(t)u(t)] = \int_{0-}^{\infty} e^{-at}f(t)e^{-st} \, dt = \int_{0-}^{\infty} f(t)e^{-(s+a)t} \, dt = F(s+a) \qquad (8\text{-}71)$$

Thus, the result of multiplying a time function by e^{-at} is a translation of value a in the complex frequency variable s. This result is a very useful one in permitting us to readily extend our list of transform pairs. For example, from Eq. (8-56), since the Laplace transformation of $\sin bt \, u(t)$ is $b/(s^2 + b^2)$, we see that the transform of $e^{-at} \sin bt \, u(t)$ is simply found to be $b/[(s + a)^2 + b^2]$. Other useful results are readily obtained. We may summarize the above discussion as follows:

SUMMARY 8-3.2

Complex Frequency Variable Translation Theorem. The Laplace transformation of a function $f(t)u(t)$ which is multiplied by the term e^{-at} is equal to $F(s + a)$, where $F(s)$ is the transform of the function $f(t)u(t)$ alone. ∎

One other procedure involving the Laplace transform that is of interest in studies of circuit theory is the transformation of functions in the time domain which have been subjected to a time scaling. For our purposes here, we may define "scaling" as multiplying all values of the variable t by some positive constant a. Thus, given an unscaled function $f(t)u(t)$, let us investigate the Laplace transformation of a time-

TABLE 8-3.1   Basic Laplace Transform Pairs

Number	$f(t)$†	$F(s)$
1	$\delta(t)$	1
2	$\delta^{(n)}(t)$‡	s^n
3	$u(t)$	$\dfrac{1}{s}$
4	t	$\dfrac{1}{s^2}$
5	t^n	$\dfrac{n!}{s^{n+1}}$
6	$e^{-at} \quad a > 0$	$\dfrac{1}{s+a}$
7	$te^{-at} \quad a > 0$	$\dfrac{1}{(s+a)^2}$
8	$\sin \omega t$	$\dfrac{\omega}{s^2+\omega^2}$
9	$\cos \omega t$	$\dfrac{s}{s^2+\omega^2}$
10	$\sin(\omega t + \theta)$	$\dfrac{s \sin \theta + \omega \cos \theta}{s^2+\omega^2}$
11	$\cos(\omega t + \theta)$	$\dfrac{s \cos \theta - \omega \sin \theta}{s^2+\omega^2}$
12	$e^{-at} \sin \omega t \quad a > 0$	$\dfrac{\omega}{(s+a)^2+\omega^2}$
13	$e^{-at} \cos \omega t \quad a > 0$	$\dfrac{s+a}{(s+a)^2+\omega^2}$
14	$e^{-at} \sin(\omega t + \theta) \quad a > 0$	$\dfrac{(s+a)\sin \theta + \omega \cos \theta}{(s+a)^2+\omega^2}$
15	$e^{-at} \cos(\omega t + \theta) \quad a > 0$	$\dfrac{(s+a)\cos \theta - \omega \sin \theta}{(s+a)^2+\omega^2}$
16	$2|K|e^{-at} \cos(\omega t + \text{Arg } K) \quad a > 0$	$\dfrac{K}{s+a-j\omega} + \dfrac{K*}{s+a+j\omega}$

†All of the functions $f(t)$ are assumed to be zero for $t < 0$, i.e., they may be considered as being multiplied by $u(t)$.

‡The function $\delta^{(n)}(t)$ is defined as $d^n\delta(t)/dt^n$, i.e., the nth derivative of $\delta(t)$. Thus $\delta^{(1)}(t)$ is a doublet, $\delta^{(2)}(t)$ is a triplet, etc.

scaled function $f(at)u(at) = f(at)u(t)$. We may write

$$\mathcal{L}[f(at)u(t)] = \int_{0-}^{\infty} f(at)e^{-st}\, dt = \int_{0-}^{\infty} f(at)e^{-s(at)/a}\, \frac{d(at)}{a}$$

$$= \frac{1}{a}\int_{0-}^{\infty} f(at)e^{-at(s/a)}\, d(at) = \frac{1}{a}F\left(\frac{s}{a}\right) \tag{8-72}$$

This result may be summarized as follows:

The Laplace Transformation   Chap. 8

TABLE 8-3.2 Basic Laplace Transform Operations

Number	$f(t)$†	$F(s)$
1	$f_1(t) \pm f_2(t)$	$F_1(s) \pm F_2(s)$
2	$Kf(t)$	$KF(s)$
3	$\dfrac{d}{dt} f(t)$	$sF(s) - f(0-)$
4	$\dfrac{d^2}{dt^2} f(t)$	$s^2 F(s) - sf(0-) - \dfrac{df}{dt}(0-)$
5	$\dfrac{d^3}{dt^3} f(t)$	$s^3 F(s) - s^2 f(0-)$ $- s\dfrac{df}{dt}(0-) - \dfrac{d^2 f}{dt^2}(0-)$
6	$g(t) = \displaystyle\int_0^t f(\tau)\,d\tau + g(0-)$	$\dfrac{F(s)}{s} + \dfrac{g(0-)}{s}$
7	$u(t - a)$	e^{-as}/s
8	$f(t - a)u(t - a)$	$e^{-as} F(s)$
9	$e^{-at} f(t)$	$F(s + a)$
10	$f(at)$	$\dfrac{1}{a} F\left(\dfrac{s}{a}\right)$
11	$tf(t)$	$-\dfrac{d}{ds} F(s)$
12	$t^n f(t)$	$(-1)^n \dfrac{d^n}{ds^n} F(s)$
13‡	$\lim_{t \to 0} f(t)$	$\lim_{s \to \infty} sF(s)$
14‡	$\lim_{t \to \infty} f(t)$	$\lim_{s \to 0} sF(s)$

†All of the functions $f(t)$ are assumed to be zero for $t < 0$, i.e., they may be considered as being multiplied by $u(t)$.

‡This operation is defined in Sec. 8-4.

SUMMARY 8-3.3

Scaling Theorem. If a function $f(t)u(t)$ has its real variable t multiplied by a positive constant a, then the Laplace transform of the function will have its complex frequency variable s multiplied by $1/a$. In addition, the entire transform is multiplied by $1/a$. ■

The result given above is quite reasonable when one considers that, in general, frequency and time are reciprocal quantities and, thus, multiplying time by a scale factor should correspondingly divide frequency by the same factor.

The various Laplace transform pairs which have been derived in this chapter, as well as some other ones which are commonly encountered, are listed for reference in Table 8-3.1. The properties of the Laplace transformation which have been developed in this chapter are listed in Table 8-3.2. These tables will be of use to the reader when applications of the Laplace transformation are discussed in future sections of this chapter.

8-4 THE INVERSE LAPLACE TRANSFORMATION— PARTIAL-FRACTION EXPANSIONS

Let us summarize what we have accomplished so far in this chapter. In Sec. 8-1 we developed an expression for a mathematical transformation which changed functions of a real variable t into functions of a complex frequency variable s, and which had the property that the operations of differentiation and integration in the time domain were changed to algebraic operations in the frequency domain. In Sec. 8-2 we illustrated how integro-differential time-domain equations could be transformed into algebraic frequency-domain equations involving transformed variables and in Sec. 8-3 we showed how specific time functions could be similarly transformed. We are now in a position to attack our final goal, namely, reclaiming the time-domain solutions for the actual physical network variables from the transformed equations. The mathematical process by means of which this is accomplished is referred to as an *inverse transformation*. Such a process will be the major subject of this section.

The *inverse Laplace transformation* is defined by a complex *inversion integral* which may be applied to a function $F(s)$ to generate a function $f(t)$. The integral itself is somewhat formidable, and its rigorous application involves a knowledge of complex variable theory that is beyond the scope of our text.[7] Fortunately, however, for our applications, there is no need to become involved in such a complicated process. The reason for this is the *uniqueness* property of the Laplace transformation. This may be defined as follows:

SUMMARY 8-4.1

Uniqueness of the Laplace Transformation.　For a given $F(s)$, there is one and only one function $f(t)$, defined for $t \geq 0$, which is related to $F(s)$ by the Laplace transformation.[8] Because of this uniqueness, we may write

$$\mathcal{L}[f(t)] = F(s) \qquad \mathcal{L}^{-1}[F(s)] = f(t) \tag{8-73}$$

where $\mathcal{L}^{-1}[F(s)]$ may be read as "the inverse transformation of $F(s)$." ∎

---

[7]The complex inversion integral defining the inverse Laplace transformation is defined as

$$f(t) = \frac{1}{2\pi j} \int_{\sigma_0 - j\infty}^{\sigma_0 + j\infty} F(s) e^{st} \, ds$$

where the path of integration is defined as the *Bromwich path*, and is chosen so as to achieve the necessary convergence of the integral.

[8]It should be noted that this uniqueness only applies to the Laplace transformation which is defined as an integral with limits of integration from $0-$ to infinity. This transformation is sometimes referred to as the *single-sided* Laplace transformation. Another integral transformation, the *double-sided* Laplace transformation, has limits of integration from minus infinity to infinity. Such a transformation may be applied to functions which have a non-zero value for $t < 0$. In such a case, the uniqueness property described here does not apply.

As a most important application of the conclusion given above, consider the transform pairs given in Table 8-3.1. These pairs were originally derived "from left to right," i.e., given the time-domain functions $f(t)$, we found the corresponding functions $F(s)$. Since the Laplace transformation of these functions is unique, we may also use this table of transform pairs "from right to left," i.e., given any of the tabulated frequency-domain functions $F(s)$ or any linear combination of them, we may directly write the corresponding functions $f(t)$ in the time domain, thus obtaining the desired inverse transform. An example of this procedure follows.

Example 8-4.1

As an example of the procedure for finding the inverse transform, consider the following function:

$$F(s) = \frac{3s - 8}{s^2 + 4} = \frac{3(s)}{s^2 + 4} - \frac{4(2)}{s^2 + 4}$$

Here in the right member, we have separated the function into two parts which have the general form of items 8 and 9 in Table 8-3.1. From the table, we may write

$$f(t) = \mathcal{L}^{-1}[F(s)] = \mathcal{L}^{-1}\left[\frac{3(s)}{s^2 + 4} - \frac{4(2)}{s^2 + 4}\right] = 3 \cos 2t - 4 \sin 2t$$

$$= 5 \cos (2t + 53.13°)$$

where the last result is obtained by the methods outlined in Sec. 7-1.

From the above example, it should be apparent that a simple means of determining the inverse transform of a complicated function is to rewrite the function as a sum of terms, with each term having the form of one of the transforms given in Table 8-3.1. We shall now describe a general procedure for writing a rational function in such a way. The first step of this procedure is to evaluate the behavior of the function $F(s)$ in the limit as s approaches infinity. There are three possible situations that may occur. These are:

1. $\lim_{s \to \infty} F(s) = Ks^m \qquad (m = 1, 2, 3, \ldots)$

2. $\lim_{s \to \infty} F(s) = K$ \hfill (8-74)

3. $\lim_{s \to \infty} F(s) = \dfrac{K}{s^m} \qquad (m = 1, 2, 3, \ldots)$

Before describing the above situations, let us introduce the concept of a point in the complex frequency plane at which the magnitude of the function $F(s)$ goes to infinity. Such a point is called a *pole* of $F(s)$. Thus, we may describe the first situation defined in Eq. (8-74) by saying that $F(s)$ has an *mth-order pole at infinity*. This requires that the degree of the numerator polynomial of $F(s)$ be m greater than the degree of the denominator polynomial. An example of such a function is

$$F(s) = \frac{2s^4 + 1}{s^3 + 3s^2 + 4s + 1} \qquad \lim_{s \to \infty} F(s) = 2s \qquad (8-75)$$

For this function, we say that $F(s)$ has a first-order pole at infinity. To describe the second situation listed in Eq. (8-74), we may say that $F(s)$ is constant at s equals infinity. This requires that the degree of the numerator and denominator polynomials be equal. An example of such a function is

$$F(s) = \frac{3s^2 + 2s + 1}{s^2 + 4s + 4} \qquad \lim_{s \to \infty} F(s) = 3 \qquad (8\text{-}76)$$

To describe the third situation listed in Eq. (8-74), we may introduce the concept of a *zero* to indicate a place in the complex frequency plane at which the magnitude of $F(s)$ goes to zero. Thus, we may describe this case by saying that $F(s)$ has an *mth-order zero at infinity*. This requires that the degree of the numerator of $F(s)$ be m lower than the degree of the denominator. An example of a function which has a second-order zero at infinity is

$$F(s) = \frac{7}{s^2 + 3s + 2} \qquad \lim_{s \to \infty} F(s) = \frac{7}{s^2} \qquad (8\text{-}77)$$

In evaluating the behavior of a given function at infinity, if the first or second situation listed in Eq. (8-74) occurs, then the first step in putting $F(s)$ in a form such that the inverse transform is easily determined is to separate the terms defining the behavior at infinity from the rest of the rational function. This is simply done by long division. For example, consider

$$F(s) = \frac{s^3 + 5s^2 + 11s + 9}{s^2 + 3s + 2} = s + 2 + \frac{3s + 5}{s^2 + 3s + 2} \qquad (8\text{-}78)$$

where the right member is obtained as follows:

$$
\begin{array}{r}
s^2 + 3s + 2\,|\,s^3 + 5s^2 + 11s + 9\,|s + 2 \\
\underline{s^3 + 3s^2 + 2s} \\
2s^2 + 9s + 9 \\
\underline{2s^2 + 6s + 4} \\
3s + 5 \qquad \text{Remainder}
\end{array}
\qquad (8\text{-}79)
$$

Comsidering (8-78), we see that we may write from Table 8-3.1

$$f(t) = \mathcal{L}^{-1}[F(s)] = \delta^{(1)}(t) + 2\delta(t) + \mathcal{L}^{-1}\left[\frac{3s + 5}{s^2 + 3s + 2}\right] \qquad (8\text{-}80)$$

where the inverse transform of the last term in the right member is yet to be found. The results given above are easily generalized as follows:

SUMMARY 8-4.2

The Inverse Transform of the Infinite Frequency Characteristics of a Rational Function. A rational function $f(s) = C(s)/B(s)$, in which the degree of the numerator polynomial $C(s)$ is m greater ($m = 0, 1, 2, \ldots$) than the degree of the denominator polynomial $B(s)$, may be written in the form

$$F(s) = g_1 + g_2 s + g_3 s^2 + \cdots + g_{m+1} s^m + \frac{A(s)}{B(s)}$$

where the degree of $A(s)$ is less than the degree of $B(s)$. The inverse transform of this expression is given as

$$f(t) = \mathcal{L}^{-1}[F(s)] = g_1\delta(t) + g_2\delta^{(1)}(t) + g_3\delta^{(2)}(t) + \cdots$$

$$+ g_{m+1}\delta^{(m)}(t) + \mathcal{L}^{-1}\left[\frac{A(s)}{B(s)}\right]$$

where the quantities $\delta(t)$, $\delta^{(1)}(t)$, $\delta^{(2)}(t)$, ... are, respectively, an impulse, a doublet, a triplet, ... occurring at $t = 0$. ∎

In the above paragraphs we have seen how to transform the infinite frequency characteristics of a rational function into time-domain behavior. In most network situations, the degree of the numerator will never be more than 1 greater than the degree of the denominator; thus, only the impulse and the doublet will normally appear in the time function. Now let us consider the inverse transform of the remaining portion of the rational function, i.e., the part in which the degree of the numerator is less than the degree of the denominator.[9] Let $F(s) = A(s)/B(s)$ be such a function, and let us assume that any non-unity leading coefficient of the denominator polynomial $B(s)$ has been moved to the numerator. If we now factor both the numerator and the denominator polynomials of $F(s)$, we may write the function in the form

$$F(s) = \frac{A(s)}{B(s)} = \frac{K(s - z_1)(s - z_2)(s - z_3)\cdots(s - z_m)}{(s - p_1)(s - p_2)(s - p_3)\cdots(s - p_n)}$$

$$= \frac{K\prod_{i=1}^{m}(s - z_i)}{\prod_{i=1}^{n}(s - p_i)} \qquad m < n \qquad (8\text{-}81)$$

The quantities p_i represent locations in the finite s plane where the magnitude of $F(s)$ goes to infinity. Thus, they are referred to as the *poles of* $F(s)$. Similarly, the quantities z_i represent places in the finite complex plane where the magnitude of $F(s)$ is zero. They are called the *zeros of* $F(s)$. The constant K is sometimes referred to as the overall multiplicative constant or, more simply, as the *gain constant*. It should be apparent that the number of finite zeros of $F(s)$ plus the number of zeros at infinity is equal to the number of finite poles of $F(s)$, i.e., *the total number of poles of a rational function equals the total number of zeros.*

There are three cases which we can consider for functions which are of the general type shown in Eq. (8-81). The first of these is the case in which all the poles of $F(s)$ are real and simple. Thus, in this case, all the quantities p_i in Eq. (8-81) will be real numbers, and no two will have the same value. Now let us assume that we can find a method for writing $F(s)$ in the form

$$F(s) = \frac{K_1}{s - p_1} + \frac{K_2}{s - p_2} + \frac{K_3}{s - p_3} + \cdots + \frac{K_n}{s - p_n} = \sum_{i=1}^{n}\frac{K_i}{s - p_i} \qquad (8\text{-}82)$$

If this can be done, then, from the transform pairs given in Table 8-3.1 we see that the inverse transform may be immediately written as

[9]Such a function is sometimes called a *proper* rational function.

$$f(t) = \mathcal{L}^{-1}[F(s)] = K_1 e^{p_1 t} + K_2 e^{p_2 t} + K_3 e^{p_3 t} + \cdots + K_n e^{p_n t}$$

$$= \sum_{i=1}^{n} K_i e^{p_i t} \tag{8-83}$$

An expansion of a rational function $F(s)$ in the form indicated in Eq. (8-82) is called a *partial-fraction expansion*. The constants associated with each of the terms of the partial-fraction expansion are called the *residues*. More specifically, K_i is called *the residue of $F(s)$ at its pole at p_i*. If $F(s)$ has only two or three poles, then the method of *undetermined coefficients* is easily used to find the values of the K_i. As an example of this, consider the function

$$F(s) = \frac{3s + 5}{s^2 + 3s + 2} = \frac{3s + 5}{(s + 1)(s + 2)} \tag{8-84}$$

The poles are $p_1 = -1$, and $p_2 = -2$. Putting $F(s)$ in the form of a partial-fraction expansion, we obtain

$$F(s) = \frac{K_1}{s + 1} + \frac{K_2}{s + 2} = \frac{K_1(s + 2) + K_2(s + 1)}{(s + 1)(s + 2)} = \frac{s(K_1 + K_2) + (2K_1 + K_2)}{s^2 + 3s + 2} \tag{8-85}$$

If we now compare the numerators of the right members of (8-84) and (8-85) and equate corresponding coefficients of s, we obtain the equations

$$K_1 + K_2 = 3 \qquad 2K_1 + K_2 = 5 \tag{8-86}$$

Solving these equations, we find that $K_1 = 2$, $K_2 = 1$. Thus, $F(s)$ of (8-84) may be written in the form

$$F(s) = \frac{3s + 5}{s^2 + 3s + 2} = \frac{2}{s + 1} + \frac{1}{s + 2} \tag{8-87}$$

The inverse transform of $F(s)$ of Eq. (8-87) may be found by taking the inverse transform of each term in the right member. Thus, from Table 8-3.1 we obtain

$$f(t) = \mathcal{L}^{-1}[F(s)] = [2e^{-t} + e^{-2t}]u(t) \tag{8-88}$$

The method of undetermined coefficients outlined above for finding the residues K_i requires the solution of a set of simultaneous equations of order n, where n is the number of poles. Such an approach becomes unwieldy for values of n greater than 2 or 3. Thus, it is desirable to consider an alternative method for finding the residues. To see how this alternative approach works, let us again consider the function $F(s)$ defined in Eq. (8-81). If we multiply $F(s)$ by the factor $(s - p_1)$, we obtain

$$(s - p_1)F(s) = \frac{(s - p_1)K \prod_{i=1}^{m} (s - z_i)}{\prod_{i=1}^{n} (s - p_i)} = \frac{K \prod_{i=1}^{m} (s - z_i)}{\prod_{i=2}^{n} (s - p_i)} \tag{8-89}$$

Note that in the right member of (8-89), we have cancelled the factor $(s - p_1)$ in the numerator and denominator, thus the index of the product symbol in the denominator goes from 2 to n, rather than from 1 to n. Now let us also multiply the partial-fraction expansion form of $F(s)$ as given in Eq. (8-82) by the factor $(s - p_1)$. Doing this, we may write

$$(s - p_1)F(s) = K_1 + (s - p_1)\left[\frac{K_2}{s - p_2} + \frac{K_3}{s - p_3} + \cdots + \frac{K_n}{s - p_n}\right] \tag{8-90}$$

Now consider what happens if we evaluate Eqs. (8-89) and (8-90) at $s = p_1$. From the right member of Eq. (8-89) we see that even though p_1 is a pole of $F(s)$ [and therefore $F(p_1)$ is infinite], the product $(s - p_1)F(s)$ will be finite when evaluated at $s = p_1$, since the factor $(s - p_1)$ cancels in the numerator and the denominator. If we now evaluate Eq. (8-90) at $s = p_1$, the right member is simply K_1. Combining these results, we obtain

$$(s - p_1)F(s)|_{s=p_1} = K_1 \qquad (8\text{-}91)$$

This procedure is easily extended to determine the other residues, since, if we multiply $F(s)$ by the factor $(s - p_i)$, we see that the residue K_i is determined by the expression

$$K_i = (s - p_i)F(s)|_{s=p_i} \qquad (8\text{-}92)$$

Thus, we may apply this procedure to find the residue at any pole p_i. An example follows.

Example 8-4.2

It is desired to find the partial-fraction expansion and the inverse transform of the following function.

$$F(s) = \frac{6s^2 + 25s + 23}{s^3 + 6s^2 + 11s + 6} = \frac{6s^2 + 25s + 23}{(s + 1)(s + 2)(s + 3)}$$

The partial-fraction expansion for this function will have the form

$$F(s) = \frac{K_1}{s + 1} + \frac{K_2}{s + 2} + \frac{K_3}{s + 3}$$

The residues K_1, K_2, and K_3 may be found using Eq. (8-92) as follows:

$$K_1 = (s + 1)F(s)\Big|_{s=-1} = \frac{6s^2 + 25s + 23}{(s + 2)(s + 3)}\Big|_{s=-1} = 2$$

$$K_2 = (s + 2)F(s)\Big|_{s=-2} = \frac{6s^2 + 25s + 23}{(s + 1)(s + 3)}\Big|_{s=-2} = 3$$

$$K_3 = (s + 3)F(s)\Big|_{s=-3} = \frac{6s^2 + 25s + 23}{(s + 1)(s + 2)}\Big|_{s=-3} = 1$$

Thus, we may write $F(s)$ as the following partial-fraction expansion.

$$F(s) = \frac{2}{s + 1} + \frac{3}{s + 2} + \frac{1}{s + 3}$$

Since, from Table 8-3.1, the inverse transform of a term of the type $k/(s + a)$ is $ke^{-at}u(t)$, the inverse transform of $F(s)$ is readily see to be

$$f(t) = \mathcal{L}^{-1}[F(s)] = [2e^{-t} + 3e^{-2t} + e^{-3t}]u(t)$$

The results given above are summarized as follows:

SUMMARY 8-4.3

The Inverse Transform of a Rational Function Containing Only Simple Real Poles. A rational function $F(s)$ which contains only n simple real poles, and in which the degree of the numerator polynomial is less than the degree of the denominator polynomial, may always be expressed as a partial-fraction expansion having the form

$$F(s) = \sum_{i=1}^{n} \frac{K_i}{s - p_i}$$

where the K_i are the residues and the p_i are the poles. The residues are found as

$$K_i = (s - p_i) F(s)|_{s=p_i}$$

The inverse transform is given as

$$f(t) = \mathcal{L}^{-1}[F(s)] = \sum_{i=1}^{n} K_i e^{p_i t} \qquad \blacksquare$$

The second case which we shall discuss for determining a partial-fraction expansion of a function $F(s)$ in which the degree of the numerator polynomial is less than the degree of the denominator polynomial is the one in which the function has only real poles, but in which one or more of these poles is non-simple, i.e., of higher order than first. For simplicity, let us assume that there is only one such pole, located at $s = p_0$, and that it is of order n ($n = 2, 3, 4, \ldots$). We shall also assume that there are m other poles, and that these are all simple. Such a function may be written in the form

$$F(s) = \frac{A(s)}{(s - p_0)^n \displaystyle\prod_{i=1}^{m} (s - p_i)} \qquad (8\text{-}93)$$

where it is assumed that the degree of the numerator polynomial is less than $n + m$, i.e., the degree of the denominator polynomial. Now let us assume that we can find a method for writing such a function in the form

$$F(s) = \frac{K_{01}}{s - p_0} + \frac{K_{02}}{(s - p_0)^2} + \frac{K_{03}}{(s - p_0)^3} + \cdots + \frac{K_{0n}}{(s - p_0)^n}$$

$$+ \frac{K_1}{s - p_1} + \frac{K_2}{s - p_2} + \cdots + \frac{K_m}{s - p_m}$$

$$= \sum_{i=1}^{n} \frac{K_{0i}}{(s - p_0)^i} + \sum_{i=1}^{m} \frac{K_i}{s - p_i} \qquad (8\text{-}94)$$

If this can be done, then, from the transform pairs given in Table 8-3.1, we see that the inverse transform may be immediately written as

$$f(t) = \mathcal{L}^{-1}[F(s)] = K_{01} e^{p_0 t} + K_{02} t e^{p_0 t} + \frac{K_{03} t^2 e^{p_0 t}}{2} + \cdots + \frac{K_{0n} t^{n-1} e^{p_0 t}}{(n - 1)!}$$

$$+ K_1 e^{p_1 t} + K_2 e^{p_2 t} + \cdots + K_m e^{p_m t}$$

$$= e^{p_0 t} \sum_{i=1}^{n} \frac{K_{0i} t^{i-1}}{(i - 1)!} + \sum_{i=1}^{m} K_i e^{p_i t} \qquad (8\text{-}95)$$

In the partial-fraction expansion given in Eq. (8-94), the determination of the residues K_i associated with the m simple poles is accomplished by the procedure given in Summary 8-4.3. Such a determination presents no new problems. The determination of the constants K_{0i} associated with the various powers of the non-simple pole, however, requires a slight modification of the basic technique.[10] To see this, consider the function

[10] Of the constants K_{0i} ($i = 1, 2, 3, \ldots, n$) of Eq. (8-94), only K_{01} is formally called the residue.

$$F(s) = \frac{4s^2 + 11s + 9}{s^3 + 4s^2 + 5s + 2} = \frac{4s^2 + 11s + 9}{(s+1)^2(s+2)} \tag{8-96}$$

The partial-fraction expansion for this function will have the form

$$F(s) = \frac{K_{01}}{s+1} + \frac{K_{02}}{(s+1)^2} + \frac{K_1}{s+2} \tag{8-97}$$

Let us first see how to find the constant K_{02} associated with the highest degree term representing the second-order pole at $s = -1$. If we multiply both sides of Eq. (8-97) by the denominator of the term containing K_{02}, i.e., the term $(s+1)^2$, and rearrange terms, we obtain

$$(s+1)^2 F(s) = K_{02} + K_{01}(s+1) + \frac{K_1(s+1)^2}{s+2} \tag{8-98}$$

If we now evaluate both sides of (8-98) at $s = -1$, the last two terms in the right member will disappear, since they are multiplied by zero. Since the left member also contains a term $(s+1)^2$ in the denominator of $F(s)$, however, this member will not be zero. Therefore, we may write

$$K_{02} = (s+1)^2 F(s) \Big|_{s=-1} = \frac{4s^2 + 11s + 9}{s+2} \Big|_{s=-1} = 2 \tag{8-99}$$

Now let us consider how the term K_{01} may be found. As a first step, let us differentiate both sides of Eq. (8-98). Thus, we obtain

$$\frac{d}{ds}[(s+1)^2 F(s)] = K_{01} + K_1 \left[\frac{2(s+1)}{s+2} + (s+1)^2 \frac{d}{ds}\frac{1}{s+2} \right] \tag{8-100}$$

Note that in the left member of the above, the factor $(s+1)^2$ will be cancelled by an identical factor in the denominator of $F(s)$ *before* performing the differentiation. Now let us evaluate both sides of Eq. (8-100) at $s = -1$. All the terms in the right member go to zero except the quantity K_{01}. Thus, we may write

$$K_{01} = \frac{d}{ds}[(s+1)^2 F(s)] \Big|_{s=-1} = \frac{d}{ds}\left[\frac{4s^2 + 11s + 9}{s+2} \right] \Big|_{s=-1}$$

$$= \left[\frac{8s+11}{s+2} - \frac{4s^2 + 11s + 9}{(s+2)^2} \right] \Big|_{s=-1} = 1 \tag{8-101}$$

Thus, we have found K_{01} and K_{02} of Eq. (8-97). The residue K_1 is easily shown to be 3, therefore, we may write

$$F(s) = \frac{1}{s+1} + \frac{2}{(s+1)^2} + \frac{3}{s+2} \tag{8-102}$$

From Table 8-3.1 we find that the inverse transform of this function is

$$f(t) = \mathcal{L}^{-1}[F(s)] = [e^{-t} + 2te^{-t} + 3e^{-2t}]u(t) \tag{8-103}$$

The results given above are easily generalized as follows:

SUMMARY 8-4.4

The Inverse Transform of a Rational Function Containing Non-Simple Real Poles. If a rational function $F(s)$ contains a non-simple real pole of order n at $s = p_0$, then there will be a sum of terms of the form $K_{0i}/(s - p_0)^i$ $(i = 1, 2, \ldots, n)$ in the partial-fraction expansion for the function. The multiplicative constants K_{0i} of these terms are found by the relation[11]

$$K_{0i} = \frac{1}{(n-i)!} \frac{d^{n-i}}{ds^{n-i}} (s-p_0)^n F(s) \Big|_{s=p_0}$$

where the constants are most easily evaluated in the order $i = n, n-1, n-2, \ldots,$ 2, 1. There will also be a series of terms in the inverse transform corresponding with such a non-simple pole. These will have the form $e^{p_0 t} K_{0i} t^{i-1}/(i-1)!$ ∎

The third case which we shall discuss for determining a partial-fraction expansion of a rational function $F(s)$ is the case in which the function contains a pair of complex conjugate poles. For simplicity, we shall consider that the order of the poles is simple and that they are located at p_0 and p_0^*. Thus, we may write $F(s)$ in the form

$$F(s) = \frac{A(s)}{(s-p_0)(s-p_0^*)Q(s)} \tag{8-104}$$

where the degree of the numerator polynomial $A(s)$ is less than the degree of the denominator [the degree of $Q(s)$ plus 2], and where $Q(s)$ contains all the other poles of $F(s)$. Now let us separate the partial-fraction expansion terms involving the complex conjugate poles from the rest of the function. Thus, we may write

$$F(s) = \frac{K_1}{s-p_0} + \frac{K_2}{s-p_0^*} + \frac{R(s)}{Q(s)} \tag{8-105}$$

where $R(s)/Q(s)$ is the partial-fraction expansion for all the other poles of $F(s)$. The residues K_1 and K_2 are easily found as follows:

$$K_1 = (s-p_0)F(s)|_{s=p_0} \qquad K_2 = (s-p_0^*)F(s)|_{s=p_0^*} \tag{8-106}$$

From these equations we may note that, since the evaluation of the right members is made at a complex value of the argument, the residues K_1 and K_2 will, in general, be complex. In addition, we may note that $K_2 = K_1^*$. Let us define the real and imaginary parts of K_1 and p_0 by the expressions

$$K_1 = k_r + jk_i \qquad p_0 = \sigma_0 + j\omega_0 \tag{8-107}$$

The related conjugate terms then are given as

$$K_2 = K_1^* = k_r - jk_i \qquad p_0^* = \sigma_0 - j\omega_0 \tag{8-108}$$

Substituting the expressions of (8-107) and (8-108) in the general form for $F(s)$ given in (8-105) and rearranging terms, we obtain

[11]Many alternative methods have been proposed for finding the terms in the partial-fraction expansion for poles which are of higher order than the first. A sampling of some of the literature on this subject may be found in the following:

S. Karni, "Easy partial-fraction expansion with multiple poles," *Proc. IEEE*, vol. 57, No. 2, pp. 231–232, February 1969.

M. F. Moad, "On partial-fraction expansion with multiple poles through derivatives," *Proc. IEEE*, Vol. 57, No. 11, pp. 2056–2058, November 1969.

K. R. Rao and N. Ahmed, "Recursive techniques for obtaining the partial-fraction expansion of a rational function, *IEEE Trans. on Education*, Vol. E-11, No. 2, pp. 152–154, June 1968.

$$F(s) = \frac{2k_r s - 2k_r \sigma_0 - 2k_i \omega_0}{(s - \sigma_0)^2 + \omega_0^2} + \frac{R(s)}{Q(s)} \qquad (8\text{-}109)$$

We conclude that in the partial-fraction expansion the terms produced by a simple pair of complex conjugate poles with complex residues may be combined to form an expression of the form $(as + b)/[(s - \sigma_0)^2 + \omega_0^2]$ where the coefficients a, b, σ_0, and ω_0 are all real. Frequently, for low-order rational functions, the coefficients a and b can be found directly, using the method of undetermined coefficients. This is usually easier than using the complex arithmetic required when determining the residues directly by Eq. (8-106). As an example of this, consider the function

$$F(s) = \frac{4s^2 + 7s + 13}{(s^2 + 2s + 5)(s + 2)} = \frac{as + b}{s^2 + 2s + 5} + \frac{K}{s + 2} \qquad (8\text{-}110)$$

The residue K for the pole at -2 is easily shown to be 3. Using this value and equating the numerator of both sides of Eq. (8-110), we obtain

$$4s^2 + 7s + 13 = (as + b)(s + 2) + 3(s^2 + 2s + 5)$$
$$= s^2(a + 3) + s(2a + b + 6) + (2b + 15) \qquad (8\text{-}111)$$

Equating the coefficients of like powers of s, we find that $a = 1$ and $b = -1$.[12] Thus, we see that

$$F(s) = \frac{s - 1}{s^2 + 2s + 5} + \frac{3}{s + 2} = \frac{s - 1}{(s + 1)^2 + (2)^2} + \frac{3}{s + 2} \qquad (8\text{-}112)$$

The form used for the complex conjugate pole pair in the denominator of the right member of the above equation provides the key to one of the simplest methods of taking the inverse transform of such a term. To see this, let us use the transforms of the sine and cosine terms, multiplied by an exponential as given in Table 8-3.1. These are

$$\mathcal{L}[e^{-\sigma t} \sin \omega t \, u(t)] = \frac{\omega}{(s + \sigma)^2 + \omega^2}$$
$$\mathcal{L}[e^{-\sigma t} \cos \omega t \, u(t)] = \frac{s + \sigma}{(s + \sigma)^2 + \omega^2} \qquad (8\text{-}113)$$

Now let us apply these transform pairs to the first term of the right member of Eq. (8-112). In doing this, we must first separate this term into the sum of two terms, such that each is separately transformable, using the transform pairs of Eq. (8-113). Thus, we must find constants k_1 and k_2 such that

$$\frac{k_1(s + 1)}{(s + 1)^2 + (2)^2} + \frac{k_2(2)}{(s + 1)^2 + (2)^2} = \frac{s - 1}{(s + 1)^2 + (2)^2} \qquad (8\text{-}114)$$

Equating coefficients in the numerators of the left and right members of Eq. (8-114),

[12]It should be noted that, since there are three coefficients of the various powers of s (including s^0), performing this operation provides three equations, with only two unknowns to be solved for. The reason for this is that there are really three unknowns, namely, a, b, and K as given in Eq. (8-110). We have already inserted the value of one of the unknowns, namely K, into the expression. Thus, any two of the three equations may be used to find a and b. In general, it is easier to find the residues for real poles directly, and thus to reduce the order of the set of simultaneous equations which must be solved to find the coefficients associated with complex conjugate pairs of poles.

we find that $k_1 = 1$ and $k_2 = -1$. Thus we may write $F(s)$ of (8-112) in the form

$$F(s) = \frac{s + 1}{(s + 1)^2 + (2)^2} + \frac{-1(2)}{(s + 1)^2 + (2)^2} + \frac{3}{s + 2} \qquad (8\text{-}115)$$

The inverse transforms of the first two terms in the right member of this equation are readily found from Eq. (8-113). Thus, the entire inverse transform for the function $F(s)$ given in (8-112) is

$$f(t) = \mathcal{L}^{-1}[F(s)] = [e^{-t} \cos 2t - e^{-t} \sin 2t + 3e^{-2t}]u(t)$$
$$= [\sqrt{2}\, e^{-t} \cos (2t + 45°) + 3e^{-2t}]u(t) \qquad (8\text{-}116)$$

where the second form of $f(t)$, in which the two sinusoidal terms have been combined, is formed by following the methods given in Sec. 7-1. The results given above are easily generalized as follows:

SUMMARY 8-4.5

The Inverse Transform of Complex Conjugate Pole Pairs. If a rational function $F(s)$ contains a pair of complex conjugate simple poles located at $s = -\sigma_0 \pm j\omega_0$, then there will be a term of the form $(as + b)/[(s + \sigma_0)^2 + \omega_0^2]$ in the partial-fraction expansion for the function. Such a term may always be modified to have the form

$$\frac{k_1(s + \sigma_0)}{(s + \sigma_0)^2 + \omega_0^2} + \frac{k_2\omega_0}{(s + \sigma_0)^2 + \omega_0^2}$$

The inverse transform of this pair of terms is

$$e^{-\sigma_0 t}(k_1 \cos \omega_0 t + k_2 \sin \omega_0 t)u(t) \qquad \blacksquare$$

The above results are easily modified to treat the case where multiple-order complex conjugate poles are encountered. The method is similar to that described in Summary 8-4.4, except that it is necessary to use complex arithmetic. The details are left to the reader as an exercise. When the denominator polynomial of the rational function is of third or higher order, and is unfactored, finding its roots is most easily done using the digital computer. A subroutine for performing such a computation is given in Appendix C.

Frequently, it is useful to be able to determine the *initial* value of an inverse Laplace transform without actually performing the inversion process. To see how this may be done, let us first consider the expression for the Laplace transform of the derivative of a time function $f(t)$. This has been shown to be

$$\int_{0-}^{\infty} f'(t)e^{-st}\, dt = sF(s) - f(0-) \qquad (8\text{-}117)$$

Now let us modify this result by dividing the integral in the left member of the above expression into two parts as follows:

$$\int_{0-}^{0+} f'(t)e^{-st}\, dt + \int_{0+}^{\infty} f'(t)e^{-st}\, dt = sF(s) - f(0-) \qquad (8\text{-}118)$$

The first integral in the left member of the above is taken over a range of time from

The Laplace Transformation   Chap. 8

0— to 0+. In this range, the quantity e^{-st} is simply equal to unity. Thus, this first integral may simply be evaluated as follows:

$$\int_{0-}^{0+} f'(t)e^{-st}\, dt = \int_{0-}^{0+} f'(t)\, dt = f(0+) - f(0-) \tag{8-119}$$

Now let us insert the above result in Eq. (8-118) and take the limit of both sides of the equation as s approaches infinity. We obtain

$$f(0+) - f(0-) + \lim_{s\to\infty} \int_{0+}^{\infty} f'(t)e^{-st}\, dt = \lim_{s\to\infty} sF(s) - f(0-) \tag{8-120}$$

Since the limits of integration are from 0+ to infinity in the integral in the left member of the above expression, the variable of integration t is always positive. Thus, as s approaches infinity, the integrand approaches 0 [we require that $f'(t)$ be transformable for this result to hold]. As a result, the value of the integral is also zero and, adding $f(0-)$ to both sides of (8-120), we obtain

$$f(0+) = \lim_{s\to\infty} sF(s) \tag{8-121}$$

This result may be summarized as follows:

SUMMARY 8-4.6

The Initial-Value Theorem. The initial value $f(0+)$ of the inverse transform $f(t)$ of a function $F(s)$ may be found by evaluating the quantity $sF(s)$ in the limit as s approaches infinity.[13] ∎

It is also possible to directly determine the value of $f(t)$ as t approaches *infinity* from the Laplace transform without actually performing the inversion process. To see this, we may let s approach zero in the expression for the Laplace transformation of a derivative. This may be written

$$\lim_{s\to 0} \int_{0-}^{\infty} \left[\frac{df}{dt}\right] e^{-st}\, dt = \lim_{s\to 0} [sF(s) - f(0-)] \tag{8-122}$$

The left member of this equation may be rewritten as follows:

$$\lim_{s\to 0} \int_{0-}^{\infty} \left[\frac{df}{dt}\right] e^{-st}\, dt = \int_{0-}^{\infty} \left[\frac{df}{dt}\right] dt = \lim_{t\to\infty} \int_{0-}^{t} \left[\frac{df}{d\tau}\right] d\tau = \lim_{t\to\infty} [f(t) - f(0-)] \tag{8-123}$$

Equating the right members of (8-122) and (8-123), since $f(0-)$ is not a function of t or s, it may be subtracted from both members. Thus, we obtain

$$\lim_{t\to\infty} f(t) = \lim_{s\to 0} sF(s) \tag{8-124}$$

This result requires that the function $sF(s)$ have no poles on the imaginary axis or in the right half-plane, since otherwise the final value of the related time function $f(t)$

[13]The result given above assumes that $f(t)$ does not have any impulses or derivatives of impulses at the origin. If it does, the result of Summary 8-4.6 must be modified. A discussion of this case may be found in P. M. Chirlian's *Basic Circuit Theory*, Chap. 5, McGraw-Hill Book Company, New York, 1969.

is indeterminate. For example, if $sF(s) = s/(s^2 + 1)$ (which has poles on the $j\omega$ axis), the relation of Eq. (8-124) does not apply. This is clearly true, since $F(s) = 1/(s^2 + 1)$ and, thus, $f(t) = \sin t$, which is determinate for $t = \infty$. The result given above may be summarized as follows:

SUMMARY 8-4.7

The Final-Value Theorem.   If $F(s)$ is the transform of a function $f(t)$, the final value of $f(t)$ as t approaches infinity may be found by evaluating the quantity $sF(s)$ as s approaches zero [provided that $sF(s)$ has no $j\omega$ axis or right half-plane poles]. ∎

In the preceding sections of this chapter we have shown how an integro-differential equation in the time domain may be transformed into an algebraic equation in the frequency domain using the Laplace transformation. This procedure, coupled with the techniques for finding the inverse transformation by means of the partial-fraction expansion, provides a complete method for the analysis of networks based on the use of the Laplace transformation. A flow chart illustrating the general procedure is given in Fig. 8-4.1.

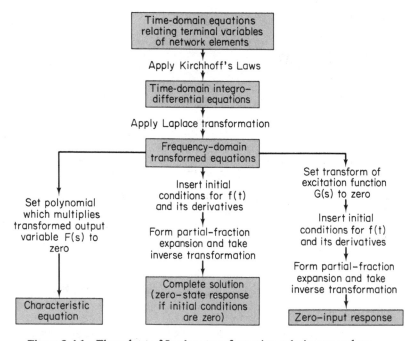

Figure 8-4.1   Flow chart of Laplace transformation solution procedure.

8-5 IMPEDANCE AND ADMITTANCE—NETWORK FUNCTIONS

In this section we shall apply the methods for finding transforms and inverse transforms developed in the preceding sections of this chapter to the general problem of analyzing a network. As a simple introduction, consider the *RLC* network shown in Fig. 8-5.1. The integro-differential equation found by applying KVL to this circuit

Figure 8-5.1 Series *RLC* circuit.

may be written in the form

$$v_S(t) = L\frac{di}{dt} + \frac{1}{C}\int_{0-}^{t} i(\tau)\,d\tau + v_C(0-) + v_R(t) \tag{8-125}$$

Transforming this equation, we obtain

$$V_S(s) = L[sI(s) - i(0-)] + \frac{1}{sC}I(s) + \frac{v_C(0-)}{s} + V_R(s) \tag{8-126}$$

Now let us assume that we desire to solve for $V_R(s)$. The variable $I(s)$ may be eliminated from Eq. (8-126) by substituting the transformed form of Ohm's law, namely, $I(s) = V_R(s)/R$. Making this substitution and rearranging terms, we obtain

$$V_S(s) = \left[\frac{s^2 + s(R/L) + (1/LC)}{s(R/L)}\right]V_R(s) - Li(0-) + \frac{v_C(0-)}{s} \tag{8-127}$$

Solving this equation for $V_R(s)$, we obtain

$$V_R(s) = \frac{s(R/L)V_S(s)}{s^2 + s(R/L) + (1/LC)} + \frac{sRi(0-) - v_C(0-)(R/L)}{s^2 + s(R/L) + (1/LC)} \tag{8-128}$$

Thus, we have found a transformed equation which explicitly gives $V_R(s)$, i.e., the transform of the response variable $v_R(t)$, in terms of the transform of the excitation variable $v_S(t)$, and the initial conditions on the inductor current $i(0-)$ and the capacitor voltage $v_C(0-)$. Specifying these quantities will permit us to determine $V_R(s)$ explicitly, and, by taking the inverse transform, we are thus able to find $v_R(t)$. Note that in Eq. (8-128) we have written the expression for $V_R(s)$ in such a way as to separate the term involving $V_S(s)$ from the term involving the initial conditions. Thus, the inverse transform of the first term in the right member of Eq. (8-128) will give the *zero-state response* for the network, i.e., the response that results from an excitation which is applied when all the initial conditions are zero. The poles of this term come from two places: (1) the network equation [the zeros of the quantity $s^2 + s(R/L) + (1/LC)$]; and (2) the poles of the function $V_S(s)$. Thus, the resulting inverse transform will have terms which are the result of the element values of the network, as well as terms which are the result of the particular form of $v_S(t)$. The situation will be identical with that shown in the "Zero-State Response" column of the "Second Decomposition of Response" section of Table 6-4.1. Whether Case I, II, or III applies will depend on whether the zeros of $s^2 + s(R/L) + (1/LC)$ are real and simple, real and double, or complex, respectively.

Now let us consider the second term in the right member of Eq. (8-128). The inverse transform of this term will give the *zero-input response* for the network, i.e., the response that results from initial conditions when the excitation $v_S(t)$ is zero. Thus, the inverse transform of this term will have elements whose basic behavior is the result of the network elements, and whose magnitude and/or phase is the result of the relative magnitude of the two initial conditions. This situation will be identical with that shown in the "Zero-Input Response" column of Table 6-4.1.

As has been discussed above, the expression for $V_R(s)$ given in Eq. (8-128) has been written in such a form as to agree with the entries entitled "Second Decomposition of Response" in Table 6-4.1. It is also possible to put the transformed expression given in Eq. (8-128) in a form so that the resulting inverse transform has the general appearance given in the columns entitled "First Decomposition of Response" in Table 6-4.1. To do this, let us define $V_S(s) = N(s)/D(s)$. Equation (8-128) may now be written in the form

$$V_R(s) = \frac{as + b}{s^2 + s(R/L) + (1/LC)} + \frac{P(s)}{D(s)} \tag{8-129}$$

where the coefficients a and b, and the polynomial $P(s)$ are readily found from Eq. (8-128). The inverse transform of the first term in the right member of Eq. (8-129) will be the transient response, i.e., the complementary solution for the network. The inverse transform of the second term will be the forced response, i.e., the particular solution. Thus, these terms give the transformed equivalent of the entries in the first two columns of Table 6-4.1. A numerical example illustrating the principles described above follows.

Example 8-5.1

For the *RLC* network shown in Fig. 8-5.1, the two-terminal elements have the values $R = 1\ \Omega$, $C = 1$ F, and $L = 0.5$ H. It is desired to find the output voltage $v_R(t)$ under the conditions that a unit ramp $v_S(t) = tu(t)$ V is applied as an excitation. It is assumed that the initial conditions in the network are zero. Since the transform of $tu(t)$ is $1/s^2$, we may substitute the given element values directly in (8-128) to obtain an expression for $V_R(s)$. This is

$$V_R(s) = \frac{2}{s(s^2 + 2s + 2)} = \frac{K_1}{s} + \frac{as + b}{(s + 1)^2 + (1)^2}$$

where

$$K_1 = s \left[\frac{2}{s(s + 2s + 2)} \right]\Bigg|_{s=0} = 1$$

Now, applying the method of undetermined coefficients to find a and b, we obtain

$$\frac{2}{s(s^2 + 2s + 2)} = \frac{1(s^2 + 2s + 2)}{s(s^2 + 2s + 2)} + \frac{s(as + b)}{s(s^2 + 2s + 2)} = \frac{s^2(1 + a) + s(2 + b) + 2}{s(s^2 + 2s + 2)}$$

Equating the coefficients of like powers of s in the first and last members of the above equation we find $a = -1$ and $b = -2$. The expanded form for $V_R(s)$ can now be written

$$V_R(s) = \frac{1}{s} - \frac{s+2}{(s+1)^2 + (1)^2} = \frac{1}{s} - \frac{s+1}{(s+1)^2 + (1)^2} - \frac{1}{(s+1)^2 + (1)^2}$$

The inverse transform of each of the terms in the right member of the above equation is easily found using the methods of Sec. 8-3. Thus, we find that

$$v_R(t) = [1 - e^{-t}(\cos t + \sin t)]u(t) = [1 - \sqrt{2}e^{-t}\cos(t - 45°)]u(t)$$

is the desired response. The first term inside the brackets is readily seen to be the particular response, corresponding with the form of the particular response given as $A + Bt$ in Table 5-2.1. In this case, $B = 0$, since the zero at the origin of the network function [the term s in the numerator of the left term in the right member of (8-128)] has cancelled one of the poles at the origin of the transform $1/s^2$ of $v_S(t)$. The second term inside the brackets is the transient response, the form of which is due to the natural frequencies of the network.

The fact that the transformed expression for the response variable includes the initial conditions explicitly makes it easy not only to observe the effect of these initial conditions, but also to select them in such a way as to achieve some desired effect on the response. An example of such an application follows.

Example 8-5.2

For the *RLC* network and the excitation used in Example 8-5.1, it is desired to specify a set of non-zero initial conditions in such a way that the phase of the sinusoidal term in the response $v_R(t)$ is zero, i.e., in such a way that the inverse transform of the term representing the complex pole pair is of the form $Ke^{-t}\cos t$. To do this, using the results found in Example 8-5.1, we may write the general expression for the response given in (8-128) as

$$V_R(s) = \frac{1}{s} - \frac{s+2}{(s+1)^2 + (1)^2} + \frac{si(0-) - 2v_C(0-)}{(s+1)^2 + (1)^2}$$

It is readily observed that if we specify $i(0-) = 0$, and $v_C(0-) = -1/2$, then

$$V_R(s) = \frac{1}{s} - \frac{s+1}{(s+1)^2 + (1)^2}$$

Taking the inverse transform, we obtain

$$v_R(t) = (1 - e^{-t}\cos t)u(t)$$

as was desired.

The techniques illustrated above are readily applied to the solution of networks containing more than a single independent variable. As an example of this, consider the network shown in Fig. 8-5.2. In addition to the two independent sources specified as $i_1(t)$ and $i_2(t)$, this circuit also has a dependent VCIS connected to the second node

mhos, henries, farads

Figure 8-5.2 Circuit with a dependent source.

and controlled by the voltage at the first node. The time-domain equations for this circuit are easily written by applying KCL at the two nodes. Thus, we obtain

$$i_1(t) = \frac{dv_1}{dt} + 2v_1(t) - v_2(t) \qquad t \geq 0$$

$$3v_1(t) + i_2(t) = 4 \int_{0-}^{t} v_2(\tau)d\tau + i_L(0-) + 4v_2(t) - v_1(t) \qquad t \geq 0 \tag{8-130}$$

Combining the $v_1(t)$ terms in the left and right members of the second equation and transforming both of the equations, we obtain

$$I_1(s) = sV_1(s) - v_1(0-) + 2V_1(s) - V_2(s)$$

$$I_2(s) = \frac{4}{s} V_2(s) + \frac{i_L(0-)}{s} + 4V_2(s) - 4V_1(s) \tag{8-131}$$

Now let us write the equations in matrix form. If we move the initial condition terms to the left member at the same time, we obtain

$$\begin{bmatrix} I_1(s) + v_1(0-) \\ I_2(s) - \dfrac{i_L(0-)}{s} \end{bmatrix} = \begin{bmatrix} s+2 & -1 \\ -4 & \dfrac{4}{s}+4 \end{bmatrix} \begin{bmatrix} V_1(s) \\ V_2(s) \end{bmatrix} \tag{8-132}$$

To solve for such a set of equations, we need merely multiply both members of the equation by the inverse of the square matrix. The inverse is readily found using the techniques given in Sec. 3-1. Thus, we obtain

$$\begin{bmatrix} V_1(s) \\ V_2(s) \end{bmatrix} = \frac{s/4}{s^2 + 2s + 2} \begin{bmatrix} \dfrac{4}{s}+4 & 1 \\ 4 & s+2 \end{bmatrix} \begin{bmatrix} I_1(s) + v_1(0-) \\ I_2(s) - \dfrac{i_L(0-)}{s} \end{bmatrix} \tag{8-133}$$

Let us assume that we desire to find an expression for the transform of $V_2(s)$. From the matrix equation given above, we may write

$$V_2(s) = \frac{sI_1(s)}{s^2 + 2s + 2} + \frac{(s/4)(s+2)I_2(s)}{s^2 + 2s + 2} + \frac{sv_1(0-) - i_L(0-)(s+2)/4}{s^2 + 2s + 2} \tag{8-134}$$

Obviously, if desired, a similar expression can be written for $V_1(s)$. In the right member of the expression given above, we have indicated three separate terms. The first term is the transform of the zero-state response for the case in which only $i_1(t)$ is non-zero. The second term is the transform of the zero-state response for the case in which only $i_2(t)$ is non-zero. It is readily observed that superposition applies, namely, the transform of the zero-state response for the case for which both $i_1(t)$ and $i_2(t)$ are non-zero is simply the sum of these two terms. Finally, the third term in the right member of Eq. (8-134) is the transform of the zero-input response, i.e., the response determined only by initial conditions. An example of the use of Eq. (8-134) to find the output $v_2(t)$ follows.

Example 8-5.3

For the network shown in Fig. 8-5.2, it is desired to find $v_2(t)$ when excitation currents $i_1(t) = tu(t)$ and $i_2(t) = e^{-2t}u(t)$ A are applied. The initial conditions are specified as $v_1(0-) = \frac{1}{4}$ V and $i_L(0-) = 0$. The transforms of $i_1(t)$ and $i_2(t)$ are $1/s^2$ and $1/(s+2)$, respectively. Inserting these values in (8-134), we obtain

$$V_2(s) = \frac{1/s}{s^2 + 2s + 2} + \frac{(s/4)(s + 2)/(s + 2)}{s^2 + 2s + 2} + \frac{s/4}{s^2 + 2s + 2}$$

Combining terms and expressing the result in the form of a partial-fraction expansion, we obtain

$$V_2(s) = \frac{(s^2/2) + 1}{s(s^2 + 2s + 2)} = \frac{K_1}{s} + \frac{as + b}{s^2 + 2s + 2}$$

where

$$K_1 = s \left[\frac{(s^2/2) + 1}{s(s^2 + 2s + 2)} \right]\bigg|_{s=0} = \frac{1}{2}$$

Using undetermined coefficients we obtain

$$\frac{(s^2/2) + 1}{s(s^2 + 2s + 2)} = \frac{\frac{1}{2}(s^2 + 2s + 2)}{s(s^2 + 2s + 2)} + \frac{s(as + b)}{s(s^2 + 2s + 2)} = \frac{s^2(\frac{1}{2} + a) + s(1 + b) + 1}{s(s^2 + 2s + 2)}$$

Equating coefficients of the numerators of the first and last members we find $a = 0$ and $b = -1$. Thus we may write

$$V_2(s) = \frac{\frac{1}{2}}{s} - \frac{1}{(s + 1)^2 + (1)^2}$$

Taking the inverse transform, we find that

$$v_2(t) = [\tfrac{1}{2} - e^{-t} \sin t]u(t)$$

The developments given above are quite general, in the sense that they may be used to determine the *complete* response (i.e., the sum of the zero-state response and the zero-input response) of an arbitrary network. In many network situations, however, the initial conditions may be specified as zero. In such a situation, it is only necessary to find the zero-state response. In this case, the determination of the transformed network equations can be considerably simplified. To see this, let us consider the transformed equation for the terminal variables of a capacitor for the case where the initial condition $v(0-)$ is zero and the excitation is assumed to be a voltage source. Using the differential equation $i(t) = C \, dv/dt$, the transformed relation may be written in the form

$$\frac{I(s)}{V(s)} = \frac{\text{Response}}{\text{Excitation}} = sC \tag{8-135}$$

In a manner similar to that used when we considered phasors in Sec. 7-7, we may define an expression of this type which relates a transformed response variable to a transformed excitation variable as a *network function*. In the case given in Eq. (8-135), since the network function is the ratio of a transformed current variable to a transformed voltage variable, and since both variables are defined at the same pair of terminals, the expression sC is called the *driving-point admittance*, or, more simply, the *admittance* of a capacitor of value C farads. Note that if we let $s = j\omega$, then we obtain the expression for the admittance of a capacitor as used in phasor analysis in Sec. 7-5.

Now consider a capacitor excited by a current source with the variables related by the integral equation $v(t) = (1/C) \int i(t) \, dt + v(0-)$. If we again consider the case where the initial condition is zero, then the transform of this equation may be written in the form

$$\frac{V(s)}{I(s)} = \frac{\text{Response}}{\text{Excitation}} = \frac{1}{sC} \qquad (8\text{-}136)$$

Since this network function is the ratio of a transformed voltage variable to a transformed current variable, it defines the *driving-point impedance* of a capacitor. It is readily observed from Eqs. (8-135) and (8-136) that driving-point impedances and admittances are reciprocal quantities. Just as was the case for the sinusoidal steady state, we may use the term *immittance* to collectively refer to impedance and admittance.[14]

The immittance of other network elements is easily determined by transforming the time-domain equations relating their terminal behavior, and by assuming zero initial conditions. For example, the driving-point impedance of an inductor is sL; correspondingly, its admittance is $1/sL$. It is easily shown also that the impedance of a resistor is R, and its admittance is $1/R$. Note that there is a significant difference between immittances as defined in the complex frequency domain in this chapter and those defined for sinusoidal steady-state conditions in Chap. 7. The immittances defined here may be evaluated for any value of the complex frequency variable s. Thus, the impedance of an inductance, which is sL, will be real when s takes on real values, imaginary when s takes on imaginary values, and complex when s takes on complex values. However, the impedance of an inductor under sinusoidal steady-state conditions, is $j\omega L$, which is always imaginary for all values of ω, since ω is always real.

All the terminology derived with respect to network functions for sinusoidal steady-state conditions in Sec. 7-7 is directly applicable to the network functions which have been defined in this section as a ratio of transformed variables. Thus, in addition to the driving-point immittances introduced above, we may consider *transfer immittances* and *dimensionless transfer ratios*. Table 7-7.1 may thus be used as a basis for defining network functions in the complex frequency variable as well as a basis for defining network functions under sinusoidal steady-state conditions. This conclusion may be summarized as follows:

SUMMARY 8-5.1

Network Functions for Transformed Network Variables. Let $E(s)$ be the transform of some excitation variable $e(t)$ for a given network. Similarly, let $R(s)$ be the transform of some response variable $r(t)$. If we transform the integro-differential equation relating $e(t)$ and $r(t)$, and if all the initial conditions are set to zero, then the remaining equation may be used to define a network function $N(s)$ relating the transformed response variable to the transformed excitation variable. Specifically,

---

[14]The terms *network function, impedance, admittance,* etc. used in this section are identical with those used for the sinusoidal steady-state case in Sec. 7-7. Such a usage follows common practice, the differentiation between the two usages being provided by the functional notation (s) or $(j\omega)$.

$N(s) = R(s)/E(s)$. Such a network function may be a driving-point immittance, a transfer immittance, or a dimensionless transfer ratio.　■

The concept of a network function developed in the preceding discussion applies equally well to circuits in which mutual inductance and/or active elements are present. Such cases are illustrated in the following examples.

Example 8-5.4

It is desired to find the network function relating the transformed excitation voltage $V_1(s) = \mathcal{L}[v_1(t)]$ and the transformed output voltage $V_2(s) = \mathcal{L}[v_2(t)]$ in the network shown in Fig. 8-5.3 .The time-domain loop equations for this circuit are

Figure 8-5.3 Network for Example 8-5.4.

$$v_1(t) = R_1 i_1(t) + L_1 \frac{di_1}{dt} + M \frac{di_2}{dt}$$

$$0 = M \frac{di_1}{dt} + L_2 \frac{di_2}{dt} + R_2 i_2(t)$$

The Laplace transformation of these relations yields the following matrix equation:

$$\begin{bmatrix} V_1(s) \\ 0 \end{bmatrix} = \begin{bmatrix} R_1 + sL_1 & sM \\ sM & R_2 + sL_2 \end{bmatrix} \begin{bmatrix} I_1(s) \\ I_2(s) \end{bmatrix} - \begin{bmatrix} L_1 i_1(0-) + M i_2(0-) \\ L_2 i_2(0-) + M i_1(0-) \end{bmatrix}$$

The network function $V_2(s)/V_1(s)$ is readily found by setting the initial conditions to zero, inverting the square matrix, solving for $I_2(s)$, and applying the relation $V_2(s) = R_2 I_2(s)$. The details are left as an exercise for the reader.

Example 8-5.5

It is desired to find the network transfer function relating the transformed excitation voltage $V_1(s)$ and the transformed output voltage $V_2(s)$ in the network shown in Fig. 7-9.6 in which an active element modeled by an ideal VCVS is used. Assuming a normalized design in which $R_1 = R_3 = C_2 = C_4 = 1$, the voltage transfer function is readily shown to be [see Example 6-5.1 and also Eq. (7-200)]

$$\frac{V_2(s)}{V_1(s)} = \frac{K}{s^2 + (3 - K)s + 1}$$

This is the desired voltage transfer function.

There is a very close tie between the network functions defined in this chapter and those defined in Chap. 7. In this chapter, we have seen in Summary 8-2.3 that, when initial conditions are not considered, network functions which relate transformed network variables may be found from the original equations by replacing differentiation by multiplication by s and integration by division by s. These, however, are the identical replacements used in Chap. 7. We may conclude that the network functions (of $j\omega$) determined for sinusoidal steady-state or phasor analysis and the network functions (of s) determined in this chapter for transformed variables are identical if we set $j\omega = s$.

As an example of this, consider the series RLC network shown in Fig. 8-5.1. From Eq. (8-128), for the zero-state case we may write

$$\frac{V_R(s)}{V_S(s)} = \frac{\text{Response}}{\text{Excitation}} = \frac{sRC}{s^2LC + sRC + 1} \qquad (8\text{-}137)$$

If we substitute $s = j\omega$ in the above equation, we obtain

$$\left.\frac{V_R(s)}{V_S(s)}\right|_{s=j\omega} = \frac{j\omega RC}{-\omega^2 LC + j\omega RC + 1} \qquad (8\text{-}138)$$

The right member of this expression is identical with the one given in Eq. (7-143). Thus, we see that by letting $s = j\omega$ in a network function in the complex frequency variable s, we obtain the network function which relates excitation and response phasors for the sinusoidal steady-state case. We may conclude that, if we know the network function which relates a transformed response variable to a transformed excitation variable (and which may thus be used to determine the zero-state response for any excitation), letting $s = j\omega$ yields the network function relating the corresponding response phasor to the excitation phasor (which may thus be used to determine the particular solution for sinusoidal steady-state conditions). Note that the phasor method only determines the *particular solution component* of the zero-state response, and not the transient component. The transform network function method, however, determines *both* components of the zero-state response, namely, the particular solution *and the transient solution*. Furthermore it accomplishes this for any input excitation function, not merely for the sinusoidal case. These differences may be observed in the following example.

Example 8-5.6

The excitation $v_S(t) = \sqrt{2}\,\cos{(2t + 45°)}u(t) = [\cos 2t - \sin 2t]u(t)$ V is applied to the simple RLC network shown in Fig. 8-5.1 and used in Example 8-5.1. If $v_R(t)$ is considered as the response variable, then the transfer function for the transformed variables is

$$\frac{V_R(s)}{V_S(s)} = \frac{2s}{s^2 + 2s + 2}$$

To find the sinusoidal steady-state response using phasors, we may defined $\mathcal{V}_R$ as the response phasor and $\mathcal{V}_S$ as the excitation phasor. Substituting $s = j\omega$ in the network function given above, we obtain

$$\frac{\mathcal{V}_R}{\mathcal{V}_S} = \frac{j2\omega}{2 - \omega^2 + j2\omega}$$

The angular frequency of the sinusoidal excitation, however, is 2 rad/s. Substituting $\omega = 2$ in the preceding relation, we obtain

$$\frac{\mathcal{V}_R}{\mathcal{V}_S} = \frac{j4}{-2 + j4} = \frac{4 - j2}{5}$$

The phasor representation for $v_S(t)$ is $1 + j1$. Inserting this value in the last relation, we find that the phasor representation for $\mathcal{V}_R$ is $(6 + j2)/5$. Thus, the sinusoidal steady-state component of $v_R(t)$ is $\frac{6}{5} \cos 2t - \frac{2}{5} \sin 2t$. Now let us compare this with the result obtained using transform techniques. Using Table 8-3.1, we find that the transform of $v_S(t)$ is

$$V_S(s) = \frac{s}{s^2 + 4} - \frac{2}{s^2 + 4} = \frac{s - 2}{s^2 + 4}$$

Thus, using the transfer function given at the beginning of this example, we find that the transform for $V_R(s)$ is

$$V_R(s) = \frac{2s}{s^2 + 2s + 2} \frac{s - 2}{s^2 + 4}$$

The inverse transform of this expression can be easily found if it is put in the form

$$V_R(s) = \frac{as + b}{s^2 + 2s + 2} + \frac{cs + d}{s^2 + 4}$$

Using undetermined coefficients we may write

$$\frac{2s(s - 2)}{(s^2 + 2s + 2)(s^2 + 4)} = \frac{(as + b)(s^2 + 4) + (cs + d)(s^2 + 2s + 2)}{(s^2 + 2s + 2)(s^2 + 4)}$$

Therefore, equating numerators and rearranging terms,

$$2s^2 - 4s = s^3(a + c) + s^2(b + 2c + 2d) + s(4a + 2c + 2d) + (4b + 2d)$$

Since the s^3 and the s^0 coefficients are zero in the left member of the above equation, from the right member we have $a = -c$ and $d = -2b$. Inserting these relations into the s^2 and s^1 coefficients we obtain

$$2 = b + 2c + d = b + 2c - 2b = -b + 2c$$

$$-4 = 4a + 2c + 2d = -4c + 2c - 4b = -4b - 2c$$

Solving these simultaneous equations we obtain $b = \frac{2}{5}$ and $c = \frac{6}{5}$. Thus, from the above relations, $a = -\frac{6}{5}$ and $d = -\frac{4}{5}$. Using these values, the expression for $V_R(s)$ may be written

$$V_R(s) = \frac{-\frac{6}{5}s + \frac{2}{5}}{s^2 + 2s + 2} + \frac{\frac{6}{5}s - \frac{4}{5}}{s^2 + 4}$$

$$= \frac{-\frac{6}{5}(s + 1)}{(s + 1)^2 + (1)^2} + \frac{\frac{8}{5}}{(s + 1)^2 + (1)^2} + \frac{\frac{6}{5}s}{s^2 + 4} - \frac{\frac{2}{5}2}{s^2 + 4}$$

where, in the last expression, the terms have been decomposed in such a way as to facilitate taking the inverse transform. From the entries in Table 8-3.1, we readily find that the inverse transform is

$$V_R(t) = \{e^{-t}[-\frac{6}{5} \cos t + \frac{8}{5} \sin t] + \frac{6}{5} \cos 2t - \frac{2}{5} \sin 2t\}u(t)$$

The last two terms in the right member of the above equation are, of course, identical with the sinusoidal steady-state solution found earlier in this example by the use of phasor methods. Thus, they represent the particular solution component of the zero-state response. The first two terms of the right member of the expression for $v_R(t)$, however, cannot be found by the phasor method, since they represent the transient component of the zero-state solution. Thus, we see that the transformed network function provides us with a means of readily finding the entire zero-state response of a network, namely, both the particular solution and the transient response.

The results given above are readily summarized as follows:

The Relation Between Network Functions Used in Sinusoidal Steady-State Analysis and Network Functions Used in Transform Analysis.   Consider a network in which a response variable $r(t)$ and an excitation variable $e(t)$ have been defined. Let $\Re$ and $\mathcal{E}$ be the phasor representations for these variables, and let $R(s)$ and $E(s)$ be the Laplace transforms of these variables. If we define the network functions $N_1(j\omega)$ and $N_2(s)$ by the relations

$$N_1(j\omega) = \frac{\Re}{\mathcal{E}} \qquad N_2(s) = \frac{R(s)}{E(s)}$$

respectively, then $N_1(j\omega)$ and $N_2(s)$ are related as follows

$$N_1(j\omega)|_{j\omega=s} = N_2(s) \qquad N_2(s)|_{s=j\omega} = N_1(j\omega)$$

Thus, we see that the functions N_1 and N_2 are identical.   ■

The concepts of impedance and admittance, as developed for phasors, are identical with those developed for transformed variables. Therefore, all the manipulations made with respect to impedances and admittances for networks under conditions of sinusoidal steady-state excitations (such as the addition of impedances for elements connected in series and the addition of admittances for network elements connected in parallel) apply to networks described by transformed equations. In the next section we shall see how to apply these rules to directly write the transformed equations and determine any desired network functions.

In this and the preceding sections of this chapter we have developed a powerful network analysis tool, the Laplace transformation. This transformation defines a complex frequency domain in which the integral and differential operations which relate the actual time-domain network variables are replaced by the algebraic operations of division and multiplication. Note that, for the same network, two different sets of initial condition terms are possible. On the one hand, if we transform the time-domain equations relating the terminal variables of the network elements and then apply Kirchhoff's laws, we obtain frequency-domain equations *with the initial conditions specified in terms of the terminal variables of the network elements.* On the other hand, if we first apply Kirchhoff's laws in the time domain to obtain a set of integro-differential equations, and then apply the Laplace transformation, we obtain frequency-domain equations *in which the initial conditions are specified in terms of the node voltage (or mesh current) variables.* In either case, the sets of frequency domain equations may be solved for some desired response variable and any of the various time-domain response components may be obtained.

8-6 MESH AND NODE EQUATIONS FOR TRANSFORMED NETWORKS

In the preceding section we showed how to apply transform methods to find the solution of networks by transforming the time-domain equations describing the network variables. Starting from these time-domain equations, we showed that if the initial conditions were specified as zero, we could then define network functions specifying a ratio of transformed network variables. Frequently, it is more convenient to determine the network functions directly from the network itself, rather than from the time-domain equations. In order to do this, in this section we will develop the concept of a *transformed network*. Such a concept is one of the most important and useful results of Laplace transformation theory.

To begin our study of transformed networks, consider the series connection of two-terminal network elements shown in Fig. 8-6.1(a). For this connection, applying

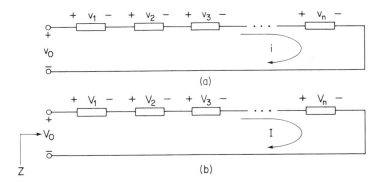

Figure 8-6.1 Series connection of two-terminal network elements.

KVL, we may write

$$v_0(t) = v_1(t) + v_2(t) + v_3(t) + \cdots + v_n(t) \qquad (8\text{-}139)$$

If we transform both sides of Eq. (8-139), we obtain

$$V_0(s) = V_1(s) + V_2(s) + V_3(s) + \cdots + V_n(s) \qquad (8\text{-}140)$$

where $V_i(s) = \mathcal{L}[v_i(t)]$. We conclude that *transformed voltage variables satisfy KVL*. In a similar fashion, we may show that *transformed current variables satisfy KCL*. These conclusions are similar to those given in Summary 7-5.1 with respect to phasors. Thus, we may redraw the circuit of Fig. 8-6.1(a) as shown in Fig. 8-6.1(b), where the reference polarities of the transformed voltage variables $V_i(s)$ follow those of the original variables $v_i(t)$. Now let $I(s)$ be the transform of $i(t)$. *If we assume that the initial conditions are zero*, we may express each of the transformed voltages in terms of the transformed current $I(s)$ and the individual impedances $Z_i(s)$ by the relations $V_i(s) = Z_i(s)I(s)$. Substituting these relations in Eq. (8-140) and rearranging terms, we obtain

$$V_0(s) = [Z_1(s) + Z_2(s) + Z_3(s) + \cdots + Z_n(s)]I(s) \qquad (8\text{-}141)$$

Since the network function defined as $V_0(s)/I(s)$ is the driving-point impedance of the entire circuit, we conclude that *the impedance of a series connection of impedances is the sum of the individual impedances*. In a similar fashion, we may show that KCL is satisfied by a set of transformed current variables, and that *the admittance of a parallel connection of admittances is equal to the sum of the individual admittances*. These conclusions are similar to those reached with respect to phasor immittances in Sec. 7.5. Considering the results given above, we see that we may define a *transformed network*, in which all the variables are the transforms of the original time-domain network variables, and in which the elements are defined in terms of their immittances. In such a transformed network, all the usual rules followed for forming network equations may be applied directly to the elements of the network. As an example of this, consider the network with time-domain variables shown in Fig. 8-6.2(a) and the corresponding transformed network shown in Fig. 8-6.2(b). Applying the voltage-divider law directly to the impedances of the transformed network, we may define the network function

$$\frac{V_L(s)}{V_s(s)} = \frac{sL}{R + (1/sC) + sL} \tag{8-142}$$

(a)   (b)   **Figure 8-6.2**   Voltage-divider network.

Thus, using the techniques developed in the preceding section, we may directly find the zero-state response for $v_L(t)$ for any arbitrary excitation $v_s(t)$ by inserting the transformed expression for the latter in Eq. (8-142) and finding the inverse transform for $V_L(s)$. These conclusions may be summarized as follows:

SUMMARY 8-6.1

The Transformed Network.   If a network configuration is specified in the time domain, and if it is in the zero state, a transformed network may be defined in the complex-frequency domain by replacing all time-domain variables by their equivalent transformed variables and by specifying all the network elements in terms of their immittances. KVL and KCL may be directly applied to the transformed network variables. Thus, elements in series have their impedances add, and elements connected in parallel have their admittances add.   ■

Just as was true for the phasor case, it should be noted that when specifying the element values for the transformed network, some identification must be made to indicate whether impedances or admittances are being specified.

The transformed network may be used to actually find the integro-differential equation relating a response and an excitation variable in a given network. To illustrate this approach, consider the network shown in Fig. 8-6.2(b). Let us assume that it is desired to find the integro-differential equation relating $v_L(t)$ and $v_s(t)$. To begin, let us put the network function relating $V_L(s)$ and $V_s(s)$ given in (8-142) in the following form.

$$sLV_s(s) = \left[R + \frac{1}{sC} + sL\right]V_L(s) \tag{8-143}$$

Since multiplication by s is the transformed equivalent of differentiation, and since division by s is similarly related to integration, the original integro-differential equation must have the form

$$L\frac{d}{dt}v_S(t) = Rv_L(t) + \frac{1}{C}\int_{0-}^{t} v_L(\tau)\,d\tau + L\frac{d}{dt}v_L(t) \qquad (8\text{-}144)$$

In general, it is easier to apply this method than it is to derive the integro-differential equation directly in the time domain. An example follows.

Example 8-6.1

It is desired to determine the integro-differential equation relating $v(t)$ and $i_1(t)$ in the network shown in Fig. 8-6.3(a). To do this, we may first draw the transformed network as shown in Fig. 8-6.3(b). Applying KVL around the two meshes, we obtain

$$V(s) = (2 + 3s)I_1(s) - 3sI_2(s)$$

$$0 = -3sI_1(s) + \left(4s + \frac{2}{s}\right)I_2(s)$$

ohms, henries, farads

impedances

Figure 8-6.3 Network for Example 8-6.1.

In matrix form, these equations may be written

$$\begin{bmatrix} V(s) \\ 0 \end{bmatrix} = \begin{bmatrix} 2 + 3s & -3s \\ -3s & 4s + \dfrac{2}{s} \end{bmatrix} \begin{bmatrix} I_1(s) \\ I_2(s) \end{bmatrix}$$

Solving these equations using the method of Sec. 3-1, we obtain

$$\begin{bmatrix} I_1(s) \\ I_2(s) \end{bmatrix} = \frac{s}{3s^3 + 8s^2 + 6s + 4} \begin{bmatrix} 4s + \dfrac{2}{s} & 3s \\ 3s & 2 + 3s \end{bmatrix} \begin{bmatrix} V(s) \\ 0 \end{bmatrix}$$

Thus, the network function relating $I_1(s)$ and $V(s)$ is

$$\frac{I_1(s)}{V(s)} = \frac{4s + 2}{3s^3 + 8s^2 + 6s + 4}$$

The original differential equation is easily seen to be

$$4v''(t) + 2v(t) = 3i_1'''(t) + 8i_1''(t) + 6i_1'(t) + 4i_1(t)$$

where the primes indicate differentiation with respect to t. To appreciate the simplicity and generality of the transformed network approach, the reader should compare the process of finding the differential equation using the transformed network in the preceding example with the process of solving the original time-domain network for the desired differential equation.

Finding the Integro-Differential Equations for a Time-Domain Network from the Network Function. If $R(s)$ and $E(s)$ are the transforms of a response variable $r(t)$ and an excitation variable $e(t)$ respectively, and, for the zero-state case, are related by the network function

$$\frac{R(s)}{E(s)} = \frac{a_0 + a_1 s + a_2 s^2 + \cdots}{b_0 + b_1 s + b_2 s^2 + \cdots}$$

where the a_i and the b_i are real coefficients, then the differential equation relating $r(t)$ and $e(t)$ is

$$a_0 e(t) + a_1 e'(t) + a_2 e''(t) + \cdots = b_0 r(t) + b_1 r'(t) + b_2 r''(t) + \cdots$$

where the primes indicate differentiation with respect to t. ∎

The basic techniques which were introduced in Chap. 3 for writing mesh and node equations for resistance networks, and which were used in Chap. 7 to write similar equations for complex immittances, may be directly applied to determine the network equations for a transformed network. Thus, in general, the transformed *mesh impedance equations* will have the form

$$\mathbf{V}(s) = \mathbf{Z}(s)\mathbf{I}(s) \tag{8-145}$$

where the rules for finding the elements of the component matrices are summarized as follows:

Rules for Forming the Mesh Impedance Equations for a Transformed Network. The elements of the matrices of Eq. (8-145) defining the mesh impedance equations for a transformed network may be found as follows:

$V_i(s)$ the sum of the transforms of all the excitation voltage sources which are present in the ith mesh, with sources whose reference polarity represents a rise when traversing the mesh in the positive reference current direction treated as positive

$z_{ij}(s)$ $(i = j)$ the sum of the transformed impedances which are present in the ith mesh

$z_{ij}(s)$ $(i \neq j)$ the negative of the sum of all the transformed impedances which are common to the ith and jth meshes

$I_i(s)$ the transform of the ith mesh current

These rules assume that all the mesh currents have the same reference direction, and that no element is traversed by more than two mesh currents. ∎

An example of the application of these rules may be found in the matrix expression given in Example 8-6.1.

A similar development may be easily made to show that node equations may be written by inspection for the transformed network. Such a set of transformed *node admittance equations* has the general form

$$\mathbf{I}(s) = \mathbf{Y}(s)\mathbf{V}(s) \qquad (8\text{-}146)$$

where the elements of the component matrices may be found by the following rules:

SUMMARY 8-6.4

Rules for Forming the Node Admittance Equations for a Transformed Network. The elements of the matrices of (8-146) defining the node admittance equations for a transformed network may be found as follows:

$I_i(s)$ — the sum of the transforms of all the excitation current sources which are connected to the ith node, with sources whose reference polarity is into the node treated as positive

$y_{ij}(s)$ $(i = j)$ — the sum of all the transformed admittances which are connected to the ith node

$y_{ij}(s)$ $(i \neq j)$ — the negative of the sum of any transformed admittances which are connected between nodes i and j

$V_i(s)$ — the transform of the ith node voltage

These rules assume that the negative reference polarity of all the transformed node voltages are located at a common ground terminal. ■

An example of the application of these rules to directly determine the node equations follows.

Example 8-6.2

It is desired to write the node admittance equations for the network shown in Fig. 8-6.4. Note that this network contains a VCIS, i.e., a dependent voltage-controlled current source as well as two independent current sources. For convenience, assume

Figure 8-6.4 Network for Example 8-6.2.

that all the elements have unity value. Following the rules given above, we obtain

$$
\begin{bmatrix} I_1(s) \\ I_2(s) \\ 3V_2(s) \\ -I_2(s) \end{bmatrix} = \begin{bmatrix} 1+\dfrac{1}{s} & -\dfrac{1}{s} & 0 & 0 \\ -\dfrac{1}{s} & 2s+1+\dfrac{1}{s} & -s & 0 \\ 0 & -s & s+2 & -1 \\ 0 & 0 & -1 & s+1 \end{bmatrix} \begin{bmatrix} V_1(s) \\ V_2(s) \\ V_3(s) \\ V_4(s) \end{bmatrix}
$$

Obviously, the term $3V_2(s)$ in the left member of the equation, which is due to the presence of the VCIS, can be moved into the square matrix, thus changing the "32" element from $-s$ to $-(s+3)$.

The techniques which have been presented in this section for using transformed networks to directly determine network functions, and thus to permit the solution for the zero-state response of time-domain network variables, are easily adapted to include the case in which it is desired to retain the simplicity of the transformed network concept but also to include the effects of initial conditions. As an introduction to the procedure for accomplishing this, consider the network consisting of an uncharged capacitor and a shunt current source as shown in Fig. 8-6.5. The relation

Figure 8-6.5 Capacitor excited by a current source.

for the terminal variables is

$$V(s) = \frac{1}{sC}I(s) \tag{8-147}$$

Now let the current source provide an excitation which is an impulse function of strength KC, i.e., let $i(t) = KC\delta(t)$. Then $I(s) = KC$, and $V(s) = K/s$. Thus, at $t = 0+$, the capacitor voltage is K volts, and we have established an initial condition on the originally uncharged capacitor. A similar procedure may be used to establish an initial condition on an inductor. Consider the series connection of an uncharged inductor and a voltage source shown in Fig. 8-6.6. The relation for the terminal

Figure 8-6.6 Inductor excited by a voltage source.

variables is

$$I(s) = \frac{1}{sL}V(s) \tag{8-148}$$

If the voltage source now provides an impulse of strength KL at $t = 0$, then we obtain $I(s) = K/s$. Thus, at $t = 0+$, the inductor current has the value K A, and an initial condition has been established through the use of an external source. The use of impulse generators to establish initial conditions described above is further illustrated by the following example.

Example 8-6.3

For the transformed network shown in Fig. 8-6.7(a), it is desired to use the concept of a transformed network to find the transformed response for $I_2(s)$ for the case in which the initial conditions on the inductor currents $i_1(0-)$ and $i_2(0-)$ are not zero. The first

(a)   (b)

Figure 8-6.7 Network for Example 8-6.3.

step is to redraw the network inserting voltage sources in series with the inductors as shown in Fig. 8-6.7(b). As indicated in the figure, the transformed output of the voltage source which has been placed in series with the 3-H inductor in the first mesh in $3K_1$ (an impulse of strength $3K_1$), while the transformed output of the voltage source which has been placed in series with the $\frac{1}{2}$-H inductor in the second mesh is $K_2/2$. The mesh impedance equations for the network may now be written by applying KVL. Thus, we obtain

$$\begin{bmatrix} V_1(s) + 3K_1 \\ \dfrac{K_2}{2} \end{bmatrix} = \begin{bmatrix} 3s + 1 & -1 \\ -1 & \dfrac{s}{2} + 1 \end{bmatrix} \begin{bmatrix} I_1(s) \\ I_2(s) \end{bmatrix}$$

Before solving these equations, let us first write the time-domain equations for the original network shown in Fig. 8-6.7(a). Applying KVL, we obtain

$$v_1(t) = 3\frac{di_1}{dt} + i_1(t) - i_2(t)$$

$$0 = \frac{1}{2}\frac{di_2}{dt} + i_2(t) - i(t_1)$$

Transforming these equations, we see that we may write

$$V_1(s) = 3sI_1(s) - 3i_1(0-) + I_1(s) - I_2(s)$$
$$0 = \tfrac{1}{2}sI_2(s) - \tfrac{1}{2}i_2(0-) + I_2(s) - I_1(s)$$

Rearranging the terms and putting the equations in matrix form, we obtain

$$\begin{bmatrix} V_1(s) + 3i_1(0-) \\ \dfrac{i_3(0-)}{2} \end{bmatrix} = \begin{bmatrix} 3s + 1 & -1 \\ -1 & \dfrac{s}{2} + 1 \end{bmatrix} \begin{bmatrix} I_1(s) \\ I_2(s) \end{bmatrix}$$

Comparing these equations with those derived directly from the transformed network with the impulse generators shown in Fig. 8-6.7(b), we conclude that if we set K_1 and K_2 equal to the desired initial conditions $i_1(0-)$ and $i_2(0-)$, respectively, then the equations are identical for all values of $t > 0$. Thus, either set of equations may be solved to give an expression for the transformed current $I_2(s)$. The result is

$$I_2(s) = \frac{2V_1(s) + 6K_1 + K_2(3s + 1)}{3s^2 + 7s} = \frac{2V_1(s) + 6i_1(0-) + i_2(0-)(3s + 1)}{3s^2 + 7s}$$

Using Impulses to Establish Initial Conditions in a Transformed Network. Non-zero initial conditions may be established in a transformed network by shunting capacitors with current sources and by placing voltage sources in series with inductors. If the outputs of these sources are all impulses occurring at $t = 0$, then their only effect is to establish initial conditions on the network elements, and they have no effect on the network at any other time. ∎

A second method for establishing initial conditions in the capacitors and inductors of a transformed network may be developed from a consideration of the transform of the integral relations for the terminal variables of these elements. For example, for capacitors, we may write

$$V(s) = \frac{1}{sC} I(s) + \frac{v(0-)}{s} \tag{8-149}$$

Thus, a voltage source in series with an uncharged capacitor may be used as an equivalent circuit for a capacitor with a non-zero initial condition of voltage. The output of the voltage source must be a step function of magnitude $v(0-)$ starting at $t = 0$. Similarly, for an inductor, the transform of the integral expression for the terminal variables is

$$I(s) = \frac{1}{sL} V(s) + \frac{i(0-)}{s} \tag{8-150}$$

Thus, a current source in shunt with an uncharged inductor may be used as an equivalent circuit for an inductor with a non-zero initial condition of current. The output of the current source must be a step function of magnitude $i(0-)$ starting at $t = 0$. It is important to observe the difference between the approach based on the use of transformed integral relations described above and the approach given in Summary 8-6.5. The method described in the summary was based on the use of impulses. Thus, the only function of the added sources was to establish the initial conditions on the uncharged network elements. For all other values of time, the sources had no effect on the circuit; thus, they could be ignored. In the method described in this paragraph, however, the uncharged element *remains initially uncharged*. Thus, the effect of the sources must be considered for all values of time. This approach may be summarized as follows:

SUMMARY 8-6.6

Using Step Functions to Establish Initial Conditions in a Transformed Network. Non-zero initial conditions may be established in a transformed network by connecting voltage sources in series with capacitors and by shunting inductors with current sources. If the output of these sources are step functions starting at $t = 0$, then each combination of a source and an uncharged reactive element is equivalent to a reactive element which has a non-zero initial condition. ∎

In deciding which of the above methods to apply in analyzing a given network, several factors must be taken into consideration. For example, if there is already some independent excitation voltage source in series with a reactive element, then the simplest approach is usually to use another series voltage source (with an impulse output for an inductor and a step output for a capacitor) to establish the initial condition. Similarly, if there is already an independent excitation current source in shunt with a reactive element, it is usually most convenient to use an additional shunt current source (with an impulse output for a capacitor and a step output for an inductor) to establish the initial condition.

As an example of the use of step and impulse functions to establish initial conditions, consider the following.

Example 8-6.4

The network shown in Fig. 8-6.8(a) is at rest with the switch S1 in the open position. The current that flows through the first of the coupled set of coils (after all the transients have disappeared) is 5 A. Since the energy associated with this current is stored in the magnetic field, and since this magnetic field links both windings, it is more meaningful to use a T equivalent circuit for the set of coupled coils (see Sec. 4-6). With such a modification, the circuit may be redrawn as shown in Fig. 8-6.8(b). In this circuit, two of the inductors in the T have an initial current and one does not. We may now add two shunt current sources each with a step output current of 5 A starting at $t = 0$ to the network as shown in Fig. 8-6.8(c). With such a representation, all three of the inductors may be treated as being uncharged. Thus, the current sources provide the initial conditions. If the switch $S1$ is closed at $t = 0$, the transformed network representation for $t > 0$ is as shown in Fig. 8-6.8(d). In this figure the initial conditions on all three of the inductors are zero. An alternative transformed representation using impulse generators is shown in Fig. 8-6.8(e). It is left as an exercise to the reader to verify that the network equations which result from the transformed networks shown in Fig. 8-6.8(d) and (e) both have the same solution.

The various cases of generating initial conditions which are most conveniently applied to the mesh and node analysis situations are summarized in Table 8-6.1. It should be noted that the two types of excitation used to generate initial conditions

TABLE 8-6.1 Generating Initial Conditions in a Transformed Network

Type of Analysis	Type of Element	
	Capacitor	Inductor
Mesh	Series voltage source with step output $v(0)u(t)$	Series voltage source with impulse output $i(0)L\delta(t)$
Node	Shunt current source with impulse output $v(0)C\delta(t)$	Shunt current source with step output $i(0)u(t)$

Figure 8-6.8. Network for Example 8-6.4.

for either an inductor or a capacitor are easily related to each other by the use of Thevenin and/or Norton equivalent circuits. The proof of this is left to the reader as an exercise. In the next section we shall see that the use of voltage sources in series with inductors and current sources in shunt with capacitors to establish initial conditions is an example of a more general principle which makes it possible to add sources to networks in such a way that the natural frequencies of the network are unchanged.

8-7 NATURAL FREQUENCIES OF NETWORKS

In the preceding sections of this chapter we defined the concept of a transformed network, and (in Summaries 8-6.3 and 8-6.4) we showed how the mesh impedance and node admittance equations could be directly written from the transformed network. The solution of such sets of equations follows the general rules for the solution of a set of simultaneous equations given in Sec. 3-3. For example, consider the mesh impedance equations given in (8-145). In expanded form, such a set of equations may be written

$$
\begin{bmatrix} V_1(s) \\ V_2(s) \\ \cdot \\ \cdot \\ \cdot \\ V_n(s) \end{bmatrix} = \begin{bmatrix} z_{11}(s) & z_{12}(s) & \cdots & z_{1n}(s) \\ z_{21}(s) & z_{22}(s) & \cdots & z_{2n}(s) \\ \cdot & \cdot & \cdots & \cdot \\ \cdot & \cdot & \cdots & \cdot \\ \cdot & \cdot & \cdots & \cdot \\ z_{n1}(s) & z_{n2}(s) & \cdots & z_{nn}(s) \end{bmatrix} \begin{bmatrix} I_1(s) \\ I_2(s) \\ \cdot \\ \cdot \\ \cdot \\ I_n(s) \end{bmatrix} \tag{8-151}
$$

where the elements of the various matrices are defined in Summary 8-6.3, and where the quantities $V_i(s)$ may include the effects of impulse generators which have been used to establish initial conditions, as well as other types of excitations. The solution to this set of equations will have the form

$$
\begin{bmatrix} I_1(s) \\ I_2(s) \\ \cdot \\ \cdot \\ \cdot \\ I_n(s) \end{bmatrix} = \frac{1}{\det \mathbf{Z}(s)} \begin{bmatrix} Z_{11}(s) & Z_{21}(s) & \cdots & Z_{n1}(s) \\ Z_{12}(s) & Z_{22}(s) & \cdots & Z_{n2}(s) \\ \cdot & \cdot & \cdots & \cdot \\ \cdot & \cdot & \cdots & \cdot \\ \cdot & \cdot & \cdots & \cdot \\ Z_{1n}(s) & Z_{2n}(s) & \cdots & Z_{nn}(s) \end{bmatrix} \begin{bmatrix} V_1(s) \\ V_2(s) \\ \cdot \\ \cdot \\ \cdot \\ V_n(s) \end{bmatrix} \tag{8-152}
$$

where the quantities $Z_{ij}(s)$ are the cofactors of the $\mathbf{Z}(s)$ matrix. Note that, as specified by Cramer's rule (see Sec. 3-3), the "12" element in the inverse matrix is the $Z_{21}(s)$ cofactor, etc. The linear nature of the transformed network is readily apparent from the equations given in (8-152). For example, consider the transform for the mesh current $I_j(s)$. From (8-152), this may be written as

$$
I_j(s) = \frac{Z_{1j}(s)}{\det \mathbf{Z}(s)} V_1(s) + \frac{Z_{2j}(s)}{\det \mathbf{Z}(s)} V_2(s) + \cdots + \frac{Z_{nj}(s)}{\det \mathbf{Z}(s)} V_n(s) \tag{8-153}
$$

From this equation we may note that the transformed response $I_j(s)$ is given as the sum of the effects produced by the individual excitations $V_1(s), V_2(s), \ldots, V_n(s)$. Thus, the principle of *superposition* applies. Similarly, if all the excitations except one, say $V_k(s)$, are zero, then multiplying this excitation by a factor k must necessarily multiply the response $I_j(s)$ by the same factor. Thus, the principle of *homogeneity* applies. These two properties, of course, define linearity.

Another very significant property of linear networks is readily observed from Eq. (8-153). This is the fact that the quantity $\det \mathbf{Z}(s)$ appears in the denominator of each term in the right member. Thus, the zeros of the determinant of $\mathbf{Z}(s)$ are the poles of all network functions which have $I_j(s)$ as a response variable, independent of which variable is the excitation variable. Thus, we may conclude that *the zeros of the determinant of the mesh impedance matrix $\mathbf{Z}(s)$ are the natural frequencies of the network.*

A development similar to that given above may be used to show that if a transformed network is described by a set of node admittance equations having the form given in Eq. (8-146), where the elements of the various matrices are defined in Summary 8-6.4, then the solution for the jth transformed node voltage variable will be

$$V_j(s) = \frac{Y_{1j}(s)}{\det \mathbf{Y}(s)} I_1(s) + \frac{Y_{2j}(s)}{\det \mathbf{Y}(s)} I_2(s) + \cdots + \frac{Y_{nj}(s)}{\det \mathbf{Y}(s)} I_n(s) \qquad (8\text{-}154)$$

where the quantities $Y_{ij}(s)$ are the cofactors of the matrix $\mathbf{Y}(s)$. Thus, we see that the poles of all network functions which have the transformed node voltages as their response variables are the zeros of $\det \mathbf{Y}(s)$. Thus, we may conclude *that the zeros of the determinant of the node admittance matrix $\mathbf{Y}(s)$ are the natural frequencies of the network.* The above conclusions, of course, assume that none of the zeros of the determinants are cancelled by corresponding zeros in the cofactors. Such a case is sometimes referred to as a *degenerate* case. The conclusions given above may be summarized as follows:

SUMMARY 8-7.1

Natural Frequencies of a Network. The natural frequencies of a network are the zeros of the determinant of the matrix $\mathbf{Z}(s)$ formed by writing the mesh impedance equations for the transformed network. The natural frequencies of a network are also the zeros of the determinant of the matrix $\mathbf{Y}(s)$ formed by writing the node admittance equations for the network. It is possible for some of these natural frequencies to be cancelled by a corresponding zero of a cofactor. ∎

The natural frequencies of a network are determined by several factors: the type of network elements, the values of elements, and the topology used to interconnect them. The natural frequencies specify the general form of the zero-input response of the network, i.e., the general form of the time-domain behavior of the network variables which results from excitation by initial conditions. Thus, the addition (or deletion) of independent sources to a network will leave its natural frequencies invariant if these sources are added in such a way so as not to change the topology of the network that results when all the independent sources are set to zero. Specifically, this is true for independent *current* sources which are connected in *parallel* with a given element or elements, and independent *voltage* sources which are connected in *series* with a given element. The latter connection is made by breaking one of the leads to the given element and connecting the two ends of the broken lead to the voltage source. As an example of this conclusion, consider the transformed network shown in Fig. 8-7.1(a). It is claimed that the natural frequencies of this network are located at $s = -1 \pm j1$. To verify this claim, let us add independent current sources $I_1(s)$ and $I_2(s)$ and define transformed node voltages $V_1(s)$ and $V_2(s)$ as shown in Fig. 8-7.1(b) Applying KCL to this network, we obtain the following node admittance equations:

$$\begin{bmatrix} I_1(s) \\ I_2(s) \end{bmatrix} = \begin{bmatrix} s+2 & -2 \\ -2 & \dfrac{4}{s}+4 \end{bmatrix} \begin{bmatrix} V_1(s) \\ V_2(s) \end{bmatrix} \tag{8-155}$$

Setting the determinant of the admittance matrix of Eq. (8-155) to zero, we obtain

$$s^2 + 2s + 2 = (s+1+j1)(s+1-j1) = 0 \tag{8-156}$$

We conclude that the natural frequencies of the network shown in Fig. 8-7.1(a) are the values of s which satisfy Eq. (8-156), namely, $s = -1 \pm j1$, as originally claimed. The network natural frequencies, of course, will be the same whether mesh or node analysis is used. As an example of this, consider the circuit shown in Fig. 8-7.1(c). This is the original transformed network as shown in Fig. 8-7.1(a), which has been

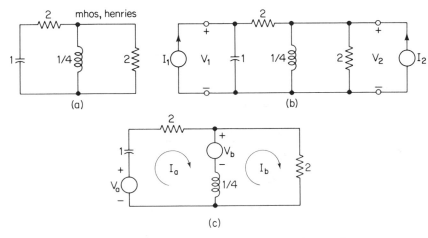

Figure 8-7.1   Networks with identical natural frequencies.

modified by inserting independent voltage sources $V_a(s)$ and $V_b(s)$ in series with the 1-F capacitor and the $\frac{1}{4}$-H inductor. Defining mesh currents $I_a(s)$ and $I_b(s)$ as shown and applying KVL, we obtain the mesh impedance equations

$$\begin{bmatrix} V_a(s) - V_b(s) \\ V_b(s) \end{bmatrix} \begin{bmatrix} \dfrac{1}{s} + \dfrac{s}{4} + \dfrac{1}{2} & -\dfrac{s}{4} \\ -\dfrac{s}{4} & \dfrac{s}{4} + \dfrac{1}{2} \end{bmatrix} \begin{bmatrix} I_a(s) \\ I_b(s) \end{bmatrix} \tag{8-157}$$

Setting the determinant of the impedance matrix of Eq. (8-157) to zero, we again obtain Eq. (8-156), verifying our claim that the natural frequencies of the network are located at $s = -1 \pm j1$. From the above example, we may conclude that the natural frequencies of a network are independent of the type of excitation (i.e., whether by independent current sources or independent voltage sources), the number of such excitation sources, and their location in the network, as long as the network is not changed when such sources are set to zero. Thus, for example, all the networks shown in Fig. 8-7.1 have the same natural frequencies. As a counter example, consider the network shown in Fig. 8-7.2(a). This network does not have the same natural

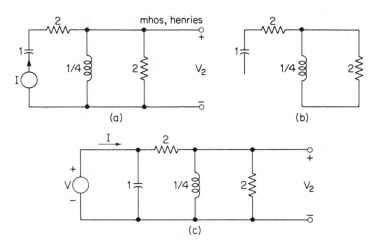

Figure 8-7.2   Networks which do not have identical natural frequencies.

frequencies as the ones shown in Fig. 8-7.1, since the capacitor and resistor which are in series with the current source do not affect the natural frequencies of the network. As an example of this, let us determine the transfer function $V_2(s)/I(s)$. This may be done directly by noting that the impedance $Z(s)$ of the parallel combination of the inductor and the resistor is

$$Z(s) = \frac{1}{(4/s) + 2} = \frac{s}{2s + 4} = \frac{s}{2(s + 2)} \tag{8-158}$$

Since $V_2(s)$ is simply $Z(s)$ times $I(s)$, we may directly write

$$\frac{V_2(s)}{I(s)} = \frac{s}{2(s + 2)} \tag{8-159}$$

thus, the network has a natural frequency at $s = -2$. This result is easily generalized to show that elements which are *in series with a current source*, such as the resistor and capacitor shown in Fig. 8-7.2(a), will not affect the natural frequencies of a network. This is done by noting that when the current source is set to zero and replaced by an open circuit, as shown in Fig. 8-7.2(b), such elements are completely superfluous as far as making any contribution to the properties of the zero-input network response, since no current flows through them. It should be noted, however, that such elements, namely, those which are in series with a current source, *will always have an effect on the natural frequencies of the driving-point impedance of the network* as seen at the terminals of the current source. For example, for the network shown in Fig. 8-7.2(a), the driving-point impedance $Z_s(s)$ seen at the terminals of the current source is the sum of the impedance $Z(s)$ of Eq. (8-158) and the impedance of the series-connected resistor and capacitor. Thus, we may write

$$Z_s(s) = \frac{V_1(s)}{I(s)} = \frac{1}{s} + \frac{1}{2} + \frac{s}{2(s+2)} = \frac{s^2 + 2s + 2}{s(s+2)} \qquad (8\text{-}160)$$

Thus, the driving-point impedance has a pole at $s = 0$, which is not one of the natural frequencies of the network. It should be noted that driving-point functions are independent of the type of source which is used to excite the network at the given pair of terminals at which the driving-point function is defined. As an example of this, consider again the network shown in Fig. 8-7.1(c). The relation for $I_a(s)/V_a(s)$ [when $V_b(s)$ is set to zero] defines a driving-point admittance at the same pair of terminals and for the same network as the driving-point impedance $Z_s(s)$ defined in Eq. (8-160). Thus, solving Eq. (8-157) for $I_a(s)/V_a(s)$, we obtain

$$\frac{I_a(s)}{V_a(s)} = \frac{s(s+2)}{s^2 + 2s + 2} \qquad (8\text{-}161)$$

This is, of course, the reciprocal of the expression obtained in (8-160).

As another example of using an excitation source in such a way as to change the natural frequencies of a network, consider the circuit shown in Fig. 8-7.2(c). The capacitor which is in parallel with the voltage source will not affect the natural frequencies of the network. It is readily shown, for example, that the voltage transfer function $V_2(s)/V(s)$ is

$$\frac{V_2(s)}{V(s)} = \frac{s}{2(s+1)} \qquad (8\text{-}162)$$

Thus, the network has a single natural frequency at $s = -1$ independent of the capacitor value. The capacitor, of course, does affect the driving-point functions of the network. For example, we may readily show that the driving-point admittance at the terminals of the voltage source is

$$\frac{I(s)}{V(s)} = \frac{s^2 + 2s + 2}{s + 1} \qquad (8\text{-}163)$$

Thus, the driving-point admittance has an additional pole at s equals infinity which is not a natural frequency of the network. The admittance is, of course, the reciprocal of the driving-point impedance, found by exciting the network at the same pair of terminals with a current source. That is, it is the reciprocal of the network function $V_1(s)/I_1(s)$ for the network shown in Fig. 8-7.1(b), and thus the numerator of Eq. (8-163) is the same as the left member of (8-156).

The results given above are easily generalized. Let a network be excited at a pair of terminals by a current source as shown in Fig. 8-7.3(a), and let the driving-point impedance have the form

(a)

(b)

Figure 8-7.3 Open-circuit and short-circuit natural frequencies.

$$\frac{V(s)}{I(s)} = \frac{A(s)}{B(s)} \tag{8-164}$$

where $A(s)$ and $B(s)$ are polynomials with real coefficients. In general, the zeros of $B(s)$ will *include* the natural frequencies of the network, and, if there are no elements in series with the current source, then *all the zeros* of $B(s)$ will be the natural frequencies of the network. If the current source is set to zero, the given network terminals are effectively open-circuited, thus these natural frequencies are frequently referred to as the *open-circuit natural frequencies* of the network. It should be noted that the term "open-circuit" refers to the condition at the indicated pair of terminals. Now let the same network be excited by a voltage source as shown in Fig. 8-7.3(b). Since the driving-point admittance must always be the reciprocal of the driving-point impedance, we see that

$$\frac{I(s)}{V(s)} = \frac{B(s)}{A(s)} \tag{8-165}$$

In this case, the zeros of $A(s)$ will *include* the natural frequencies of the network, and, if there are no elements in shunt with the voltage source, then *all the zeros of* $A(s)$ will be the natural frequencies of the network. Since, if the voltage source is set to zero, the given network terminals are effectively short-circuited, the network natural frequencies in this case are referred to as the *short-circuit natural frequencies* of the network with respect to the given terminal pair. The results given above may be summarized as follows:

SUMMARY 8-7.2

The Effect of Independent Excitation Sources on the Natural Frequencies of a Network.   The addition or deletion of current sources in parallel with network elements, or voltage sources in series with network elements, does not change the natural frequencies of the network. Correspondingly, network elements which are in series with current sources or in parallel with voltage sources have no effect on the natural frequencies of a network, although they do affect the driving-point immittance as seen from the terminals of such sources.   ■

Now let us investigate some characteristics of *transfer* functions. Specifically, let us determine the effect on the natural frequencies of a given network when we interchange the point at which an excitation is applied to a network and the point at which a response is measured. Since there are two types of excitations (independent voltage and current sources) and two types of response (voltage and current), there are four possible cases. Let the first case involve a voltage excitation source and a current response (a transfer admittance). Thus, we have the situation shown in Fig. 8-7.4(a). The interchange of the points of excitation and response produces the situation shown in Fig. 8-7.4(b). In both of these circuits, setting the source to zero produces the network shown in Fig. 8-7.4(c). Thus, we may conclude that the natural frequencies of the network are unchanged when the points of excitation and response are interchanged. As a second case, consider the use of a current excitation source and

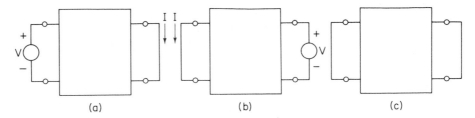

Figure 8-7.4 Network connections with the same natural frequencies.

a voltage response (a transfer impedance) as shown in Fig. 8-7.5(a). Interchanging these quantities yields the situation shown in Fig. 8-7.5(b). In both of these circuits, setting the source to zero yields the network shown in Fig. 8-7.5(c). Thus, in this case, the natural frequencies are again unaffected by the interchange. Let the third case be the one involving the use of a voltage excitation source and a voltage response (a transfer voltage ratio) as shown in Fig. 8-7.6(a). Interchanging these quantities produces the situation shown in Fig. 8-7.6(b). Comparing these two figures, we see that

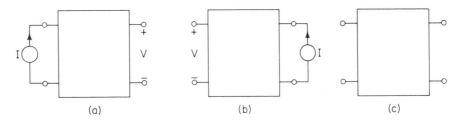

Figure 8-7.5 Network connections with the same natural frequencies.

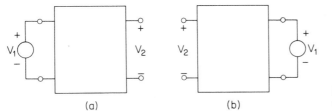

Figure 8-7.6 Network connections with different natural frequencies.

setting the sources to zero in the two figures produces different terminations on the network terminals. Thus, we conclude that the interchange of source and excitation necessarily produces a different set of network natural frequencies in this case. The final case, consisting of a current excitation source and a current response (a transfer current ratio), is shown in Fig. 8-7.7(a). The interchange of variables produces the

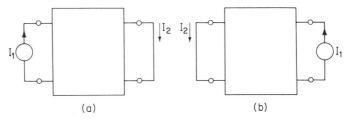

Figure 8-7.7 Network connections with different natural frequencies.

Sec. 8-7   Natural Frequencies of Networks

447

situation shown in Fig. 8-7.7(b). Again we see that the two networks will have different natural frequencies. These conclusions may be summarized as follows:

SUMMARY 8-7.3

Interchange of Excitation and Response. If a network function as defined for a given network is a transfer admittance or a transfer impedance, the interchange of the points at which the excitation is applied and the response is measured will not affect the natural frequencies of the network. If a network function is a voltage transfer ratio or a current transfer ratio, however, the interchange of excitation and response will change the natural frequencies of the network. ∎

In the first two cases we defined above, namely, the transfer immittance cases, we have seen that the natural frequencies, i.e., the *poles* of the network functions, are not affected when the points of excitation and response are interchanged. Now let us consider the *zeros* of the network functions for these two cases. For each case, there are two possibilities. The first is that the zeros will also remain unchanged. If such a result is true for any point of excitation and any point of response (excluding driving-point situations), then the network is said to be *reciprocal*. If there are one or more transfer functions for which the zeros are different when such an interchange is made, then the network is correspondingly said to be *non-reciprocal*. A direct result of the reciprocal or non-reciprocal nature of a network is seen in the mesh impedance formulation given in Eq. (8-151). If the matrix $\mathbf{Z}(s)$ is symmetric, i.e., if $z_{ij}(s) = z_{ji}(s)$, then the inverse matrix given in Eq. (8-152) will also be symmetric. Thus, the effect on the ith mesh current $I_i(s)$ due to a voltage source $V_j(s)$ in the jth mesh, i.e., $Z_{ji}(s)/$ det $\mathbf{Z}(s)$, will be the same as the effect on the jth mesh current $I_j(s)$ produced by a voltage source $V_i(s)$ in the ith mesh, i.e., $Z_{ij}(s)/\det \mathbf{Z}(s)$. Thus, *if $\mathbf{Z}(s)$ is symmetric, the network is reciprocal.* A similar conclusion holds for the node admittance matrix given in Eq. (8-146). That is, if $\mathbf{Y}(s)$ is symmetric, the network is reciprocal. The converse is readily shown to hold, namely, if $\mathbf{Z}(s)$ is non-symmetric, the network is non-reciprocal. Similarly, if $\mathbf{Y}(s)$ is non-symmetric, the network is non-reciprocal.

This result is summarized as follows:

SUMMARY 8-7.4

Reciprocity. A reciprocal network is one in which the interchange of excitation and response in any transfer immittance leaves the given transfer immittance invariant. A non-reciprocal network is one in which this is not true. Reciprocal networks have symmetric mesh impedance and node admittance matrices. ∎

If the various example networks which have been analyzed in this chapter are examined, it is fairly simple to make some general observations concerning the type of networks which are reciprocal and those which are non-reciprocal. In general, networks comprised only of two-terminal elements such as resistors, capacitors, and inductors will always be reciprocal, while networks containing controlled sources will usually be non-reciprocal.[15] This is true independent of whether the two-terminal elements are active or passive.

Example 8-7.1

As an example of the application of the concepts developed in this section, consider a network with two pairs of terminals which is tested in three ways as shown in Fig. 8-7.8. Each test defines a network function with the indicated numerator and denomina-

$$\frac{I_1(s)}{V_1(s)} = \frac{A(s)}{B(s)} \qquad \frac{V_2(s)}{I_1(s)} = \frac{C(s)}{D(s)} \qquad \frac{V_1(s)}{I_2(s)} = \frac{E(s)}{F(s)}$$

Figure 8-7.8 Natural frequencies of a network.

tor polynomials. It is desired to predict any equalities that may exist among the polynomials. To do this we first note that, from test 1, with terminals 1-1′ shorted and terminals 2-2′ open circuited, the natural frequencies of the network are given by $B(s)$ (the short-circuit natural frequencies). Since test 1 defines a driving-point admittance at terminals 1-1′, if we replace the voltage source with a current source, we obtain the (reciprocal) driving-point impedance function $V_1(s)/I_1(s) = B(s)/A(s)$. Thus, with terminals 1-1′ and 2-2′ open circuited, the natural frequencies of the network are given by $A(s)$ (the open-circuit natural frequencies). In test 2 the natural frequencies are given by $D(s)$. Since terminals 1-1′ and 2-2′ are open circuited in this test, we conclude that $D(s) = A(s)$. In test 3 the natural frequencies are given by $F(s)$. Since terminals 1-1′ and 2-2′ are also open circuited in this test, we conclude that $F(s) = D(s) = A(s)$. In addition, since test 2 and 3 both define a transfer impedance, and since they also represent the interchange of the points of excitation and response, if the network is reciprocal we have $C(s) = E(s)$.

8-8 LOCATIONS OF CRITICAL FREQUENCIES OF NETWORK FUNCTIONS

In the preceding sections of this chapter we have discussed two types of critical frequencies for network functions. These are *poles*, which are defined as points in the complex frequency plane (or at infinity) where the magnitude of a given network

---

[15]If the controlling variable *and* the output variable are both defined at the same branch, then the controlled source acts as a single two-terminal element, and the network containing such elements is reciprocal.

function goes to infinity, and *zeros* which are places where the magnitude goes to zero. One very useful way of visualizing such critical frequencies of a network function is by drawing a three-dimensional representation in which the real and the imaginary axes of the complex frequency plane are plotted in two directions, and the magnitude of the network function is plotted in the third direction. In such a representation, the locus of all points defined by the network function will generate a three-dimensional surface specifying the magnitude of the function at every point on the complex plane. Such a surface is illustrated in Fig. 8-8.1 for the network function $F(s)$, where

$$F(s) = \frac{s}{(s + 1 + j1)(s + 1 - j1)} \tag{8-166}$$

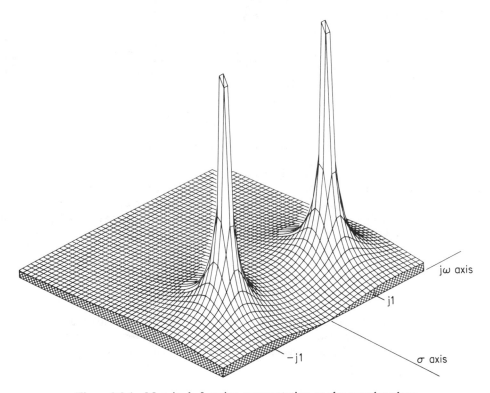

Figure 8-8.1 Magnitude function representation on the complex plane.

From the figure, we see that as the points $-1 \pm j1$ are approached on the s plane, the magnitude surface goes to infinity, thus visually as well as mathematically producing a "pole." Similarly, at the point $s = 0$, the magnitude surface goes to zero. In this section we shall present a summary of the basic principles which limit the locations of critical frequencies, i.e., the poles and zeros, for various classes of network functions, and for various types of networks. Because of the complexity of the underlying mathematics, our treatment will be non-rigorous. The results, however, are quite interesting, and well worth including as part of our study of the properties of networks.

To begin our study of the locations of the critical frequencies of network functions, let us define a network function $F(s)$ as

$$F(s) = \frac{R(s)}{E(s)} = \frac{A(s)}{B(s)} = \frac{a_0 + a_1 s + a_2 s^2 + \cdots + a_m s^m}{b_0 + b_1 s + b_2 s^2 + \cdots + b_n s^n} \tag{8-167}$$

where $R(s)$ and $E(s)$ are rational functions representing the transformed response and excitation, and $A(s)$ and $B(s)$ are polynomials. If we let the excitation to the network be a unit impulse, then, for all $t > 0$, there is no excitation to the network. Thus, the resulting response is the zero-input response, i.e., the response which results from the initial conditions which are generated by the excitation impulse. The situation is similar to that described in Summary 8-6.5. For this case, $E(s)$ is the transform of a unit impulse, i.e., $E(s) = 1$. Thus, we may write

$$F(s) = R(s) = \frac{A(s)}{B(s)} \tag{8-168}$$

We see that, *for a unit impulse excitation, the network function equals $R(s)$, the transform of the response variable*. The resulting response is appropriately referred to as the *impulse response*. Thus, we may determine all the properties of the network function by examining the properties of $R(s)$ for an applied impulse of excitation. As an example of the determination of such a property, let us consider the bounds on network variables. For a passive network under zero-input conditions, the total energy stored in the network will either be constant (for a lossless or LC network) or decreasing (if dissipation, i.e., resistance, is present). Thus the zero-input response for any network variable must be bounded. Thus, pole locations which represent time-domain behavior which grows indefinitely (and thus is not bounded) are not possible for such a passive network. For example, poles which are in the right half-plane (RHP) represent a time-domain positive exponential behavior. Thus, they violate the requirement for boundedness and they are not possible for passive networks. Similarly, poles on the $j\omega$ axis of higher than first order represent nonbounded time-domain behavior, thus they will not occur in passive networks. Simple poles on the $j\omega$ axis, however, represent bounded behavior and thus they are permissible. Poles in the left half-plane (LHP) of any order are permitted. For example, for a pole of order m on the negative real axis at $s = -a \, (a > 0)$, the corresponding time-domain behavior is $Kt^{m-1}e^{-at}$. By applying l'Hospital's rule, it is easy to show that

$$\lim_{t \to \infty} Kt^{m-1}e^{-at} = 0 \qquad a > 0 \tag{8-169}$$

for all finite m. This is due to the fact that the negative exponential term dominates for large t. It may similarly be shown that *network function poles of multiple order anywhere in the LHP* represent bounded time-domain zero-input behavior. A summary of the impulse response characteristics for various pole locations on the complex frequency plane is given in Fig. 8-8.2. From this figure and the discussion given above, we conclude the following:

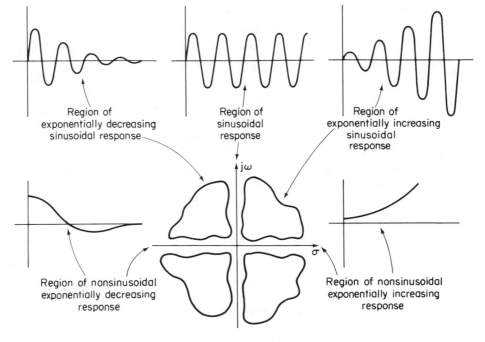

Region of
exponentially decreasing
sinusoidal response

Region of
sinusoidal
response

Region of
exponentially increasing
sinusoidal
response

$j\omega$

σ

Region of nonsinusoidal
exponentially decreasing
response

Region of nonsinusoidal
exponentially increasing
response

Figure 8-8.2   Impulse response characteristics.

SUMMARY 8-8.1

Natural Frequencies of Passive Networks.   The natural frequencies of a network are the poles of the transform of the impulse response. These are the same as the poles of the network function. The poles of network functions for passive networks must be in the LHP or on the $j\omega$ axis. Any poles on the $j\omega$ axis must be simple.   ∎

In general, no simple conclusion can be made concerning the poles of *active* networks, since it is equally possible for these to be in the RHP or the LHP. As an example of this consider the network shown in Fig. 8-8.3. This network contains

Figure 8-8.3   Active *RC* network.

passive resistors and capacitors and a VCVS. Let us investigate the possible locations for the poles of the voltage transfer function of this network. In Example 8-5.5 the voltage transfer function for this network was shown to be

$$\frac{V_2(s)}{V_1(s)} = \frac{K}{s^2 + s(3 - K) + 1} \tag{8-170}$$

Obviously, from this equation, the poles of the voltage transfer function, i.e., the natural frequencies of the network, are a function of the value of K, i.e., the gain of the VCVS. A plot of the locus of these poles as K is varied is given in Fig. 8-8.4. Such

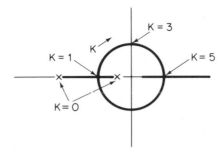

Figure 8-8.4 Root-locus plot.

a plot is frequently referred to as a *root-locus plot*, since it shows the locus of the roots of the denominator polynomial of the network function $V_2(s)/V_1(s)$ as they are affected by the value of K. From the figure we note that if $0 \leq K \leq 3$, the poles of the voltage transfer function are in the LHP or on the $j\omega$ axis. For this range of K, the impulse response of the network is bounded, and the network is said to be *stable*. For values of $K > 3$, however, the poles are in the RHP, and the impulse response is unbounded. For such values of K, the network is said to be *unstable*. Thus, we may conclude that active networks may have stable or unstable characteristics, depending on the particular configuration and on the specific values of the elements. On the other hand, passive networks, i.e., networks comprised only of passive network elements, are always stable.

Now let us investigate some other restrictions on the poles of various network functions. First, let us consider networks comprised only of inductors and capacitors, i.e., LC networks. Since there is no resistance in such circuits, there is no dissipation of any energy present in the circuit. Thus, the variables of such a circuit will oscillate sinusoidally as energy is transferred back and forth between the capacitors and inductors of the circuit. The amplitude of each sinusoidal oscillation component will remain fixed. The frequency-domain location for the poles of the network function corresponding with such time-domain behavior is the $j\omega$ axis. Thus, we conclude that LC networks have their poles restricted to the $j\omega$ axis. For boundedness of the response, such poles must be simple. Now let us consider RC circuits, i.e., circuits comprised only of resistors and capacitors. Since there is only one type of energy-storage element in such a circuit, i.e., capacitors, there can be no sinusoidal component in the zero-input response. In addition, since there is dissipation in the circuit, the values of all variables must eventually go to zero. The location of the poles in the complex frequency plane that correspond with such time-domain behavior is the negative-real axis. We conclude that RC networks have their poles only on the negative-real axis. Although not required by boundedness, it can be shown that these poles must also be simple. Such networks may also have a simple pole at the origin. The driving-point impedance of a capacitor is an example of this latter case. Similar logic leads us to the conclusion that the poles of network functions for RL networks, i.e., networks comprised only of resistors and inductors, will have poles only on the negative-real axis (or at the origin), and that these poles will be simple. The results given above may be summarized as follows:

SUMMARY 8-8.2

Critical Frequencies of LC, RC, and RL Networks. The poles of network functions for *LC* networks are restricted to the *jω* axis (and the origin) of the complex frequency plane, while those for *RC* and *RL* networks are restricted to the negative-real axis (and the origin). All the poles are simple. ∎

The restrictions given above apply to any type of network function, i.e., driving-point functions *or* transfer functions. If we now restrict our attention to driving-point functions, we may make some observations about the zeros of the three classes of networks discussed above. Since the zeros of driving-point functions become the poles when the excitation is changed from a voltage source to a current source or vice versa, the restrictions given in Summary 8-8.2 apply equally well to the zeros of driving-point functions. In addition, it may be shown that the alternation of poles and zeros along the negative-real axis (for *RC* and *RL* networks), and along the *jω* axis (for *LC* networks) is a necessary condition for driving-point immittances.

A precise determination of the permissible locations of the poles of driving-point immittances for passive *RLC* networks is considerably more complicated than those for the two-element type of networks outlined above. As such, it is beyond the scope of this text. In general, however *the poles of RLC networks may be located anywhere in the LHP and on the jω axis.* On the *jω* axis, boundedness of the network response requires that they be simple.

Now let us consider the permissible locations for the poles and zeros of *transfer functions*. The poles of such functions are, of course, the natural frequencies of the network. Thus, they are determined by the same restrictions given above for the poles of driving-point functions. The zeros, however, are almost completely unrestricted, and, in general, they may occur in either the LHP or the RHP and they need not necessarily be simple. Certain network configurations do, however, place restrictions on the location of the zeros. As an example of such a network configuration, consider the voltage divider shown in Fig. 8-8.5(a). The network function for the voltage transfer ratio is

$$\frac{V_2(s)}{V_1(s)} = \frac{Z_2(s)}{Z_1(s) + Z_2(s)} \tag{8-171}$$

(a)   (b)

Figure 8-8.5 Voltage divider.

where $Z_1(s)$ is called a *series* element (it is in series with the source), and $Z_2(s)$ is called a *shunt* element (it is in shunt with the output terminals). As a more specific example, consider the network shown in Fig. 8-8.5(b). The impedance $Z_1(s)$ of the series LC element is $s/(s^2 + 1)$ and the impedance $Z_2(s)$ of the shunt RC element is $(s + 2)/s$. From (8-171), the network function for the voltage transfer ratio is

$$\frac{V_2(s)}{V_1(s)} = \frac{(s + 2)/s}{[s/(s^2 + 1)] + [(s + 2)/s]} \tag{8-172}$$

Now let us consider what happens if the excitation $v_1(t)$ is chosen so that its transform has a pole at the identical frequency at which the impedance $Z_2(s)$ of the shunt RC element *has a zero*, namely, at $s = -2$. For this, from Table 8-3.1, the excitation will have the form $v_1(t) = Ke^{-2t}$ and its transform will be $V_1(s) = K/(s + 2)$. Substituting this expression in Eq. (8-172), we obtain

$$V_2(s) = \frac{K/s}{[s/(s^2 + 1)] + [(s + 2)/s]} = \frac{K(s^2 + 1)}{s^3 + 3s^2 + s + 2} \tag{8-173}$$

From Eq. (8-173) we see that there is no pole of $V_2(s)$ at $s = -2$. Thus, even though the excitation $v_1(t)$ was of the form Ke^{-2t}, *there will not be an e^{-2t} term present in the response for $v_2(t)$.* We may summarize this result by saying that since the network function of Eq. (8-172) has a zero at $s = -2$, exciting the network at this complex frequency mathematically cancels the factor $(s + 2)$ in the numerator and denominator of the expression for $V_2(s)$. From a physical viewpoint, we may say that when a network is excited at a complex frequency at which it has a zero, then this excitation does not appear in the response. Even though the excitation frequency does not appear in the response, there will, of course, be other response terms due to the other natural frequencies of the network. These other natural frequencies are, of course, excited by the applied excitation function.

A similar situation occurs if the network shown in Fig. 8-8.5(b) is excited at some frequency at which the impedance $Z_1(s)$ of the series LC element *has a pole*, i.e., at $s = \pm j1$. Such an excitation occurs if we let $v_1(t) = K \sin t$; thus, $V_1(s) = K/(s^2 + 1)$. Substituting this expression in (8-172), we obtain

$$V_2(s) = \frac{K(s + 2)/[s(s^2 + 1)]}{[s/(s^2 + 1)] + [(s + 2)/s]} = \frac{K(s + 2)}{s^3 + 3s^2 + s + 2} \tag{8-174}$$

From Eq. (8-174) we see that there are no poles of $V_2(s)$ at $s = \pm j1$, even though the transformed excitation function has poles there. This is because the impedance of $Z_1(s)$ also has poles at $s = \pm j1$.

In the example discussed above, we showed that the zero of the driving-point impedance $Z_2(s)$ of the shunt element and the pole of the driving-point impedance $Z_1(s)$ of the series element of the network shown in Fig. 8-8.5(b) both produced a zero in the resulting network voltage transfer function. This conclusion is easily generalized. To see this, consider again the voltage divider shown in Fig. 8-8.5(a). Suppose that the driving-point impedance of $Z_2(s)$ has a zero at $s = p_0$. Thus, we may write

$$Z_2(s) = \frac{(s - p_0)A(s)}{B(s)} \tag{8-175}$$

where $A(s)$ and $B(s)$ are arbitrary polynomials, and $B(p_0) \neq 0$. The voltage transfer function for the network may now be written as

$$\frac{V_2(s)}{V_1(s)} = \frac{(s - p_0)A(s)/B(s)}{Z_1(s) + (s - p_0)A(s)/B(s)} = \frac{(s - p_0)A(s)}{Z_1(s)B(s) + (s - p_0)A(s)} \quad (8\text{-}176)$$

Evaluating the network function at $s = p_0$, we see that since $Z_2(p_0) = 0$, the network function itself is also equal to 0. Now let us assume that $V_1(s)$ has a pole at p_0. In this case, it may be written in the form

$$V_1(s) = \frac{C(s)}{(s - p_0)D(s)} \quad (8\text{-}177)$$

where $C(s)$ and $D(s)$ are arbitrary polynomials. The response $V_2(s)$ must now have the form

$$V_2(s) = \frac{(s - p_0)A(s)}{Z_1(s)B(s) + (s - p_0)A(s)} \frac{C(s)}{(s - p_0)D(s)}$$

$$= \frac{A(s)C(s)}{[Z_1(s)B(s) + (s - p_0)A(s)]D(s)} \quad (8\text{-}178)$$

From Eq. (8-178) we see that $V_2(s)$ does not have a pole at p_0. Thus, we may conclude that, in general, when $Z_2(s)$ has a zero at some complex frequency, any excitation which is applied at that frequency produces no output voltage (at that frequency). In effect, we can think of the impedance $Z_2(s)$ as being a "short circuit" at that value of *complex* frequency, and thus shorting such an excitation frequency to ground. The input will, of course, excite the other natural frequencies of the network. Now let us consider a different case for the same network. For this case, let us assume that $Z_1(s)$ has a pole at $s = p_0$. Thus, we may write

$$Z_1(s) = \frac{A(s)}{(s - p_0)B(s)} \quad (8\text{-}179)$$

where $A(s)$ and $B(s)$ are arbitrary polynomials, and $A(p_0) \neq 0$. The voltage transfer function will now have the form

$$\frac{V_2(s)}{V_1(s)} = \frac{(s - p_0)B(s)Z_2(s)}{(s - p_0)B(s)Z_2(s) + A(s)} \quad (8\text{-}180)$$

If we evaluate this expression at $s = p_0$, then the network function itself will be equal to 0. Again we may conclude that if $V_1(s)$ contains an excitation frequency at p_0, there will be no output frequency p_0 present in $V_2(s)$. In this case, we may think of $Z_1(s)$ as being an "open circuit" at the complex frequency p_0; thus, it blocks the transmission of any input at that frequency.

The conclusions given above for a voltage divider are easily generalized to include a ladder network of the form shown in Fig. 8-8.6. As such they apply to any

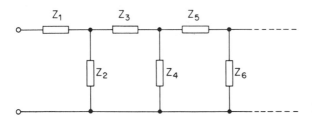

Figure 8-8.6 Ladder network.

transfer function defined from one end of the network to the other. At any value of complex frequency at which any of the impedances $Z_1(s)$, $Z_3(s)$, $Z_5(s)$, etc., has a pole, there will be no transmission, i.e., the transfer function will have a zero. Similarly, at any value of complex frequency at which any of the impedances $Z_2(s)$, $Z_4(s)$, $Z_6(s)$, etc., has a zero, there will also be a zero in the transfer function. Since *RLC* networks can only have poles and zeros in the LHP, we may conclude that *the transfer functions of ladder networks can only have zeros in the LHP and on the jω axis.* Such conclusions make it easy to directly evaluate some of the characteristics of frequently encountered ladder networks. For example, consider the network shown in Fig. 8-8.7. The poles of

Figure 8-8.7 Low-pass ladder network.

the series impedances (the inductors) are all at $s = \infty$. Similarly, the zeros of the shunt impedances (the capacitors) are at $s = \infty$. Thus, the transmission zeros of the network will all be at $s = \infty$. We conclude that the network is a low-pass network. Similar reasoning shows that the network shown in Fig. 8-8.8 is a high-pass network. These results may be summarized as follows:

Figure 8-8.8 High-pass ladder network.

SUMMARY 8-8.3

Transmission Zeros of Ladder Networks. The zeros of transfer functions for ladder networks will occur at the complex frequencies at which the driving-point impedances of any of the series elements have poles and at the complex frequencies at which the driving-point impedances of any of the shunt elements have zeros. Thus the zeros can only occur in the LHP or on the *jω* axis.[16] ∎

8-9 POLES AND ZEROS AND FREQUENCY RESPONSE

In Summary 8-5.2 we showed that there is a direct relation between the type of network function which relates *transformed excitation and response variables*, and the type of network function which relates *excitation and response phasors*. This latter network

---

[16]For a shunt element to produce a zero of transmission at a given frequency, the impedance of the remainder of the network as seen from the element must *not* also be zero at that frequency. Similarly, for a series element to produce a zero of transmission, at a given frequency, the impedance of the rest of the network must *not* be infinite at that frequency.

function, of course, determines the sinusoidal steady-state response of a given network. Specifically, we showed that the two network functions are related by a simple argument change, namely, the substitution of $j\omega$ for s and vice versa. In this section we shall illustrate a graphical method of relating these types of network functions, i.e., of determining the sinusoidal steady-state response of a network directly from the locations of the poles and zeros in the complex frequency plane. The method has the advantage that it is readily visualized. In addition, it provides valuable insight into the properties of various classes of network functions.

To begin our discussion, let us consider a network function $F(s)$ written in the form

$$F(s) = K\frac{(s - z_1)(s - z_2)\cdots(s - z_m)}{(s - p_1)(s - p_2)\cdots(s - p_n)} \tag{8-181}$$

where the quantities z_i and p_i are the zeros and the poles of the network function, respectively, and where K is a multiplicative constant. Let us now consider what happens when we evaluate the function $F(s)$ by letting $s = j\omega_0$, where ω_0 is some specific value of ω, i.e., some specific point on the $j\omega$ axis. For such a situation, Eq. (8-181) may be written

$$F(j\omega_0) = K\frac{(j\omega_0 - z_1)(j\omega_0 - z_2)\cdots(j\omega_0 - z_m)}{(j\omega_0 - p_1)(j\omega_0 - p_2)\cdots(j\omega_0 - p_n)} \tag{8-182}$$

Now let us consider a typical numerator term of the form $(j\omega_0 - z_i)$ where $j\omega_0$ and z_i are specific points on the complex frequency plane. Such a pair of points is shown in Fig. 8-9.1(a). Since these points are complex quantities, they may be represented by

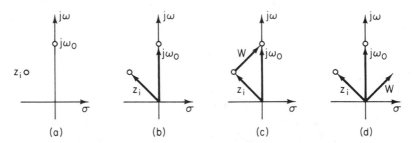

Figure 8-9.1 Graphical evaluation of a factor of a network function.

vectors as shown in Fig. 8-9.1(b). Now let us construct a vector W going from z_i to $j\omega_0$ as shown in Fig. 8-9.1(c). By simple vector addition, we see that $j\omega_0 = W + z_i$ or $W = j\omega_0 - z_i$. Thus we may conclude that the complex quantity $(j\omega_0 - z_i)$ is equal to the (complex) value of a vector drawn from the point z_i to the point $j\omega_0$ on the complex plane. The actual magnitude and phase of such a vector are, of course, more readily visualized if the vector is shifted to the origin as shown in Fig. 8-9.1(d). Thus we have developed a graphical means of evaluating numerator factors of Eq. (8-182) having the form $(j\omega_0 - z_i)$. Obviously, denominator factors of Eq. (8-182) which have the form $(j\omega_0 - p_i)$ may be evaluated in the same manner.

The graphical evaluation technique described above is readily applied to determine the magnitude and phase of a given network function at any specified point on the $j\omega$ axis. As an example, consider the network function

$$F(s) = 3\frac{s(s+1)}{s^2 + 2s + 2} = 3\frac{s(s+1)}{(s+1-j1)(s+1+j1)} \tag{8-183}$$

The zeros of this function are $z_1 = 0$ and $z_2 = -1$, and the poles are $p_1 = -1 + j1$ and $p_2 = -1 - j1$. In addition, the multiplicative constant is $K = 3$. Now let us suppose that it is desired to evaluate $F(s)$ of Eq. (8-183) for $s = j\omega_0 = j1$. We may do this by the process shown in Fig. 8-9.2. The first numerator factor s, evaluated at

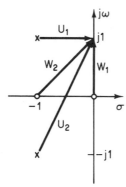

Figure 8-9.2 Graphical evaluation of a network function.

$s = j1$ is simply $j1$, i.e., the vector W_1 of length 1 and angle 90° shown in the figure. The second numerator factor $(s+1)$, evaluated at $j1$, is the vector W_2. This has length $\sqrt{2}$ and angle 45°. Similarly, the denominator factors may be represented by the vectors U_1 and U_2 shown in the figure. In terms of these four vectors, we see that

$$F(j1) = 3\frac{W_1 W_2}{U_1 U_2} = 3\frac{1e^{j90°}\sqrt{2}e^{j45°}}{1e^{j0}\sqrt{5}e^{j63.43°}} = \frac{3\sqrt{2}}{\sqrt{5}}e^{j71.57°} \tag{8-184}$$

The results given above are easily generalized as follows:

SUMMARY 8-9.1

Use of Poles and Zeros in Determination of Frequency Response. A network function $F(s)$ having the form

$$F(s) = K\frac{(s-z_1)(s-z_2)\cdots(s-z_m)}{(s-p_1)(s-p_2)\cdots(s-p_n)} \tag{8-185}$$

may be evaluated for sinusoidal steady-state behavior for any value of frequency ω_0 by constructing a set of vectors with complex values W_i from the various zeros z_i to the point $j\omega_0$, and by constructing a set of vectors of values U_i from the various poles p_i to the same point. The process is illustrated in Fig. 8-9.3. In terms of these vectors, we may write

$$|F(j\omega)| = K\frac{\prod_{i=1}^{m}|W_i|}{\prod_{i=1}^{n}|U_i|} \tag{8-186}$$

$$\text{Arg } F(j\omega) = \sum_{i=1}^{m}\text{Arg } W_i - \sum_{i=1}^{n}\text{Arg } U_i \qquad \blacksquare$$

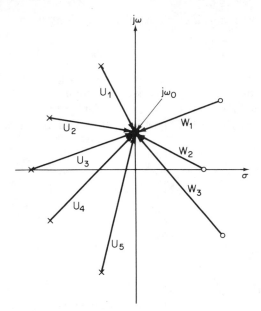

Figure 8-9.3 Graphical evaluation of frequency response.

The results given above may be used to determine many useful results concerning the sinusoidal steady-state response of various networks. For example, in the limit as ω approaches infinity, the vectors from all the finite poles and zeros will approach angles of 90°. Thus, the phase of the network function in this limit is simply equal to the number of finite zeros minus the number of finite poles times 90°. As another example, it is readily shown that for values of frequency above and below a zero on the $j\omega$ axis, the phase must differ by 180°.

Example 8-9.1

As an example of the use of graphical techniques to make an approximate evaluation of the sinusoidal steady-state magnitude and phase of a network function, consider the function

$$F(s) = \frac{1}{s^2 + 0.4s + 1.04}$$

with the poles shown in Fig. 8-9.4(a). To evaluate $F(0)$ we draw vectors as shown in Fig. 8-9.4(b). Since both vectors have an approximate magnitude of unity, $|F(0)| \approx 1$. Since the arguments of the two vectors are the negative of each other, their sum gives Arg $F(0) = 0$. As ω increases, the magnitude of the vector from $s = -0.2 + j1$ to the $j\omega$ axis decreases more rapidly than the magnitude of the vector from $s = -0.2 - j1$ increases. Thus, the product of the magnitudes of these vectors decreases, and the magnitude of $F(j\omega)$ increases. The maximum magnitude of $F(j\omega)$ occurs approximately at the frequency where the magnitude of the vector from $s = -0.2 + j1$ has its minimum value, i.e., at $\omega = 1$. To evaluate $F(j1)$ we draw vectors as shown in Fig. 8-9.4(c). From these vectors we see that $|F(j1)| \approx 1/(2 \times 0.2) = 2.5$, and Arg $F(j1) \approx 90°$. For higher values of ω, $|F(j\omega)|$ decreases. At $\omega = \infty$, both vectors are of infinite length and have an argument of 90° as shown in Fig. 8-9.4(d). Thus we have $|F(\infty)| = 0$, and Arg $F(\infty) = -180°$. Using the data given above, the approximate magnitude and phase plots of the network function have the form shown in Fig. 8-9.5.

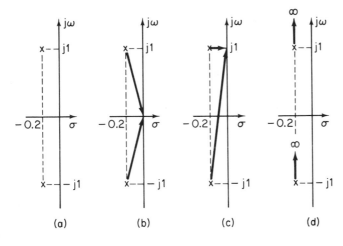

Figure 8-9.4 Evaluating sinusoidal steady-state magnitude and phase.

(a) (b) (c) (d)

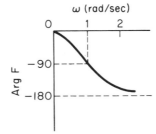

Figure 8-9.5 Magnitude and phase plots for Example 8-9.1.

As a final topic in this section we now show that there is a direct relation between the quality factor Q introduced in Sec. 7-7 and the pole positions of a given network function. As an example of this relation, consider the band-pass network function

$$N(s) = \frac{Ks}{s^2 + as + b} \qquad (8\text{-}187)$$

Evaluating this for $s = j\omega$, we obtain

$$N(j\omega) = \frac{j\omega K}{b - \omega^2 + j\omega a} = \frac{K}{a + j[\omega - (b/\omega)]} \qquad (8\text{-}188)$$

At resonance the denominator must have its smallest value. This occurs when the imaginary part of the denominator is zero. Thus, letting ω_0 be the resonant frequency, we see that the following relation must be satisfied:

$$\omega_0 - \frac{b}{\omega_0} = 0 \quad \text{or} \quad \omega_0 = \sqrt{b} \qquad (8\text{-}189)$$

At resonance, i.e., when $s = j\omega_0$, the network function is real and has the value

$$N(j\omega_0) = \frac{K}{a} \qquad (8\text{-}190)$$

Now let us compute the value of the frequencies at which the magnitude of the network function $|N(j\omega)|$ is 0.707 of its value at resonance, i.e., the 3-dB frequencies. From Eqs. (8-188) and (8-190), these values of frequency ω must satisfy the relation

$$\frac{K}{\sqrt{[\omega - (b/\omega)]^2 + a^2}} = \frac{K}{\sqrt{2}\,a} \tag{8-191}$$

Solving this equation for ω we obtain

$$\omega = \pm\frac{a}{2} \pm \sqrt{\left(\frac{a}{2}\right)^2 + b} \tag{8-192}$$

Choosing the appropriate combination of signs, we see that

$$\omega_{\text{upper 3 dB}} = \frac{a}{2} + \sqrt{\left(\frac{a}{2}\right)^2 + b}$$
$$\omega_{\text{lower 3 dB}} = -\frac{a}{2} + \sqrt{\left(\frac{a}{2}\right)^2 + b} \tag{8-193}$$

Thus, we find that the bandwidth B is

$$B = \omega_{\text{upper 3 dB}} - \omega_{\text{lower 3 dB}} = a \tag{8-194}$$

In Sec. 7-7 we showed that Q was defined as $Q = \omega_0/B$. Using the relations of Eqs. (8-189) and (8-194) in this expression, we obtain the following expression for Q in terms of the coefficients of the network function given in Eq. (8-187). This is

$$Q = \frac{\sqrt{b}}{a} \tag{8-195}$$

Thus, we conclude that for a second-order band-pass network function having the form given in Eq. (8-187), the Q is simply expressed as the ratio of the square root of the zero-degree coefficient to the first-degree coefficient. Since the square root of the zero-degree coefficient is equal to the magnitude of a vector drawn from the origin to the pole position and since the first-degree coefficient is -2 times the real part of the pole position, we see that an alternative definition for Q for a complex conjugate pair of poles located at $s = p_0$ and $s = p_0^*$ is

$$Q = \left|\frac{p_0}{2\,\text{Re}\,p_0}\right| \tag{8-196}$$

The relations given in Eqs. (8-195) and (8-196) provide a simple means of calculating Q in network functions which are characterized by a second-degree denominator polynomial.

8-10 THEVENIN AND NORTON EQUIVALENT TRANSFORMED NETWORKS

To this point in this chapter we have introduced the Laplace transformation and showed how it may be used to solve for the values of the variables of transformed networks. Many of the techniques for forming the equations describing the transformed network which we have introduced in this chapter are practically identical with those which were introduced in Chap. 3, where we discussed the formulation of network equations for resistance networks. For example, in Sec. 8-6, we showed that mesh impedance and node admittance equations could be written for transformed networks by following the same rules for the determination of the elements of the matrices that were used in the resistance network case. The modification of almost all the other techniques for analyzing resistance networks given in Chap. 3 (such as converting voltage sources to current sources and vice versa, adding parallel voltage sources and series current sources, etc.) so that they apply to the transformed network is a trivial task, since the only major change in the techniques is to substitute the word impedance for the word resistance, and the word admittance for the word conductance. As an example of the modification of one such technique, consider the development of a Thevenin equivalent circuit for a transformed network. As an example of this procedure, we will use the network shown in Fig. 8-10.1. The circuit consists of an

Figure 8-10.1 Simple transformed network.

independent voltage source with an output $V_s(s) = 3s/(s^2 + 4)$ [thus, the time-domain excitation is $v_s(t) = 3 \cos 2t\, u(t)$], a $\frac{1}{9}$-H inductor, and a 1-F capacitor. The Thevenin equivalent circuit will, of course, consist of an independent voltage source and a series two-terminal impedance. To determine the components of the Thevenin equivalent circuit, following the procedure outlined in Sec. 3-8, we first find the transformed open-circuit output voltage $V_0(s)$. This is easily done by considering the capacitor and the inductor as elements of a voltage divider. Thus, we may write

$$V_0(s) = \frac{1/s}{(s/9) + (1/s)} V_s(s) = \frac{9}{s^2 + 9} \frac{3s}{s^2 + 4} \tag{8-197}$$

The techniques developed in Sec. 8-4 for determining a partial-fraction expansion are readily applied to change the product form of the right member of (8-197) into a sum form. Thus, we may write

$$V_0(s) = \frac{-27s/5}{s^2 + 9} + \frac{27s/5}{s^2 + 4} \tag{8-198}$$

Considering this result, and recalling that the open-circuit voltage specifies the value of the voltage source in the equivalent Thevenin circuit, we see that the time-domain equivalent voltage source will have an output $v_0(t) = (27/5)[-\cos 3t + \cos 2t]u(t)$. This may be modeled in the transformed network as a single voltage source with an output as given by the right member of Eq. (8-197). Now let us find the second part of the equivalent circuit, the two-terminal impedance. This is found by setting the output of the independent source $V_s(s)$ to zero and determining the impedance at the terminals of the network. For the network shown in Fig. 8-10.1 this is simply the impedance of the parallel combination of the inductor and the capacitor. Thus, the Thevenin equivalent of the transformed network shown in Fig. 8-10.1 may be drawn as shown in Fig. 8-10.2.

To verify the validity of the Thevenin equivalent circuit derived above, let us determine the expression for the current $I(s)$ which flows in a 1-Ω resistor that is connected to the terminals of the original circuit shown in Fig. 8-10.1. The connection is illustrated in Fig. 8-10.3(a). For this circuit, we may find $I(s)$ by writing the mesh

$$V_S = \frac{27s}{(s^2 + 9)(s^2 + 4)}$$

henries, farads

1/9

1

henries, ohms, farads

Figure 8-10.2 Thevenin equivalent for the transformed network shown in Fig. 8-10.1.

$$V_S = \frac{3s}{s^2 + 4}$$

1/9

I_1

I

(a)

$$V_S = \frac{27s}{(s^2 + 9)(s^2 + 4)}$$

1/9

1

I

(b)

Figure 8-10.3 Testing the circuits of Figs. 8-10.1 and 8-10.2 using a one-Ω termination.

equations. Thus, we obtain

$$\begin{bmatrix} \dfrac{3s}{s^2 + 4} \\[2ex] 0 \end{bmatrix} = \begin{bmatrix} \dfrac{s}{9} + \dfrac{1}{s} & -\dfrac{1}{s} \\[2ex] -\dfrac{1}{s} & 1 + \dfrac{1}{s} \end{bmatrix} \begin{bmatrix} I_1(s) \\[2ex] I(s) \end{bmatrix} \tag{8-199}$$

Solving these equations, we find

$$I(s) = \frac{27s}{(s^2 + 4)(s^2 + s + 9)} \tag{8-200}$$

Now consider the problem of finding the current $I(s)$ in a 1-Ω resistor used as a termination for the Thevenin equivalent circuit shown in Fig. 8-10.2. The configuration is shown in Fig. 8-10.3(b). In this case, we find that the total impedance of the mesh is $Z(s) = (s^2 + s + 9)/(s^2 + 9)$. Thus, the current $I(s)$ is simply given as

$$I(s) = \frac{V_0(s)}{Z(s)} = \frac{27s}{(s^2 + 4)(s^2 + 9)} \frac{s^2 + 9}{s^2 + s + 9}$$

$$= \frac{27s}{(s^2 + 4)(s^2 + s + 9)} \tag{8-201}$$

Therefore, as is to be expected, the same results are obtained in the two cases. The procedure outlined above for finding the Thevenin equivalent circuit for a transformed network is summarized as follows:

SUMMARY 8-10.1

The Thevenin Equivalent Circuit for a Transformed Network. A transformed network representation for an actual time-domain circuit in which the initial conditions are zero may be converted at a given pair of terminals to a transformed Thevenin

The Laplace Transformation   Chap. 8

equivalent network consisting of a transformed voltage source in series with a two-terminal impedance. The output of the transformed Thevenin voltage source is simply the transformed voltage seen at the pair of terminals when these are open-circuited. The Thevenin impedance is the impedance of the network seen between the pair of terminals when all the independent sources have been set to zero. ∎

By a similar development, we may extend the procedure for finding a Norton equivalent circuit given in Sec. 3-8 to include transformed networks. The transformed equivalent circuit will consist of an independent current source and a shunt two-terminal admittance. As an example, consider finding a Norton equivalent circuit for the network shown in Fig. 8-10.1. First we will find the transformed current that will flow in a short circuit placed across the terminals of the network. This gives the value for the equivalent transformed Norton current generator. Letting $I_0(s)$ be the expression for this current, we find from Fig. 8-10.4(a) that

$$I_0(s) = \frac{3s}{s^2 + 4} \frac{9}{s} = \frac{27}{s^2 + 4} \tag{8-202}$$

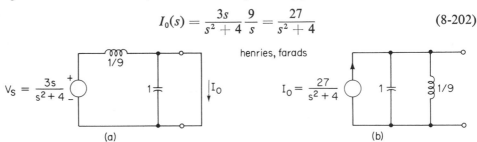

henries, farads

Figure 8-10.4 Finding a Norton equivalent circuit.

The equivalent admittance is simply the parallel connection of the inductor and the capacitor, i.e., the admittance seen at the specified pair of terminals when the independent source is set to zero. Thus, we obtain an equivalent Norton circuit as shown in Fig. 8-10.4(b). These results may be summarized as follows:

SUMMARY 8-10.2

The Norton Equivalent Circuit for a Transformed Network. A transformed network representation for an actual time-domain circuit in which the initial conditions are zero may be converted at a given pair of terminals to an equivalent transformed Norton circuit consisting of a transformed current source in shunt with a two-terminal admittance. The output of the Norton current source is simply the transformed current that flows through a short circuit placed across the pair of terminals. The Norton admittance is the admittance of the network seen between the pair of terminals when all the independent sources have been set to zero. Its value is the reciprocal of the value of the Thevenin impedance. ∎

In practice, a circuit realization for a Thevenin impedance (or a Norton admittance) is easily found directly from the transformed network without any computation.

This is so because it is simply the portion of the original transformed circuit connected to the specified terminal pair which remains after the independent transformed sources have been set to zero, i.e., after voltage sources have been replaced by a short circuit and current sources by an open circuit. An example of such a situation follows.

Example 8-10.1

The transformed network shown in Fig. 8-10.5(a) contains two independent sources and one (dependent) VCIS. It is desired to find a Thevenin equivalent circuit for the original network as seen at the pair of terminals labeled a-b. To aid in such a determination, we may first convert the branch consisting of a series-connected voltage source and an inductor to a Norton equivalent as shown in Fig. 8-10.5(b). In addition,

Figure 8-10.5 Network for Example 8-10.1.

using the techniques given in Example 3-8.4, a source producing a test current I_0 has been added at nodes a-b as shown in the figure. For this network, we obtain

$$\begin{bmatrix} I_S(s) - \dfrac{V_S(s)}{s} \\ I_0(s) + \dfrac{V_S(s)}{s} \end{bmatrix} = \begin{bmatrix} 2s + \dfrac{1}{s} & -\dfrac{1}{s} \\ -2 - \dfrac{1}{s} & 3 + \dfrac{1}{s} \end{bmatrix} \begin{bmatrix} V_1(s) \\ V_2(s) \end{bmatrix}$$

where the effect of the controlled source has been included in the node admittance matrix. Solving this matrix equation we obtain

$$\begin{bmatrix} V_1(s) \\ V_2(s) \end{bmatrix} = \frac{s}{6s^2 + 2s + 1} \begin{bmatrix} 3 + \dfrac{1}{s} & \dfrac{1}{s} \\ 2 + \dfrac{1}{s} & 2s + \dfrac{1}{s} \end{bmatrix} \begin{bmatrix} I_S(s) - \dfrac{V_S(s)}{s} \\ I_0(s) + \dfrac{V_S(s)}{s} \end{bmatrix}$$

Solving for $V_2(s)$ we obtain

$$V_2(s) = \frac{(2s + 1)I_S(s) + 2(s - 1)V_S(s)}{6s^2 + 2s + 1} + \frac{(2s^2 + 1)I_0(s)}{6s^2 + 2s + 1} = V_{Thev.}(s) + Z_{Thev.}(s)I_0(s)$$

Thus, the Thevenin voltage is a function of the two independent sources $I_S(s)$ and $V_S(s)$, and the Thevenin impedance is specified by the expression multiplying the test

The Laplace Transformation   Chap. 8

current $I_0(s)$. Alternately, the Thevenin impedance may simply be represented by the original network in which the independent sources have been set to zero. Thus, the complete Thevenin equivalent circuit is as shown in Fig. 8-10.5(c).

The procedures given above for finding the Thevenin or Norton equivalent of a transformed network in which all the initial conditions have been assumed to be zero are readily modified for the case where the effect of non-zero initial conditions is to be considered. This may be done by applying the results of Summaries 8-6.5 and 8-6.6, in which it is shown that impulse and step generators may be used to establish initial conditions in a network which otherwise would be in the zero state. An example follows.

Example 8-10.2

For the network shown in Fig. 8-10.6(a), it is desired to use the concept of a transformed network to find the Thevenin equivalent circuit as seen at terminals a-b for the case where the initial conditions $v_1(0+)$ and $i_L(0+)$ are not zero. The first step is to redraw the network inserting a current source in shunt across the capacitor and a voltage source in series with the inductor as shown in Fig. 8-10.6(b). As indicated in the

(a)

(b)

Figure 8-10.6 Network for Example 8-10.2.

figure, the output of the additional transformed current source is K_1 (an impulse of strength K_1) and the output of the new transformed voltage source is $K_2/4$. Analyzing the circuit and making the substitutions $K_1 = v_1(0+)$ and $K_2 = i_L(0+)$, then the equations describing the circuit are identical with those given in Eq. (8-132), which were derived directly from the time-domain equations for the same network (assuming that the initial conditions are the same at $0+$ as they are at $0-$). Thus, the solution for $V_2(s)$ yields the value of the equivalent Thevenin voltage generator. This is easily seen to be

$$V_2(s) = \frac{4I_1(s)}{s^2 + 2s + 2} + \frac{(s + 2)I_2(s)}{s^2 + 2s + 2} + \frac{4v_1(0+) - (s + 2)i_L(0+)/s}{s^2 + 2s + 2}$$

Note that, in the above equation, the term multiplying $I_2(s)$ is the network function which defines the Thevenin impedance of the network at the terminals a-b.

8-11 CONCLUSION

In this chapter we have introduced one of the most useful and powerful techniques that exists for the analysis of lumped linear time-invariant circuits, namely, the complex frequency variable as defined by the Laplace transformation. In addition to its use as an analysis tool, through the methods of Sec. 8-9, the complex frequency variable provides an important link between the sinusoidal steady-state characteristics of a network, and its more general time-domain properties. The complex frequency variable has another very important use. It provides the main mechanism by means

Figure 8-11.1 Flow chart for time-domain and frequency-domain analysis.

The Laplace Transformation   Chap. 8

of which synthesis operations are made possible. By synthesis, we mean an operation which starts with some desired driving-point or transfer function, and produces a network realization of that function. The subject of network synthesis is beyond the scope of this text. The titles of several books which treat this subject, however, may be found in the Bibliography at the end of the book.

As a summary of the various techniques for applying Laplace transformation methods to the analysis of networks and a comparison of those techniques with the time-domain methods introduced in Chaps. 5 and 6, we present on a single flow chart in Fig. 8-11.1 the most significant operations of both time-domain and frequency-

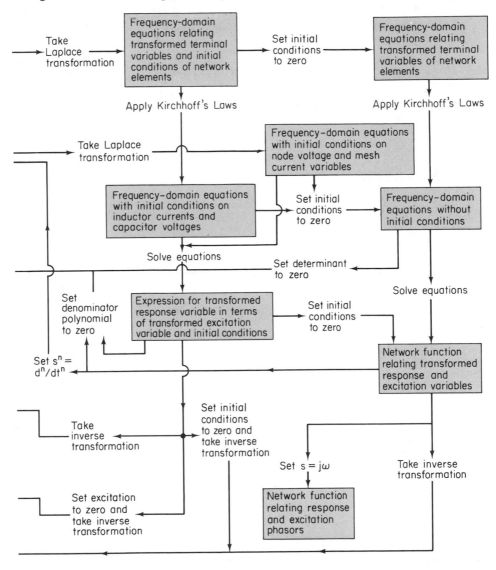

Figure 8-11.1 (Cont.)

domain analysis. At first glance, this flow chart will probably appear somewhat formidable to the reader. This shouldn't be too surprising when you consider that it summarizes several hundred pages of text material! The left half of the flow chart represents operations in the time domain, and the right half represents operations in the frequency domain. Of particular interest to the student wishing a thorough grasp of the basic fundamentals of circuit analysis is the multitude of routes that can be taken to arrive at a given answer. For example, from the flow chart it is readily apparent that the characteristic equation can be obtained from the homogeneous time-domain differential equation [by substituting $f(t) = Ke^{st}$], from the frequency-domain mesh or node equations (by setting the determinant to zero), or from the network function (by setting the denominator polynomial to zero). Many other interesting equivalences will be found in the flow chart. To effectively study the flow chart and thus to review the material of the first eight chapters of this book, it is suggested that the student start with a simple network such as may be found in any of the examples in Chap. 5, 6, or 8 and actually carry through each of the steps indicated in the flow chart. The successful completion of such a project indicates an excellent level of comprehension of the text material.

PROBLEMS

Problem 8-1 (Sec. 8-2)

For each of the circuits shown in Fig. P8-1:
(a) Find the differential equation relating the output voltage $v(t)$ and the input current $i(t)$.
(b) Find an expression for $V(s)$, the Laplace transform of $v(t)$ in terms of $I(s)$, the Laplace transform of $i(t)$, and the initial conditions on $v(t)$.
(c) Identify the characteristic polynomial.

ohms, henries, farads

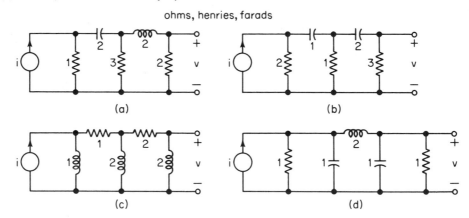

Figure P8-1

Problem 8-2 (Sec. 8-2)

For each of the circuits shown in Fig. P8-2:
(a) Find the differential equation relating the output voltage $v_2(t)$ and the input voltage $v_1(t)$.

(b) Find an expression for $V_2(s)$, the Laplace transform of $v_2(t)$, in terms of $V_1(s)$, the Laplace transform of $v_1(t)$, and the initial conditions on $v_2(t)$.

(c) Identify the characteristic polynomial.

ohms, henries, farads

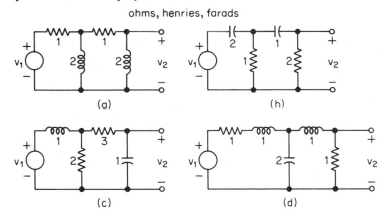

(a)

(h)

(c)

(d)

Figure P8-2

Problem 8-3 (Sec. 8-2)

Repeat Problem 8-2 for each of the networks shown in Fig. P8-3.

ohms, farads

$K = 1.586$

(a)

(b)

Figure P8-3

Problem 8-4 (Sec. 8-2)

Derive the expression for the Laplace transform of the nth-order derivative of a time-domain function $f(t)$ given in Summary 8-2.1.

Problem 8-5 (Sec. 8-3)

Find the Laplace transform of each of the functions shown in Fig. P4-13.

Problem 8-6 (Sec. 8-3)

Find the Laplace transform of each of the functions shown in Fig. P4-14. Assume that the functions are zero for $t < 0$.

Problem 8-7 (Sec. 8-3)

Derive the expression for the Laplace transform of the function $Kt^n u(t)$, where n is a positive integer, given in Table 8-3.1.

Problem 8-8 (Sec. 8-3)

Find the Laplace transform of each of the following time-domain functions.

(a) $t \cos t$ (b) $t \sin t$

(c) $e^{-t} \cos t$ (d) $e^{-t} \sin t$

Problem 8-9 (Sec. 8-4)

Find the inverse Laplace transform of each of the following frequency-domain functions.

(a) $\dfrac{3s + 2}{s^2 + 1}$ (b) $\dfrac{1}{(s + 1)^2}$

(c) $\dfrac{3s^2 + 2s + 2}{(s + 2)^2(s + 3)}$ (d) $\dfrac{s - 1}{s^2 + 2s + 2}$

(e) $\dfrac{4s^2 - 3s + 5}{s(s^2 + 2s + 5)}$ (f) $\dfrac{3s^3 + 2s^2 + s + 1}{(s^2 + 2s + 2)^2}$

Problem 8-10 (Sec. 8-4)

Find the inverse Laplace transform of each of the following frequency-domain functions.

(a) $\dfrac{s^3}{s + 1}$ (b) $\dfrac{s^3 + s^2 + 3s + 2}{s(s^2 + s + 1)}$

(c) $\dfrac{s^4 + 1}{(s + 1)^3}$ (d) $\dfrac{s}{(s^2 + 1)^2}$

Problem 8-11 (Sec. 8-4)

Find the inverse Laplace transform of the following partial-fraction expansion.

$$F(s) = \frac{j1}{s + 1 + j1} + \frac{-j1}{s + 1 - j1} + \frac{-1}{(s + 1 + j1)^2} + \frac{-1}{(s + 1 - j1)^2}$$

Problem 8-12 (Sec. 8-4)

Find the initial values for the inverse Laplace transform of each of the functions given in Problem 8-9 by applying the results of Summary 8-4.6.

Problem 8-13 (Sec. 8-4)

Find the final values for the inverse Laplace transform of each of the functions given in Problems 8-9 and 8-10 by applying the results of Summary 8-4.7.

Problem 8-14 (Sec. 8-4)

Make a plot of the values of the residues K_1 and K_2 for the function $F(s)$ given below. Use a as the independent variable over the range 0 to 1.

$$F(s) = \frac{1}{(s + 1)(s + a)} = \frac{K_1}{s + 1} + \frac{K_2}{s + a}$$

Problem 8-15 (Sec. 8-4)

(a) Determine the response to a step function $u(t)$ for each of the network functions given below. $F_1(s)$ is a low-pass function satisfying a Bessel or maximally flat delay (in the time-domain) characteristic. $F_2(s)$ is a low-pass function satisfying a Butterworth or maximally flat magnitude (as a function of a sinusoidal frequency) characteristic. Plot the responses as a function of time over a range from 0 to 10 s.

(b) Repeat the problem for the functions defined as $F_3(s)$ and $F_4(s)$.

$$F_1(s) = \frac{3}{s^2 + 3s + 3} \qquad F_2(s) = \frac{1}{s^2 + \sqrt{2}\,s + 1}$$

$$F_3(s) = \frac{15}{s^3 + 6s^2 + 15s + 15} = \frac{15}{(s + 2.322)(s^2 + 3.6778s + 6.4594)}$$

$$F_4(s) = \frac{1}{s^3 + 2s^2 + 2s + 1} = \frac{1}{(s + 1)(s^2 + s + 1)}$$

Problem 8-16 (Sec. 8-5)

For the network used in Example 8-5.1, find the voltage $v_R(t)$ across the resistor (with the positive reference polarity on the left), for the case where the input voltage $v_S(t)$ is $10 \cos t\, u(t)$ V and the initial conditions are zero.

Problem 8-17 (Sec. 8-5)

Use the Laplace transform technique to find $v_2(t)$ in the network shown in Fig. P8-17. Assume that all initial conditions are zero and that $i(t) = 2e^{-t}u(t)$ A.

Figure P8-17

Problem 8-18 (Sec. 8-5)

Use the Laplace transform technique to find an expression for $v_2(t)$ for the network shown in Fig. P8-17 if $i(t) = u(t)$ A and if the initial conditions are $v_1(0-) = 1$ V, $i_L(0-) = 2$ A.

Problem 8-19 (Sec. 8-5)

Find the initial conditions $v_1(0-)$ and $i_L(0-)$ such that the output $v_2(t) = e^{-t} \cos t$ V in the circuit shown in Fig. P8-17. Let $i(t) = u(t)$ A.

Problem 8-20 (Sec. 8-5)

Find the driving-point immittance of each of the networks shown in Fig. P8-20.

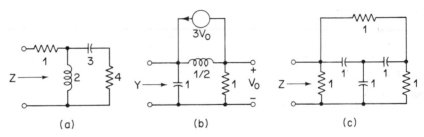

Figure P8-20

Problem 8-21 (Sec. 8-5)

In the network shown in Fig. P8-21, if the initial conditions are zero, find an expression for $v_2(t)$ for the following cases:

(a) When the excitation voltage $v_1(t)$ is $\sin t\, u(t)$ V.

(b) When the excitation voltage $v_1(t)$ is $\sin 2t\, u(t)$ V.

Figure P8-21

Problem 8-22 (Sec. 8-5)

Find an expression for the Laplace transform $V_C(s)$ of the voltage $v_C(t)$ which appears across the capacitance (with the positive reference polarity at the top) in Fig. 8-5.1 in terms of the Laplace transform $V_S(s)$ of the excitation voltage $v_S(t)$ and the initial conditions. Identify the zero-state response component and the zero-input response component of your answer.

Problem 8-23 (Sec. 8-5)

Find an expression for the Laplace transform $V_L(s)$ of the voltage $v_L(t)$ which appears across the inductance (with the positive reference polarity at the left) in Fig. 8-5.1 in terms of the Laplace transform $V_S(s)$ of the excitation voltage $v_S(t)$ and the initial conditions. Identify the zero-state response component and the zero-input response component of your answer.

Problem 8-24 (Sec. 8-5)

For the network shown in Fig. P8-24, find the expressions for the following variables in terms of the excitation current $I_1(s)$ and the initial conditions on the energy-storage elements:

(a) $V_1(s)$ (b) $V_2(s)$

(c) $V_L(s)$ (d) $I_C(s)$

Figure P8-24

Problem 8-25 (Sec. 8-5)

For the RLC network and the values used in Example 8-5.1, find a set of initial conditions such that the sinusoidal term in the response of $v_R(t)$, the voltage across the resistor, is of the form $Ke^{-t} \sin t$, where $K > 0$. The positive reference polarity $v_R(t)$ is assumed to be at the left of the resistor.

Problem 8-26 (Sec. 8-5)

Verify the solution for the constants a, b, c, and d in Example 8-5.6.

Problem 8-27 (Sec. 8-5)

Find a set of initial conditions $v_1(0-)$ and $i_L(0-)$ such that there is no sinusoidal oscillation in the output for $v_2(t)$ in the circuit shown in Fig. P8-27.

Figure P8-27

Problem 8-28 (Sec. 8-6)

In the circuit shown in Fig. P8-28, use Laplace transform techniques to find an expression for $v_2(t)$. Assume that the initial conditions are zero.

ohms, henries, farads

Figure P8-28

Problem 8-29 (Sec. 8-6)

Write the network equations for the circuit shown in Fig. P8-29. The equations should have the form $\mathbf{V}(s) = \mathbf{Z}(s)\mathbf{I}(s)$, where $\mathbf{I}(s)$ is a column matrix whose elements are $I_1(s)$ and $I_2(s)$. Assume that the initial conditions are zero.

ohms, henries, farads

Figure P8-29

Figure P8-30

Problem 8-30 (Sec. 8-6)

For the circuit shown in Fig. P8-30, write the matrix equations $\mathbf{I}(s) = \mathbf{Y}(s)\mathbf{V}(s)$, where $\mathbf{V}(s)$ is a column matrix with elements $V_1(s)$, $V_2(s)$, and $V_3(s)$. Assume that all elements have unity value and all initial conditions are zero.

Problem 8-31 (Sec. 8-6)

For the network shown in Fig. P8-31:
(a) Find an expression for $V_2(s)$ in terms of $I_1(s)$, $v_1(0)$, and $i_L(0)$.
(b) Find an expression for $V_2(s)$ in terms of $I_1(s)$, $v_2(0)$, and $v_2'(0)$.

ohm, henries, farads

Figure P8-31

Problem 8-32 (Sec. 8-6)

In the network shown in Fig. P8-32, the switch $S1$ has been closed for a long time. It is opened at $t = 0$. Find the expression for $v_2(t)$.

Figure P8-32

Figure P8-33

Problem 8-33 (Sec. 8-6)

In the circuit shown in Fig. P8-33, switch $S1$ has been closed for a long time. It is opened at $t = 0$. Find the expression for $v_2(t)$.

Problem 8-34 (Sec. 8-6)

In the circuit shown in Fig. P8-33, switch $S1$ has been open for a long time. It is closed at $t = 0$. Find the expression for $v_2(t)$.

Problem 8-35 (Sec. 8-6)

Find the integro-differential equation relating the input quantity $v_1(t)$ and the output quantity $v_2(t)$ for the network shown in Fig. P8-35.

Figure P8-35

Figure P8-36

Problem 8-36 (Sec. 8-6)

Find the integro-differential equation relating the excitation variable $i_1(t)$ and the response variable $v_2(t)$ in the network shown in Fig. P8-36.

Problem 8-37 (Sec. 8-6)

Find the differential equation which relates $v_1(t)$ and $v_2(t)$ in the network shown in Fig. P8-37.

Figure P8-37

The Laplace Transformation   Chap. 8

Problem 8-38 (Sec. 8-6)

Add independent sources to the network shown in Fig. P8-35 so that the initial conditions on the reactive elements are established by the specification of impulses on these sources, then write the mesh equations for the network.

Problem 8-39 (Sec. 8-6)

Repeat Problem 8-38 for the network shown in Fig. P8-36. Use node equations instead of mesh equations.

Problem 8-40 (Sec. 8-6)

Repeat Problem 8-38 using sources with a step function output to generate the initial conditions.

Problem 8-41 (Sec. 8-7)

For the network shown in Fig. P8-20 add voltage sources as follows:
(a) Across terminals 1-2 in Fig. P8-20(a).
(b) In series with the 1-F capacitor in Fig. P8-20(b).
(c) In series with the 1-Ω resistor on the left of Fig. P8-20(c).
Now write mesh equations for each of the networks and from these equations find the natural frequencies.

Problem 8-42 (Sec. 8-7)

For the networks shown in Fig. P8-20 add current sources as follows:
(a) Across the 1-Ω resistor in Fig. P8-20(a) (also put a short across terminals 1-2).
(b) In parallel with the 1-F capacitor in Fig. P8-20(b).
(c) In parallel with the left 1-Ω resistor in Fig. P8-20(c).
Now write node equations for each of the networks and from these find the natural frequencies.

Problem 8-43 (Sec. 8-7)

Compare the results obtained in Problems 8-41 and 8-42.

Problem 8-44 (Sec. 8-7)

(a) If a voltage source is used to excite the network shown in Fig. P8-44 at the terminals a-b, what are the natural frequencies of the network?
(b) If a current source is used, what are the natural frequencies?
(c) Should the network be called "open-circuit stable" or "short-circuit stable" (or both)?

ohms, henries, farads

Figure P8-44

Problem 8-45 (Sec. 8-7)

A set of network functions are defined by making two tests on the same network as shown in Fig. P8-45. Predict any equalities that will exist between the various polynomials $A(s)$, $B(s)$, $C(s)$, $D(s)$, $E(s)$, $F(s)$, $G(s)$, and $H(s)$.

$$\frac{V_1(s)}{I_1(s)} = \frac{A(s)}{B(s)}; \frac{V_2(s)}{V_1(s)} = \frac{C(s)}{D(s)} \qquad \frac{I_1(s)}{V_1(s)} = \frac{E(s)}{F(s)}; \frac{V_2(s)}{I_1(s)} = \frac{G(s)}{H(s)}$$

Figure P8-45

Problem 8-46 *(Sec. 8-7)*

As a physical example of the conclusions reached in Problem 8-45, find the polynomials $A(s), B(s), C(s), D(s), E(s), F(s), G(s)$, and $H(s)$ defined in that problem for the network shown in Fig. P8-46.

Figure P8-46

Problem 8-47 *(Sec. 8-7)*

A series of four tests are made on a two two-terminal-pair network as shown in Fig. P8-47. If the network is reciprocal, find any equalities that will exist between the various polynomials $A(s), B(s), C(s), D(s), E(s), F(s), G(s)$, and $H(s)$.

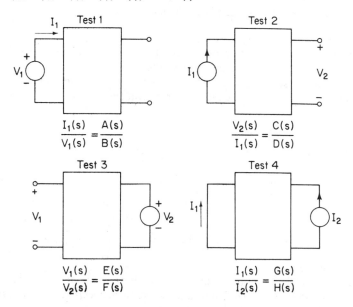

$$\frac{I_1(s)}{V_1(s)} = \frac{A(s)}{B(s)} \qquad \frac{V_2(s)}{I_1(s)} = \frac{C(s)}{D(s)}$$

$$\frac{V_1(s)}{V_2(s)} = \frac{E(s)}{F(s)} \qquad \frac{I_1(s)}{I_2(s)} = \frac{G(s)}{H(s)}$$

Figure P8-47

The Laplace Transformation   Chap. 8

Problem 8-48 (Sec. 8-7)

As a physical example of the conclusions reached in Problem 8-47, find the polynomials $A(s)$, $B(s)$, $C(s)$, $D(s)$, $E(s)$, $F(s)$, $G(s)$, and $H(s)$ defined in that problem for the network shown in Fig. P8-46.

Problem 8-49 (Sec. 8-7)

Find any relations that exist between any of the polynomials $A(s)$, $B(s)$, $C(s)$, $D(s)$, $E(s)$, and $F(s)$ in the three network functions defined by the three tests on the same network shown in Fig. P8-49.

Figure P8-49

Problem 8-50 (Sec. 8-7)

Find any relations that may exist between the various polynomials $A(s)$, $B(s)$, $C(s)$, $D(s)$, $E(s)$, $F(s)$, $G(s)$, and $H(s)$ which define the different network functions for the network shown in Fig. P8-50.

Figure P8-50

Problem 8-51 (Sec. 8-7)

    (a) Predict which of the networks shown in Fig. P8-51 are reciprocal by observing the form and the components of the networks.

    (b) Find the network functions $Y_{12}(s) = I_1(s)/V_2(s)$ [with $V_1(s) = 0$] and $Y_{21}(s) = I_2(s)/V_1(s)$ [with $V_2(s) = 0$] or the network functions $Z_{12}(s) = V_1(s)/I_2(s)$ [with $I_1(s) = 0$] and $Z_{21}(s) = V_2(s)/I_1(s)$ [with $I_2(s) = 0$] for each of the networks shown in the figure to verify the conclusions reached in part (a).

Figure P8-51

Problem 8-52 (Sec. 8-8)

    The networks shown in Fig. P8-52 are to be operated as open-circuit voltage dividers. Without determining the actual network function defining the voltage transfer ratio for these networks, predict the location of any transmission zeros. Verify your prediction by actual analysis.

Figure P8-52

The Laplace Transformation   Chap. 8

Problem 8-53 (Sec. 8-8)

Find the location and multiplicity of any zeros of transmission which are present in the ladder network shown in Fig. P8-53.

henries, farads

Figure P8-53

Problem 8-54 (Sec. 8-8)

(a) Assuming that the function given below is an impulse response, find the network function.

(b) Assuming that the function given below is a step response, find the network function

$$f(t) = e^{-t} + \cos t$$

Problem 8-55 (Sec. 8-8)

Treat the network shown in Fig. P8-55 as a ladder network and predict the general form of $v_2(t)$ for an input function $v_1(t) = e^{-t} - e^{-2t}$. Verify your prediction by analyzing the network and solving for $v_2(t)$. Assume that the initial conditions are zero.

ohms, farads

Figure P8-55

Problem 8-56 (Sec. 8-8)

In each of the networks shown in Fig. P8-56 assume that all the elements have unity value. Predict as much information as you can about the network functions defined by the indicated input and output variables, for example: order of numerator and denominator polynomials, location of natural frequencies, location and order of transmission zeros, etc. It is NOT necessary that you try to write and solve the network equations for the circuits.

(a)   (b)

Figure P8-56

(c)   (d)

Figure P8-56 (Cont.)

Problem 8-57 (Sec. 8-8)

(a) Find values of L and C for the network shown in Fig. P8-57 such that the circuit has no steady-state sinusoidal output component [of $v_2(t)$].

(b) Find the time-domain expression for the output $v_2(t)$ for the values chosen in part (a). Assume that the initial conditions are zero.

Figure P8-57

Problem 8-58 (Sec. 8-8)

The circuit shown in Fig. P8-58 is to be used to realize the open-circuit voltage transfer function

$$\frac{V_2}{V_1} = \frac{(s^2 + 1)(s^2 + 4)}{as^4 + bs^3 + cs^2 + ds + 8}$$

Find values for the elements $L_1, L_2, C_1, C_2,$ and R.

Figure P8-58

Problem 8-59 (Sec. 8-9)

Draw approximate sketches of the magnitude and phase of each of the pole-zero configurations shown in Fig. P8-59.

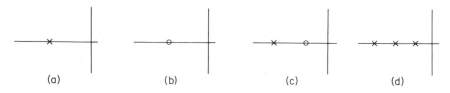

(a)   (b)   (c)   (d)

Figure P8-59

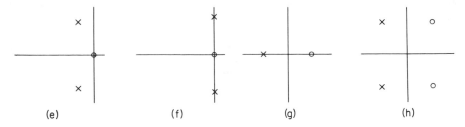

(e)   (f)   (g)   (h)

Figure P8-59 (Cont.)

Problem 8-60 (Sec. 8-9)

Sketch the magnitude and the phase response curves for each of the pole-zero configurations shown in Fig. P8-60. Give numerical values for significant points on your plots. Assume all network functions have unity gain constants.

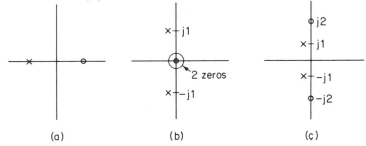

(a)   (b)   (c)

Figure P8-60

Problem 8-61 (Sec. 8-9)

Sketch magnitude and phase plots for each of the pole-zero configurations shown in Fig. P8-61. Assume that the gain constant is unity and indicate significant values on the ordinates and abscissas of your plots.

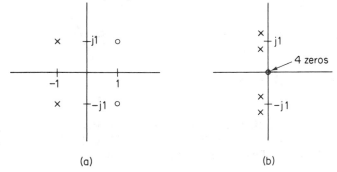

(a)   (b)   **Figure P8-61**

Problem 8-62 (Sec. 8-9)

The magnitude characteristics shown in Fig. P8-62 are each derived from a network function which has a single pole and a single zero. (Either the pole or the zero may be located at infinity.) Draw an approximate pole-zero plot which will produce each of the given characteristics.

Problem 8-63 (Sec. 8-9)

The phase characteristics shown in Fig. P8-63 are each derived from a network function which has a single pole and a single zero. (Either the pole or the zero may be located at infinity.) Draw an approximate pole-zero plot which will produce each of the given characteristics.

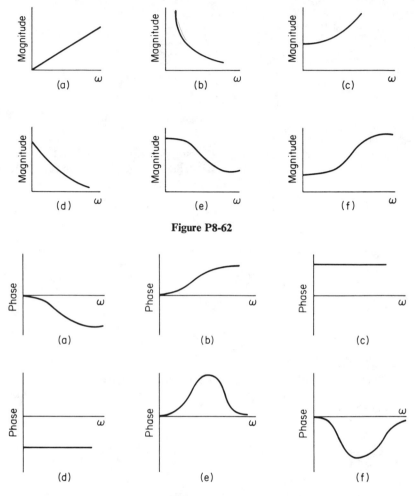

Figure P8-62

Figure P8-63

Problem 8-64 (Sec. 8-9)

The magnitude characteristics shown in Fig. P8-64 are each derived from a network function which has two complex conjugate poles and two zeros. The latter may only be located at the origin or at infinity. Draw the pole-zero plot which will produce each of the characteristics.

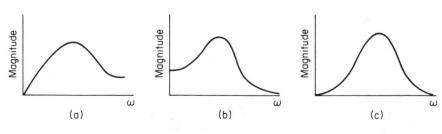

Figure P8-64

Problem 8-65 (Sec. 8-9)

(a) Draw two magnitude loci (labeled I and II) on the same plot for the pole-zero configurations shown in I and II in Fig. P8-65.

(b) Repeat for the pole-zero configurations shown in III and IV.

(c) Repeat for the pole-zero configurations shown in V and IV.

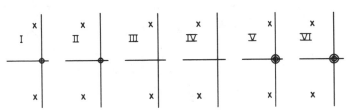

Figure P8-65

Problem 8-66 (Sec. 8-9)

For each of the pole-zero configurations labeled (a), (b), (c), and (d) shown in Fig. P8-66 identify the approximate waveform of the step response as one of those labeled (1), (2), (3), or (4).

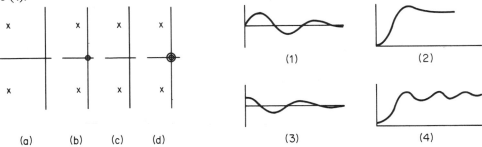

Figure P8-66

Problem 8-67 (Sec. 8-10)

Find Thevenin and Norton equivalent circuits for each of the transformed networks shown in Fig. P8-67. Assume that all the initial conditions are zero.

Problem 8-68 (Sec. 8-10)

Test the validity of the Thevenin and Norton circuits found in Problem 8-67 by calculating the transformed current that will flow in a 1-Ω resistor connected across the output terminals of the original circuit, the Thevenin equivalent circuit, and the Norton equivalent circuit.

Problem 8-69 (Sec. 8-10)

Repeat Problem 8-67 for the circuit shown in Fig. P8-67(a) under the assumption that the reactive element shown has a non-zero initial condition.

Problem 8-70 (Sec. 8-10)

Repeat Problem 8-68 for the circuit shown in Fig. P8-67(a) under the assumption that the reactive element shown has a non-zero initial condition.

Problem 8-71 (Sec. 8-10)

Find a Thevenin equivalent circuit for the network shown in Fig. 8-10.5 at the terminals where $V_1(s)$ is defined.

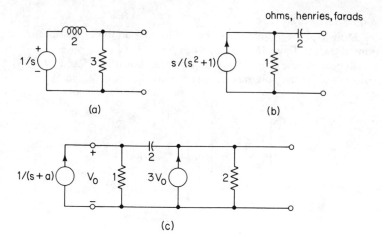

(a)

ohms, henries, farads

(b)

(c)

Figure P8-67

9

Two-Port Network

Parameters

One of the most frequent tasks to which networks are applied is the process of shaping (or filtering) some electrical signal information. In order to accomplish this filtering process, it is necessary that a specific pair of terminals in the network be designated as the place at which an excitation or input is to be applied and another pair of terminals be designated as the place where the response or output, i.e., the filtered signal, is present. A frequently used name applied to a pair of terminals used for input or output purposes is a *port*. Thus, the input terminal-pair (this is customarily drawn at the left side of the network schematic) may be referred to as the *input port* and the output terminal-pair (this is usually drawn at the right side of the network schematic) is called the *output port*. Thus, in a typical filtering application, the network may be referred to as a *two-port network*. In this chapter we shall investigate some ways of characterizing two-port networks. Before we do this, however, we must consider some of the more general details that apply to all networks which have two ports. This is done in the following section.

9-1 ONE-PORT AND TWO-PORT NETWORKS

Before we investigate some of the properties of the general two-port network, let us first consider a simpler case, namely, the one-port network. It should be noted here that all the developments which follow refer to *transformed networks* (as defined in Sec. 8-6). Thus, all the variables and all the network functions will be functions of the complex frequency variable. Generally, we may draw a one-port network as shown in Fig. 9-1.1, indicating that only two of all the network's nodes are to be made available for excitation and measurement of response. The interior of the box may contain any arbitrary collection of resistors, capacitors, inductors, and dependent sources. The network shown in Fig. 9-1.1 has two variables defined at the port. The relative reference directions for these two variables follow the convention for associated reference directions developed in Sec. 1-2. The network itself will, of course, impose a relation between these two variables, namely, the driving-point impedance $Z(s)$. Thus, we may write the expression

$$V(s) = Z(s)I(s) \tag{9-1}$$

relating the variables defined at the port. Alternately we may write

$$I(s) = Y(s)V(s) \tag{9-2}$$

Thus, we see that in a one-port network, two network functions are defined. One of these is the driving-point impedance $Z(s)$, where

$$Z(s) = \frac{V(s)}{I(s)} \tag{9-3}$$

in which (since network functions are defined as response/excitation) it is assumed that the excitation is provided by a current source as shown in Fig. 9-1.2. The other network function defined for a one-port network is $Y(s)$, where

$$Y(s) = \frac{I(s)}{V(s)} \tag{9-4}$$

and the excitation is provided by a voltage source as shown in Fig. 9-1.3. Obviously, $Y(s)$ of Eq. (9-4) equals the reciprocal of $Z(s)$ of Eq. (9-3).

Figure 9-1.1 One-port network.

Figure 9-1.2 One-port network excited by a current source.

Figure 9-1.3 One-port network excited by a voltage source.

The description of a one-port network given above assumes that *there are no independent sources* in the network, since, if such sources are present, a voltage may appear across the open-circuited terminals of the network. Thus, if independent sources are present, Eq. (9-1) will have the general form

$$V(s) = Z(s)I(s) + V_o(s) \tag{9-5}$$

where the term $V_o(s)$ represents the effects of the independent sources, i.e., when $I(s)$ is zero, $V(s) = V_o(s)$. Another effect which is observed if independent sources are present in a one-port network is that a current may flow through a short circuit placed across the terminals of the port. In this case, Eq. (9-2) must have the form

$$I(s) = Y(s)V(s) + I_o(s) \tag{9-6}$$

where the term $I_o(s)$ represents the effects of the independent sources. The above discussion may be summarized as follows:

SUMMARY 9-1.1

Definition of a One-Port Network. A one-port network is any network which does not contain independent sources and in which only a single terminal pair is to be used for the application of an excitation and the measurement of a response. It has the following properties:

1. There are two port variables, $V(s)$ and $I(s)$, with the relative reference polarities shown in Fig. 9-1.1.
2. There is one equation which relates the variables.
3. The equation may be written in either of two forms; thus, it defines two network functions $Y(s)$ and $Z(s)$ which are reciprocal to each other. ■

Now let us see how the situation changes when we have a two-port network. At each of the ports we shall now define a voltage variable and a current variable as shown in Fig. 9-1.4. For convenience, we have used the subscript 1 to refer to

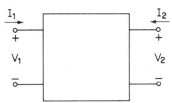

Figure 9-1.4 Variables for a two-port network.

the variables at the input port (at the left) and the subscript 2 to refer to the variables at the output port (at the right). At both of these ports the variables are defined so that their relative reference directions obey the usual convention. Since we now have four variables rather than the two used to describe the one-port network, we now require two equations to relate these variables. These equations will have the general form

$$U_1(s) = k_{11}(s)W_1(s) + k_{12}(s)W_2(s)$$
$$U_2(s) = k_{21}(s)W_1(s) + k_{22}(s)W_2(s)$$

(9-7)

where the quantities $U_1(s)$, $U_2(s)$, $W_1(s)$, and $W_2(s)$ may be any of the voltage and current variables $V_1(s)$, $V_2(s)$, $I_1(s)$, or $I_2(s)$, and the $k_{ij}(s)$ are the network functions which relate these variables. These latter are frequently referred to as *network parameters*.

Now let us consider the various ways in which we can select two variables for the left members $U_1(s)$ and $U_2(s)$ of Eq. (9-7). If we investigate the number of possible combinations in which two quantities may be selected from a total group of four, we find that there are six possibilities. Thus, there will be six different possible sets of coefficients $k_{ij}(s)$. A tabulation of these six possible combinations is given in Table 9-1.1. In this table, it will be noted that the fifth and sixth sets of network parameters have one variable, namely, $I_2(s)$, defined with a minus sign. The reason for this will be made clear in a following section. Each of the sets of parameters listed in the table has very specific properties that make it different from any other combination and make it the best one to apply to a certain class of specific network situations. These properties will be discussed in more detail in the sections that follow. Here we may note that any one of these six sets of parameters (if it exists) has the property that it completely characterizes the network. We shall find that if a given set of parameters for a network is known, then any other parameter set which exists may be found directly from the known set.

TABLE 9-1.1 The Six Sets of Network Parameters

Case	$U_1(s)$	$U_2(s)$	$W_1(s)$	$W_2(s)$
1	$V_1(s)$	$V_2(s)$	$I_1(s)$	$I_2(s)$
2	$I_1(s)$	$I_2(s)$	$V_1(s)$	$V_2(s)$
3	$I_1(s)$	$V_2(s)$	$V_1(s)$	$I_2(s)$
4	$V_1(s)$	$I_2(s)$	$I_1(s)$	$V_2(s)$
5	$V_1(s)$	$I_1(s)$	$V_2(s)$	$-I_2(s)$
6	$V_2(s)$	$I_2(s)$	$V_1(s)$	$-I_1(s)$

The general two-port network configuration shown in Fig. 9-1.4 has four individual terminals. Thus, it is theoretically possible to define four current variables rather than the two shown in the figure. To eliminate this possibility, it is customary to add the requirement to a network port that the current into one of the port terminals is, at every instant of time, equal to the current out of the other terminal of the port. This is called the *port current requirement* and it is illustrated in Fig. 9-1.5. Most of

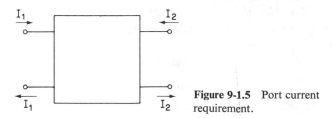

Figure 9-1.5 Port current requirement.

the situations to which two-port networks are applied automatically satisfy this requirement. For example, if a port is open-circuited, i.e., if nothing is connected to it, then obviously the currents at the two terminals defining the port are zero. Thus, the requirement is satisfied. Now consider the case where *any two-terminal element* (including a short circuit) is connected to the port. Since the currents into and out of the two-terminal element must obviously be equal, the port current requirement is always satisfied in this case. The requirement is also satisfied for the case where a group of two-port networks is connected in cascade as shown in Fig. 9-1.6. To see this, consider the first network in the cascade. The restriction on the currents is obviously satisfied at the input port of this network. Thus, the sum of the terminal currents at this port is zero. KCL now requires that the sum of the other two terminal currents also be zero, thus the port current requirement is satisfied. Obviously, since the requirement is satisfied at the *output* port of the first network, it must also be satisfied at the *input* port of the second network since the ports are directly connected. Continuing in this fashion, we conclude that the restrictions on the port currents are satisfied for all the networks in the cascade. A network configuration in which the port condition

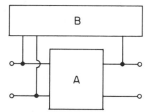

Figure 9-1.6 Cascade connection of two-port networks.

may not be satisfied is shown in Fig. 9-1.7. Depending upon the elements of the *B* network, the currents at the terminals of each of the two ports of the *A* network may be different. This type of a configuration will not be frequently encountered in this text.

Figure 9-1.7 Network interconnection which may not satisfy the port current requirement.

By far the most important subclass of two-port networks is the one in which the minus reference terminals of the input and output ports are at the same potential, i.e., a network in which a short circuit exists in the network between these terminals. Such a network can be drawn with the short circuit external to the box representing the network as shown in Fig. 9-1.8. This type of a network configuration is appropri-

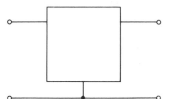

Figure 9-1.8 Three-terminal network.

ately referred to as a *three-terminal* network as well as a two-port network. Since the only modification required to change a three-terminal network to a two-port network is to connect an extra wire to the third terminal, all the developments that we will make in this chapter regarding two-port networks apply equally well to three-terminal networks.

In this section we have discussed some general considerations that are applicable to one- and two-port networks. In the following sections of this chapter we shall present a detailed treatment of each of the six possible types of two-port network parameters.

9-2 z PARAMETERS

In the last section we showed that there are six ways in which the voltage and current variables of a two-port network may be selected so as to define a set of network functions. The various cases are shown in Table 9-1.1. In this section we shall consider the first of these cases, namely, we shall choose $V_1(s)$ and $V_2(s)$ for the quantities $U_1(s)$ and $U_2(s)$ shown in Eq. (9-7), and $I_1(s)$ and $I_2(s)$ as the quantities $W_1(s)$ and $W_2(s)$. Since the left members of the resulting set of equations have the dimensions of voltage, all the terms in the right members of these equations must also have the dimensions of voltage. Therefore, since the quantities $W_1(s)$ and $W_2(s)$ have the dimensions of current, this requires that coefficients $k_{ij}(s)$ have the dimensions of impedance. Thus, the equations may be written in the form

$$V_1(s) = z_{11}(s)I_1(s) + z_{12}(s)I_2(s)$$
$$V_2(s) = z_{21}(s)I_1(s) + z_{22}(s)I_2(s)$$
(9-8)

where the quantities $z_{ij}(s)$ are called the *z parameters*. More succinctly, we may write the equations given above in matrix form as

$$\mathbf{V}(s) = \begin{bmatrix} V_1(s) \\ V_2(s) \end{bmatrix} = \begin{bmatrix} z_{11}(s) & z_{12}(s) \\ z_{21}(s) & z_{22}(s) \end{bmatrix} \begin{bmatrix} I_1(s) \\ I_2(s) \end{bmatrix} = \mathbf{Z}(s)\mathbf{I}(s)$$
(9-9)

The matrix $\mathbf{Z}(s)$ is referred to as the *z-parameter matrix*.

For a given network the z parameters may be found by applying a set of test excitations to the network. To see this, consider the first equation given in Eq. (9-8). If we let the output port (port 2) of the network be open-circuited, then the variable $I_2(s)$ must be zero. Thus, the equation will have the form

$$V_1(s) = z_{11}(s)I_1(s)|_{I_2(s)=0}$$
(9-10)

The quantity $z_{11}(s)$ is a network function. Thus, recalling that network functions are defined as response/excitation, (9-10) implies the use of a current excitation (rather than a voltage excitation) at the input port (port 1) of the network. Thus, the parameter $z_{11}(s)$ gives the driving-point impedance at port 1 when port 2 is open-circuited. More precisely, $z_{11}(s)$ is defined by the relation

$$z_{11}(s) = \frac{V_1(s)}{I_1(s)}\bigg|_{I_2(s)=0}$$
(9-11)

As an example of finding this parameter, consider the network shown in Fig. 9-2.1(a). If we apply a current source of value $I_1(s)$ at the input port, as shown in Fig. 9-2.1(b), the resulting impedance at this port is $2s + 1$. Thus, for this network $z_{11}(s)$ equals $2s + 1$.

A similar testing procedure may be applied to find the parameter $z_{12}(s)$. In this case, we must let port 1 be open-circuited, thus, $I_1(s) = 0$, and the first equation of

ohms, henries, farads

(a) (b) (c)

Figure 9-2.1 Tests for the z parameters.

Eq. (9-8) has the form

$$V_1(s) = z_{12}(s)I_2(s)\,|_{I_1(s)=0} \tag{9-12}$$

The quantity $z_{12}(s)$ thus defines the transfer impedance relating a current excitation at port 2 to a voltage response at port 1 (when port 1 is open-circuited). Thus, we may write

$$z_{12}(s) = \frac{V_1(s)}{I_2(s)}\bigg|_{I_1(s)=0} \tag{9-13}$$

An example of the determination of this parameter for the network shown in Fig. 9-2.1(a) is shown in Fig. 9-2.1(c). From this figure, we see that $V_1 = 2sI_2(s)$. Thus, $z_{12}(s)$ equals $2s$.

Test excitations similar to those described above may be applied so that the second equation of Eq. (9-8) may be used to find the parameters $z_{21}(s)$ and $z_{22}(s)$. For example, to find $z_{21}(s)$, we open-circuit port 2 and apply a current excitation at port 1. Thus, for the example network shown in Fig. 9-2.1(a), we may use the testing configuration shown in Fig. 9-2.1(b). The quantity $z_{21}(s)$ is now seen to be the transfer impedance from port 1 to port 2 when port 2 is open-circuited. Thus, in general, it is defined as

$$z_{21}(s) = \frac{V_2(s)}{I_1(s)}\bigg|_{I_2(s)=0} \tag{9-14}$$

From Fig. 9-2.1(b), we see that $z_{21}(s)$ equals $2s$. Finally, to find $z_{22}(s)$, we opencircuit port 1 and apply a test excitation current as shown in Fig. 9-2.1(c). Thus, $z_{22}(s)$ is the driving-point impedance at port 2 with port 1 open-circuited. From the second equation of (9-8), we see that

$$z_{22}(s) = \frac{V_2(s)}{I_2(s)}\bigg|_{I_1(s)=0} \tag{9-15}$$

For the example network, $z_{22}(s)$ equals $2s + (3/s)$. To summarize the above development, we see that the z-parameter matrix for the network shown in Fig. 9-2.1(a) is

$$\mathbf{Z}(s) = \begin{bmatrix} 2s + 1 & 2s \\ 2s & 2s + \dfrac{3}{s} \end{bmatrix} \tag{9-16}$$

In the applications of the test excitations described above, we note that in every case the port at which the response is measured is *open-circuited*. In addition, since the impedance of the current sources used to apply the test excitations is infinite, the port to which such an excitation is applied also sees an infinite impedance termination. Thus, from an impedance viewpoint, the excitation port may also be considered as being open-circuited, although, of course, an excitation current does flow into it. Considering the above, we see that another name which is also applicable to the z parameters is the *open-circuit impedance parameters.*

All the z parameters described above are defined as a ratio of voltage to current. In determining these parameters, rather than considering the excitation current as a variable and writing it explicitly, as was done above, it is frequently convenient to simply set the output of the excitation current source equal to unity. Since we are considering transformed variables, this implies that the physical current is set equal to an impulse [the inverse transform of 1 is $\delta(t)$]. As was pointed out in Sec. 8-8, when the transformed excitation is set to unity, *the network function is identically equal to the response.* Thus, if we set the excitation currents to unity, the resulting expressions for the response voltages directly specify the z parameters. This is usually a convenient technique for finding the z parameters. An example of this procedure follows.

Example 9-2.1

It is desired to find the z parameters for the network shown in Fig. 9-2.2(a). First, let us set up a testing condition which will permit us to determine the parameters $z_{11}(s)$ and $z_{21}(s)$. This is shown in Fig. 9-2.2(b). Note that the transformed current excitation applied at port 1 is unity. Since the port current $I_1(s)$ is also equal to the controlling current for the ICIS, the current through the resistor will be 4 while the current through the inductor is 1. Thus, applying KVL, $V_1(s) = 2s + 4$ and $V_2(s) = 4$. Since, for unity excitation, the response $V_1(s)$ equals the network function $z_{11}(s)$ and the response $V_2(s)$ equals $z_{21}(s)$, we conclude that $z_{11}(s)$ equals $2s + 4$ and $z_{21}(s)$ equals 4. A similar procedure is used to find the remaining z parameters, namely, $z_{12}(s)$ and $z_{22}(s)$. For this case, we apply an excitation current source of unity value at port 2 as shown in Fig. 9-2.2(c). Since the controlling current for the ICIS is zero, in this case we find that $V_1(s) = V_2(s) = 1$ and thus, $z_{12}(s) = z_{22}(s) = 1$. The z-parameter matrix for the network shown in Fig. 9-2.2(a) is thus

$$\begin{bmatrix} 2s + 4 & 1 \\ 4 & 1 \end{bmatrix}$$

(a)　　　　　　　　(b)　　　　　　　　(c)

Figure 9-2.2 Tests for the z parameters.

SUMMARY 9-2.1

The z Parameters of a Two-Port Network. The z parameters of a two-port network are the functions $z_{ij}(s)$ which specify the port voltages as functions of the port currents. Two of the four z parameters may be found by applying a unit test excitation current at port 1 and measuring the resulting voltage response at both ports. The other two z parameters may be found by similarly applying a unit test excitation current at the other port. The various test conditions used to define each parameter are summarized in Fig. 9-2.3. ∎

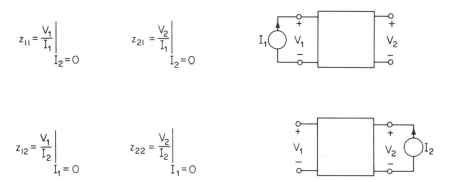

$$z_{11} = \left.\frac{V_1}{I_1}\right|_{I_2=0} \qquad z_{21} = \left.\frac{V_2}{I_1}\right|_{I_2=0}$$

$$z_{12} = \left.\frac{V_1}{I_2}\right|_{I_1=0} \qquad z_{22} = \left.\frac{V_2}{I_2}\right|_{I_1=0}$$

Figure 9-2.3 Test conditions used to define z parameters.

Now let us investigate some properties of the z parameters. In Sec. 8-7 we showed that if the points of excitation and response are interchanged, the transfer impedance of a *reciprocal* network will be unchanged, while the transfer impedance of a *non-reciprocal* network will be modified. Such an interchange of excitation and response is exactly what is done in determining the parameters $z_{12}(s)$ and $z_{21}(s)$ as shown in Fig. 9-2.3. We may conclude that a reciprocal network will, in general, have a *symmetric* z-parameter matrix, i.e., $z_{12}(s)$ will equal $z_{21}(s)$, and that a non-reciprocal network will have a non-symmetric z-parameter matrix. As an example of this the parameter matrix given in (9-16) for the network shown in Fig. 9-2.1(a) is symmetric. This is to be expected since the network is comprised only of two-terminal network elements and such networks are always reciprocal. An example of a non-symmetric z-parameter matrix is given in Example 9-2.1. This result again could be anticipated, since the network contains an active element, i.e., an ICIS.

An important property of the z parameters is that if, for a given network, the z parameters are known, we may find any other desired network function directly from them. For example, consider the determination of the open-circuit voltage transfer function $V_2(s)/V_1(s)$. By the term "open-circuit" here, we imply that the response port, i.e., the port where $V_2(s)$ is measured, is open-circuited. To find this voltage transfer function, let us rewrite the equations of (9-8) with the added restriction that $I_2(s) = 0$. Thus, we obtain

$$V_1(s) = z_{11}(s)I_1(s)$$
$$V_2(s) = z_{21}(s)I_1(s) \tag{9-17}$$

Taking the ratio of these two equations and inverting the result, we obtain

$$\frac{V_2(s)}{V_1(s)} = \frac{z_{21}(s)}{z_{11}(s)} \tag{9-18}$$

This is the desired voltage transfer function expressed in terms of the z parameters. In a similar fashion we may obtain the short-circuit current transfer function $I_2(s)/I_1(s)$. By the term "short-circuit" here, we imply that the response port, i.e., the one at which $I_2(s)$ is measured, is short-circuited. To find this function, we need simply write the second equation of (9-8) with the added restriction that $V_2(s) = 0$. Thus, we obtain

$$0 = z_{21}(s)I_1(s) + z_{22}(s)I_2(s) \tag{9-19}$$

Rearranging the terms of the above equation, we see that

$$\frac{I_2(s)}{I_1(s)} = \frac{-z_{21}(s)}{z_{22}(s)} \tag{9-20}$$

This is the desired current transfer function. As a third example of using the z parameters to find other network functions, consider the problem of finding the input impedance of a given two-port network when the output port is short-circuited. In this case, the relation between the variables $I_2(s)$ and $I_1(s)$ is given by Eq. (9-20). Substituting this result in the first equation of (9-8) to eliminate $I_2(s)$, we obtain

$$\frac{V_1(s)}{I_1(s)} = z_{11}(s) - \frac{z_{12}(s)z_{21}(s)}{z_{22}(s)} \tag{9-21}$$

This is the short-circuit driving-point impedance of a two-port network. A procedure similar to those given above may be used to find expressions for any other network function which is defined for a two-port network in terms of the z parameters of the two-port network.

In many network configurations we can simplify the determination of the z parameters by separating the network into component parts which are simpler to analyze than the original network. As an example of this, consider the case where two-port networks are connected as shown in Fig. 9-2.4. This type of a connection

Figure 9-2.4 Series connection of two two-port networks.

is frequently referred to as a *series* connection of two-port networks. For the A network and the voltage and current variables shown in the figure, we may define a set of z parameters $z_{ij}^a(s)$ by the relations

$$\mathbf{V}^a(s) = \begin{bmatrix} V_1^a(s) \\ V_2^a(s) \end{bmatrix} = \begin{bmatrix} z_{11}^a(s) & z_{12}^a(s) \\ z_{21}^a(s) & z_{22}^a(s) \end{bmatrix} \begin{bmatrix} I_1^a(s) \\ I_2^a(s) \end{bmatrix} = \mathbf{Z}^a(s)\mathbf{I}^a(s) \tag{9-22}$$

Similarly, for the B network we may define a set of parameters $z_{ij}^b(s)$ by the relations

$$\mathbf{V}^b(s) = \begin{bmatrix} V_1^b(s) \\ V_2^b(s) \end{bmatrix} = \begin{bmatrix} z_{11}^b(s) & z_{12}^b(s) \\ z_{21}^b(s) & z_{22}^b(s) \end{bmatrix} \begin{bmatrix} I_1^b(s) \\ I_2^b(s) \end{bmatrix} = \mathbf{Z}^b(s)\mathbf{I}^b(s) \qquad (9\text{-}23)$$

The series connection of the separate two-port networks labeled A and B can be considered as still another two-port network as shown in Fig. 9-2.5. Using the voltage

Figure 9-2.5 Series connection of two two-port networks.

and current variables shown in the figure, we may define a third set of parameters $z_{ij}(s)$ by the relations

$$\mathbf{V}(s) = \begin{bmatrix} V_1(s) \\ V_2(s) \end{bmatrix} = \begin{bmatrix} z_{11}(s) & z_{12}(s) \\ z_{21}(s) & z_{22}(s) \end{bmatrix} \begin{bmatrix} I_1(s) \\ I_2(s) \end{bmatrix} = \mathbf{Z}(s)\mathbf{I}(s) \qquad (9\text{-}24)$$

Applying KVL, we see that the sum of the port voltages of the A and B networks equals the port voltages of the overall network. Similarly, it is readily noted that the port currents at the input and output ports are the same for all three networks. Thus, the following relations must hold between the three sets of voltage and current variables for the various two-port networks.

$$\mathbf{V}(s) = \begin{bmatrix} V_1(s) \\ V_2(s) \end{bmatrix} = \begin{bmatrix} V_1^a(s) + V_1^b(s) \\ V_2^a(s) + V_2^b(s) \end{bmatrix} = \mathbf{V}^a(s) + \mathbf{V}^b(s) \qquad (9\text{-}25)$$

$$\mathbf{I}(s) = \begin{bmatrix} I_1(s) \\ I_2(s) \end{bmatrix} = \begin{bmatrix} I_1^a(s) \\ I_2^a(s) \end{bmatrix} = \begin{bmatrix} I_1^b(s) \\ I_2^b(s) \end{bmatrix} = \mathbf{I}^a(s) = \mathbf{I}^b(s) \qquad (9\text{-}26)$$

Combining the results from Eqs. (9-24), (9-25), and (9-26), we obtain

$$\mathbf{V}(s) = \mathbf{V}^a(s) + \mathbf{V}^b(s) = \mathbf{Z}^a(s)\mathbf{I}^a(s) + \mathbf{Z}^b(s)\mathbf{I}^b(s)$$

$$= [\mathbf{Z}^a(s) + \mathbf{Z}^b(s)]\mathbf{I}(s) = \mathbf{Z}(s)\mathbf{I}(s) \qquad (9\text{-}27)$$

Thus, we may conclude that, for a two-port network which is formed as a series connection of two separate two-port networks, the z parameters of the overall network may be found by adding the z parameters of the component networks. An example follows.

Example 9-2.2

In Fig. 9-2.6(a), a network consisting of three resistors and an ICVS is shown. To simplify the determination of the z parameters of this network it may be separated into two component networks connected in series as shown in parts (b) and (c) of the figure. For these networks, the z-parameter matrices are readily shown to be

Figure 9-2.6 Series connection of two-port networks in which the z parameters add.

$$\begin{bmatrix} 3 & 0 \\ K & 2 \end{bmatrix} \quad \text{and} \quad \begin{bmatrix} 1 & 1 \\ 1 & 1 \end{bmatrix}$$

If we now add the two z-parameter matrices given above, we obtain

$$\begin{bmatrix} 4 & 1 \\ 1+K & 3 \end{bmatrix}$$

These are the z parameters for the overall network shown in Fig. 9-2.6(a).

It should be noted that, in making a series connection of two two-port networks, the result given above, namely, that the z parameters add, does not apply if the port current requirement given in Sec. 9-1 is violated. As an example of this, consider the network shown in Fig. 9-2.7(a) [with the test current $I_1(s)$ applied]. This may be

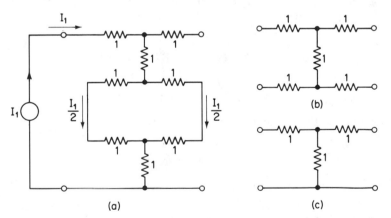

Figure 9-2.7 Series connection of two-port networks in which the z parameters do NOT add.

considered as a series connection of the two component networks shown in parts (b) and (c) of the figure. For these networks, the z-parameter matrices are, respectively,

$$\begin{bmatrix} 3 & 1 \\ 1 & 3 \end{bmatrix} \quad \text{and} \quad \begin{bmatrix} 2 & 1 \\ 1 & 2 \end{bmatrix} \tag{9-28}$$

The series connection of the two networks, shown in Fig. 9-2.7(a), however, has z parameters

$$\begin{bmatrix} 4 & 3 \\ 3 & 4 \end{bmatrix} \tag{9-29}$$

Obviously, in this case, the z parameters of the overall network are not equal to the sum of the z parameters of the component networks. To see why this is so, consider the path of the test current $I_1(s)$ shown in Fig. 9-2.7(a). At the indicated interior node this current branches into two equal currents $I_1(s)/2$ as shown. Thus, for the upper two port component network, the current into one of the terminals of its port 1 is $I_1(s)$, and the current out of the other terminal of this port is $I_1(s)/2$. This clearly violates the port current condition, and, as a result, it invalidates the z parameters for the component networks.

The results given above may be summarized as follows:

SUMMARY 9-2.2

Properties of the z Parameters of a Two-Port Network. The four z parameters of a given two-port network completely determine the properties of the network in the sense that they may be used to find any other network function that is defined for the two-port network. If two two-port networks are connected in series, the sum of their individual z parameters gives the z parameters for the resulting overall two-port network (unless the connection violates the port current requirement on the port currents of the component two-port networks). ∎

In the preceding paragraphs descriptions of various techniques for finding the z parameters of a specific network have been given. Now let us consider the inverse problem: Given a set of z parameters, how do we find a two-port network which has such a set of parameters? This problem is basically a synthesis problem, and there is, in general, no unique solution to synthesis problems. There are, however, some useful network forms which may be used as circuit representations for a given set of z parameters. One of the most common of these forms is the "T" network representation. This is applicable to all reciprocal networks, i.e., ones in which $z_{12}(s)$ is equal to $z_{21}(s)$. To see how this T representation may be developed, let us consider first of all the T configuration of two-terminal impedances shown in Fig. 9-2.8. The z-parameter matrix for this network is readily shown to be

$$\begin{bmatrix} z_{11}(s) & z_{12}(s) \\ z_{12}(s) & z_{22}(s) \end{bmatrix} = \begin{bmatrix} Z_a(s) + Z_c(s) & Z_c(s) \\ Z_c(s) & Z_b(s) + Z_c(s) \end{bmatrix} \tag{9-30}$$

Figure 9-2.8   T network.

Equating like terms in the above equation, we find the relations giving the network elements $Z_a(s)$, $Z_b(s)$, and $Z_c(s)$ as functions of the network parameters are

$$Z_a(s) = z_{11}(s) - z_{12}(s)$$
$$Z_b(s) = z_{22}(s) - z_{12}(s)$$
$$Z_c(s) = z_{12}(s)$$

(9-31)

Thus, the T network may be redrawn with the expressions specifying the two-terminal impedances given directly in terms of the z parameters as shown in Fig. 9-2.9. As an

Figure 9-2.9 T network realization of z parameters.

example of finding a network representation for a set of z parameters, consider the z-parameter matrix

$$\begin{bmatrix} 7 & 3 \\ 3 & 5 \end{bmatrix}$$

(9-32)

Applying the relations given in (9-31), we see that the T configuration for a two-port network which realizes these parameters is as shown in Fig. 9-2.10. It should be noted

Figure 9-2.10 T network realizing the z parameters of Eq. (9-32).

that when the relations given in Eq. (9-31) are applied to the case where the z parameters are complicated functions of the complex frequency variable, due to the subtraction of rational functions required in the right members of these equations, the network functions describing the impedances $Z_a(s)$, $Z_b(s)$, and $Z_c(s)$ may be difficult to synthesize.

In order to realize non-symmetric z-parameter matrices, we must use network representations which are non-reciprocal. One of the most useful representations for such a set of z parameters utilizes two controlled sources of the ICVS type. The representation has the form shown in Fig. 9-2.11. Applying KVL at each of the ports shown in this figure, it is readily verified that this representation satisfies the original equations defining the z parameters given in Eq. (9-8). Another equivalent representation for a set of z parameters for a non-reciprocal network is the one shown in Fig. 9-2.12. This representation uses only a single controlled source. If this representation is applied to a reciprocal network in which $z_{21}(s) = z_{12}(s)$, it will simply appear as the T configuration shown in Fig. 9-2.9.

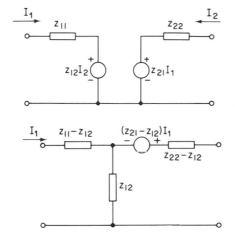

Figure 9-2.11 Two-controlled-source realization of a non-reciprocal set of z parameters.

Figure 9-2.12 One-controlled-source realization of a non-reciprocal set of z parameters.

In this section we have investigated the use of z parameters to describe two-port networks. In the following sections we shall make a similar investigation of the other types of parameters which are categorized in Table 9.1.1.

9-3 y PARAMETERS

In this section we shall discuss the second of the six ways in which the voltage and current variables of a two-port network may be selected so as to define a set of network functions. The case is the second one shown in Table 9-1.1, namely, we will choose $I_1(s)$ and $I_2(s)$ for the quantities $U_1(s)$ and $U_2(s)$ shown in (9-7) and $V_1(s)$ and $V_2(s)$ as the quantities $W_1(s)$ and $W_2(s)$. For this case, the left members of the resulting set of equations have the dimensions of current. Therefore, all the terms in the right members of these equations must also have the dimensions of current. Therefore, the coefficients $k_{ij}(s)$ of (9-7) must have the dimensions of admittance. Thus, the equations may be written in the form

$$I_1(s) = y_{11}(s)V_1(s) + y_{12}(s)V_2(s)$$
$$I_2(s) = y_{21}(s)V_1(s) + y_{22}(s)V_2(s)$$
(9-33)

where the quantities $y_{ij}(s)$ are called the *y parameters*. The equations given above may also be written in matrix form as

$$\mathbf{I}(s) = \begin{bmatrix} I_1(s) \\ I_2(s) \end{bmatrix} = \begin{bmatrix} y_{11}(s) & y_{12}(s) \\ y_{21}(s) & y_{22}(s) \end{bmatrix} \begin{bmatrix} V_1(s) \\ V_2(s) \end{bmatrix} = \mathbf{Y}(s)\mathbf{V}(s)$$
(9-34)

The matrix $\mathbf{Y}(s)$ is referred to as the *y-parameter matrix*.

For a given network, the y parameters may be found in a manner similar to that used for the z parameters in the previous section. In this case, however, the test excitations will be applied by voltage sources rather than by current sources. For example, consider the first equation given in (9-33). If we short-circuit port 2, then the variable $V_2(s)$ must be 0. Thus, the network function $y_{11}(s)$ is seen to be the driving-point admittance at port 1 when port 2 is short-circuited, and it is defined by the relation

$$y_{11}(s) = \frac{I_1(s)}{V_1(s)}\bigg|_{V_2(s)=0} \qquad (9\text{-}35)$$

In a similar manner, if we consider the second equation of (9-33) and the same terminating condition, namely, $V_2(s) = 0$, then the network function $y_{21}(s)$ which is the transfer admittance from port 1 to port 2 is defined by the relation

$$y_{21}(s) = \frac{I_2(s)}{V_1(s)}\bigg|_{V_2(s)=0} \qquad (9\text{-}36)$$

As an example of finding these two parameters, consider the network shown in Fig. 9-3.1(a). If we apply a voltage source $V_1(s)$ at port 1 and short-circuit the terminals of port 2 as shown in Fig. 9-3.1(b), then we see from the figure that $I_1(s) = (2s^2 + 1)V_1(s)/2s$ and $I_2(s) = -V_1(s)/2s$. Thus, for this network $y_{11}(s) = (2s^2 + 1)/2s$ and $y_{21}(s) = -1/2s$. Note that the minus sign in the expression for $y_{21}(s)$ occurs because for a positive $V_1(s)$, the actual current direction in the short circuit flows opposite to the assumed positive reference direction for $I_2(s)$.

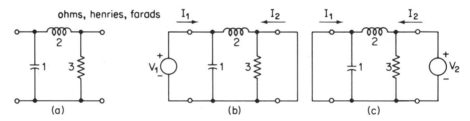

Figure 9-3.1 Tests for the y parameters.

A similar testing procedure may be applied to find the parameters $y_{12}(s)$ and $y_{22}(s)$. In this case, let us short-circuit port 1. Thus, the variable $V_1(s)$ must equal 0. From the first equation of (9-33), we find that $y_{12}(s)$, the transfer admittance from port 2 to port 1, is defined by the relation

$$y_{12}(s) = \frac{I_1(s)}{V_2(s)}\bigg|_{V_1(s)=0} \qquad (9\text{-}37)$$

Similarly, from the second equation of (9-33) the parameter $y_{22}(s)$, the driving-point admittance at port 2 when port 1 is short-circuited, is defined by the relation

$$y_{22}(s) = \frac{I_2(s)}{V_2(s)}\bigg|_{V_1(s)=0} \qquad (9\text{-}38)$$

As an example of the determination of these last two parameters, the network shown in Fig. 9-3.1(a) may be excited by a voltage source at port 2 with port 1 short-circuited as shown in Fig. 9-3.1(c). From this figure, we see that $I_1(s) = -V_2(s)/2s$ and $I_2(s) = (2s + 3)V_2(s)/6s$. Thus, we conclude that $y_{12}(s) = -1/2s$ and $y_{22}(s) = (2s + 3)/6s$. Combining the above results, we may write the y-parameter matrix for the network shown in Fig. 9-3.1(a) as

$$\mathbf{Y}(s) = \begin{bmatrix} \dfrac{2s^2 + 1}{2s} & -\dfrac{1}{2s} \\[3mm] -\dfrac{1}{2s} & \dfrac{(2s/3) + 1}{2s} \end{bmatrix} \qquad (9\text{-}39)$$

In the applications of the test excitations described above, we note that in each case the port at which the response is measured is short-circuited. Also, since the internal impedance of the voltage sources used to apply the test excitations is zero, the port to which these excitations are applied in effect sees a zero impedance termination. Thus, from an impedance standpoint, the excitation port may also be considered as being short-circuited, although, of course, there will be a voltage present across its terminals. Considering the above, we see that another equally descriptive name which may be applied to the *y* parameters is *short-circuit admittance parameters*. The same remarks which were made in the preceding section with respect to the use of unit-valued excitations to directly determine the network parameters also applies to the *y*-parameter case. An example of this procedure follows.

Example 9-3.1

It is desired to find the *y* parameters for the network shown in Fig. 9-3.2(a). In order to find $y_{11}(s)$ and $y_{21}(s)$, we may apply a unit-valued test excitation voltage at port 1 and a short circuit at port 2, as shown in Fig. 9-3.2(b). For this excitation the (transformed) output of the VCIS is 4. The current $I_1(s)$ consists of two components. The first of these is the current (downward) through the 1-mho resistor. Since 1 V is applied across the resistor, the current through it is 1 A. The second component of $I_1(s)$ is the current through the capacitor. Since the right end of the capacitor is grounded by the short circuit at port 2, 1 V also appears across it and the current through it (from left to right) is $2s$. Adding these two components we find $I_1(s) = y_{11}(s) = 2s + 1$. The current $I_2(s)$ also consists of two components. The first of these is the current through the capacitor. This is $-2s$ [the minus sign occurs since the actual current is in a direction opposite to the reference direction for $I_2(s)$]. The second component of $I_2(s)$ comes from the VCIS and has the value -4. Thus $I_2(s) = y_{21}(s) = -2s - 4$. Note that the voltage across the 3-mho resistor is zero, thus no current flows through it. To find the remaining *y* parameters, we apply a unit-valued test excitation voltage at port 2 and a short circuit at port 1 as shown in Fig. 9-3.2(c). In this case, since $V_1(s) = 0$, the output of the VCIS is zero. It is easily shown that $I_1(s) = y_{12}(s) = -2s$ and $I_2(s) = y_{22}(s) = 2s + 3$. Thus, the *y* parameters for the network shown in Fig. 9-3.2(a) are

$$\mathbf{Y}(s) = \begin{bmatrix} 2s + 1 & -2s \\ -2s - 4 & 2s + 3 \end{bmatrix}$$

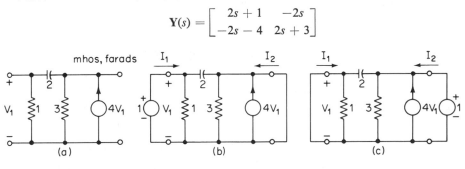

Figure 9-3.2 Tests for the *y* parameters.

The y Parameters of a Two-Port Network. The y parameters of a two-port network are the network functions $y_{ij}(s)$ which specify the port currents as functions of the port voltages. Two of the four y parameters may be found by applying a unit test excitation voltage at port 1, short-circuiting port 2, and measuring the resulting current response at the two ports. The other two y parameters may be found by similarly applying a unit test excitation voltage at port 2 and short-circuiting port 1. The various excitations used to determine the y parameters are summarized in Fig. 9-3.3. ∎

$$y_{11} = \frac{I_1}{V_1}\bigg|_{V_2=0} \qquad y_{21} = \frac{I_2}{V_1}\bigg|_{V_2=0}$$

$$y_{12} = \frac{I_1}{V_2}\bigg|_{V_1=0} \qquad y_{22} = \frac{I_2}{V_2}\bigg|_{V_1=0}$$

Figure 9-3.3 Test conditions used to determine y parameters.

In the preceding section it was shown that the z parameters for a two-port network were characterized by a matrix equation having the form

$$\mathbf{V}(s) = \mathbf{Z}(s)\mathbf{I}(s) \tag{9-40}$$

where $\mathbf{Z}(s)$ is the z-parameter matrix. Now let us pre-multiply both sides of this expression by the y-parameter matrix $\mathbf{Y}(s)$. We obtain

$$\mathbf{Y}(s)\mathbf{V}(s) = \mathbf{Y}(s)\mathbf{Z}(s)\mathbf{I}(s) \tag{9-41}$$

The left member of this equation, however, from (9-34), simply defines the column matrix $\mathbf{I}(s)$. Thus, considering the right member of the equation, we see that the matrix product $\mathbf{Y}(s)\mathbf{Z}(s)$ can only be equal to the identity matrix. We conclude that the y-parameter matrix is the inverse of the z-parameter matrix, namely, that the relations

$$\mathbf{Y}(s) = [\mathbf{Z}(s)]^{-1} \qquad \mathbf{Z}(s) = [\mathbf{Y}(s)]^{-1} \tag{9-42}$$

are satisfied. Since this is true, we may directly apply the relations for the inverse of a second-order matrix developed in Sec. 3-1 to determine the elements of the y-parameter matrix in terms of the elements of the z-parameter matrix. Thus, we may write

$$\begin{bmatrix} y_{11}(s) & y_{12}(s) \\ y_{21}(s) & y_{22}(s) \end{bmatrix} = \frac{1}{\det \mathbf{Z}(s)} \begin{bmatrix} z_{22}(s) & -z_{12}(s) \\ -z_{21}(s) & z_{11}(s) \end{bmatrix} \tag{9-43}$$

Similarly, the elements of the z-parameter matrix may be found from the elements of the y-parameter matrix by the relations

$$\begin{bmatrix} z_{11}(s) & z_{12}(s) \\ z_{21}(s) & z_{22}(s) \end{bmatrix} = \frac{1}{\det \mathbf{Y}(s)} \begin{bmatrix} y_{22}(s) & -y_{12}(s) \\ -y_{21}(s) & y_{11}(s) \end{bmatrix} \qquad (9\text{-}44)$$

Since the inverse of a symmetric matrix is still a symmetric matrix, the conclusions relating symmetricity of the parameter matrix and reciprocity of the network made for the z parameters also apply to the y parameters; namely, a reciprocal network will have a symmetric y-parameter matrix and the y-parameter matrix for a non-reciprocal network will not be symmetric. It is also easily shown that, as was the case for the z parameters, once the y parameters for a network are known, any other network function may be expressed in terms of them. The procedures used for determining the other network functions are similar to those presented for the z parameters.

In many network configurations, we can simplify the determination of the y parameters by separating the network configuration into component parts which are simpler to analyze than the original network. For example, consider the case where two two-port networks are connected as shown in Fig. 9-3.4. This type of a connec-

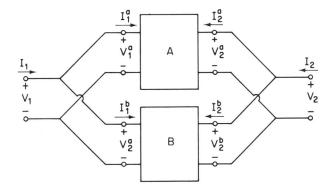

Figure 9-3.4 Parallel connection of two two-port networks.

tion is called a *parallel* connection of two two-port networks. For the A network and the voltage and current variables shown in the figure, we may define a set of y parameters $y_{ij}^a(s)$ by the relations

$$\mathbf{I}^a(s) = \begin{bmatrix} I_1^a(s) \\ I_2^a(s) \end{bmatrix} = \begin{bmatrix} y_{11}^a(s) & y_{12}^a(s) \\ y_{21}^a(s) & y_{22}^a(s) \end{bmatrix} \begin{bmatrix} V_1^a(s) \\ V_2^a(s) \end{bmatrix} = \mathbf{Y}^a(s)\mathbf{V}^a(s) \qquad (9\text{-}45)$$

Similarly, for the B network we may define a set of parameters $y_{ij}^b(s)$ by the relations

$$\mathbf{I}^b(s) = \begin{bmatrix} I_1^b(s) \\ I_2^b(s) \end{bmatrix} = \begin{bmatrix} y_{11}^b(s) & y_{12}^b(s) \\ y_{21}^b(s) & y_{22}^b(s) \end{bmatrix} \begin{bmatrix} V_1^b(s) \\ V_2^b(s) \end{bmatrix} = \mathbf{Y}^b(s)\mathbf{V}^b(s) \qquad (9\text{-}46)$$

The parallel connection of the separate two-port networks A and B can be considered as still another two-port network as shown in Fig. 9-3.4. The voltage and current variables specified in the figure for this overall two-port network define a third set of parameters $y_{ij}(s)$ by the relations

$$\mathbf{I}(s) = \begin{bmatrix} I_1(s) \\ I_2(s) \end{bmatrix} = \begin{bmatrix} y_{11}(s) & y_{12}(s) \\ y_{21}(s) & y_{22}(s) \end{bmatrix} \begin{bmatrix} V_1(s) \\ V_2(s) \end{bmatrix} = \mathbf{Y}(s)\mathbf{V}(s) \qquad (9\text{-}47)$$

Applying KCL, we see that the sum of the port currents of the A and B networks equals the port currents of the overall network. Similarly, it is readily noted that the port voltages at the input and output ports are the same for all three networks. Thus, by a proof analogous to that used for the z-parameter case, we can show that the y parameters of the overall network may be found by adding the y parameters of the component networks. This result is expressed by the following relation:

$$\mathbf{Y}(s) = \mathbf{Y}^a(s) + \mathbf{Y}^b(s) \qquad (9\text{-}48)$$

An example follows.

Example 9-3.2

As an example of a situation in which the y parameters of component paralleled networks may be added to find the parameters of an overall two-port network, consider the network shown in Fig. 9-3.5(a). This may be treated as the parallel combination of the networks shown in Fig. 9-3.5(b) and (c) with y-parameter matrices

$$\begin{bmatrix} 1.05s + 0.1 & -0.05s \\ -0.05s & 0.55s + 0.2 \end{bmatrix} \quad \text{and} \quad \begin{bmatrix} 0 & 0 \\ G & 0 \end{bmatrix}$$

The sum of the two matrices given above is the y-parameter matrix for the overall two-port network shown in Fig. 9-3.5(a). It is

$$\begin{bmatrix} 1.05s + 0.1 & -0.05s \\ G - 0.05s & 0.55s + 0.2 \end{bmatrix}$$

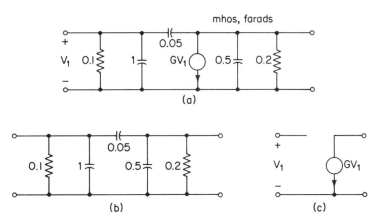

Figure 9-3.5 Parallel connection of two-port networks in which the y parameters add.

The same caution which was presented in the preceding section regarding the possible violation of the port current requirement by the interconnection of certain types of networks also applies to the y-parameter case. It may be shown, however, that when *three-terminal networks* (*treated as two-port networks*) *are connected in parallel, the port current requirement is automatically satisfied.* Thus, the y parameters of three-terminal networks may always be added, when such networks are connected in parallel, to determine the y parameters of the resulting two-port network. The results given above may be summarized as follows:

SUMMARY 9-3.2

Properties of the y Parameters of a Two-Port Network. The four y parameters of a two-port network have the following properties:

1. They completely determined the characteristics of the network in the sense that they may be used to find any other network function which is defined for the network.
2. The y-parameter matrix is the inverse of the z-parameter matrix.
3. If two two-port networks are connected in parallel, the sum of their individual y parameters gives the y parameters for the resulting overall two-port network (unless the connection violates the port current requirement on the port currents of the component two-port networks).
4. When three-terminal networks are connected in parallel, the port current requirement is always satisfied. ■

In the preceding paragraphs a description of various techniques for finding the y parameters of a specific network have been discussed. Now let us consider the inverse problem. Given a set of y parameters, how do we find a two-port network which realizes such a set of parameters? One of the most common of the network forms for doing this is the "π" network representation. This is applicable to all reciprocal networks, i.e., networks for which $y_{12}(s)$ is equal to $y_{21}(s)$. The π representation is developed in a manner similar to that used for the determination of the T representation for the z parameters. To see this, consider the π configuration of two-terminal admittances shown in Fig. 9-3.6. The y-parameter matrix for this network is readily

Figure 9-3.6 π network.

shown to be

$$\begin{bmatrix} y_{11}(s) & y_{12}(s) \\ y_{12}(s) & y_{22}(s) \end{bmatrix} = \begin{bmatrix} Y_a(s) + Y_c(s) & -Y_c(s) \\ -Y_c(s) & Y_b(s) + Y_c(s) \end{bmatrix} \qquad (9\text{-}49)$$

Equating like terms in the above equation, we find the relations giving the network elements $Y_a(s)$, $Y_b(s)$, and $Y_c(s)$ as functions of the network parameters are

$$Y_a(s) = y_{11}(s) + y_{12}(s)$$
$$Y_b(s) = y_{22}(s) + y_{12}(s) \qquad (9\text{-}50)$$
$$Y_c(s) = -y_{12}(s)$$

Thus, the π network may be redrawn with the expressions specifying the two-terminal admittances given directly in terms of the y parameters as shown in Fig. 9-3.7. As an example of finding a network representation for a set of y parameters, consider the y-parameter matrix

$$\begin{bmatrix} 3s + 2 & -1 \\ -1 & 2s + 3 \end{bmatrix} \qquad (9\text{-}51)$$

Applying the relations given in (9-50), we see that the π configuration for a two-port network which realizes these parameters is as shown in Fig. 9-3.8. It should be noted

Figure 9-3.7 π network realization of y parameters.

Figure 9-3.8   Network realizing the y parameters of Eq. (9-51).

that, in general, the synthesis of the two-terminal admittances may not be as simple as it is in this example if the expressions for the $y_{ij}(s)$ are complicated functions.

In order to realize non-symmetrical y-parameter matrices, one of the most useful representations is one utilizing two controlled sources of the VCIS type. This representation has the form shown in Fig. 9-3.9. If we apply KCL at each of the

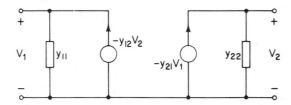

Figure 9-3.9   Two-controlled-source realization of a non-reciprocal set of y parameters.

ports shown in this figure, it is readily verified that this representation satisfies the original equations defining the y parameters given in (9-33). Another equivalent representation for a set of y parameters for a non-reciprocal network is the one shown in Fig. 9-3.10. This representation utilizes only a single controlled source. For a reciprocal network, i.e., for the case where $y_{12}(s) = y_{21}(s)$, this representation simply becomes the π configuration shown in Fig. 9-3.7.

The z- and y-parameter sets may not both exist for a given network configuration. For example, consider a two-port network consisting of a single series element as shown in Fig. 9-3.11. The y-parameter matrix for this network is readily shown

Figure 9-3.10 One-controlled-source realization of a non-reciprocal set of y parameters.

to be

$$\begin{bmatrix} Y(s) & -Y(s) \\ -Y(s) & Y(s) \end{bmatrix} \qquad (9\text{-}52)$$

Clearly, the determinant of this matrix equals zero. Therefore, the inverse matrix, and consequently the z parameters as determined by (9-44), do not exist. As another example of a network which does not have both z and y parameters, consider a two-port network consisting of a single shunt element as shown in Fig. 9-3.12. The z-parameter matrix for this network is

$$\begin{bmatrix} Z(s) & Z(s) \\ Z(s) & Z(s) \end{bmatrix} \qquad (9\text{-}53)$$

Again, since the matrix is singular, the inverse matrix, and therefore the y parameters as determined by (9-43), do not exist for this network. For some networks, neither the z nor the y parameters exist. As an example of this, consider the ideal transformer shown in Fig. 9-3.13. The equations defining this two-port network are

Figure 9-3.11 Single series element.

Figure 9-3.12 Single shunt element.

Figure 9-3.13 Ideal transformer.

$$V_1(s) = n V_2(s) \qquad (9\text{-}54\text{a})$$

$$I_1(s) = -\frac{1}{n} I_2(s) \qquad (9\text{-}54\text{b})$$

Since these equations require that the voltage at one port be specified as a function of the voltage at the other port and similarly that the current be expressed as a function of the other port current, neither the z parameters nor the y parameters can be used to describe the characteristics of this network. We shall see, however, that some of the other representations which we shall discuss may be used to characterize it. In general, except for some very unusual degenerate networks, there will be at least one parameter set which will exist for any given network configuration.

9-4 HYBRID PARAMETERS

In the last two sections we discussed the use of z and y parameters to represent two-port networks. These parameters are characterized by the fact that they have the dimensions of impedance and admittance, respectively. Thus, collectively they may be referred to as *immittance parameters*. In this section we shall introduce two more sets of parameters which are quite different in nature. These are the third and fourth types given in the list in Table 9-1.1 of the six possible parameter sets for a two-port network. They are named the g parameters and the h parameters and are collectively referred to as *hybrid parameters*. The name is appropriate because the individual elements of both of these types of parameters do not have a common dimension.

The first set of hybrid parameters is defined by the third case in Table 9-1.1, namely, the variables $I_1(s)$ and $V_2(s)$ are chosen for the quantities $U_1(s)$ and $U_2(s)$ in Eq. (9-7), and the variables $V_1(s)$ and $I_2(s)$ are chosen for the quantities $W_1(s)$ and $W_2(s)$. The equations have the form

$$I_1(s) = g_{11}(s)V_1(s) + g_{12}(s)I_2(s)$$
$$V_2(s) = g_{21}(s)V_1(s) + g_{22}(s)I_2(s)$$

(9-55)

where the quantities $g_{ij}(s)$ are called the g *parameters*. In the first equation, the left member has the dimensions of current. Therefore, in the right member, $g_{11}(s)$ must have the dimensions of admittance and $g_{12}(s)$ must be dimensionless. In the second equation, the left member has the dimensions of voltage; therefore, in the right member $g_{21}(s)$ must be dimensionless and $g_{22}(s)$ must have the dimensions of impedance. The matrix representation for the g parameters will have the form

$$\begin{bmatrix} I_1(s) \\ V_2(s) \end{bmatrix} = \begin{bmatrix} g_{11}(s) & g_{12}(s) \\ g_{21}(s) & g_{22}(s) \end{bmatrix}\begin{bmatrix} V_1(s) \\ I_2(s) \end{bmatrix} = \mathbf{G}(s)\begin{bmatrix} V_1(s) \\ I_2(s) \end{bmatrix}$$

(9-56)

where the matrix $\mathbf{G}(s)$ is called the *g-parameter matrix*.

Test excitations similar to those defined for the z and y parameters may be used to determine the g parameters. For example, if we open-circuit port 2, thus setting $I_2(s) = 0$, and apply a voltage test excitation at port 1, from (9-55) we obtain the relations

$$I_1(s) = g_{11}(s)V_1(s) \tag{9-57a}$$

$$V_2(s) = g_{21}(s)V_1(s) \tag{9-57b}$$

Thus, $g_{11}(s)$ is the open-circuit driving-point admittance at port 1 of the network and $g_{21}(s)$ is the open-circuit voltage transfer function from port 1 to port 2. As an example of the determination of these parameters, consider the network shown in Fig. 9-4.1(a). The test conditions are shown in Fig. 9-4.1(b). From this figure, it is easily seen that $g_{11}(s) = (2s^2 + 3s + 1)/(2s + 3)$, and $g_{21}(s) = 3/(2s + 3)$. In a similar manner, we may find the remaining two g parameters by short-circuiting port 1, thus setting $V_1(s) = 0$, and applying a test excitation current at port 2. From Eq. (9-55), the relations defining the parameters are

$$I_1(s) = g_{12}(s)I_2(s) \tag{9-58a}$$

$$V_2(s) = g_{22}(s)I_2(s) \tag{9-58b}$$

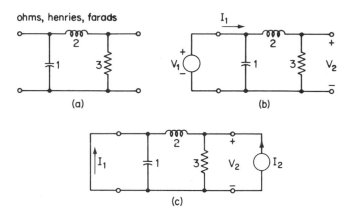

Figure 9-4.1 Tests for the g parameters.

Thus, $g_{12}(s)$ is the short-circuit current transfer ratio from port 2 to port 1 and $g_{22}(s)$ is the short-circuit driving-point impedance at port 2. For the network shown in Fig. 9-4.1(a), the test conditions are shown in Fig. 9-4.1(c). From the figure we find that $g_{12}(s) = -3/(2s + 3)$. Note that the minus sign occurs because for a positive current $I_2(s)$ the actual current flows opposite to the assumed positive reference direction for $I_1(s)$. From Fig. 9-4.1(c) we also see that $g_{22}(s) = 6s/(2s + 3)$. Thus, the g-parameter matrix for the example network shown in Fig. 9-4.1(a) is

$$\mathbf{G}(s) = \begin{bmatrix} \dfrac{2s^2 + 3s + 1}{2s + 3} & \dfrac{-3}{2s + 3} \\ \dfrac{3}{2s + 3} & \dfrac{6s}{2s + 3} \end{bmatrix} \tag{9-59}$$

It should be apparent that we may use unit input excitations to determine the g parameters just as we did in the case of z and y parameters. The results given above may be summarized as follows:

SUMMARY 9-4.1

The g **Parameters of a Two-Port Network.** The g parameters of a two-port network are the network functions $g_{ij}(s)$ which specify the current at port 1 and the voltage at port 2 as functions of the other port variables. The exact relations are given in Eq. (9-55). Two of the four g parameters may be found by applying a unit test excitation voltage at port 1 and open-circuiting port 2. The other two may be found by applying a unit test excitation current at port 2 and short-circuiting port 1. The various test conditions used to define each parameter are summarized in Fig. 9-4.2. ■

In the last section we showed how the y parameters could be found directly from the z parameters and vice versa. This conclusion is readily extended to show that, in general, from any given set of parameters we can find any other set of param-

$$g_{11} = \frac{I_1}{V_1}\bigg|_{I_2=0} \qquad g_{21} = \frac{V_2}{V_1}\bigg|_{I_2=0}$$

$$g_{12} = \frac{I_1}{I_2}\bigg|_{V_1=0} \qquad g_{22} = \frac{V_2}{I_2}\bigg|_{V_1=0}$$

Figure 9-4.2 Test conditions used to define g parameters.

eters (if such a representation exists). As an example of this, let us develop the relationship between the g parameters and the y parameters. From (9-33), the second equation which defines the y parameters is

$$I_2(s) = y_{21}(s)V_1(s) + y_{22}(s)V_2(s) \qquad (9\text{-}60)$$

Solving this equation for $V_2(s)$, we obtain

$$V_2(s) = \frac{1}{y_{22}(s)}I_2(s) - \frac{y_{21}(s)}{y_{22}(s)}V_1(s) \qquad (9\text{-}61)$$

Comparing this equation with (9-55), we see that it has the same form as the second equation for the g parameters, namely, it expresses $V_2(s)$ as a function of $V_1(s)$ and $I_2(s)$. To derive the first g-parameter equation in terms of the y parameters, let us consider the first equation for the y parameters given in (9-33). This is

$$I_1(s) = y_{11}(s)V_1(s) + y_{12}(s)V_2(s) \qquad (9\text{-}62)$$

If we substitute (9-61) in this equation and rearrange terms, we obtain

$$I_1(s) = \frac{\det \mathbf{Y}(s)}{y_{22}(s)}V_1(s) + \frac{y_{12}(s)}{y_{22}(s)}I_2(s) \qquad (9\text{-}63)$$

where $\det \mathbf{Y}(s) = y_{11}(s)y_{22}(s) - y_{12}(s)y_{21}(s)$. This equation has the form of the first equation of the g parameters, namely, it expresses $I_1(s)$ as a function of $V_1(s)$ and $I_2(s)$. We may summarize the results given in (9-61) and (9-63) by the matrix equality

$$\begin{bmatrix} g_{11}(s) & g_{12}(s) \\ g_{21}(s) & g_{22}(s) \end{bmatrix} = \frac{1}{y_{22}(s)}\begin{bmatrix} \det \mathbf{Y}(s) & y_{12}(s) \\ -y_{21}(s) & 1 \end{bmatrix} \qquad (9\text{-}64)$$

Thus, we have developed the relations giving the g parameters as functions of the y parameters. A series of steps similar to the above may be used to obtain the inverse relations, namely, to specify the y parameter in terms of the g parameters. The results of these steps is the matrix equality

$$\begin{bmatrix} y_{11}(s) & y_{12}(s) \\ y_{21}(s) & y_{22}(s) \end{bmatrix} = \frac{1}{g_{22}(s)}\begin{bmatrix} \det \mathbf{G}(s) & g_{12}(s) \\ -g_{21}(s) & 1 \end{bmatrix} \qquad (9\text{-}65)$$

where $\det \mathbf{G}(s) = g_{11}(s)g_{22}(s) - g_{12}(s)g_{21}(s)$. It should be noted that, implicit in Eqs. (9-64) and (9-65), are the conditions under which it is possible to find one set of parameters from another. For example, in Eq. (9-64) we see that every element of the right member of the equation is divided by $y_{22}(s)$. We may conclude that, for a two-port network, if the y parameters exist, the g parameters will also exist if and only if $y_{22}(s) \neq 0$. Similarly, from Eq. (9-65), if a set of g parameters for a two-port network exists, the y parameters will exist if and only if $g_{22}(s) \neq 0$.

By an extension of the method given above, it is readily verified that all the parameter sets are related. Thus, any properties of a two-port network which are expressed in terms of one set of parameters may be easily related to another set of parameters. For example, it has been shown that a reciprocal network is one in which $y_{12}(s)$ equals $y_{21}(s)$. From Eq. (9-65) we see that the equivalent equality in terms of the g parameters is that $g_{12}(s)/g_{22}(s) = -g_{21}(s)/g_{22}(s)$. Thus, the condition for reciprocity in term of the g parameters is simply

$$g_{12}(s) = -g_{21}(s) \tag{9-66}$$

By a similar substitution, any desired network function may be found in terms of the g parameters. For example, for the open-circuit voltage transfer function, we readily see that

$$\frac{V_2(s)}{V_1(s)} = \frac{-y_{21}(s)}{y_{22}(s)} = g_{21}(s) \tag{9-67}$$

Just as was the case for the z and y parameters, it is possible to specify an interconnection of two-port networks such that the g parameters add. Such an interconnection, however, is seldom encountered in practice. The details concerning it are left as an exercise for the reader.

The second set of hybrid parameters, which is given as the fourth case in Table 9-1.1, is the one in which $U_1(s)$ in Eq. (9-7) is set equal to $V_1(s)$ and $U_2(s)$ is set equal to $I_2(s)$. Similarly, $W_1(s)$ and $W_2(s)$ are set equal to $I_1(s)$ and $V_2(s)$, respectively. The resulting equations have the form

$$\begin{aligned} V_1(s) &= h_{11}(s)I_1(s) + h_{12}(s)V_2(s) \\ I_2(s) &= h_{21}(s)I_1(s) + h_{22}(s)V_2(s) \end{aligned} \tag{9-68}$$

where the quantities $h_{ij}(s)$ are called the h parameters. The left member of the first equation given in Eq. (9-68) has the dimensions of voltage. Thus, we see that $h_{11}(s)$ has the dimensions of impedance and $h_{12}(s)$ is dimensionless. Similarly, since the left member of the second equation has the dimensions of current, $h_{21}(s)$ must be dimensionless and $h_{22}(s)$ must have the dimensions of admittance. If the equations are written in matrix form, we obtain

$$\begin{bmatrix} V_1(s) \\ I_2(s) \end{bmatrix} = \begin{bmatrix} h_{11}(s) & h_{12}(s) \\ h_{21}(s) & h_{22}(s) \end{bmatrix} \begin{bmatrix} I_1(s) \\ V_2(s) \end{bmatrix} = \mathbf{H}(s) \begin{bmatrix} I_1(s) \\ V_2(s) \end{bmatrix} \tag{9-69}$$

where the square matrix $\mathbf{H}(s)$ is referred to as the h-parameter matrix. If we compare the form of the equation in (9-69) with that given in (9-56), we see that the h-parameter matrix and the g-parameter matrix are related by

$$\mathbf{H} = \mathbf{G}^{-1} \qquad \mathbf{G} = \mathbf{H}^{-1} \tag{9-70}$$

The application of the test excitations to determine the h parameters is similar to that given for the g parameters. The results are summarized as follows:

The h **Parameters of a Two-Port Network.** The h parameters of a two-port network are the network functions $h_{ij}(s)$ which specify the voltage variable at port 1 and the current variable at port 2 as functions of the other port variables. The exact relations are given in Eq. (9-68). Two of the four h parameters may be found by applying a unit test excitation current at port 1 with port 2 short-circuited. The other two h parameters may be found by applying a unit test excitation voltage at port 2 with port 1 open-circuited. The various test conditions used to define each of the h parameters are summarized in Fig. 9-4.3. ■

$$h_{11} = \left.\frac{V_1}{I_1}\right|_{V_2=0} \qquad h_{21} = \left.\frac{I_2}{I_1}\right|_{V_2=0}$$

$$h_{12} = \left.\frac{V_1}{V_2}\right|_{I_1=0} \qquad h_{22} = \left.\frac{I_2}{V_2}\right|_{I_1=0}$$

Figure 9-4.3 Test conditions used to define h parameters.

Remarks similar to those made for the g parameters also apply to the h parameters, namely, reciprocity implies that $h_{12}(s)$ equals $-h_{21}(s)$, and, in general, any other set of parameters or any network function may be expressed directly in terms of the h parameters. A summary of the interrelations between the various parameter sets is given in Table 9-7.1.

Just as was the case for the z and y parameters, there are several network representations which may be used to model a given set of hybrid parameters. As an example, let us again consider the g parameters. The first equation of (9-55) says that $I_1(s)$ is equal to the sum of a term which is proportional to $V_1(s)$ (such a term may be physically represented by a shunt admittance) and a term which is proportional to $I_2(s)$ (which may be represented by a controlled source). Similarly, from the second equation of (9-55) we see that $V_2(s)$ is equal to the sum of a term which is proportional to $V_1(s)$ (represented by a VCVS) and a term which is proportional to $I_2(s)$ (which is represented by a series impedance). Combining these results, we see that the set of equations given in Eq. (9-55) may be modeled by the network representation shown in Fig. 9-4.4. If $g_{12}(s)$ equals zero, this model appears as shown in Fig. 9-4.5. Thus, there is no transmission from port 2 to port 1 and the network may be considered as a *unilateral* one. A representation of the type shown in Fig. 9-4.5 is useful for modeling many active devices. An example follows.

Figure 9-4.4 Two-controlled-source realization of a set of *g* parameters.

Figure 9-4.5 One-controlled-source realization for a unilateral network.

Example 9-4.1

As an example of the use of *g* parameters to model an active element, consider a unilateral voltage amplifier which is specified as having an input resistance of 100 kΩ, and a no-load low-frequency gain of 100 which decreases at a rate of 6 dB/octave starting at 1000 rad/s. In addition, suppose that the no-load output voltage at any frequency is halved if a 1-kΩ load is added. Such a device may be directly modeled using the *g* parameters. Since it is unilateral, we conclude that $g_{12}(s)$ equals 0. The parameter $g_{11}(s)$ equals 1/100,000, namely, the reciprocal of the input impedance. To meet the no-load specifications the transfer admittance $g_{21}(s)$ must have a pole at $s = -1000$. In addition, $g_{21}(0)$ must equal 100. Thus, $g_{21}(s)$ equals $10^5/(s + 10^3)$. Finally, the network load characteristic specifies $g_{22}(s)$ equals 10^3 Ω, since this "internal" impedance must equal the external impedance which causes the output voltage to be halved. Thus, the amplifier, as specified, may be modeled by the following set of *g* parameters.

$$\begin{bmatrix} 10^{-5} & 0 \\ \dfrac{10^5}{s + 10^3} & 10^3 \end{bmatrix}$$

The *g*-parameter model for this amplifier is shown in Fig. 9-4.6.

Figure 9-4.6 Network realization for Example 9-4.1.

A network model which may be used to represent a set of *h* parameters is easily derived following logic similar to that given above. It serves as a useful model for the case where the output of an active element has the basic characteristics of a current source. Thus, it is common to use *h* parameters to represent transistors. The model is shown in Fig. 9-4.7.

Figure 9-4.7 Two-controlled-source realization of a set of *h* parameters.

Example 9-4.2

As an example of the application of the h parameters, consider the dc model of a grounded emitter transistor with elements r_b (base resistance), r_e (emitter resistance), r_c (collector resistance), and β (current gain) as shown in Fig. 9-4.8(a). Applying the test conditions given in Fig. 9-4.3, we obtain the h parameters

$$\begin{bmatrix} r_b + \dfrac{r_c r_e (1 + \beta)}{r_c + r_e} & \dfrac{r_e}{r_c + r_e} \\ \beta - \dfrac{r_e(1 + \beta)}{r_c + r_e} & \dfrac{1}{r_c + r_e} \end{bmatrix} = \begin{bmatrix} h_{ie} & h_{re} \\ h_{fe} & h_{oe} \end{bmatrix}$$

where the subscripts in the right matrix are the ones usually found in transistor literature to specify the various h parameters. These have the following meaning: $i =$ input, $r =$ reverse, $f =$ forward, and $o =$ output. In addition, e stands for emitter, indicating that the parameters are for the transistor in the grounded emitter configuration. In the limit, for large values of r_c, the parameters approach the values

$$\begin{bmatrix} r_b + r_e(1 + \beta) & 0 \\ \beta & 0 \end{bmatrix}$$

Thus, using these values in the circuit shown in Fig. 9-4.7, we obtain the transistor model shown in Fig. 9-4.8(b). Note that in this circuit, the input resistance is $r_b + r_e(1 + \beta)$. Thus, the resistance r_e has been multiplied by the factor $1 + \beta$, providing an increase in resistance caused by the ICIS. This increase in resistance is frequently referred to as the *Miller effect*.

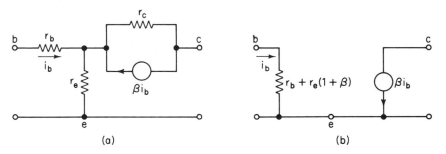

Figure 9-4.8 Use of h parameters to model a grounded-emitter transistor.

9-5 ABCD PARAMETERS

All the types of two-port network parameters which have been introduced so far in this chapter have had the property that the quantities $U_1(s)$ and $U_2(s)$ shown in Eq. (9-7) have included one variable from each of the two ports. Thus, for the z parameters these quantities have included a voltage variable from each port; similarly, for the y parameters they have included a current variable from each port; finally, for the hybrid parameters they have included a voltage variable from one port and a current variable from another. In this section we shall introduce two sets of network parameters which are quite different in that the quantities $U_1(s)$ and $U_2(s)$ will include both the voltage *and* the current variable from a single port. The first of these parameter sets covers the situation shown as the fifth case in Table 9-1.1. Namely, the variables $V_1(s)$ and

$I_1(s)$ are used as the quantities $U_1(s)$ and $U_2(s)$ and the variables $V_2(s)$ and $-I_2(s)$ are used as the quantities $W_1(s)$ and $W_2(s)$. Note that this last assignment of variable includes a minus sign. The reason for this will become apparent later in this section. For the coefficients k_{ij} in Eq. (9-7), we shall use the quantities $A(s)$, $B(s)$, $C(s)$, and $D(s)$. Thus, the equations for this case may be written in the form

$$V_1(s) = A(s)V_2(s) - B(s)I_2(s)$$
$$I_1(s) = C(s)V_2(s) - D(s)I_2(s)$$

(9-71)

The quantities $A(s)$, $B(s)$, $C(s)$, and $D(s)$ defined for the above equations are collectively referred to as the *ABCD parameters*. For convenience of notation, we may write these in the form

$$\begin{bmatrix} V_1(s) \\ I_1(s) \end{bmatrix} = \begin{bmatrix} A(s) & B(s) \\ C(s) & D(s) \end{bmatrix} \begin{bmatrix} V_2(s) \\ -I_2(s) \end{bmatrix} = \mathbf{A}(s) \begin{bmatrix} V_2(s) \\ -I_2(s) \end{bmatrix}$$

(9-72)

where the matrix $\mathbf{A}(s)$ is called the *ABCD parameter matrix*.

For a given network the *ABCD* parameters may be found by applying a set of test excitations to the network in a manner similar to that done for the other parameters which have been introduced in this chapter. To see this, consider the first equation given in (9-71). If we let port 2 of the network be open-circuited, then the variable $I_2(s)$ must be zero. For such a termination, the equation becomes

$$V_1(s) = A(s)V_2(s)$$

(9-73)

To set up a test situation which may be used to implement this equation, we note that $V_2(s)$ *cannot* be used as an excitation variable, since we have already postulated that $I_2(s) = 0$. Therefore, we conclude that $V_1(s)$ must be the excitation variable and $V_2(s)$ must be the response variable. Since we have followed the practice of writing network functions as the ratio of response to excitation (rather than the ratio of excitation to response), it is more consistent to actually define the parameter $A(s)$ by the following expression:

$$\frac{1}{A(s)} = \frac{V_2(s)}{V_1(s)} \bigg|_{I_2(s)=0}$$

(9-74)

Thus, we see that the parameter $A(s)$ is the reciprocal of the open-circuit voltage transfer ratio from port 1 to port 2 of a two-port network. As an example of the determination of this parameter, consider the network consisting of a single shunt element $Y(s)$ as shown in Fig. 9-5.1. If a test voltage excitation is applied at port 1 and port 2 is open-circuited, as shown in Fig. 9-5.2, then $V_1(s) = V_2(s)$ and, from Eq. (9-74), we conclude that $A(s) = 1$. Now consider the determination of the $B(s)$ parameter. If we set $V_2(s)$ to zero by short-circuiting port 2, the first equation of (9-71)

Figure 9-5.1 Single shunt element.

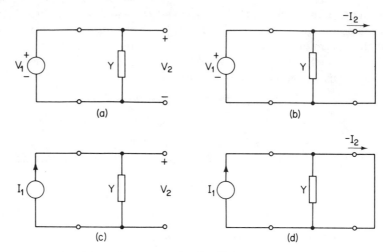

Figure 9-5.2 Tests for the $ABCD$ parameters.

becomes

$$V_1(s) = -B(s)I_2(s) \qquad (9\text{-}75)$$

Following the logic given above, we see that $V_1(s)$ is the excitation variable and $-I_2(s)$ is the response variable. Thus, the parameter $B(s)$ is defined by the relation

$$\frac{1}{B(s)} = \frac{-I_2(s)}{V_1(s)}\bigg|_{V_2(s)=0} \qquad (9\text{-}76)$$

We conclude that $B(s)$ is the reciprocal of the short-circuit transfer impedance from port 1 to port 2. As an example of finding this parameter, again consider the network shown in Fig. 9-5.1. The test conditions with a short circuit on port 2 are shown in Fig. 9-5.2(b). Note that, since the short circuit is effectively applied across the voltage source, and since the impedance of a voltage source is zero, the magnitude of $I_2(s)$ will be infinite for any non-zero voltage excitation. Therefore, the magnitude of $1/B(s)$ is also infinite, and we see that the parameter $B(s) = 0$.

Now consider the second equation of (9-71). If we open-circuit port 2 to set $I_2(s)$ to zero, then, considering the remaining variables in this equation, we see that $I_1(s)$ is the excitation variable and $V_2(s)$ is the response variable. Following the procedure used above, we may define the network function $C(s)$ by the relation

$$\frac{1}{C(s)} = \frac{V_2(s)}{I_1(s)}\bigg|_{I_2(s)=0} \qquad (9\text{-}77)$$

From the above we see that the network parameter $C(s)$ is the reciprocal of the open-circuit current transfer admittance from port 1 to port 2. As an example of the determination of this parameter for the network shown in Fig. 9-5.1, in Fig. 9-5.2(c) a current test excitation is shown being applied to port 1 and port 2 has been open-circuited. Obviously, for this figure, $V_2(s) = I_1(s)/Y(s)$ and from (9-77) we see that $C(s) = Y(s)$. Now consider the determination of $D(s)$. If we short-circuit port 2, then $V_2(s)$ is zero. Considering the remaining variables in the second equation of (9-71), we see that $I_1(s)$ is the excitation variable and $-I_2(s)$ is the response variable.

Thus, we may define the parameter $D(s)$ by the relation

$$\frac{1}{D(s)} = \frac{-I_2(s)}{I_1(s)}\Bigg|_{V_2(s)=0} \tag{9-78}$$

We see that the parameter $D(s)$ is the reciprocal of the short-circuit current transfer ratio from port 1 to port 2. As an example of finding this parameter, for the network shown in Fig. 9-5.1, the test conditions are shown in Fig. 9-5.2(d). Obviously, from this figure $D(s) = 1$. Considering the above development, we see that the complete $ABCD$ parameter matrix for the network shown in Fig. 9-5.1 is

$$\begin{bmatrix} 1 & 0 \\ Y(s) & 1 \end{bmatrix} \tag{9-79}$$

In all the descriptions of the $ABCD$ parameters given above, we see that each parameter defines a *transfer* function rather than a driving-point function. Thus, all these parameters are concerned with the *transmission* of a signal from port 1 to port 2 for various types of excitation and termination. Considering this, we see that another name which is equally applicable to these parameters (and one which is frequently used) is *transmission parameters*.

All the $ABCD$ parameters described above are readily determined by applying unity-valued excitations following the methods used in previous sections. Thus, we may summarize the discussion given above as follows:

SUMMARY 9-5.1

The $ABCD$ Parameters of a Two-Port Network. The $ABCD$ parameters of a two-port network are the quantities $A(s)$, $B(s)$, $C(s)$, and $D(s)$ which specify the variables at port 1 as functions of the variables at port 2. The exact relations are given in Eq. (9-72). Two of the four $ABCD$ parameters may be found by applying a unit test excitation voltage at port 1 and measuring the resulting open-circuit and short-circuit responses at port 2. The other two $ABCD$ parameters may be found by applying a unit test excitation current at port 1 and similarly measuring the responses at port 2. The parameters are the reciprocal of the network functions thus determined. The various test conditions used to define the parameters are summarized in Fig. 9-5.3. ∎

Now let us investigate some properties of the $ABCD$ parameters. In general, these are similar to the properties of the other parameter sets discussed in previous sections. For example, any network function which is defined for a two-port network may be expressed in terms of the $ABCD$ parameters. Similarly, any other set of parameters may be expressed in terms of the $ABCD$ parameters. As an example of such a determination, if we solve the second equation of (9-71) for the variable $V_2(s)$, we obtain

$$V_2(s) = \frac{1}{C(s)}I_1(s) + \frac{D(s)}{C(s)}I_2(s) \tag{9-80}$$

Substituting this result in the first equation of (9-71) and rearranging terms, we obtain

$$A = \frac{V_1}{V_2}\bigg|_{I_2=0}$$

$$B = \frac{V_1}{-I_2}\bigg|_{V_2=0}$$

$$C = \frac{I_1}{V_2}\bigg|_{I_2=0}$$

$$D = \frac{I_1}{-I_2}\bigg|_{V_2=0}$$

Figure 9-5.3 Test conditions used to define the *ABCD* parameters.

$$V_1(s) = \frac{A(s)}{C(s)}I_1(s) + \left[\frac{A(s)D(s)}{C(s)} - B(s)\right]I_2(s) \tag{9-81}$$

The two equations given above may be written in the following form:

$$\begin{bmatrix} V_1(s) \\ V_2(s) \end{bmatrix} = \frac{1}{C(s)}\begin{bmatrix} A(s) & \det \mathbf{A}(s) \\ 1 & D(s) \end{bmatrix}\begin{bmatrix} I_1(s) \\ I_2(s) \end{bmatrix} \tag{9-82}$$

Comparing this with (9-9), we see that we may write the following matrix equality:

$$\begin{bmatrix} z_{11}(s) & z_{12}(s) \\ z_{21}(s) & z_{22}(s) \end{bmatrix} = \frac{1}{C(s)}\begin{bmatrix} A(s) & \det \mathbf{A}(s) \\ 1 & D(s) \end{bmatrix} \tag{9-83}$$

Thus, we have defined the z parameters in terms of the *ABCD* parameters. Other expressions determining the relations between various parameter sets are easily derived. They are summarized in Table 9-7.1. As another example of the properties of the *ABCD* parameters, let us consider the restriction that reciprocity places on these parameters. In Sec. 9-2 it was shown that a reciprocal network is characterized as one in which the z-parameter matrix is symmetric, i.e., $z_{12}(s) = z_{21}(s)$. From Eq. (9-83) we see that this requires that

$$\det \mathbf{A}(s) = A(s)D(s) - B(s)C(s) = 1 \tag{9-84}$$

We conclude that the determinant of the *ABCD* parameter matrix is unity for a reciprocal network.

In previous sections we showed how the z-parameter matrices of component two-port networks are added when such networks are connected in series and considered as a new two-port network, and how the y-parameter matrices similarly add when networks are connected in parallel. Thus, we may anticipate that there is also some connection of two-port networks such that the *ABCD* parameters of the component networks may be used to find the $ABCD$ parameters of some overall two-port network. There is, indeed, such a connection; however, it involves matrix *multiplication* rather than matrix *addition*. To see how this comes about, consider the connection of two two-port networks shown in Fig. 9-5.4. This is called a *cascade* connection of

Figure 9-5.4 Cascade connection of two two-port networks.

networks. For the A network and the voltage and current variables shown in the figure, we may define a set of $ABCD$ parameters $A^a(s)$, $B^a(s)$, $C^a(s)$, and $D^a(s)$ by the relations

$$\begin{bmatrix} V_1^a(s) \\ I_1^a(s) \end{bmatrix} = \begin{bmatrix} A^a(s) & B^a(s) \\ C^a(s) & D^a(s) \end{bmatrix} \begin{bmatrix} V_2^a(s) \\ -I_2^a(s) \end{bmatrix} \tag{9-85}$$

Similarly, for the B network we may define a set of parameters $A^b(s)$, $B^b(s)$, $C^b(s)$, and $D^b(s)$ by the relations

$$\begin{bmatrix} V_1^b(s) \\ I_1^b(s) \end{bmatrix} = \begin{bmatrix} A^b(s) & B^b(s) \\ C^b(s) & D^b(s) \end{bmatrix} \begin{bmatrix} V_2^b(s) \\ -I_2^b(s) \end{bmatrix} \tag{9-86}$$

The quantities $V_1^b(s)$ and $I_1^b(s)$ are obviously equal to the quantities $V_2^a(s)$ and $-I_2^a(s)$, respectively. Thus, we may combine Eqs. (9-85) and (9-86) to obtain

$$\begin{bmatrix} V_1^a(s) \\ I_1^a(s) \end{bmatrix} = \begin{bmatrix} A^a(s) & B^a(s) \\ C^a(s) & D^a(s) \end{bmatrix} \begin{bmatrix} A^b(s) & B^b(s) \\ C^b(s) & D^b(s) \end{bmatrix} \begin{bmatrix} V_2^b(s) \\ -I_2^b(s) \end{bmatrix} \tag{9-87}$$

This result, of course, occurs since the port 2 current variable is defined with a minus sign. Now let us consider the overall two-port network formed as the cascade of networks A and B with port variables defined as shown in Fig. 9-5.4. Obviously, the variables at port 1 of the overall network are the same as those at port 1 of network A. Similarly, the variables at port 2 of the overall network are the same as those at port 2 of network B. Thus, we may write Eq. (9-87) in terms of the variables of the overall network. We obtain

$$\begin{bmatrix} V_1(s) \\ I_1(s) \end{bmatrix} = \begin{bmatrix} A^a(s) & B^a(s) \\ C^a(s) & D^a(s) \end{bmatrix} \begin{bmatrix} A^b(s) & B^b(s) \\ C^b(s) & D^b(s) \end{bmatrix} \begin{bmatrix} V_2(s) \\ -I_2(s) \end{bmatrix} \tag{9-88}$$

The product of the two component $ABCD$ parameter matrices clearly defines a new second-order matrix which gives the $ABCD$ parameters of the overall network formed by the cascade connection of the two component networks. Note that since matrix multiplication is not commutative, i.e., in general, if $\mathbf{E}(s)$ and $\mathbf{F}(s)$ are two square matrices, the product $\mathbf{E}(s)\mathbf{F}(s)$ is not equal to the product $\mathbf{F}(s)\mathbf{E}(s)$, the component parameter matrices must be multiplied in the same left-to-right order as the component networks appear in the cascade connection.

As an example of the application of the results given above, let us determine the transmission parameters of the "L" network shown in Fig. 9-5.5(a). This may be considered as a cascade connection of two component networks A and B as shown in Fig. 9-5.5(b). For the A network, the $ABCD$ parameter matrix has already been given

Network A   Network B

(a)   (b)

Figure 9-5.5 Shunt element cascaded with a series element.

in Eq. (9-79) [with $Y(s)$ replaced by $Y_a(s)$]. For the B network, the $ABCD$ parameter matrix is readily shown to be

$$\begin{bmatrix} 1 & Z_b(s) \\ 0 & 1 \end{bmatrix} \tag{9-89}$$

Thus, the $ABCD$ parameters for the overall network are defined by the equation

$$\begin{bmatrix} 1 & 0 \\ Y_a(s) & 1 \end{bmatrix}\begin{bmatrix} 1 & Z_b(s) \\ 0 & 1 \end{bmatrix} = \begin{bmatrix} 1 & Z_b(s) \\ Y_a(s) & 1 + Y_a(s)Z_b(s) \end{bmatrix} \tag{9-90}$$

The results given above are easily generalized to cover a cascade of an arbitrary number of two-port networks as shown in Fig. 9-5.6. For such a cascade, the trans-

Figure 9-5.6 Cascade of n two-port networks.

mission parameters are found by multiplying the transmission matrices of the individual component networks in the proper order. Thus, if $\mathbf{A}^{(i)}(s)$ $(i = 1, 2, \ldots, n)$ is the transmission parameter matrix for the ith component network, we may write

$$\mathbf{A}(s) = \mathbf{A}^{(1)}(s)\mathbf{A}^{(2)}(s)\mathbf{A}^{(3)}(s) \cdots \mathbf{A}^{(n)}(s) \tag{9-91}$$

where $\mathbf{A}(s)$ is the transmission parameter matrix for the overall two-port network. Note that in a cascade connection of two-port networks such as is shown in Fig. 9-5.6, we need not concern ourselves with whether or not the port current requirement is satisfied, since, as was shown in Sec. 9-1, this requirement is always satisfied for a cascade connection of networks. The results of Eqs. (9-79) and (9-89) provide a simple procedure for finding the $ABCD$ parameters of a ladder network. An example follows.

Example 9-5.1

The ladder network shown in Fig. 9-5.7 may be considered as a ladder network consisting of two series branches of impedance 1 and $4s$, and two shunt branches consisting of admittances $(6s^2 + 1)/2s$ and 1. Thus, using Eqs. (9-79) and (9-89), the transmission parameters for the entire network may be found by evaluating the matrix product

$$\begin{bmatrix} 1 & 1 \\ 0 & 1 \end{bmatrix} \begin{bmatrix} 1 & 0 \\ \dfrac{6s^2 + 1}{2s} & 1 \end{bmatrix} \begin{bmatrix} 1 & 4s \\ 0 & 1 \end{bmatrix} \begin{bmatrix} 1 & 0 \\ 1 & 1 \end{bmatrix}$$

Performing the matrix multiplication, we obtain

$$\begin{bmatrix} \dfrac{24s^3 + 14s^2 + 8s + 1}{2s} & \dfrac{24s^3 + 8s^2 + 6s}{2s} \\ \dfrac{24s^3 + 6s^2 + 6s + 1}{2s} & \dfrac{24s^3 + 6s}{2s} \end{bmatrix}$$

Figure 9-5.7 Ladder network.

SUMMARY 9-5.2

Properties of the *ABCD* Parameters of a Two-Port Network. The four *ABCD* parameters of a given two-port network completely determine the properties of the network in the sense that they may be used to find any other network function that is defined for the two-port network. If a set of two-port networks are connected in cascade, the product of their individual *ABCD* parameter matrices, in the same order from left to right in which the component networks are cascaded, gives the *ABCD* parameters for the resulting overall two-port network. ∎

The transmission parameters of a two-port network may be modeled by a network consisting of three controlled sources and one impedance as shown in Fig. 9-5.8. Applying KVL at port 1 and KCL at port 2, it is readily shown that this network

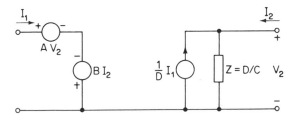

Figure 9-5.8 Controlled source realization for a set of *ABCD* parameters.

satisfies the relations of (9-71). As an example of the use of this model, consider the ideal transformer shown in Fig. 9-5.9(a). It is easily shown that the *ABCD* parameters for this two-port network are

$$\begin{bmatrix} n & 0 \\ 0 & \dfrac{1}{n} \end{bmatrix} \tag{9-92}$$

Thus, the ideal transformer may be modeled with two controlled sources as shown in Fig. 9-5.9(b).

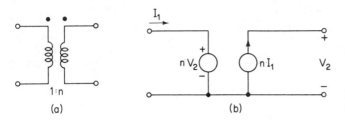

Figure 9-5.9 Controlled source realization for an ideal transformer.

(a)

(b)

The last remaining set of network parameters for a two-port network to be considered is defined as the sixth case in Table 9-1.1. This parameter set is usually referred to as the $\mathcal{ABCD}$ *parameters* (read "script" ABCD parameters). It is defined by the matrix equation

$$\begin{bmatrix} V_2(s) \\ I_2(s) \end{bmatrix} = \begin{bmatrix} \mathcal{A}(s) & \mathcal{B}(s) \\ \mathcal{C}(s) & \mathcal{D}(s) \end{bmatrix} \begin{bmatrix} V_1(s) \\ -I_1(s) \end{bmatrix} \tag{9-93}$$

where the square matrix is called the $\mathcal{ABCD}$ *parameter matrix* or the *inverse transmission parameter matrix*. Its characteristics differ only slightly from those of the *ABCD* parameters. As a result, it is used very infrequently.

9-6 APPLICATIONS OF TWO-PORT NETWORK PARAMETERS

In the preceding sections of this chapter we have defined various sets of network parameters and illustrated how they may be applied to characterize a wide range of two-port networks. Additional applications of these parameters can be made in many typical circuit situations. One such application frequently encountered is the use of a two-port network to serve as a signal transmitting device. As such, it is usually excited by some non-ideal signal source, and transmits the signal to some load impedance. In addition to transmitting the signal, the two-port network is frequently called upon to provide isolation between the excitation and the load, as well as to amplify, filter, or modify the signal in some way. The basic configuration is shown in Fig. 9-6.1. In this figure, a box is used to represent the two-port network. The load and source

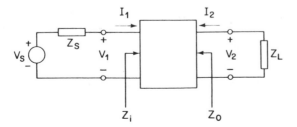

Figure 9-6.1 Two-port network with terminating impedances.

impedances $Z_L(s)$ and $Z_S(s)$ are sometimes called *terminations*. There are many network functions which are of interest for such a configuration. In this section we shall develop specific equations for several of these. In determining the network functions, we shall assume that the two-port network is specified in terms of one of the parameter sets defined in the previous sections. Expressions for the desired network function in terms of other parameter sets are readily obtained using the relations given in Table 9-7.1.

The first network function for the configuration shown in Fig. 9-6.1 that we shall determine is $Z_i(s)$, i.e., the impedance seen looking into the input port of the two-port network when a load impedance $Z_L(s)$ is connected to the output port. Such a network function specifies the impedance seen by the source which drives the network. Thus, it permits an evaluation of the degree of impedance match which is obtained at the input port and thus (as was discussed in Sec. 7-10) an evaluation of the efficiency of the power transfer. In defining the input impedance $Z_i(s)$, it is assumed that the effect of the source impedance $Z_S(s)$ is not considered. Thus, the actual test to determine $Z_i(s)$ is as shown in Fig. 9-6.2. Comparing this circuit with the one shown in Fig. 9-6.1, we see that the arrow used in Fig. 9-6.1 to define $Z_i(s)$ may be interpreted as implying that all circuit components behind the arrowhead are ignored.

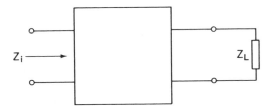

Figure 9-6.2 Definition of Z_i.

To determine $Z_i(s)$, we first note that the presence of the load impedance $Z_L(s)$ requires that the transformed variables $V_2(s)$ and $I_2(s)$ satisfy the relation

$$I_2(s) = -\frac{V_2(s)}{Z_L(s)} \qquad (9\text{-}94)$$

If we substitute this into the equations defining the y parameters of a two-port network given in (9-33), we may solve for the relation between $V_1(s)$ and $I_1(s)$. Thus we obtain the desired expression for the input impedance $Z_i(s)$ in terms of the y parameters of the two-port network

$$Z_i(s) = \frac{V_1(s)}{I_1(s)} = \frac{y_{22}(s) + Y_L(s)}{\det \mathbf{Y}(s) + y_{11}(s)Y_L(s)} \qquad (9\text{-}95)$$

where $\det \mathbf{Y}(s)$ is the determinant of the y-parameter matrix and where $Y_L(s) = 1/Z_L(s)$. Expressions for the input impedance $Z_i(s)$ in terms of the other two-port network parameters may be found in Table 9-6.1.

As an example of the use of the above expression for $Z_i(s)$, consider a circuit in which the two-port network consists of a transistor in a common emitter configuration as shown in Fig. 9-6.3(a). If we use a simple T model for the small-signal behavior of the transistor, the overall circuit may be redrawn as shown in Fig. 9-6.3(b). The z parameters of the transistor are easily determined to be

$$\begin{bmatrix} r_e + r_b & r_e \\ t_e - \alpha r_c & r_e + r_c(1 - \alpha) \end{bmatrix} \qquad (9\text{-}96)$$

TABLE 9-6.1 Two-Port Network Properties

	Z	**Y**	**G**	**H**	**A**	**$\mathscr{A}$**
Z_i	$\dfrac{\det \mathbf{Z} + z_{11}Z_L}{z_{22} + Z_L}$	$\dfrac{y_{22} + Y_L}{\det \mathbf{Y} + y_{11}Y_L}$	$\dfrac{g_{22} + Z_L}{\det \mathbf{G} + g_{11}Z_L}$	$\dfrac{\det \mathbf{H} + h_{11}Y_L}{h_{22} + Y_L}$	$\dfrac{AZ_L + B}{CZ_L + D}$	$\dfrac{\mathscr{D}Z_L + \mathscr{B}}{\mathscr{C}Z_L + \mathscr{A}}$
Z_0	$\dfrac{\det \mathbf{Z} + z_{22}Z_S}{z_{11} + Z_S}$	$\dfrac{y_{11} + Y_S}{\det \mathbf{Y} + y_{22}Y_S}$	$\dfrac{\det \mathbf{G} + g_{22}Y_S}{g_{11} + Y_S}$	$\dfrac{h_{11} + Z_S}{\det \mathbf{H} + h_{22}Z_S}$	$\dfrac{DZ_S + B}{CZ_S + A}$	$\dfrac{\mathscr{A}Z_S + \mathscr{B}}{\mathscr{C}Z_S + \mathscr{D}}$
$A_i = -\dfrac{I_2}{I_1}$	$\dfrac{z_{21}}{z_{22} + Z_L}$	$\dfrac{-y_{21}Y_L}{\det \mathbf{Y} + y_{11}Y_L}$	$\dfrac{g_{21}}{\det \mathbf{G} + g_{11}Z_L}$	$\dfrac{-h_{21}Y_L}{h_{22} + Y_L}$	$\dfrac{1}{D + CZ_L}$	$\dfrac{\det \mathscr{A}}{\mathscr{A} + \mathscr{C}Z_L}$
$A_v = \dfrac{V_2}{V_1}$	$\dfrac{z_{21}Z_L}{\det \mathbf{Z} + z_{11}Z_L}$	$\dfrac{-y_{21}}{y_{22} + Y_L}$	$\dfrac{g_{21}Z_L}{g_{22} + Z_L}$	$\dfrac{-h_{21}}{\det \mathbf{H} + h_{11}Y_L}$	$\dfrac{Z_L}{B + AZ_L}$	$\dfrac{\det \mathscr{A}}{\mathscr{B} + \mathscr{D}Z_L}$

Substituting the values $r_e = 25\ \Omega$, $r_b = 100\ \Omega$, $r_c = 10^6\ \Omega$, and $\alpha = 0.99$, the approximate z parameters become

$$\begin{bmatrix} 125 & 25 \\ -10^6 & 10^4 \end{bmatrix} \qquad (9\text{-}97)$$

Applying the relations given in Table 9-6.1, we readily find that the input impedance $Z_i(s)$ with a load impedance $Z_L(s)$ of $1000\ \Omega$ is $2390\ \Omega$. This is the impedance with which the non-ideal source is loaded.

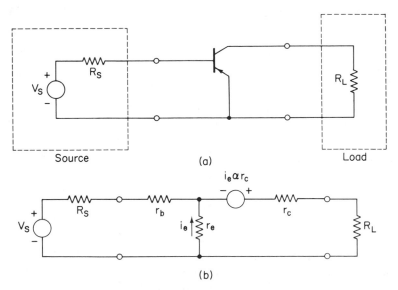

Figure 9-6.3 Simple transistor circuit.

The second network function which we shall consider is $Z_0(s)$, i.e., the impedance seen looking into the output terminals of a two-port network when a source which has an internal impedance $Z_S(s)$ is connected to the input port. Such a network function

Two-Port Network Parameters   Chap. 9

gives us the impedance seen by a circuit connected to the output port. Thus, it permits us to evaluate the impedance match at the output port. The test for determining $Z_0(s)$ is shown in Fig. 9-6.4. In this figure it should be noted that the source voltage $V_S(s)$

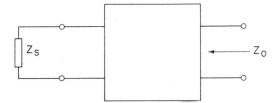

Figure 9-6.4 Definition of Z_0.

has been set to zero. This is test condition implied by the arrow defining $Z_0(s)$ shown in Fig. 9-6.1. In the determination of the network function $Z_0(s)$, we note from Fig. 9-6.4 that the transformed input variables $V_1(s)$ and $I_1(s)$ are related as follows:

$$V_1(s) = -Z_S(s)I_1(s) \tag{9-98}$$

Again substituting into the equations defining the y parameters and solving for the ratio of $V_2(s)$ to $I_2(s)$, we obtain

$$Z_0(s) = \frac{V_2(s)}{I_2(s)} = \frac{y_{11}(s) + Y_S(s)}{\det \mathbf{Y}(s) + y_{22}(s)Y_S(s)} \tag{9-99}$$

where $Y_S(s) = 1/Z_S(s)$. Expressions for the output impedance $Z_0(s)$ in terms of other sets of two-port network parameters are readily obtained and are summarized in Table 9-6.1. As an example of the determination of $Z_0(s)$, let us consider again the circuit shown in Fig. 9-6.3(b). Using the z parameters for the transistor given in (9-97) and the expression given in Table 9-6.1, we find that the output impedance $Z_0(s)$ for a source impedance $Z_S(s)$ of 600 Ω is 44.5 kΩ. Thus any network connected to the output terminals of this circuit will be driven by an impedance of 44.5 kΩ.

The third network function that we shall consider is the current transfer function $A_i(s)$ for the terminated circuit. This network function is defined in terms of the variables shown in Fig. 9-6.1 as

$$A_i(s) = \frac{-I_2(s)}{I_1(s)} \tag{9-100}$$

It should be noted that, for convenience, we define the output current as $-I_2(s)$ where $I_2(s)$ is the conventional two-port output current variable. To determine this network function, using the defining equations for the z parameters and Eq. (9-94), we obtain

$$A_i(s) = \frac{z_{21}(s)}{z_{22}(s) + Z_L(s)} \tag{9-101}$$

Alternative expressions for $A_i(s)$ in terms of other sets of two-port network parameters are given in Table 9-6.1. As an example of the determination of $A_i(s)$, consider again the circuit shown in Fig. 9-6.3(b). Using the z parameters given in Eq. (9-97) and a load impedance $Z_L(s)$ of 1000 Ω, we find $A_i(s) = -190$.

The final network function that we shall consider in this section is the voltage transfer function $A_v(s)$. This network function is defined in terms of the variables shown in Fig. 9-6.1 as

$$A_v(s) = \frac{V_2(s)}{V_1(s)} \tag{9-102}$$

Using Eq. (9-94), and the defining equations for the y parameters, this network function is readily shown to be

$$A_v(s) = \frac{-y_{21}(s)}{y_{22}(s) + Y_L(s)} \tag{9-103}$$

where $Y_L(s) = 1/Z_L(s)$. This is the desired network function for the voltage gain of the terminated network. Alternative expressions for $A_v(s)$ in terms of other sets of two-port network parameters are given in Table 9-6.1. As an example of the determination of this network function, consider again the circuit shown in Fig. 9-6.3(b). Using the z parameters given in Eq. (9-97) and the expression for $A_v(s)$ in terms of the z parameters given in Table 9-6.1, and assuming a load impedance $Z_L(s)$ of 1000 Ω, we find $A_v(s) = -38$.

TABLE 9-7.1 Two-Port

	Z		**Y**		**G**	
Z	z_{11}	z_{12}	$\dfrac{y_{22}}{\det \mathbf{Y}}$	$-\dfrac{y_{12}}{\det \mathbf{Y}}$	$\dfrac{1}{g_{11}}$	$-\dfrac{g_{12}}{g_{11}}$
	z_{21}	z_{22}	$-\dfrac{y_{21}}{\det \mathbf{Y}}$	$\dfrac{y_{11}}{\det \mathbf{Y}}$	$\dfrac{g_{21}}{g_{11}}$	$\dfrac{\det \mathbf{G}}{g_{11}}$
Y	$\dfrac{z_{22}}{\det \mathbf{Z}}$	$-\dfrac{z_{12}}{\det \mathbf{Z}}$	y_{11}	y_{12}	$\dfrac{\det \mathbf{G}}{g_{22}}$	$\dfrac{g_{12}}{g_{22}}$
	$-\dfrac{z_{21}}{\det \mathbf{Z}}$	$\dfrac{z_{11}}{\det \mathbf{Z}}$	y_{21}	y_{22}	$-\dfrac{g_{21}}{g_{22}}$	$\dfrac{1}{g_{22}}$
G	$\dfrac{1}{z_{11}}$	$-\dfrac{z_{12}}{z_{11}}$	$\dfrac{\det \mathbf{Y}}{y_{22}}$	$\dfrac{y_{12}}{y_{22}}$	g_{11}	g_{12}
	$\dfrac{z_{21}}{z_{11}}$	$\dfrac{\det \mathbf{Z}}{z_{11}}$	$-\dfrac{y_{21}}{y_{22}}$	$\dfrac{1}{y_{22}}$	g_{21}	g_{22}
H	$\dfrac{\det \mathbf{Z}}{z_{22}}$	$\dfrac{z_{12}}{z_{22}}$	$\dfrac{1}{y_{11}}$	$-\dfrac{y_{12}}{y_{11}}$	$\dfrac{g_{22}}{\det \mathbf{G}}$	$-\dfrac{g_{12}}{\det \mathbf{G}}$
	$-\dfrac{z_{21}}{z_{22}}$	$\dfrac{1}{z_{22}}$	$\dfrac{y_{21}}{y_{11}}$	$\dfrac{\det \mathbf{Y}}{y_{11}}$	$-\dfrac{g_{21}}{\det \mathbf{G}}$	$\dfrac{g_{11}}{\det \mathbf{G}}$
A	$\dfrac{z_{11}}{z_{21}}$	$\dfrac{\det \mathbf{Z}}{z_{21}}$	$-\dfrac{y_{22}}{y_{21}}$	$-\dfrac{1}{y_{21}}$	$\dfrac{1}{g_{21}}$	$\dfrac{g_{22}}{g_{21}}$
	$\dfrac{1}{z_{21}}$	$\dfrac{z_{22}}{z_{21}}$	$-\dfrac{\det \mathbf{Y}}{y_{21}}$	$-\dfrac{y_{11}}{y_{21}}$	$\dfrac{g_{11}}{g_{21}}$	$\dfrac{\det \mathbf{G}}{g_{21}}$
$\mathcal{A}$	$\dfrac{z_{22}}{z_{12}}$	$\dfrac{\det \mathbf{Z}}{z_{12}}$	$-\dfrac{y_{11}}{y_{12}}$	$\dfrac{1}{y_{12}}$	$-\dfrac{\det \mathbf{G}}{g_{12}}$	$-\dfrac{g_{22}}{g_{12}}$
	$\dfrac{1}{z_{12}}$	$\dfrac{z_{11}}{z_{12}}$	$-\dfrac{\det \mathbf{Y}}{y_{12}}$	$\dfrac{y_{22}}{y_{12}}$	$-\dfrac{g_{11}}{g_{12}}$	$-\dfrac{1}{g_{12}}$

9-7 CONCLUSION

When the Laplace transformation was introduced in Chap. 8, we employed the concept of a transformed network and the network functions describing it as a tool for the analysis of lumped linear time-invariant circuits. In this chapter we have seen an additional application of such an approach, namely, the use of a set of four parameters, each of which is a network function, to characterize a given two-port network (Table 9-7.1). Such a characterization provides a convenient technique for decomposing a given system into simpler subcomponents as an aid to its analysis. Such a characterization also provides a means of defining the properties of a network in a manner which is independent of the planned use of the network or the type of termination which may be connected to it. Network parameters are widely used to define the properties of transistors and other electronic devices, and a set of such parameters will usually be found on the manufacturer's specification sheets for such devices.

Parameter Relations

H		A		$\mathscr{A}$	
$\dfrac{\det \mathbf{H}}{h_{22}}$	$\dfrac{h_{12}}{h_{22}}$	$\dfrac{A}{C}$	$\dfrac{\det \mathbf{A}}{C}$	$\dfrac{\mathscr{D}}{\mathscr{C}}$	$\dfrac{1}{\mathscr{C}}$
$-\dfrac{h_{21}}{h_{22}}$	$\dfrac{1}{h_{22}}$	$\dfrac{1}{C}$	$\dfrac{D}{C}$	$\dfrac{\det \mathscr{A}}{\mathscr{C}}$	$\dfrac{\mathscr{A}}{\mathscr{C}}$
$\dfrac{1}{h_{11}}$	$-\dfrac{h_{12}}{h_{11}}$	$\dfrac{D}{B}$	$-\dfrac{\det \mathbf{A}}{B}$	$\dfrac{\mathscr{A}}{\mathscr{B}}$	$-\dfrac{1}{\mathscr{B}}$
$\dfrac{h_{21}}{h_{11}}$	$\dfrac{\det \mathbf{H}}{h_{11}}$	$-\dfrac{1}{B}$	$\dfrac{A}{B}$	$-\dfrac{\det \mathscr{A}}{\mathscr{B}}$	$\dfrac{\mathscr{D}}{\mathscr{B}}$
$\dfrac{h_{22}}{\det \mathbf{H}}$	$-\dfrac{h_{12}}{\det \mathbf{H}}$	$\dfrac{C}{A}$	$-\dfrac{\det \mathbf{A}}{A}$	$\dfrac{\mathscr{C}}{\mathscr{D}}$	$-\dfrac{1}{\mathscr{D}}$
$-\dfrac{h_{21}}{\det \mathbf{H}}$	$\dfrac{h_{11}}{\det \mathbf{H}}$	$\dfrac{1}{A}$	$\dfrac{B}{A}$	$\dfrac{\det \mathscr{A}}{\mathscr{D}}$	$\dfrac{\mathscr{B}}{\mathscr{D}}$
h_{11}	h_{12}	$\dfrac{B}{D}$	$\dfrac{\det \mathbf{A}}{D}$	$\dfrac{\mathscr{B}}{\mathscr{A}}$	$\dfrac{1}{\mathscr{A}}$
h_{21}	h_{22}	$-\dfrac{1}{D}$	$\dfrac{C}{D}$	$\dfrac{\det \mathscr{A}}{\mathscr{A}}$	$\dfrac{\mathscr{C}}{\mathscr{A}}$
$-\dfrac{\det \mathbf{H}}{h_{21}}$	$-\dfrac{h_{11}}{h_{21}}$	A	B	$\dfrac{\mathscr{D}}{\det \mathscr{A}}$	$\dfrac{\mathscr{B}}{\det \mathscr{A}}$
$-\dfrac{h_{22}}{h_{21}}$	$-\dfrac{1}{h_{21}}$	C	D	$\dfrac{\mathscr{C}}{\det \mathscr{A}}$	$\dfrac{\mathscr{A}}{\det \mathscr{A}}$
$\dfrac{1}{h_{12}}$	$\dfrac{h_{11}}{h_{12}}$	$\dfrac{D}{\det \mathbf{A}}$	$\dfrac{B}{\det \mathbf{A}}$	$\mathscr{A}$	$\mathscr{B}$
$\dfrac{h_{22}}{h_{12}}$	$\dfrac{\det \mathbf{H}}{h_{12}}$	$\dfrac{C}{\det \mathbf{A}}$	$\dfrac{A}{\det \mathbf{A}}$	$\mathscr{C}$	$\mathscr{D}$

PROBLEMS

Problem 9-1 (Sec. 9-2)

Find the z parameters for each of the two-port networks shown in Fig. P9-1.

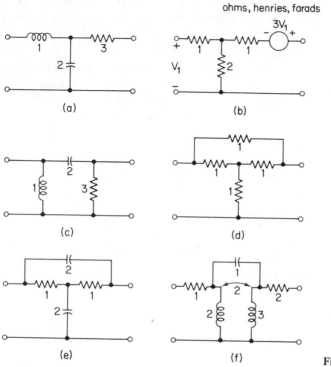

ohms, henries, farads

(a)

(b)

(c)

(d)

(e)

(f)

Figure P9-1

Problem 9-2 (Sec. 9-2)

Find the z parameters of the two-port network shown in Fig. P9-2.

Figure P9-2

Problem 9-3 (Sec. 9-2)

Find the z parameters for the networks shown in Fig. P9-3 by:
(a) Applying the usual testing conditions.
(b) Separating the network into two component subnetworks connected in series and finding the z parameters of each subnetwork separately.

Problem 9-4 (Sec. 9-2)

Find a network which realizes each of the following sets of z parameters.

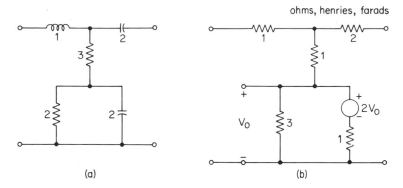

(a)   (b)

Figure P9-3

$$\text{(a)} \begin{bmatrix} 3 & 1 \\ 1 & 2 \end{bmatrix} \quad \text{(b)} \begin{bmatrix} 1 + \dfrac{4}{s} & \dfrac{2}{s} \\ \dfrac{2}{s} & 3 + \dfrac{2}{s} \end{bmatrix} \quad \text{(c)} \begin{bmatrix} 3 & 2 \\ -4 & 4 \end{bmatrix}$$

Problem 9-5 (Sec. 9-3)

Find the y parameters for each of the two-port networks shown in Fig. P9-5.

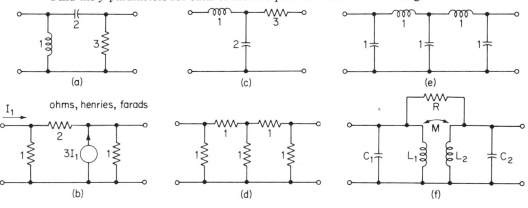

Figure P9-5

Problem 9-6 (Sec. 9-3)

Find the y parameters for each of the networks shown in Fig. P9-6 by

Figure P9-6

(a) Applying the usual test conditions.

(b) Separating the network into two component subnetworks connected in parallel and finding the y parameters of each subnetwork separately.

Problem 9-7 (Sec. 9-3)

Find a network which realizes each of the following sets of y parameters.

$$\text{(a)} \begin{bmatrix} 3 & -1 \\ -1 & 2 \end{bmatrix} \quad \text{(b)} \begin{bmatrix} 1 + \dfrac{4}{s} & -\dfrac{2}{s} \\ -\dfrac{2}{s} & 3 + \dfrac{2}{s} \end{bmatrix} \quad \text{(c)} \begin{bmatrix} 3 & -2 \\ 4 & 4 \end{bmatrix}$$

Problem 9-8 (Sec. 9-3)

Show that the product of the z-parameter matrix for the network shown in Fig. P9-1(a) and the y-parameter matrix for the network shown in Fig. P9-5(c) yields the identity matrix.

Problem 9-9 (Sec. 9-3)

Show that the product of the z-parameter matrix for the network shown in Fig. P9-1(c) and the y-parameter matrix for the network shown in Fig. P9-5(a) gives the identity matrix.

Problem 9-10 (Sec. 9-3)

To find the y parameters of the network shown in Fig. P9-10(a), a student decides to find the y parameters of the two networks shown in parts (b) and (c) of the figure separately and add the result. Is his reasoning correct? Test your conclusion by performing the analysis.

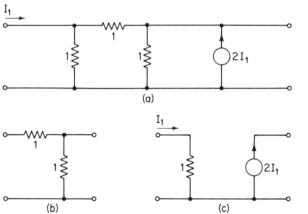

Figure P9-10

Problem 9-11 (Sec. 9-4)

Find the g parameters for each of the two-port networks shown in Fig. P9-11.

Problem 9-12 (Sec. 9-4)

Show an interconnection of two two-port networks such that their g parameters add. Use blocks to represent the two-port networks and assume that the port condition is not violated.

Problem 9-13 (Sec. 9-4)

Find the h parameters for each of the two-port networks shown in Fig. P9-11.

Problem 9-14 (Sec. 9-4)

Show an interconnection of two two-port networks such that their h parameters add. Use blocks to represent the two-port networks and assume that the port condition is not violated.

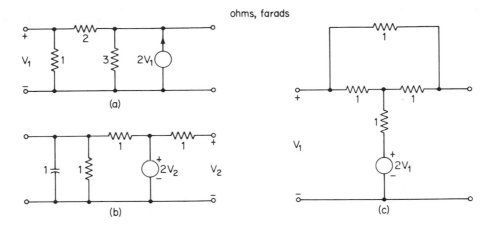

Figure P9-11

Problem 9-15 (Sec. 9-4)
Find a network which realizes the following *g* parameters.

$$\begin{bmatrix} s+4 & 0 \\ 10 & 4 \end{bmatrix}$$

Problem 9-16 (Sec. 9-5)
Find the transmission parameters for each of the two-port networks shown in Fig. P9-16.

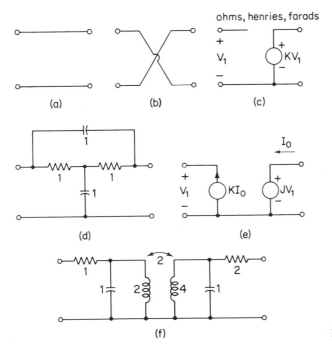

Figure P9-16

Problem 9-17 (Sec. 9-5)
Find the *y* parameters of the network shown in Fig. P9-17 by first finding the transmission parameters of the transformer and the series resistors, converting these into *y* parameters,

and then adding the y parameters of the capacitor. The transformer is assumed to be ideal. Test the result by finding the y parameters directly using the usual test conditions.

ohms, farads

Figure P9-17

Problem 9-18 (Sec. 9-5)

Find the z parameters of the networks shown in Fig. P9-18 by first finding the transmission parameters of each element, then using these to find the overall transmission parameters of the cascade of elements, and finally converting the result to z parameters. Check your answer by directly applying the usual tests to determine the z parameters of the networks.

ohms, henries

(a)   (b)   Figure P9-18

Problem 9-19 (Sec. 9-5)

Find all possible sets of parameters for each of the active devices shown in Fig. P9-19.

(a)   (b)

(c)   (d)   Figure P9-19

Problem 9-20 (Sec. 9-5)

Two three-port networks with variables defined as shown in Fig. P9-20 are each specified by a set of parameters having the variables indicated below. Show an interconnection of the ports of the two networks such that their individual parameters add to form the parameters of the overall three-port network.

$$\begin{bmatrix} V_1 \\ I_2 \\ V_3 \end{bmatrix} = [\text{Set of parameters}] \begin{bmatrix} I_1 \\ V_2 \\ I_3 \end{bmatrix}$$

Figure P9-20

Problem 9-21 (Sec. 9-6)

Find the network functions $Z_i(s)$, $Z_0(s)$, $A_i(s)$, and $A_v(s)$ for the circuit shown in Fig. 9-6.3(a), with the transistor in a common base configuration rather than a common emitter configuration. Use the transistor model shown in Fig. 9-6.3(b), and assume that $r_b = 100\ \Omega$, $r_e = 25\ \Omega$, $r_c = 10^6\ \Omega$, $Z_S = 100\ \Omega$, $\alpha = 0.99$, and $Z_L = 10^4\ \Omega$.

Problem 9-22 (Sec. 9-6)

Using the relations given in Table 9-6.1, we may develop the concepts of image impedances. These are the impedances $Z_{I1}(s)$ and $Z_{I2}(s)$, which, when connected respectively at ports 1 and 2 of a two-port network, simultaneously provide an impedance match at both ports. Find expressions for $Z_{I1}(s)$ and $Z_{I2}(s)$ in terms of the *ABCD* parameters.

Problem 9-23 (Sec. 9-6)

Find the image impedances defined in Problem 9-22 for the symmetric lattice network with impedances $Z_a(s)$ and $Z_b(s)$ shown in Fig. P9-23. Show that when $Z_a(s) = 1/Z_b(s)$, the lattice has image impedances of 1 Ω. (In such a case it is called a constant resistance lattice.)

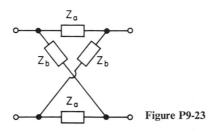

Figure P9-23

10

The Fourier Series

and the Fourier Integral

In Chap. 7 we discussed the determination of the steady-state response of networks which were excited only by sinusoidal functions. As part of our development, we introduced the concept of a network function and showed that the product of an input sinusoidal function (represented as a phasor) and the network function (a function of $j\omega$) gave us the output or response sinusoid (also represented as a phasor). A most important extension of this approach allows similar techniques to be applied to find the steady-state response of a network which is excited by periodic functions *which are not sinusoids*. Basically, the method consists of using a series of sinusoids to represent the non-sinusoidal periodic input and output waveforms. The basic concepts of such an approach can also be extended to cover non-periodic functions. In this context, the ideas of this chapter provide an important link between the sinusoidal steady-state material introduced in Chap. 7 and the concepts of a transformed network presented in Chap 8. The details of this connection will become apparent in the material which follows.

10-1 NON-SINUSOIDAL PERIODIC FUNCTIONS

In this section we shall introduce the concept of using a series of sinusoidal functions to represent a non-sinusoidal periodic function. First of all, let us review some terminology associated with a sinusoidal function. If we let $f(t)$ be a sinusoidal waveform, then we may write

$$f(t) = F_0 \sin(\omega_0 t + \phi) \tag{10-1}$$

where F_0 is the peak value or magnitude of the sinusoidal function, ω_0 is the frequency of oscillation in radians per second, and ϕ is the phase angle (or argument) in radians. Two other quantities of interest which are used in connection with such a sinusoidal function are f_0, the frequency of oscillation in hertz (where $f_0 = \omega_0/2\pi$), and T, the period in seconds (where $T = 1/f_0$). For purposes of sinusoidal steady-state analysis it is assumed that a function such as is given by Eq. (10-1) is defined over all t, i.e., over the interval $-\infty \le t \le \infty$. A sinusoidal function of this type has the property that

$$f(t) = f\left(t \pm \frac{2\pi}{\omega_0}\right) = f(t \pm T) \tag{10-2}$$

for all t.

Let us now see how a series of sinusoidal functions as described above can be utilized to describe non-sinusoidal periodic functions. Such non-sinusoidal functions are of considerable importance since they are frequently encountered in network applications. For example, the square wave shown in Fig. 10-1.1(a) is available as an output from most signal generators used to test electronic circuits. Other examples of non-sinusoidal periodic functions are the triangular wave shown in Fig. 10-1.1(b), the train of pulses shown in Fig. 10-1.1(c), and the irregular waveform shown in Fig. 10-1.1(d). In general, any function that satisfies the relation

$$f(t) = f(t + T) \tag{10-3}$$

for a given value of T, and for $-\infty \le t \le \infty$ is said to be a *periodic function* with a period of T seconds. To see how (10-3) applies, consider the square wave shown in Fig. 10-1.1(a). The period of this square wave is 2. Thus, if we let $f(t)$ designate this function and if we choose $t = 1.1$ as a "test" value of t, we see that $f(1.1) = -1$. From the figure we also see that $f(3.1) = -1$, $f(5.1) = -1$, etc., as required by (10-3).

To see how a square wave such as is shown in Fig. 10-1.1(a) may be treated as a series of sinusoids, let us start with a sinusoid $f_1(t)$ which has the same period, namely

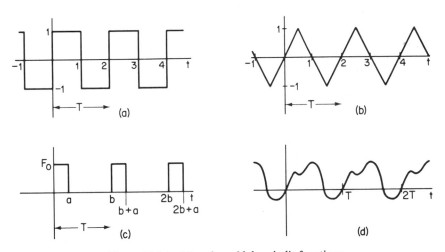

Figure 10-1.1   Non-sinusoidal periodic functions.

$T = 2$. Thus, $\omega_0 = 2\pi/T = \pi$. In addition, let $f_1(t)$ have a peak magnitude of $4/\pi$. Thus, we may write

$$f_1(t) = \frac{4}{\pi} \sin \pi t \qquad (10\text{-}4)$$

This function is plotted in Fig. 10-1.2(a). It obviously doesn't look very much like a square wave. To this sinusoid let us now add a second sinusoid with a frequency three times that of $f_1(t)$ and a magnitude $\frac{1}{3}$ as great. In the next section we shall see how this second sinusoid was chosen. Let us call the resulting sum $f_2(t)$. It may be expressed as

$$f_2(t) = \frac{4}{\pi} \left(\sin \pi t + \frac{1}{3} \sin 3\pi t \right) \qquad (10\text{-}5)$$

A plot of this function is shown in Fig. 10-1.2(b). Now let us add a third sinusoid which has a frequency five times as great as $f_1(t)$ and a magnitude $\frac{1}{5}$ as large. If we call the result $f_3(t)$, we may write

$$f_3(t) = \frac{4}{\pi} \left(\sin \pi t + \frac{1}{3} \sin 3\pi t + \frac{1}{5} \sin 5\pi t \right) \qquad (10\text{-}6)$$

A plot of this function is shown in Fig. 10-1.2(c). Continuing in this fashion by adding a sinusoid of frequency seven times as large as $f_1(t)$ and a magnitude $\frac{1}{7}$ as large, we

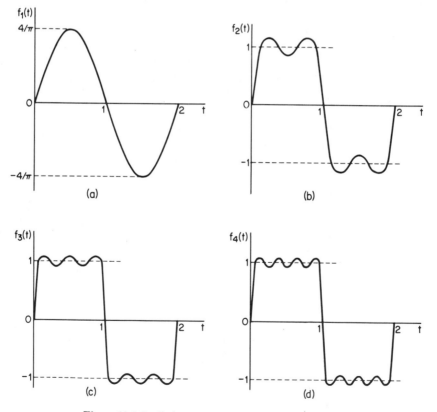

Figure 10-1.2 Series representation of a square wave.

The Fourier Series and the Fourier Integral   Chap. 10

obtain the function $f_4(t)$ shown in Fig. 10-1.2(d). This latter function obviously begins to approach the general appearance of a square wave. In the next section we shall show that in the limit as the number of sinusoidal terms becomes infinite, their sum generates a waveform identical with the one shown in Fig. 10-1.1(a). Thus, we may write

$$f(t) = \frac{4}{\pi} \sum_{k=1}^{\infty} \frac{1}{2k-1} \sin (2k-1)\pi t \tag{10-7}$$

where $f(t)$ is the square wave shown in Fig. 10-1.1(a). A similar procedure may be followed for other periodic waveforms. For example, consider

$$f(t) = \frac{8}{\pi^2} \left(\sin \pi t - \frac{1}{9} \sin 3\pi t + \frac{1}{25} \sin 5\pi t \right) \tag{10-8}$$

A plot of this function is shown in Fig. 10-1.3. It is easily observed that this waveform is quite similar to the original triangular waveform shown in Fig. 10-1.1(b). In the above discussion we have demonstrated that we can add and subtract sinusoidal functions in such a way as to produce functions whose waveforms approach those of

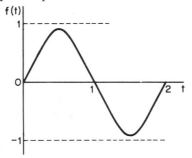

Figure 10-1.3 Series representation of a triangular wave.

non-sinusoidal periodic functions. In addition, we have shown that the more sinusoidal terms which we add, the closer the resemblance becomes. Thus, we have used a series representation of sinusoids to represent non-sinusoidal periodic functions. Such series are called *Fourier series*. In the next section we shall show how to find the elements of such series, i.e., how to find the magnitude, frequency, and phase of the component sinusoidal terms. Before proceeding to that task, however, we will determine some general properties of periodic functions.

The examples of using a series of sinusoidal functions to represent a non-sinusoidal periodic function which were given above all used sine functions in the resulting expressions. Other periodic non-sinusoidal functions may require the use of cosine functions. For example, consider the square wave shown in Fig. 10-1.4(a). We will see shortly that such a function may be represented by a Fourier series similar to (10-7) in which the sine functions are replaced by cosine functions. Thus, for the function $f(t)$ shown in this figure, we may write

$$f(t) = \frac{4}{\pi} \left(\cos \pi t - \frac{1}{3} \cos 3\pi t + \frac{1}{5} \cos 5\pi t - \cdots \right) \tag{10-9}$$

Similarly, a series which involves only cosine terms may be written for the triangular wave shown in Fig. 10-1.4(b). The above examples which use either sine or cosine functions are actually special cases of a more general class of functions which require the use of both sine and cosine functions for their representation. These will be covered in more detail in the following section.

Figure 10-1.4 Waveforms represented by series of cosine terms.

The functions expressed by the sine and cosine series given above are examples of two special types of functions which are frequently encountered in treating periodic functions. To emphasize their properties, we will define these types more fully in the following paragraphs. The definitions apply to all periodic functions (sinusoidal and non-sinusoidal) and have to do with the symmetry of the function. The first special type is defined by the relation

$$f(t) = f(-t) \qquad (10\text{-}10)$$

where $f(t)$ is a periodic function. If this relation holds for all values of t, then $f(t)$ is said to be an *even function*. As an example of such a function, consider the square wave shown in Fig. 10-1.4(a). It is readily shown that this function satisfies Eq. (10-10). Similarly, the triangular wave shown in Fig. 10-1.4(b) is readily observed to be an even function. One of the best known even functions is the cosine function itself. We shall make reference to its even properties frequently in the sections that follow. In general, a function whose waveform is symmetrical with respect to the ordinate is an even function. Frequently, such a function is referred to as being "symmetrical about the origin." It should be noted that even functions may have a non-zero average value.

The second case that we shall be concerned with regarding the symmetry of periodic functions is defined by the relation

$$f(t) = -f(-t) \qquad (10\text{-}11)$$

where $f(t)$ is a periodic function. If this relation holds for all t, then $f(t)$ is said to be an *odd function*. As an example of such a function, consider the square wave shown in Fig. 10-1.1(a). It is readily shown that this waveform satisfies Eq. (10-11) and, thus, is an odd function. Similarly, the triangular wave shown in Fig. 10-1.1(b) is an odd function. Another important odd function which we will use frequently in the sections that follow is the sine function. A function which satisfies the defining relation given in (10-11) is frequently said to be "skew-symmetric about the origin." Unlike even functions, the average value of an odd function is always zero.

We may now develop some properties which have to do with even and odd functions which will be of importance in the sections that follow. As a first property, let us consider the sum of a series of even functions $f_i^{(e)}(t)$ $(i = 1, 2, \ldots, n)$. If we let the sum of such a series of functions be $f(t)$, we may write

$$f(t) = \sum_{i=1}^{n} f_i^{(e)}(t) \qquad (10\text{-}12)$$

Now let us replace the argument t by the argument $-t$ in the above equation. Since the functions in the right member have been defined as being even, from Eq. (10-10) such a replacement does not change the value of the right member. Therefore, the left member of the equation must also be unchanged, i.e., $f(-t) = f(t)$. We may conclude that the sum of a series of even functions is also an even function. A similar analysis may be used to show that the sum of a series of odd functions is also an odd function.

Now let us evaluate the situation that occurs when even and odd functions are multiplied. First let us consider the product of two even functions. Let $f_1^{(e)}$ and $f_2^{(e)}(t)$ be two even functions and let $f(t)$ be their product. Thus, we may write

$$f(t) = f_1^{(e)}(t) \times f_2^{(e)}(t) \qquad (10\text{-}13)$$

If we replace t by $-t$ in both members of the above equation, we obtain

$$f(-t) = f_1^{(e)}(-t) \times f_2^{(e)}(-t) \qquad (10\text{-}14)$$

Now let us apply Eq. (10-10) to the even functions in the right member of the above. We obtain

$$f(-t) = f_1^{(e)}(t) \times f_2^{(e)}(t) \qquad (10\text{-}15)$$

Comparing Eqs. (10-13) and (10-15), we conclude that the product of two even functions is also an even function. In a similar manner, we may show that the product of two odd functions is an *even* function. Finally, let us consider the case where an even function $f^{(e)}(t)$ and an odd function $f^{(o)}(t)$ are multiplied to produce a function $f(t)$. Thus, we may write

$$f(t) = f^{(e)}(t) \times f^{(o)}(t) \qquad (10\text{-}16)$$

Changing t to $-t$ in both members in the above equation and applying (10-10) and (10-11), we obtain

$$f(-t) = f^{(e)}(-t) \times f^{(o)}(-t) = -f^{(e)}(t) \times f^{(o)}(t) = -f(t) \qquad (10\text{-}17)$$

Thus, we conclude that the product of an even function and an odd function is an odd function.

The results given above may be summarized as follows:

SUMMARY 10-1.1

Sums and Products of Even and Odd Functions. An even periodic function is one which is symmetric about the origin, i.e., one in which $f(t) = f(-t)$. An odd periodic function is one which is skew-symmetric about the origin, i.e., one in which $f(t) = -f(-t)$. Sums and products of such functions satisfy the following characterizations.

$$\text{even} + \text{even} = \text{even}$$
$$\text{odd} + \text{odd} = \text{odd}$$
$$\text{even} \times \text{even} = \text{even}$$
$$\text{odd} \times \text{odd} = \text{even}$$
$$\text{odd} \times \text{even} = \text{odd}$$
∎

The one case of combining even and odd functions not covered in the above discussion is the case in which an even function is added to an odd function. In general, such a sum will be neither even nor odd. To see this, let $f^{(e)}(t)$ be an even function and $f^{(o)}(t)$ be an odd function. If we let $f(t)$ be their sum, we may write

$$f(t) = f^{(e)}(t) + f^{(o)}(t) \tag{10-18}$$

Now let us replace the argument t by $-t$ in the above. We obtain

$$f(-t) = f^{(e)}(-t) + f^{(o)}(-t) \tag{10-19}$$

Substituting from the defining relations for even and odd functions given in Eqs. (10-10) and (10-11), this expression may be rewritten

$$f(-t) = f^{(e)}(t) - f^{(o)}(t) \tag{10-20}$$

If we now add Eqs. (10-18) and (10-20) and rearrange terms, we obtain

$$f^{(e)}(t) = \tfrac{1}{2}[f(t) + f(-t)] \tag{10-21}$$

Similarly, subtracting (10-18) and (10-20) and rearranging terms, we obtain

$$f^{(o)} = \tfrac{1}{2}[f(t) - f(-t)] \tag{10-22}$$

From the results of Eqs. (10-21) and (10-22) given above, we obtain a most important result: If we have any arbitrary function $f(t)$ which does not have even or odd symmetricity and apply Eq. (10-21), we obtain an even function $f^{(e)}(t)$ which is the *even part of* $f(t)$. Similarly, applying Eq. (10-22) to $f(t)$, we obtain an odd function $f^{(o)}(t)$ which is called the *odd part of* $f(t)$. Thus, we may conclude that an arbitrary periodic function may always be decomposed into an even and an odd part. It is readily shown that the addition of such parts gives the original function. An example follows.

Example 10-1.1

As an example of the determination of the even and odd parts of an arbitrary periodic function, consider the periodic sequence of rectangular pulses labeled $f(t)$ in Fig. 10-1.5(a). The corresponding function $f(-t)$ is shown in Fig. 10-1.5(b). Adding these two functions and dividing by two to satisfy (10-21), we obtain a function $f^{(e)}(t)$ as shown in Fig. 10-1.5(c). Obviously this function has even symmetricity. Similarly, applying (10-22) we obtain the odd function $f^{(o)}(t)$ shown in Fig. 10-1.5(d). The sum of the even part and the odd part of $f(t)$ obviously gives us the original function $f(t)$ shown in Fig. 10-1.5(a).

The symmetry aspects of periodic functions discussed above are sometimes referred to as the properties of *full-wave symmetry*. Another type of symmetry which is frequently found to be useful in discussing periodic functions is *half-wave symmetry*.

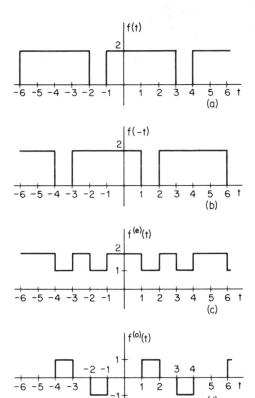

Figure 10-1.5 Even and odd parts of a periodic function.

There are two types, odd and even. *Odd half-wave symmetry* is defined by the relation

$$f(t) = -f\left(t + \frac{T}{2}\right) \tag{10-23}$$

for all t over the range $-\infty \le t \le \infty$, where T is the period. Thus, the shape of the waveform in the second half of any given period is the negative of the waveform in the first half. As an example of odd half-wave symmetry, consider the periodic function shown in Fig. 10-1.6. It is readily observed that this waveform satisfies the relation

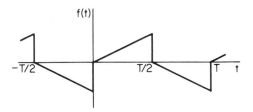

Figure 10-1.6 Periodic function with half-wave symmetry.

given in Eq. (10-23). The other type of half-wave symmetry that might be considered is *even half-wave symmetry*. This would be defined by the relation

$$f(t) = f\left(t + \frac{T}{2}\right) \tag{10-24}$$

where T is the period. Actually, this is not a special case, since, from Eq. (10-2) this relation merely defines a function with a period $T/2$ rather than a period T. Henceforth, when we refer to half-wave symmetry, we shall imply *odd* half-wave symmetry as defined by Eq. (10-23).

10-2 THE FOURIER SERIES

In the first section of this chapter we presented some specific examples of how series of sinusoidal terms called Fourier series could be formed such that in the limit as the number of terms was increased the waveform of the series approached that of a specified non-sinusoidal periodic function. In this section, we shall investigate the general form of such a Fourier series and show how to determine it for any arbitrary periodic non-sinusoidal function.

Let us assume $f(t)$ is an arbitrary periodic function of period T. To avoid any possibility of trying to obtain a Fourier series for "weird" functions, we shall place some restrictions on $f(t)$. First, let us assume that it has only a finite number of maxima or minima in any one period, and that these maxima and minima are finite in magnitude. Second, let us assume that there are only a finite number of jumps or discontinuities in any one period. Finally, let us assume that the integral of the absolute value of the function taken over one period is also finite, i.e.,

$$\int_0^T |f(t)| \, dt < \infty \tag{10-25}$$

These conditions are called the *Dirichlet conditions*, and they are sufficient to ensure that the Fourier series of $f(t)$ exists. An example of a function which does not satisfy the Dirichlet conditions is $\sin(1/t)$. In practice, all the functions that are normally encountered satisfy these conditions. For such a function it may be shown that the Fourier series has the form

$$f(t) = a_0 + \sum_{i=1}^{\infty} (a_i \cos i\omega_0 t + b_i \sin i\omega_0 t) \tag{10-26}$$

In this equation $\omega_0 \ (= 2\pi/T)$ is called the *fundamental frequency* in radians per second of the periodic function and each of the quantities $i\omega_0 \ (= i2\pi/T)$ is referred to as the *ith harmonic* of the fundamental frequency. These quantities can, of course, also be expressed in hertz, in which case $f_0 \ (= 1/T)$ is the fundamental frequency and $if_0 \ (= i/T)$ is the *i*th harmonic of the fundamental. The quantities a_i and b_i are called the *Fourier coefficients*. In this section we shall see how they may be determined. Before doing this, however, let us review some properties of the functions $\cos n\omega_0 t$ and $\sin m\omega_0 t$ (where n and m are positive integers). We may first note that the integral covering one period of the fundamental of a sine or a cosine function or any of its harmonics is zero. Thus, we may write

$$\int_0^T \cos n\omega_0 t \, dt = 0 \qquad \int_0^T \sin m\omega_0 t \, dt = 0 \tag{10-27}$$

These relations are easily seen to be true, since, over one period, the integrands generate equal positive- and negative-valued areas. Now let us note that over one period of the fundamental the integral of either the sine squared or the cosine squared of the fundamental or any harmonic is equal to $T/2$. Thus, we may write

$$\int_0^T \cos^2 n\omega_0 t\, dt = \frac{T}{2} \qquad \int_0^T \sin^2 m\omega_0 t\, dt = \frac{T}{2} \tag{10-28}$$

As a final property of the integrals of sine and cosine functions, we note that the integrals of products of sine and cosine terms over one period, as given in the following relations, are zero.

$$\int_0^T \cos n\omega_0 t \sin m\omega_0 t\, dt = 0 \qquad \text{all } m, n$$

$$\int_0^T \cos n\omega_0 t \cos m\omega_0 t\, dt = 0 \qquad m \neq n \tag{10-29}$$

$$\int_0^T \sin n\omega_0 t \sin m\omega_0 t\, dt = 0 \qquad m \neq n$$

The proofs of the relations given above may be found in any basic text on integral calculus. In the expressions given in Eqs. (10-27), (10-28), and (10-29), it should be noted that, although the limits of integration are 0 and T, in general, the theorems hold for any limits of integration which define one complete period. Thus, we may replace the lower limit of the integrals by t_0 and the upper limit by $t_0 + T$ in any of the above relations without changing the results. The relations of Eq. (10-29) may be collectively summarized by saying that the functions $\cos n\omega_0 t$ and $\sin m\omega_0 t$ are *orthogonal* with respect to integration from t_0 to $t_0 + T$, i.e., integration over one period.

Using the relations given in Eqs. (10-27), (10-28), and (10-29), we may now present a general process for determining the Fourier coefficients a_i and b_i. As an example of this procedure, let us first consider the details of obtaining the coefficient a_0. If we integrate both sides of (10-26) from 0 to T, we obtain

$$\int_0^T f(t)\, dt = \int_0^T a_0\, dt + \int_0^T \sum_{i=1}^{\infty} (a_i \cos i\omega_0 t + b_i \sin i\omega_0 t)\, dt \tag{10-30}$$

It may be shown that the operations of integration and summation indicated in the right term of the right member of (10-30) can be interchanged. Doing this, we may write

$$\int_0^T f(t)\, dt = \int_0^T a_0\, dt + \sum_{i=1}^{\infty} \left[\int_0^T a_i \cos i\omega_0 t\, dt + \int_0^T b_i \sin i\omega_0 t\, dt \right] \tag{10-31}$$

Integrating the first term in the right member of Eq. (10-31), we obtain $a_0 T$. Integrating the terms in the summation in the right member, we find that all these terms are zero. Thus, we have derived an integral relation which only involves the original function $f(t)$ and the coefficient a_0, and which may be written in the form

$$a_0 = \frac{1}{T} \int_0^T f(t)\, dt \tag{10-32}$$

From this relation, we see that a_0 gives the average value of the periodic function $f(t)$ over any single period. Obviously, since the function is periodic, this average value will be the same for any other period and we may conclude that a_0 is the average value of the non-sinusoidal periodic function $f(t)$. As an example of determining such a coefficient, consider the square wave shown in Fig. 10-2.1. The period T is obviously 2 s and the integral of the function over one period (the area under one square pulse) is 2. Applying (10-32), we conclude that $a_0 = 1$.

Figure 10-2.1 Square wave.

In determining the other Fourier coefficients a procedure quite similar to the one described above is followed. The only major difference is that we must multiply both sides of Eq. (10-26) by an appropriate sinusoidal function before integrating. Thus, the orthogonality relations of Eq. (10-29) may be used to isolate any single Fourier coefficient. As an example of this procedure, consider the determination of the coefficient a_1. Multiplying both members of (10-26) by the term $\cos \omega_0 t$ and integrating from 0 to T, we obtain

$$\int_0^T f(t) \cos \omega_0 t \, dt = \int_0^T a_0 \cos \omega_0 t \, dt + \int_0^T \cos \omega_0 t \sum_{i=1}^\infty (a_i \cos i\omega_0 t + b_i \sin i\omega_0 t) \, dt$$

$$(10\text{-}33)$$

Again the operations of integration and summation may be interchanged. Doing this and applying the orthogonality relations of (10-29), all terms in the right member of the above equation except the term involving the coefficient a_1 become zero. Thus, we obtain

$$\int_0^T f(t) \cos \omega_0 t \, dt = \int_0^T a_1 \cos^2 \omega_0 t \, dt = a_1 \frac{T}{2} \qquad (10\text{-}34)$$

Rearranging the terms, we see that

$$a_1 = \frac{2}{T} \int_0^T f(t) \cos \omega_0 t \, dt \qquad (10\text{-}35)$$

We conclude that the relation given above may be applied to any periodic function to find the Fourier coefficient a_1. As an example of such an application, let us determine the coefficient a_1 for the square wave shown in Fig. 10-2.1. From Eq. (10-35) we obtain

$$a_1 = \frac{2}{2} \int_0^1 2 \cos \pi t \, dt = \frac{2 \sin \pi t}{\pi} \Big|_0^1 = 0 \qquad (10\text{-}36)$$

The approach used above to determine a_1 is easily extended to provide a general method for obtaining any of the Fourier coefficients a_i. The result is

$$a_i = \frac{2}{T} \int_0^T f(t) \cos i\omega_0 t \, dt \qquad i = 1, 2, 3, \ldots \qquad (10\text{-}37)$$

This result is easily verified using the orthogonality relations of Eq. (10-29). Applying this relation to the square wave shown in Fig. 10-2.1, we find that all the coefficients a_i are zero.

The Fourier Series and the Fourier Integral   Chap. 10

Now let us consider the determination of the coefficients associated with the sine terms in the Fourier series. To determine the coefficient b_1 we may multiply both members of Eq. (10-26) by the factor $\sin \omega_0 t$ and integrate from 0 to T. Thus, we obtain

$$\int_0^T f(t) \sin \omega_0 t \, dt = \int_0^T a_0 \sin \omega_0 t \, dt + \int_0^T \sin \omega_0 t \sum_{i=1}^{\infty} (a_i \cos i\omega_0 t + b_i \sin i\omega_0 t) \, dt \tag{10-38}$$

Interchanging the operations of integration and summation and applying the orthogonality relations of Eq. (10-29), all the terms in the right member except the one involving b_1 become zero. Thus, we conclude that

$$b_1 = \frac{2}{T} \int_0^T f(t) \sin \omega_0 t \, dt \tag{10-39}$$

Extending this procedure, we obtain the following general relation for determining the coefficients b_i of a Fourier series.

$$b_i = \frac{2}{T} \int_0^T f(t) \sin i\omega_0 t \, dt \tag{10-40}$$

As an example of the application of this relation, consider the square wave shown in Fig. 10-2.1. For the coefficient b_1 we obtain

$$b_1 = \frac{2}{2} \int_0^1 2 \sin \pi t \, dt = -\frac{2 \cos \pi t}{\pi} \Big|_0^1 = \frac{4}{\pi} \tag{10-41}$$

Similarly, for the other coefficients b_i we find that

$$b_i = \begin{cases} 0 & i \text{ even} \\ 4/i\pi & i \text{ odd} \end{cases} \tag{10-42}$$

Comparing the results given above for the determination of the coefficients a_i and b_i of the square wave shown in Fig. 10-2.1, we find that this square wave may be expressed as the following Fourier series.

$$f(t) = 1 + \frac{4}{\pi} \left(\sin \pi t + \frac{1}{3} \sin 3\pi t + \frac{1}{5} \sin 5\pi t + \cdots \right) \tag{10-43}$$

Thus, we have verified (with the addition of a constant term) the assumed form given for the series representation of a square wave in Eq. (10-7). If we apply the procedure given above to the square wave shown in Fig. 10-2.2, then we obtain

$$f(t) = 1 + \frac{4}{\pi} \left(\cos \pi t - \frac{1}{3} \cos 3\pi t + \frac{1}{5} \cos 5\pi t - \cdots \right) \tag{10-44}$$

This is similar to the result obtained in Eq. (10-9). The results given above may be summarized as follows:

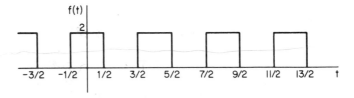

Figure 10-2.2   Square wave.

The Fourier Series for Periodic Functions. Any periodic function $f(t)$ which satisfies the Dirichlet conditions may be expressed as a Fourier series having the form

$$f(t) = a_0 + \sum_{i=1}^{\infty} (a_i \cos i\omega_0 t + b_i \sin i\omega_0 t)$$

The quantities a_i and b_i are called the Fourier coefficients and are found using the relations summarized in Table 10-3.1. ∎

The general expression for a Fourier series given in Eq. (10-26) is only one of several forms which may be used to express such a series. As an example of another form, let us recall from Sec. 7-1 that when two sinusoids having the same frequency are added, the result is another sinusoid of the same frequency. Thus, each of the terms in the summation in the right member of Eq. (10-26) may be expressed as a single sinusoidal term which includes a phase angle. An alternative form for a Fourier series may thus be written

$$f(t) = c_0 + \sum_{i=1}^{\infty} c_i \cos (i\omega_0 t + \phi_i) \tag{10-45}$$

where

$$c_0 = a_0 \qquad c_i = \sqrt{a_i^2 + b_i^2} \qquad \phi_i = \tan^{-1} \frac{-b_i}{a_i} \tag{10-46}$$

The inverse relations for the quantities a_i and b_i in terms of c_i and ϕ_i are

$$a_i = c_i \cos \phi_i \qquad b_i = -c_i \sin \phi_i \tag{10-47}$$

The form of the Fourier series given in Eq. (10-45) is especially useful in many applications since it directly specifies the magnitude of each of the harmonic frequency components of a given periodic non-sinusoidal function. Thus, for such a function, we may construct a *magnitude line spectrum* in which lines are used to specify the magnitude of these various frequency components. For example, for the square wave shown in Fig. 10-2.1, using the series given in Eq. (10-43), the magnitude line spectrum is as shown in Fig. 10-2.3. Similarly, a *phase spectrum* may be constructed in which lines are used to specify the value of the phase for the various frequency components.

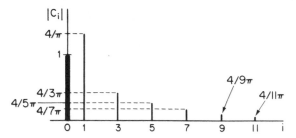

Figure 10-2.3   Magnitude line spectrum.

Now let us consider the effect of changing the reference time $t = 0$ on a Fourier series. As an example of such a change of reference, consider the square waves shown in Figs. 10-2.1 and 10-2.2. The waveform of these square waves is identical but the reference point $t = 0$ on the t axis has been shifted by $\frac{1}{4}$ of the period. The result of this is to change the form of the series from one involving only sine terms as given in Eq. (10-43) to one involving only cosine terms as given in Eq. (10-44). Comparing these two series and using the representation form given in Eq. (10-45), we see that the magnitude coefficients c_i for the two series are the same, thus they have the same magnitude line spectrum. In considering the effect on the phase angles ϕ_i when a series such as (10-43) is written in the form of Eq. (10-45), we note that a change of θ radians in the phase of the fundamental frequency produces a change of $n\theta$ radians in the phase of the nth harmonic. This conclusion may be seen more clearly by considering Fig. 10-2.4. In this figure, a fundamental sinusoidal function $f_1(t)$ and a

Figure 10-2.4 Relations for phase in harmonic waveforms.

third harmonic sinusoidal function $f_2(t)$ are shown. If we choose point 1 as the reference point for $t = 0$, then clearly $f_1(t)$ is a sine function and $f_2(t)$ is a sine function with a frequency three times as great, i.e., a third harmonic frequency. On the other hand, if we use point 2 as a reference point for $t = 0$, then we have changed the phase of $f_1(t)$ by $\frac{1}{4}$ of its period. Correspondingly, we have changed the phase of the function $f_2(t)$ by $\frac{3}{4}$ of one of its periods. Applying this result to the square wave shown in Fig. 10-2.1 with the Fourier series given in Eq. (10-43), we see that we may write such a series in the form

$$f(t) = 1 + \frac{4}{\pi} \sum_{i=1,3,5}^{\infty} \frac{1}{i} \cos\left(i\pi t - \frac{i\pi}{2}\right) \tag{10-48}$$

Comparing this series with the one given in Eq. (10-44) for the square wave shown in Fig. 10-2.2, we see that the difference between the two square waves is simply one of phase. The results given above may be summarized as follows:

SUMMARY 10-2.2

Magnitude and Phase Representation of a Fourier Series. A Fourier series for any function $f(t)$ which satisfies the Dirichlet conditions may be written in the form

$$f(t) = c_0 + \sum_{i=1}^{\infty} c_i \cos(i\omega_0 t + \phi_i)$$

where the coefficients c_i are defined in Eqs. (10-46) and (10-47). Changing the reference point $t = 0$ which has been chosen for such a function does not affect the values of the magnitude coefficients c_i although it will change the phase angles ϕ_i. ∎

An interesting result of the properties described above is that an even function with a zero average value may be changed to an odd function by choosing a different point for the reference time. The two square waves shown in Figs. 10-1.1(a) and 10-1.4(a) illustrate this.

In the preceding paragraphs we have discussed the effect of shifting a periodic function along the horizontal (or t) axis, i.e., selecting different reference points for $t = 0$. Now let us consider the effect of shifting a function along the vertical axis. This simply adds a constant term to all values of the function. Thus, it affects only the term a_0 in the Fourier series. As an example of this conclusion, consider the square wave shown in Fig. 10-2.5. This differs from the square wave shown in Fig. 10-2.2 only in that it has an average value which is easily seen to be 4. Thus, we may obtain its Fourier series directly from Eq. (10-44) by changing the constant term to a value of 4.

$$= \frac{4}{T} \int_0^{\frac{1}{2}} \frac{1}{2} 5 \cos i w_0 t \, dt$$

Figure 10-2.5 Square wave.

As a final topic for this section, in the following paragraphs we consider the question of how well the Fourier series performs its task of representing a given non-sinusoidal periodic function. Such information is necessarily of great interest, especially with respect to the problem of evaluating the accuracy of the representation as a function of the number of terms which are used in the series. To investigate this, suppose we let $f(t)$ be a periodic function and let $f_n(t)$ be its Fourier series which includes only the first n harmonic terms. The truncation error caused by not including the remaining terms may be defined as $e_n(t)$, where

$$e_n(t) = f(t) - f_n(t) \tag{10-49}$$

In evaluating this error, it is usually helpful to use the function $e_n(t)$ to define some scalar error constant from which the time dependence has been eliminated. One of the most frequently used means of doing this is to define a mean-square error E_n by the relation

$$E_n = \int_0^T e_n^2(t) \, dt \tag{10-50}$$

This expression in effect sums the time dependent error $e_n(t)$ over an entire period. In addition, the use of the square relation weights the larger errors more heavily than the smaller ones and also prevents any false minimization that might occur as a result of positive and negative polarity errors tending to cancel. It may be shown that a

Fourier series with coefficients a_i and b_i as determined by Eqs. (10-32), (10-37), and (10-40) is optimum in the sense that there are no other values of the coefficients that will give a lower value of E_n for any specified value of n.[1] This property of the Fourier series is referred to as the *least squares error property*. It is one of the reasons that the Fourier series is such a valuable tool for the representation of non-sinusoidal periodic waveforms.

In general, the type of function for which the Fourier series representation is the poorest is a periodic function in which discontinuities occur. It will be recalled that the Dirichlet conditions permit a finite number of such discontinuities. The first problem associated with a discontinuity is the question of what value the series will assign to the function at the time at which the discontinuity occurs. For example, suppose a function $f(t)$ has a discontinuity at $t = t_0$. This requires that $f(t_0-)$ is not equal to $f(t_0+)$.[2] The truncated series $f_n(t)$, being composed of a sum of continuous terms, obviously cannot be discontinuous at $t = t_0$. Instead, it may be shown that it provides the value $f(t_0)$ which satisfies the relation

$$f(t_0) = \frac{f(t_0+) + f(t_0-)}{2} \qquad (10\text{-}51)$$

i.e., the series provides the average of the values of the function which precede and follow the discontinuity. As higher numbers of terms are added to the series $f_n(t)$, it will tend to overshoot the values of $f(t_0+)$ and $f(t_0-)$ in its effort to approximate a discontinuity. A good example of this is a square wave. The Fourier series for such a wave produces a characteristic overshoot and damped oscillatory decay which approaches 8.95 percent of the total magnitude of the discontinuity. This phenomenon is illustrated in Fig. 10-2.6. As n is increased, the frequency of oscillation increases;

Figure 10-2.6 Gibbs phenomenon.

however, the magnitude of the overshoot still remains the same. This tendency of a truncated Fourier series $f_n(t)$ to exhibit a damped oscillatory overshoot in its representation of a discontinuity in a function $f(t)$ is called the *Gibbs phenomenon*. A more specific example of this phenomenon is shown in Fig. 10-2.7. Here we show a computer-plotted reconstruction of the first half-cycle of a square wave with a peak magnitude of plus and minus unity, and a period of two seconds. One of the loci shows the

[1]See P. M. Chirlian, *Basic Network Theory*, Sec. 11.3, McGraw-Hill Book Company, New York, 1969, for a development which indicates the proof of this statement.

[2]The symbols t_0- and t_0+ are explained in Sec. 4-3.

waveform that results when terms through the eleventh harmonic are used. The second locus shows the results of using terms through the fifty-first harmonic. It is readily observed that an overshoot of $2 \times 0.0895 = 0.179$ occurs which is independent of the number of terms used in the series. (The "2" occurs since the magnitude of the discontinuity is from -1 to $+1$, i.e., a total discontinuity of 2.)

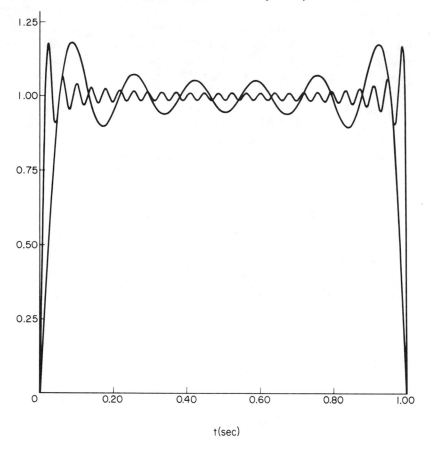

t(sec)

Figure 10-2.7 Gibbs phenomenon.

10-3 HOW THE FORM OF A PERIODIC FUNCTION AFFECTS THE COEFFICIENTS OF THE FOURIER SERIES

In this section we shall discuss some of the most significant ways in which the form of a non-sinusoidal periodic function affects the various coefficients a_i and b_i of the Fourier series as defined by Eq. (10-26). We shall investigate two major types of such phenomena. The first of these is the effect that the various symmetry properties discussed in Sec. 10-1 have on the coefficients. The second is the effect that discontinuities in the function being represented have on the coefficients.

To begin, let $f(t)$ be an arbitrary periodic function with an even part $f^{(e)}(t)$ and

The Fourier Series and the Fourier Integral   Chap. 10

an odd part $f^{(o)}(t)$. Now let us use Eq. (10-37) to find the coefficients a_i. Since the functions $f^{(e)}(t)$ and $f^{(o)}(t)$ specify symmetry properties with respect to the origin, we will change the limits of integration in the integral defining the coefficients a_i from zero and T to $-T/2$ and $T/2$, respectively, so that the integration interval is symmetrical with respect to the origin. Thus, we obtain

$$a_i = \frac{2}{T} \left[\int_{-T/2}^{T/2} f^{(e)}(t) \cos i\omega_0 t \, dt + \int_{-T/2}^{T/2} f^{(o)}(t) \cos i\omega_0 t \, dt \right] \qquad (10\text{-}52)$$

In the right member of the above equation the first integrand contains a product of even functions; therefore, from Summary 10-1.1 it must be an even function, i.e., one in which $f(t) = f(-t)$. Thus, the integral may be evaluated by integrating over the limits zero to $T/2$ and multiplying the result by two. The second integrand contains the product of an even and an odd function. Therefore, the integrand is an odd function in which $f(t) = -f(-t)$. The integrand of such a function taken over limits which are symmetrical with respect to the origin must be zero. Considering the above, we see that the expression for the coefficients a_i may be written

$$a_i = \frac{4}{T} \int_0^{T/2} f^{(e)}(t) \cos i\omega_0 t \, dt \qquad (10\text{-}53)$$

where $f^{(e)}(t)$ is the even part of some periodic function $f(t)$. A most important conclusion of the above is that only the even part of an arbitrary periodic function $f(t)$ determines the coefficients a_i. Thus, the Fourier series of a purely odd function will contain no cosine terms. It should be noted that since an even function can have a nonzero average value, there may also be a constant a_0 as determined by Eq. (10-32) for such a case. As an example of this conclusion, consider the expression given in Eq. (10-44) for the square wave shown in Fig. 10-2.2. The series obviously contains only cosine terms plus a constant term and the square wave is clearly an even function.

Now let us make a development similar to the above for the Fourier sine coefficients b_i. Using Eq. (10-40) and changing the limits of integration, we may write

$$b_i = \frac{2}{T} \left[\int_{-T/2}^{T/2} f^{(e)}(t) \sin i\omega_0 t \, dt + \int_{-T/2}^{T/2} f^{(o)}(t) \sin i\omega_0 t \, dt \right] \qquad (10\text{-}54)$$

In the right member of the above equation the first integrand contains a product of an even and an odd function. Therefore, it is odd and the resulting integral is zero. Similarly, the second integrand contains the product of two odd functions; therefore, it is even and its integral may be evaluated by taking the integral from zero to $T/2$ and multiplying by two. As a result, the expression for determining the coefficients b_i may be written in the form

$$b_i = \frac{4}{T} \int_0^{T/2} f^{(o)}(t) \sin i\omega_0 t \, dt \qquad (10\text{-}55)$$

where $f^{(o)}(t)$ is the odd part of some periodic function $f(t)$. We conclude that the Fourier series of an even function will contain no sine terms. As an example of this, consider the Fourier series given in Eq. (10-7) for the square wave shown in Fig. 10-1.1(a). The series obviously contains only sine terms and the square wave clearly has odd symmetry. Since the average value of an odd function is zero, the coefficient a_0 will always be zero for such a function.

Now let us consider the effect of half-wave symmetry as defined by Eq. (10-23) on the Fourier coefficients. An example of a function with half-wave symmetry is shown by the solid line in Fig. 10-3.1(a). Also, on this figure we have indicated by a dashed line a cosine wave with the same period as the original function. The expression used to determine the Fourier coefficients a_i ($i = 1, 2, ..$) is

$$a_i = \frac{2}{T} \int_{-T/2}^{T/2} f(t) \cos i\omega_0 t \, dt \qquad (10\text{-}56)$$

Thus, the integrand for the constant a_1 is determined by the product of the dashed and solid functions shown in the figure. Now consider the portion of the integral over

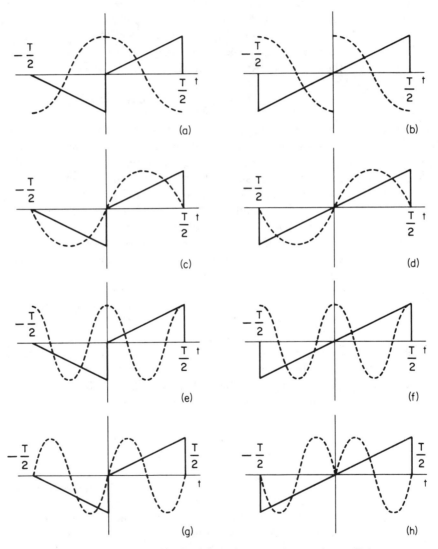

Figure 10-3.1 Effect of half-wave symmetry on the Fourier coefficients.

the limits $-T/2$ to zero. Suppose we reverse the functions indicated in the figure over this interval. The result of doing this is shown in Fig. 10-3.1(b). The value of the integral over this interval is unchanged since the total area is still the same. From Fig. 10-3.1(b), however, we see that over the entire period from $-T/2$ to $T/2$ both the dashed line and the solid line define odd functions. Therefore, the product of these two, namely, the integrand of Eq. (10-56), is even and an expression for the coefficient a_1 may be derived similar in form to the one given in (10-53), namely,

$$a_1 = \frac{4}{T} \int_0^{T/2} f(t) \cos \omega_0 t \, dt \qquad (10\text{-}57)$$

The same logic may be used to find the coefficient b_1. In this case, the integrand will be the product of the original function shown as a solid line in Fig. 10-3.1(c) and the sine function shown as a dashed line in this figure. Again reversing the functions over the interval $-T/2$ to zero, we obtain the waveforms shown in Fig. 10-3.1(d). As before, we see that both functions are odd; thus, their product is even. We conclude that a relation similar to (10-55) may be used to determine the b_1 coefficient, namely,

$$b_1 = \frac{4}{T} \int_0^{T/2} f(t) \sin \omega_0 t \, dt \qquad (10\text{-}58)$$

Now let us consider the integrands used in determining the coefficients a_2 and b_2. The original function and the second harmonic cosine function are shown by solid and dashed lines, respectively, in Fig. 10-3.1(e). Reversing the functions over the interval $-T/2$ to zero, we obtain the situation shown in Fig. 10-3.1(f). The solid line now represents an odd function but the dashed line is an even function. Therefore, the product of these two functions, i.e., the integrand of Eq. (10-56), will be odd and the integral over the limits $-T/2$ to $T/2$ will be zero. Thus, the coefficient a_2 is zero. Similarly, considering the sinusoidal second harmonic component of the Fourier series and of the original function as shown in Fig. 10-3.1(g), and reversing the waveshapes over the period $-T/2$ to zero as shown in Fig. 10-3.1(h), we may conclude that the coefficient b_2 is zero. If this evaluation procedure is extended and generalized, we reach the conclusion that all the even coefficients a_i $(i = 0, 2, 4, \ldots)$ and b_i $(i = 2, 4, \ldots)$ are zero for a periodic function which has half-wave symmetry. Thus, such a function will have only odd harmonic terms. The reader should be careful to distinguish between a function of *odd symmetry* and a function which contains only *odd harmonic terms*. The former has only sine terms in its series expansion. The latter will, in general, have both sine and cosine terms but only the odd harmonic ones.

The results given above are useful in two major ways. First, by inspection, they permit us to easily determine some advanced knowledge about the coefficients of the Fourier series for a non-sinusoidal periodic function. Second, they permit us to find the Fourier coefficients of complicated functions by first finding the coefficients for simpler functions and summing the results. As an example of the application of these results, consider the periodic function shown in Fig. 10-3.2. It is readily seen that this function has half-wave symmetry. Therefore, all the even coefficients of the Fourier series must be zero. If we now split the function into its even and odd parts, the even part $f^{(e)}(t)$ is a square wave identical with the one shown in Fig. 10-1.4(a). Its Fourier series is given in Eq. (10-9). Similarly, the odd part $f^{(o)}(t)$ is a triangular wave identical with the one shown in Fig. 10-1.1(b), which has a Fourier series whose first terms are given in Eq. (10-8). The Fourier series of the function given in Fig. 10-3.2 is now simply found by summing the two series. Thus, we obtain

$$f(t) = \frac{4}{\pi} \cos \pi t + \frac{8}{\pi^2} \sin \pi t - \frac{4}{3\pi} \cos 3\pi t - \frac{8}{9\pi^2} \sin 3\pi t + \cdots \qquad (10\text{-}59)$$

The results given above may be summarized as follows:

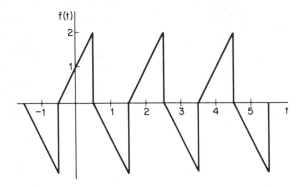

Figure 10-3.2 Periodic function.

SUMMARY 10-3.1

Effect of Waveform Symmetry on the Fourier Series Coefficients. The various waveform symmetries which a periodic function may have affect the Fourier series coefficients in the following ways:

1. An even function will have a Fourier series consisting only of cosine terms and possibly a constant. Thus, the coefficients b_i $(i = 1, 2, \ldots)$ of the general series given in Eq. (10-26) are zero.
2. An odd function will have a Fourier series consisting only of sine terms. Thus, the coefficients a_i $(i = 0, 1, 2, \ldots)$ are zero.
3. A function with half-wave symmetry will have only odd harmonic terms. Thus, the coefficients a_i $(i = 0, 2, 4, \ldots)$ and b_i $(i = 2, 4, \ldots)$ are zero. A summary of the various symmetry relations and the equations for finding the Fourier coefficients are given in Tables 10-3.1 and 10-3.2. ∎

TABLE 10-3.1 Expressions for Fourier Coefficients

Type of symmetry	Condition for symmetry	a_0	a_i ($i = 1, 2, 3, \ldots$)	b_i ($i = 1, 2, 3, \ldots$)
none	—	$\dfrac{1}{T}\int_0^T f(t)\,dt$	$\dfrac{2}{T}\int_0^T f(t) \cos i\omega_0 t\,dt$	$\dfrac{2}{T}\int_0^T f(t) \sin i\omega_0 t\,dt$
even	$f(t) = f(-t)$	$\dfrac{2}{T}\int_0^{T/2} f(t)\,dt$	$\dfrac{4}{T}\int_0^{T/2} f(t) \cos i\omega_0 t\,dt$	0
odd	$f(t) = -f(-t)$	0	0	$\dfrac{4}{T}\int_0^{T/2} f(t) \sin i\omega_0 t\,dt$

TABLE 10-3.2 Summary of Symmetry Properties

Coefficient	Description	Symmetry of $f(t)$		
		Even $f(t) = f(-t)$	Odd $f(t) = -f(-t)$	Half-wave $f(t) = -f\left(t + \frac{2}{T}\right)$
a_0	dc term	may be non-zero	zero	zero
$a_i\ (i = 1, 3, \ldots)$	odd harmonic cosine terms	may be non-zero	zero	may be non-zero
$a_i\ (i = 2, 4, \ldots)$	even harmonic cosine terms	may be non-zero	zero	zero
$b_i\ (i = 1, 3, \ldots)$	odd harmonic sine terms	zero	may be non-zero	may be non-zero
$b_i\ (i = 2, 4, \ldots)$	even harmonic sine terms	zero	may be non-zero	zero

The second relationship between the form of a non-sinusoidal periodic function and the resulting coefficients of the Fourier series that we shall investigate in this section is the effect of discontinuities in the function. At the end of Sec. 10-2, we showed that the Fourier series for a function containing a discontinuity will produce an overshoot referred to as the Gibbs phenomenon. Therefore, we might very well expect that such a function will require large numbers of high-frequency harmonics in its Fourier series representation to achieve reasonable accuracy. Correspondingly, we might expect that functions which are well behaved in the sense that they are not discontinuous might require considerably fewer terms. Such a conjecture is indeed well founded. In fact, it may be shown that the number of terms required in a Fourier series for a given accuracy of representation has a direct relation to the number of times the function being considered may be differentiated without producing any discontinuities. As an example of this consider three periodic functions which have a cycle as shown in Fig. 10-3.3. In part (a) of this figure, we show a function which has a parabolic behavior. It is defined by the relations

$$f_a(t) = \begin{cases} \dfrac{F_0 t^2}{2} - \dfrac{F_0 T}{4} t & 0 \leq t \leq \dfrac{T}{2} \\[2ex] \dfrac{3F_0 T}{4} t - \dfrac{F_0 t^2}{2} - \dfrac{F_0 T^2}{4} & \dfrac{T}{2} \leq t \leq T \end{cases} \qquad (10\text{-}60)$$

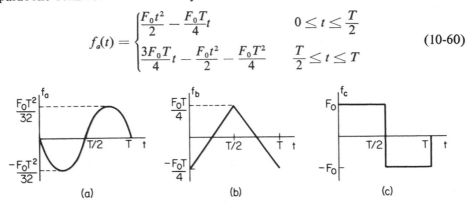

Figure 10-3.3 Periodic functions related by differentiation.

If we differentiate this parabolic function, we obtain the relations

$$f_b(t) = \begin{cases} F_0 t - F_0 \dfrac{T}{2} & 0 \le t \le \dfrac{T}{2} \\[2mm] \dfrac{3F_0 T}{4} - F_0 t & \dfrac{T}{2} \le t \le T \end{cases} \tag{10-61}$$

These equations define the triangular periodic function shown in Fig. 10-3.3(b). Similarly, a second differentiation of the parabolic waveform produces the relations

$$f_c(t) = \begin{cases} F_0 & 0 < t < \dfrac{T}{2} \\[2mm] -F_0 & \dfrac{T}{2} < t < T \end{cases} \tag{10-62}$$

These relations define a discontinuous function, namely, the square wave shown in Fig. 10-3.3(c). Now let us consider the Fourier series for the three functions. We have already encountered the square and triangular waveforms in previous discussions. From series similar to those given in Eqs. (10-7) and (10-8), we see that for the square wave the magnitude of the non-zero coefficients decreases in direct proportion to $1/n$, where n is the number of the harmonic term which the coefficient multiplies. The magnitudes of the coefficients for the continuous triangular wave (which if differentiated *once* leads to the discontinuous square wave), however, are seen to be proportional to $1/n^2$. Now consider the parabolic function shown in Fig. 10-3.3(a). This function must be differentiated *twice* before a discontinuous function (a square wave) results. If we determine the Fourier series for this function, we find that the coefficients of the various terms decrease in direct proportion to $1/n^3$. Comparing the results for the three waveforms given in Fig. 10-3.3, we may conclude that the more times a function must be differentiated before discontinuities are produced, the more rapidly the magnitudes of the coefficients of the Fourier series decrease. This conclusion is an example of a general law regarding the manner in which the magnitudes of the Fourier coefficients decrease as determined by the nature of the function which is being represented. Such a general law may be summarized as follows:

SUMMARY 10-3.2

The Rate at Which the Higher-Order Harmonic Coefficients of a Fourier Series Decrease. If $f(t)$ is a non-sinusoidal periodic function, and if k differentiations of this function are required before a discontinuous function is produced, then the magnitude of the Fourier coefficients a_i and b_i associated with the ith harmonic term satisfy the inequalities

$$|a_i| \le \frac{M}{i^{k+1}} \qquad |b_i| \le \frac{M}{i^{k+1}} \tag{10-63}$$

where the value of the constant M depends on the specific function $f(t)$ being analyzed.[3]

■

[3] R. J. Schwarz and B. Friedland, *Linear Systems*, p. 144, McGraw-Hill Book Company, New York, 1965.

The relations given above are of considerable value in anticipating the form of the magnitude spectrum of Fourier coefficients that will characterize a given periodic function. They are also of value in helping the engineer to anticipate the form of a periodic function from its magnitude spectrum. For example, from Eq. (10-63) it is readily verified that any *continuous function* will have terms in its Fourier series whose magnitude decreases at least as rapidly as $1/n^2$. Such a conclusion provides valuable information on the number of terms which must be included when a Fourier series is used to analyze a network for a given input waveform. Such an application will be considered in the following section.

10-4 NETWORK ANALYSIS USING THE FOURIER SERIES

In the preceding sections of this chapter we have described the use of the Fourier series in the representation of non-sinusoidal periodic functions. Now let us see how these representations may be used to make a steady-state analysis of a network excited by such a function. Note that here we say "steady-state" rather than "*sinusoidal* steady-state" since we are considering *non*-sinusoidal periodic wave forms. The basic technique is simply to determine the network function (as a function of $j\omega$) by the methods described in Chap 7. The Fourier series representation of the non-sinusoidal periodic waveform may then be treated as a series of separate sinusoidal excitations at different frequencies, each of which produces its own sinusoidal output. Because of the linear nature of the network, the superposition of the input waveforms (which yields the Fourier series representation of the non-sinusoidal periodic input function) produces a corresponding superposition of the output waveforms. This latter gives the desired response of the network. As an example of the technique, consider the network shown in Fig. 10-4.1(a). As indicated, the excitation voltage $v_s(t)$ is a square wave with a period $T = 4\pi$. Thus, the fundamental frequency is 0.5 rad/s. For simplicity, let us consider only the first three harmonic terms. Thus, the Fourier series of the input waveform is readily found to be [see Eq. (10-9)]

$$v_s(t) = 4(\cos \tfrac{1}{2}t - \tfrac{1}{3} \cos \tfrac{3}{2}t + \tfrac{1}{5} \cos \tfrac{5}{2}t) \tag{10-64}$$

Since the right member of the above is a summation of terms all of which have the dimensions of voltage, the input waveform may also be represented as a series connection of voltage sources each of which has an output voltage at one of the sinusoidal

(a)   (b)

Figure 10-4.1 Network analysis using a Fourier series.

frequencies represented in the Fourier series. Such a representation is shown in Fig. 10-4.1(b). Now let us use phasor notation for the network. Using the methods of Chap. 7, we find that the network function $N(j\omega)$ for this network is

$$N(j\omega) = \frac{\mathcal{V}_o}{\mathcal{V}_s} = \frac{1}{1 + j\omega RC} \tag{10-65}$$

where $\mathcal{V}_s$ and $\mathcal{V}_o$ are phasors. For the first or fundamental frequency (as applied by the first voltage source) we may define the phasor $\mathcal{V}_s^{(1)} = 4\underline{/0°}$. Since the frequency of this source is 0.5 rad/s, we may write

$$\mathcal{V}_o^{(1)} = N(j0.5)\mathcal{V}_s^{(1)} = \frac{4}{1 + j0.5} = 3.576\underline{/-26.57°} \tag{10-66}$$

where $\mathcal{V}_o^{(1)}$ is the phasor component of the output produced by this first harmonic term. Thus, the steady-state component of this output in the time domain may be expressed as $v_o^{(1)}(t)$, where

$$v_o^{(1)}(t) = 3.576 \cos (\tfrac{1}{2}t - 26.57°) \tag{10-67}$$

Continuing in this manner, we may compute the effect of the second voltage source. Its frequency is 1.5 rad/s and its phasor representation is $\mathcal{V}_s^{(2)} = \tfrac{4}{3}\underline{/180°}$. If we let its effect on the output be the phasor $\mathcal{V}_o^{(2)}$, we may write

$$\mathcal{V}_o^{(2)} = N(j1.5)\mathcal{V}_s^{(2)} = \frac{-\tfrac{4}{3}}{1 + j1.5} = 0.7392\underline{/123.7°} \tag{10-68}$$

The corresponding steady-state time-domain output component is $v_o^{(2)}(t)$, where

$$v_o^{(2)}(t) = 0.7392 \cos (\tfrac{3}{2}t + 123.7°) \tag{10-69}$$

Continuing, the output $v_o^{(3)}(t)$ produced by the third voltage source may be shown to be

$$v_o^{(3)}(t) = 0.2972 \cos (\tfrac{5}{2}t - 68.2°) \tag{10-70}$$

The process described above obviously may be continued for as many terms of the Fourier series as may be required. The steady-state output waveform $v_o(t)$ is then readily found by summing the individual output terms $v_o^{(i)}(t)$. For this example, using only the terms given in Eqs. (10-67), (10-69), and (10-70), we obtain

$$v_o(t) = 3.576 \cos (\tfrac{1}{2}t - 26.57°) + 0.7392 \cos (\tfrac{3}{2}t + 123.7°) + 0.2972 \cos (\tfrac{5}{2}t - 68.2°) \tag{10-71}$$

A sketch of this output waveform is shown in Fig. 10-4.2. It should be noted that this

Figure 10-4.2 Output waveform for the circuit shown in Fig. 10-4.1.

waveform represents only the steady-state component, i.e., the particular solution, of the network response to the applied square wave input. If the complete response is desired, then the time-domain techniques described in Chaps. 5 and 6 must also be applied. The procedure illustrated in the preceding paragraphs may be summarized as follows:

SUMMARY 10-4.1

Analysis of a Network Excited by a Non-Sinusoidal Periodic Waveform. If a nonsinusoidal periodic waveform $f(t)$ with fundamental frequency ω_0 and Fourier series

$$f(t) = c_0 + \sum_{k=1}^{\infty} c_k \cos(k\omega_0 t + \alpha_k)$$

is applied as an excitation to a linear network, the output function $g(t)$ will be described by a non-sinusoidal periodic waveform with Fourier series

$$g(t) = d_0 + \sum_{k=1}^{\infty} d_k \cos(k\omega_0 t + \beta_k)$$

The quantities d_0, d_k, and β_k are determined by the relations

$$d_0 = N(j0)c_0$$
$$d_k = |N(jk\omega_0)|c_k \qquad (k = 1, 2, \ldots)$$
$$\beta_k = \underline{/N(jk\omega_0)} + \alpha_k \qquad (k = 1, 2, \ldots)$$

where $N(j\omega)$ is the network function relating the output and input phasor.   ■

Fourier series techniques are of special use in evaluating the steady-state performance of filter circuits in which it is desired to select or eliminate a certain frequency component from a waveform which may contain many harmonic components. An example of such an application follows.

Example 10-4.1

The circuit shown in Fig. 10-4.3(a) consists of a full-wave bridge rectifier operated from a secondary transformer winding. The other transformer winding is connected to a sinusoidal voltage source of 60 Hz. The rectifier portion of the circuit is followed by

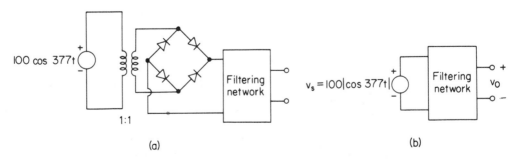

(a)   (b)

Figure 10-4.3   Circuit for Example 10-4.1.

a filtering network. This is a common configuration used in electronic circuits to provide a filtered or nearly-constant value of dc voltage from a sinusoidally varying ac supply. The full-wave rectifier portion of the circuit may be modeled by a single voltage source whose output is a rectified sine-wave voltage; thus, the circuit may be represented as shown in Fig. 10-4.3(b).[4] Such a waveform has a second half-cycle which is identical to its first half-cycle. Thus, for convenience, it may be considered as a periodic wave having twice the applied frequency. The first few terms of the Fourier series for such a waveform may be shown to be

$$v_s = \frac{200}{\pi}\left(1 - \frac{2}{3}\cos 754t - \frac{2}{15}\cos 1508t - \frac{2}{35}\cos 2262t + \cdots\right)$$

Since the first harmonic component of this series (which is the second harmonic component of the source frequency) is the largest of the harmonic terms in the series, it is desirable to design the filtering network to reduce this term as much as possible. To do this we will use the network shown in Fig. 10-4.4. The sinusoidal steady-state network

Figure 10-4.4 Filtering network for Example 10-4.1.

function for this circuit is readily found to be

$$N(j\omega) = \frac{\mathcal{V}_o}{\mathcal{V}_s} = \frac{1}{1 - \omega^2 LC + (j\omega L/R)}$$

Let us assume that circuit considerations require that the value of the resistor be fixed at 2 kΩ and the value of the capacitor at 20 μF. We may now choose the value of inductance so as to make the total peak-to-peak second-harmonic output voltage variation at 754 rad/s be some small percentage of the dc output voltage. If we write the output voltage $v_o(t)$ in the form

$$v_o(t) = V_o^{(0)} + V_o^{(1)}\cos(754t + \alpha_1) + V_o^{(2)}\cos(1508t + \alpha_2) + \cdots$$

and choose the percentage as 5 percent, we may write

$$0.05 = \frac{2V_o^{(1)}}{V_o^{(0)}} = \frac{2\left(\frac{400}{3\pi}\right)|N(j754)|}{\left(\frac{200}{\pi}\right)|N(j0)|}$$

In determining $|N(j754)|$, if we substitute the values for the resistor and capacitor given above, we obtain

$$|N(j754)| = \left|\frac{1}{1 - (754)^2(20 \times 10^{-6})L + j(754 \times 0.0005)L}\right| \approx \left|\frac{1}{1 - 11.38L}\right|$$

where we have assumed that the magnitude of the imaginary term in the denominator is smaller than the real term. Since $N(j0)$ is clearly 1, we find that L is 2.433 H. Evaluating the network function at the first, second, and third harmonics of the series (the

---

[4]It should be noted that this model is only valid if the input current to the filter does not reverse in polarity. Thus, the final result of the example must be checked to make certain that this condition is satisfied.

second, fourth, and sixth harmonics of the original sinusoidal waveform), we obtain

$$N(j754) = 0.0375\underline{/-178.0°}$$
$$N(j1508) = 0.00912\underline{/-179.0°}$$
$$N(j2262) = 0.00403\underline{/-179.4°}$$

We may now apply phasor methods to determine the output phasors $\mathcal{V}_o^{(1)}$, $\mathcal{V}_o^{(2)}$, and $\mathcal{V}_o^{(3)}$ for the first, second, and third harmonic components. Thus, we obtain

$$\mathcal{V}_o^{(1)} = N(j754)\left[\frac{400}{3\pi}\underline{/180°}\right] = 1.59\underline{/2.0°}$$

$$\mathcal{V}_o^{(2)} = N(j1508)\left[\frac{400}{15\pi}\underline{/180°}\right] = 0.0774\underline{/1.0°}$$

$$\mathcal{V}_o^{(3)} = N(j2262)\left[\frac{400}{35\pi}\underline{/180°}\right] = 0.0147\underline{/0.6°}$$

Thus, the time-domain expression for the sinusoidal output waveform becomes

$$v_o(t) = \frac{200}{\pi} + 1.59 \cos(754t + 2.0°) + 0.0774 \cos(1508t + 1.0°)$$
$$+ 0.0147 \cos(2262t + 0.6°)$$

The usefulness of such a filtering circuit in transforming the pulsating rectified sinusoidal input wave into a relatively constant dc voltage is readily apparent.

One of the most useful aspects of the treatment of non-sinusoidal periodic waveforms using the Fourier series given above is that it provides a ready tool for making computations of power dissipation and transmission through a given network. As a preliminary step to a discussion of this topic, since we have already demonstrated the utility of rms quantities in power considerations concerning sinusoidal signals, let us first consider the rms value of a *non*-sinusoidal periodic waveform. To do this, let $f(t)$ be a nonsinusoidal periodic waveform represented by the magnitude and phase form of the Fourier series. Thus, we may write

$$f(t) = c_0 + c_1 \cos(\omega_0 t + \phi_1) + c_2 \cos(2\omega_0 t + \phi_2) + \cdots \qquad (10\text{-}72)$$

The general relation for the rms value of a periodic function $f(t)$ with period T as developed in Sec. 7-1 is

$$F_{rms} = \sqrt{\frac{1}{T} \int_0^T f^2(t)\, dt} \qquad (10\text{-}73)$$

Substituting the expression for the Fourier series of $f(t)$ as given in Eq. (10-72) into this last expression, we obtain

$$F_{rms} = \sqrt{\frac{1}{T} \int_0^T [c_0 + c_1 \cos(\omega_0 t + \phi_1) + c_2 \cos(2\omega_0 t + \phi_2) + \cdots]^2\, dt} \qquad (10\text{-}74)$$

In this expression the bracketed integrand containing the infinite summation is squared. We may indicate the first terms of the resulting integrand produced when the squared expression is expanded as follows:

$$c_0^2 + c_0 c_1 \cos(\omega_0 t + \phi_1) + \cdots + c_1^2 \cos^2(\omega_0 t + \phi_1)$$
$$+ c_1 \cos(\omega_0 t + \phi_1)c_2 \cos(2\omega_0 t + \phi_2) + \cdots \qquad (10\text{-}75)$$

If we now apply the orthogonality relations of (10-27), (10-28), and (10-29), we find that when the integration indicated in (10-74) is performed, all the product terms involving components of different harmonic frequencies disappear. Thus, we may write the expression for the rms value in the form

$$F_{\text{rms}} = \sqrt{\frac{1}{T} \int_0^T \left[c_0^2 + \sum_{k=1}^{\infty} c_k^2 \cos^2 (k\omega_0 t + \phi_k) \right] dt} \qquad (10\text{-}76)$$

If we interchange the operations of integration and summation in the above equation, we obtain

$$F_{\text{rms}} = \sqrt{\frac{1}{T} \int_0^T c_0^2 \, dt + \sum_{k=1}^{\infty} \frac{1}{T} \int_0^T c_k^2 \cos^2 (k\omega_0 t + \phi_k) \, dt} \qquad (10\text{-}77)$$

The separate integrals of this expression, however, simply define the rms values of the separate harmonic components of the Fourier series. Thus, in general, we may write

$$F_{\text{rms}} = \sqrt{c_0^2 + \sum_{k=1}^{\infty} F_k^2} \qquad (10\text{-}78)$$

where the quantity F_k^2 is the square of the rms value of the kth harmonic component of the Fourier series $f(t)$ and where c_0^2 is the square of the constant component of the Fourier series. This constant, of course, is numerically equal to its own rms value. We may summarize the results given as follows:

SUMMARY 10-4.2

The rms Value of a Non-Sinusoidal Periodic Function.   The rms or effective value of a non-sinusoidal periodic function is equal to the square root of the sum composed of the squares of the rms values of the various Fourier series harmonic components and the square of the constant (or average) term.   ■

Now let us consider the power which appears at a pair of terminals at which the voltage and current waveforms are non-sinusoidal but periodic. Defining the voltage and current variables as $v(t)$ and $i(t)$ we may write the Fourier series for these quantities as

$$v(t) = v_0 + \sum_{k=1}^{\infty} v_k \cos (k\omega_0 t + \phi_k)$$

$$i(t) = i_0 + \sum_{k=1}^{\infty} i_k \cos (k\omega_0 t + \theta_k) \qquad (10\text{-}79)$$

Rather than treating the instantaneous power $p(t) = v(t)i(t)$ (which is a function of t), let us define an average power P_{avg} by the relation

$$P_{\text{avg}} = \frac{1}{T} \int_0^T p(t) \, dt = \frac{1}{T} \int_0^T v(t)i(t) \, dt \qquad (10\text{-}80)$$

Substituting the expressions for the Fourier series for $v(t)$ and $i(t)$ given in (10-79), we obtain

$$P_{\text{avg}} = \frac{1}{T} \int_0^T \left[v_0 + \sum_{k=1}^{\infty} v_k \cos (k\omega_0 t + \phi_k) \right] \left[i_0 + \sum_{k=1}^{\infty} i_k \cos (k\omega_0 t + \theta_k) \right] dt$$

(10-81)

If we again apply the orthogonality relations of Sec. 10-2 to the integrand terms that result from multiplying the bracketed terms in Eq. (10-81), we again find that the product terms involving different harmonic frequencies disappear. Thus, the expression may be simplified as follows.

$$P_{\text{avg}} = \frac{1}{T} \int_0^T v_0 i_0 \, dt + \sum_{k-1}^{\infty} \left[\frac{1}{T} \int_0^T v_k \cos (k\omega_0 t + \phi_k) i_k \cos (k\omega_0 t + \theta_k) \, dt \right] \quad (10\text{-}82)$$

The separate integrals of this expression, however, simply define the average power dissipated at each of the various harmonic frequency components of the Fourier series. Thus, in general, we may write

$$P_{\text{avg}} = P_0 + \sum_{k=1}^{\infty} P_k \tag{10-83}$$

where P_0 is the average dc power, and the quantity P_k is the average of the power being dissipated at the kth harmonic frequency component in the Fourier series. The results given above may be summarized as follows:

SUMMARY 10-4.3

Average Power at a Pair of Terminals at Which the Variables Are Periodic but Non-Sinusoidal. The average power dissipated at a pair of terminals of a network at which the variables are periodic but non-sinusoidal may be found by summing the average powers found at each of the harmonic component frequencies (including dc) of the Fourier series representation for the variables. ∎

The conclusions given above with respect to average power are readily extended to the case where complex power is considered.

Because of the complexities of the calculations required in applying the Fourier series, it is frequently desirable to implement them with digital computer techniques. Some programs which may be used for this are given in Appendix C.

10-5 THE FOURIER TRANSFORMATION

In this section we show how the Fourier series techniques developed in the preceding sections of this chapter may be extended so as to be applicable to nonperiodic functions. Such an application is made possible by the use of the *Fourier transformation* (also referred to as the *Fourier integral*). In this section we will derive the Fourier transformation as a limiting case of the Fourier series. Before treating this topic, we will first consider a new way of writing the Fourier series which will be useful in the development of the transformation.

In Sec. 10-2 we developed two forms which could be used for expressing the Fourier series. These are

$$f(t) = a_0 + \sum_{k=1}^{\infty} a_k \cos k\omega_0 t + \sum_{k=1}^{\infty} b_k \sin k\omega_0 t \tag{10-84a}$$

$$f(t) = a_0 + \sum_{k=1}^{\infty} c_k \cos (k\omega_0 t + \phi_k) \tag{10-84b}$$

where the first series given above may be referred to as the *sine and cosine* form of the Fourier series and the second may be referred to as the *magnitude and phase* form. The relations between the coefficients of the two forms of the series in Eqs. (10-84a) and (10-84b) are given in Eqs. (10-46) and (10-47). Let us now consider a third form for the Fourier series. We begin by stating Euler's relations for expressing sine and cosine quantities. These are:

$$\cos x = \frac{e^{jx} + e^{-jx}}{2} \qquad \sin x = \frac{e^{jx} - e^{-jx}}{2j}$$

$$e^{\pm jx} = \cos x \pm j \sin x \tag{10-85}$$

Inserting the first two of these relations in Eq. (10-84a) and rearranging terms, we obtain

$$f(t) = a_0 + \sum_{k=1}^{\infty} \frac{a_k - jb_k}{2} e^{jk\omega_0 t} + \sum_{k=1}^{\infty} \frac{a_k + jb_k}{2} e^{-jk\omega_0 t} \tag{10-86}$$

Now let us define the constant C_k as follows:

$$C_k = \frac{a_k - jb_k}{2} \tag{10-87}$$

Using these constants, the Fourier series given in Eq. (10-86) may be written in the form

$$f(t) = a_0 + \sum_{k=1}^{\infty} C_k e^{jk\omega_0 t} + \sum_{k=1}^{\infty} C_k^* e^{-jk\omega_0 t} \tag{10-88}$$

where C_k^* is the complex conjugate of C_k. From the defining integral relations for the Fourier coefficients a_k and b_k given in (10-37) and (10-40), we may write the following integral expression[5] for the coefficients C_k:

$$C_k = \frac{1}{T} \int_{-T/2}^{T/2} f(t)(\cos k\omega_0 t - j \sin k\omega_0 t) \, dt \tag{10-89}$$

This equation may be simplified by using the last expression given in Eq. (10-85). Thus, we obtain

$$C_k = \frac{1}{T} \int_{-T/2}^{T/2} f(t) e^{-jk\omega_0 t} \, dt \tag{10-90}$$

We note that since the cosine function is even, i.e., $\cos(-x) = \cos x$, for a typical cosine term in the series in Eq. (10-84a), we may write

$$a_k \cos k\omega_0 t = a_{-k} \cos(-k\omega_0 t) = a_{-k} \cos k\omega_0 t \tag{10-91}$$

---

[5] Since we may integrate over any full period, the upper and lower limits of integration have been changed from 0 and T to $-T/2$ and $T/2$, respectively, to facilitate future developments.

Comparing the first and last members of the above equation, we conclude that $a_k = a_{-k}$, i.e., in a Fourier series in which the integer constant k in the argument of the cosine terms is changed from k to $-k$, the coefficients a_k multiplying the cosine terms remain unchanged. Similarly, since the sine function is an odd function, i.e., $\sin(-x) = -\sin(x)$, for a typical sine term in the series of Eq. (10-84a) we may write

$$b_k \sin k\omega_0 t = b_{-k} \sin(-k\omega_0 t) = -b_{-k} \sin k\omega_0 t \qquad (10\text{-}92)$$

Considering the first and last members of the above equation, we conclude that $b_k = -b_{-k}$. That is, if we change the integer constant k in the argument of the sine terms of a Fourier series from k to $-k$ we must also change the coefficients b_k multiplying these sine terms, to the negative of their original value to keep the value of the series the same. Substituting these results in (10-87), we find that

$$C_{-k} = \frac{a_k + jb_k}{2} \qquad (10\text{-}93)$$

Comparing this expression with Eq. (10-87), we conclude that C_{-k} is the complex conjugate of C_k. We may now rewrite the exponential form of the Fourier series given in Eq. (10-88) using the constants C_k and C_{-k}. Thus, we obtain

$$f(t) = a_0 + \sum_{k=1}^{\infty} C_k e^{jk\omega_0 t} + \sum_{k=1}^{\infty} C_{-k} e^{-jk\omega_0 t} \qquad (10\text{-}94)$$

If we now substitute $-k$ for k in the second summation given in the right member of above equation, we obtain

$$f(t) = a_0 + \sum_{k=1}^{\infty} C_k e^{jk\omega_0 t} + \sum_{k=-1}^{-\infty} C_k e^{jk\omega_0 t} \qquad (10\text{-}95)$$

The only value of k not included in the two summations in the right member of the above equation is $k = 0$. To see what happens for this value, let us evaluate the integral relation given in (10-90) for the case where $k = 0$. Thus, we obtain

$$C_0 = \frac{1}{T} \int_{-T/2}^{T/2} f(t)\, dt \qquad (10\text{-}96)$$

Comparing this relation with the one given in (10-32), we see that $a_0 = C_0$. We conclude that the entire right member of Eq. (10-95) may be consolidated into a single summation term which is run through limits of minus infinity to plus infinity. The expression for $f(t)$ may now be written

$$f(t) = \sum_{k=-\infty}^{\infty} C_k e^{jk\omega_0 t} \qquad (10\text{-}97)$$

Thus, we have defined a new form of Fourier series in which the coefficients C_k are, in general, complex. Such a form is called the *exponential* form (or the *complex* form) of the Fourier series.

The relation given in Eq. (10-87) defines the manner in which the coefficients C_k of the exponential form of the Fourier series are related to the coefficients a_k and b_k of the sine and cosine form of such a series. The relations between the coefficients c_k and ϕ_k of the magnitude and phase form of the series given in Eq. (10-84b) and the coefficients C_k are also readily derived. To see this, let us first express the coefficients

C_k defined in (10-87) in polar form. Thus, we obtain for positive values of k

$$C_k = \frac{1}{2}\sqrt{a_k^2 + b_k^2}\ \underline{\bigg/ \tan^{-1}\left(\frac{-b_k}{a_k}\right)}$$ (10-98a)

Similarly, for negative values of k

$$C_k = \frac{1}{2}\sqrt{a_{|k|}^2 + b_{|k|}^2}\ \underline{\bigg/ \tan^{-1}\left(\frac{b_{|k|}}{a_{|k|}}\right)}$$ (10-98b)

Comparing this with the relations defining the coefficients c_k and ϕ_k in terms of a_k and b_k [which are defined in Eqs. (10-46) and (10-47)], we see that for positive values of k

$$|C_k| = \tfrac{1}{2}c_k \qquad \underline{/C_k} = \phi_k$$ (10-99a)

Similarly, for negative values of k

$$|C_k| = \tfrac{1}{2}c_{|k|} \qquad \underline{/C_k} = -\phi_{|k|}$$ (10-99b)

In other words, the magnitude of C_k equals the quantity c_k divided by two and the argument of C_k is numerically equal to plus or minus ϕ_k. These relations hold for all values of k except $k = 0$, since we have already shown that the coefficient C_0 is equal to a_0 which, in turn, has been shown equal to c_0. The factor of $\tfrac{1}{2}$ that relates the magnitudes of the exponential and magnitude and phase forms of the Fourier series results from the fact that the summation index k for the coefficients C_k has a range from minus infinity to plus infinity, whereas the index for the coefficients c_k only has a range from 0 to infinity. Thus, in the exponential form of the series, there will always be two terms C_k which both contribute to the same harmonic frequency. The results given above may be summarized as follows:

SUMMARY 10-5.1

Exponential Form of the Fourier Series. The Fourier series for a non-sinusoidal function $f(t)$ has the form

$$f(t) = \sum_{k=-\infty}^{\infty} C_k e^{jk\omega_0 t}$$

where the C_k are complex coefficients and $C_{-k} = C_k^*$. For positive values of k,

$$C_k = \frac{a_k - jb_k}{2} = \frac{c_k}{2}\underline{/\phi_k}$$

where the quantities a_k, b_k, c_k, and ϕ_k are defined in Eq. (10-84). ∎

An example of the determination of the exponential form of a Fourier series follows.

Example 10-5.1

It is desired to determine the exponential form of the Fourier series for the half-wave rectified sine wave shown in Fig. 10-5.1. Applying Eq. (10-90), and noting that $T/2 = \pi$ (thus $\omega_0 = 1$) and that the function is zero for $-\pi \leq t \leq 0$, we obtain

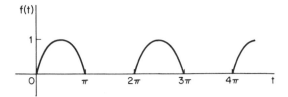

f(t)

Figure 10-5.1 Waveform for Example 10-5.1.

$$C_k = \frac{1}{2\pi} \int_0^\pi \sin t \, e^{-jkt} \, dt$$

$$= -\frac{1}{2\pi} \frac{1 + e^{-jk\pi}}{k^2 - 1}$$

This relation is easily evaluated for the various values of k. Thus, for $k = 0$,

$$C_0 = -\frac{1}{2\pi} \frac{1 + 1}{-1} = \frac{1}{\pi}$$

For $k = 1$, the general expression for C_k is indeterminate, so this value must be determined directly from the defining integral. Thus, we obtain

$$C_1 = \frac{1}{2\pi} \int_0^\pi \sin t \, e^{-jt} \, dt$$

$$= \frac{1}{2\pi} \int_0^\pi \sin t \, (\cos t - j \sin t) \, dt = -j\frac{1}{4}$$

Continuing by applying the general relation for C_k we find $C_2 = -1/3\pi$, $C_3 = 0$, $C_4 = 1/15\pi$, etc. Thus the exponential form of the series for $f(t)$ is

$$f(t) = \cdots + \frac{1}{15\pi} e^{-j4t} - \frac{1}{3\pi} e^{-j2t} + j\frac{1}{4} e^{-jt} + \frac{1}{\pi} - j\frac{1}{4} e^{jt} - \frac{1}{3\pi} e^{j2t} + \frac{1}{15\pi} e^{j4t} + \cdots$$

The corresponding magnitude and phase form of the series is readily shown to be

$$f(t) = \frac{1}{\pi} + \frac{1}{2} \cos (t - 90°) + \frac{2}{3\pi} \cos (2t - 180°) + \frac{1}{15\pi} \cos (4t - 180°) + \cdots$$

A line spectrum can be constructed for the magnitude and phase components of the exponential form of a Fourier series in the same manner as was done for the magnitude and phase form of the series. From the discussion given above and from (10-99a) and (10-99b), it is easily shown that there will be two main differences between the line spectra for the two series. These are the following:

1. All harmonic components (with the exception of the zero-frequency component) will have a magnitude one-half as large in the exponential form of the series as they do in the magnitude and phase form.
2. In the exponential form there will be a negative-frequency spectrum in addition to the positive-frequency spectrum. The negative-frequency spectrum will, of course, by symmetric about the origin with respect to the positive-frequency one.

As an example illustrating the two types of line spectra, consider the function shown in Fig. 10-5.2. The magnitude and phase form of the Fourier series for this

f(t)

F_0

$-T$ $-a/2$ $a/2$ T t

Figure 10-5.2 Rectangular pulse train.

function may be shown to be

$$f(t) = \frac{F_0 a}{T} + \sum_{k=1}^{\infty} \frac{2F_0 a}{T} \frac{\sin (k\omega_0 a/2)}{k\omega_0 a/2} \cos k\omega_0 t \tag{10-100}$$

The associated line spectrum for $T/a = 5$ is shown in Fig. 10-5.3(a). The exponental form of the Fourier series for this function may be written as

$$f(t) = \sum_{k=-\infty}^{\infty} \frac{F_0 a}{T} \frac{\sin (k\omega_0 a/2)}{k\omega_0 a/2} e^{jk\omega_0 t} \tag{10-101}$$

The line spectrum for this exponential form is shown in Fig. 10-5.3(b). A comparison of the two line spectra readily illustrates the differences between them for the given function.

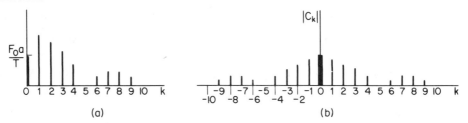

(a)   (b)

Figure 10-5.3 Magnitude line spectra for the waveform shown in Fig. 10-5.2.

Now let us see how the exponential form of the Fourier series presented above may be used to develop a Fourier transformation which may be applied to non-periodic functions. The key to our development here will be the concept of considering a non-periodic function as a limiting case of a periodic one, the limit being achieved by extending the period to infinity. Thus, over the entire range of t from minus infinity to plus infinity the non-periodic function may be considered as spanning only a single period. Now let us see how the exponential form of the Fourier series discussed above may be extended to apply to this situation. Specifically, we will be interested in examining the relations given in (10-90) and (10-97) in the limit as T approaches infinity. First let us consider the variable T itself. It is related to the fundamental harmonic frequency ω_0 by the expression

$$T = \frac{2\pi}{\omega_0} \tag{10-102}$$

The quantity ω_0 is not only the fundamental harmonic frequency, it is also the difference between adjacent harmonic frequencies. If we use the symbol $\Delta\omega$ for this difference, where $\Delta\omega = (k + 1)\omega_0 - k\omega_0 = \omega_0$, we may define T as

$$T = \frac{2\pi}{\Delta\omega} \tag{10-103}$$

Now let us define the quantity $F(jk\omega_0)$ as the coefficient C_k multiplied by T. Thus, using (10-103) and the defining relation for C_k given in (10-90), we may write

$$F(jk\omega_0) = TC_k = \frac{2\pi C_k}{\Delta\omega} = \int_{-T/2}^{T/2} f(t)e^{-jk\omega_0 t}\, dt \qquad (10\text{-}104)$$

As the period T is increased, we see from (10-102) that ω_0 becomes smaller. In the limit as T approaches infinity, the discrete set of quantities $k\omega_0$, where k is an integer, simply becomes a continuous variable which we will designate as ω. Thus, (10-104) may be written in the limit as

$$F(j\omega) = \int_{-\infty}^{\infty} f(t)e^{-j\omega t}\, dt \qquad (10\text{-}105)$$

This relation is called the *Fourier transformation* or the *Fourier integral*. As such it provides the defining expression for the quantity $F(j\omega)$ which in turn is called the *Fourier transform* or the *Fourier spectrum* of the non-periodic time-domain function $f(t)$. Since $F(j\omega)$ has frequency, namely, ω, as its variable, it is said to define a function in the *frequency domain*.

Now let us see how we may develop an expression specifying the inverse process, i.e., defining a function $f(t)$ in terms of a function $F(j\omega)$ in the frequency domain. Again we will start with the case where the period T is not infinite. From the first three members of Eq. (10-104) we may express C_k as a function of $F(jk\omega_0)$. Thus, we obtain

$$= \frac{F(jk\omega_0)}{T} = \frac{\Delta\omega\, F(jk\omega_0)}{2\pi} \qquad (10\text{-}106)$$

Now let us insert this result in the expression for $f(t)$ given in Eq. (10-97). We obtain

$$f(t) = \sum_{k=-\infty}^{\infty} \frac{\Delta\omega\, F(jk\omega_0)}{2\pi} e^{jk\omega_0 t} \qquad (10\text{-}107)$$

Let us now consider what happens as the period T becomes larger. In such a case, from Eq. (10-103), we see that $\Delta\omega$ becomes smaller. Actually, in the limit as K approaches infinity, $\Delta\omega$ becomes the differential $d\omega$, and the infinite sum of terms given in Eq. (10-107), becomes a continuous integral. As before, the discrete set of variables $k\omega_0$ becomes a continuous variable ω. Inserting these results in (10-107), we obtain

$$f(t) = \frac{1}{2\pi} \int_{-\infty}^{\infty} F(j\omega)e^{j\omega t}\, d\omega \qquad (10\text{-}108)$$

This relation defines the *inverse Fourier transformation*, i.e., it specifies how a function of the frequency variable $j\omega$ may be transformed back into a function of time. It is convenient to use the script letter $\mathcal{F}$ to symbolically indicate the Fourier transformation. Thus, the relations of Eqs. (10-105) and (10-108) may be indicated as

$$F(j\omega) = \mathcal{F}[f(t)] \qquad f(t) = \mathcal{F}^{-1}[F(j\omega)] \qquad (10\text{-}109)$$

respectively. Taken collectively, these two relations may be said to comprise a *transform pair* for any function $f(t)$ for which the Fourier transformation exists. It may be shown that the Fourier transformation produces a unique result, i.e.,

$$f(t) = \mathcal{F}^{-1}[F(j\omega)] = \mathcal{F}^{-1}\{\mathcal{F}[f(t)]\} \qquad (10\text{-}110)$$

The conditions for which the Fourier transformation of a given function may be found are similar to those enumerated for the Fourier series in Sec. 10-2. They are summarized in the following:

SUMMARY 10-5.2

The Fourier Transformation. Given a non-periodic function in the time domain $f(t)$ which has only a finite number of finite maxima, minima, and discontinuities and which satisfies the relation

$$\int_{-\infty}^{\infty} |f(t)| \, dt < \infty$$

we may determine a function $F(j\omega)$ by the Fourier transformation

$$F(j\omega) = \int_{-\infty}^{\infty} f(t)e^{-j\omega t} \, dt$$

where $F(j\omega)$ is a frequency-domain function called the Fourier transform of $f(t)$. Similarly, given a frequency-domain function $F(j\omega)$, we may apply the inverse Fourier transformation

$$f(t) = \frac{1}{2\pi} \int_{-\infty}^{\infty} F(j\omega)e^{j\omega t} \, d\omega$$

to find the time-domain function $f(t)$. ∎

As an example of the determination of the Fourier transform of a non-periodic function, consider the single pulse shown in Fig. 10-5.4. Applying Eq. (10-105), we

Figure 10-5.4 Single pulse.

obtain

$$F(j\omega) = \int_{-a/2}^{a/2} F_0 e^{-j\omega t} \, dt = F_0 \frac{e^{-j\omega t}}{-j\omega} \Big|_{-a/2}^{a/2} = F_0 \frac{e^{j\omega a/2} - e^{-j\omega a/2}}{j\omega} \qquad (10\text{-}111)$$

If we apply Euler's identity of Eq. (10-85), we see that this result may be written in the form

$$F(j\omega) = F_0 a \frac{\sin \omega a/2}{\omega a/2} \qquad (10\text{-}112)$$

A plot of the magnitude of this function is given in Fig. 10-5.5. From this plot we note that the magnitude decreases as ω increases, thus, the higher-frequency signal content is less than the lower-frequency signal content. It should be noted, however, that if a is decreased, the magnitude of the plot stays higher for a greater frequency interval. Thus, we conclude that short duration pulses will contain more signal information at higher frequencies than longer duration pulses do. In the physical world we can find many examples of this phenomenon. For instance, a lightning stroke, which is of very short duration, produces observable signal information (static) over all the communication spectrum from the relatively low frequencies used in radio reception to the considerably higher ones used in television reception.

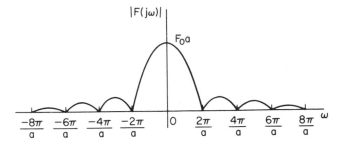

Figure 10-5.5   Fourier transform of the function shown in Fig. 10-5.4.

A spectrum or plot of the magnitude of some Fourier transform of the type shown in Fig. 10-5.5 may be considered as an extension of the discrete line spectrum discussed earlier in this section and in Sec. 10-2. It may be shown that the shape of such a continuous spectrum is the same as the *envelope* of the line spectrum that results when the identical waveform is repeated periodically. As an example of this, the periodic waveform shown in Fig. 10-5.2 is readily seen to consist of a periodic repetition of the same single pulse which was considered as a non-periodic waveform in Fig. 10-5.4. Thus the envelope of the discrete line spectrum shown in Fig. 10-5.3(b) for this periodic function has the same shape as the continuous spectrum shown in Fig. 10-5.5.

In the following section we shall develop some additional properties of the Fourier transformation, and illustrate the application of it to some basic problems in network analysis.

10-6 PROPERTIES OF THE FOURIER TRANSFORMATION

In this section we will present some properties of the Fourier transformation. The reader will note a considerable similarity between the properties developed here and those derived for the Laplace transformation in Chap. 8. These similarities are not merely coincidences. In this section we shall show that the two transformations are directly related and we shall discuss the relation in detail. Just as we did for the Laplace transformation, in all the developments of this section we shall use a capital letter to indicate the transformation of a time-domain function defined by the same letter in lower case. Thus, we will use the notation $F(j\omega) = \mathfrak{F}[f(t)]$, $G(j\omega) = \mathfrak{F}[g(t)]$, etc. As a first property of the Fourier transformation we note that it satisfies the properties of homogeneity and superposition (the proof is similar to that given in Sec. 8-2 for the Laplace transformation). Thus, *the Fourier transformation is a linear transformation.*

Now let us derive the transformation of the derivative of a function $f(t)$. As a starting point, let us consider the defining relation for the inverse Fourier transformation given in Eq. (10-108). Differentiating both sides of this relation with respect to t, we obtain

$$\frac{df}{dt} = \frac{1}{2\pi} \int_{-\infty}^{\infty} j\omega F(j\omega)e^{j\omega t}\, d\omega \qquad (10\text{-}113)$$

As indicated in the right member of the above equation, we see that the operation of differentiating a time-domain function $f(t)$ corresponds exactly with multiplication of a frequency-domain function $F(j\omega)$ by the factor $j\omega$. Thus, we may write

$$\mathcal{F}\left[\frac{df}{dt}\right] = j\omega F(j\omega) \qquad (10\text{-}114)$$

where $F(j\omega)$ is the Fourier transform of $f(t)$.

A similar conclusion is reached if we investigate the Fourier transformation of the integral of a time-domain function. To do this, let us consider the function

$$g(t) = \int_{-\infty}^{t} f(\tau)\, d(\tau)$$

Since $dg/dt = f(t)$ and if $G(j\omega)$ is the Fourier transform of $g(t)$, then from the above result on differentiation we must have $j\omega\, G(j\omega) = F(j\omega)$, where $F(j\omega)$ is the Fourier transform of $f(t)$. However, in order for the function $g(t)$ to be transformable, $g(\infty)$ must equal 0. This requires that

$$g(\infty) = \int_{-\infty}^{\infty} f(t)\, dt = F(0) = 0 \qquad (10\text{-}115)$$

In this case, we may write[6]

$$\mathcal{F}\left[\int_{-\infty}^{t} f(\tau)\, d\tau\right] = \frac{F(j\omega)}{j\omega} \qquad (10\text{-}116)$$

Thus, we conclude that integration of a function in the time domain is transformed into division by the factor $j\omega$ for the corresponding frequency-domain function.

Now let us consider some other properties of the Fourier transformation. One of the most interesting of these is the effect of changing the time scale of a function. Basically, this consists of replacing the variable t by a new variable at, where a is some positive constant. If the original function is specified as $f(t)$, the time-scaled function thus becomes $f(at)$. Taking the Fourier transformation of such a function, we obtain

$$\mathcal{F}[f(at)] = \int_{-\infty}^{\infty} f(at)e^{-j\omega t}\, dt \qquad (10\text{-}117)$$

To evaluate the integral in the right member of the above equation, let us make a substitution of variable of integration $x = at$. The differential dt now becomes dx/a. Substituting this in Eq. (10-117), we obtain

$$\mathcal{F}[f(at)] = \frac{1}{a} \int_{-\infty}^{\infty} f(x)e^{-j(\omega/a)x}\, dx = \frac{1}{a}F\left(j\frac{\omega}{a}\right) \qquad (10\text{-}118)$$

[6]If this condition is not satisfied, then the more general result is

$$\mathcal{F}\left[\int_{-\infty}^{t} f(\tau)\, d\tau\right] = \frac{F(j\omega)}{j\omega} + \pi F(0)\, \delta(\omega)$$

An example of this result may be found in (10-125).

From the above relation we see that the result of scaling the variable t of a time-domain function $f(t)$ is a reciprocal scaling of the variable ω in the transformed frequency-domain function $F(j\omega)$. In addition, there is a scaling of the magnitude of $F(j\omega)$ equal to $1/a$. This result provides a more rigorous justification for the conclusion reached in the preceding section with respect to shortening the duration of a pulse. Such a shortening can be considered as a scaling of the type indicated above in which $a > 1$. For example, a pulse $f(t)$ which occurs from 0 to 1 s may be scaled to a pulse of similar waveform but occuring from 0 to $\frac{1}{3}$ s if we express it as $f(3t)$. Correspondingly, from Eq. (10-118) the frequency spectrum of the transformed function $F(j\omega)$ will be tripled in width, and a scale factor of $\frac{1}{3}$ will be introduced.

Another interesting property of the Fourier transformation can be developed by finding the transform of the quantity $f(t) \cos \omega_0 t$. Such a time-domain quantity is frequently encountered in studies of the processes associated with radio transmission. Specifically, it is referred to as *amplitude modulation* in which the quantity $f(t)$ is a signal which is to be transmitted by modulating the amplitude of a radio-frequency carrier of ω_0 rad/s. Applying the Fourier transformation and using Euler's relation for the cosine function, we find

$$\mathscr{F}[f(t) \cos \omega_0 t] = \int_{-\infty}^{\infty} f(t) \frac{e^{j\omega_0 t} + e^{-j\omega_0 t}}{2} e^{-j\omega t} \, dt$$

$$= \frac{1}{2} \int_{-\infty}^{\infty} f(t) e^{-j(\omega - \omega_0)t} \, dt + \frac{1}{2} \int_{-\infty}^{\infty} f(t) e^{-j(\omega + \omega_0)t} \, dt$$

$$= \frac{1}{2} \{ F[j(\omega - \omega_0)] + F[j(\omega + \omega_0)] \} \tag{10-119}$$

where the quantities $F[j(\omega - \omega_0)]$ and $F[j(\omega + \omega_0)]$ are readily interpreted by recalling that $F(j\omega)$ is the Fourier transform of $f(t)$. Thus, we see that the Fourier spectrum of the resulting time function contains two distinct bands whose shapes differ only by a scale factor from the Fourier spectrum of $f(t)$ but whose positions are displaced along the frequency axis so that they are centered at $\pm j\omega_0$ rather than the origin.

As a final property of the Fourier transformation, let us consider the effect of shifting the time scale. That is, let us investigate the Fourier transformation of $f(t - a)$, namely, a function $f(t)$ which has been shifted in the direction of decreasing time by the quantity a. Applying the Fourier transformation, we obtain

$$\mathscr{F}[f(t - a)] = \int_{-\infty}^{\infty} f(t - a) e^{-j\omega t} \, dt \tag{10-120}$$

The integral of the right member of the above equation may be evaluated by defining a new variable of integration $x = t - a$. Thus, we obtain

$$\mathscr{F}[f(t - a)] = \int_{-\infty}^{\infty} f(x) e^{-j\omega(x + a)} \, dx = e^{-j\omega a} \int_{-\infty}^{\infty} f(x) e^{-j\omega x} \, dx \tag{10-121}$$

This may be written in the form

$$\mathscr{F}[f(t - a)] = e^{-j\omega a} F(j\omega) \tag{10-122}$$

where $F(j\omega) = \mathscr{F}[f(t)]$. We conclude that the effect of shifting a function in the time domain is simply multiplication of the transformed function by an exponential. If s

is substituted for $j\omega$, this result is identical with the result developed in connection with the Laplace transformation in Sec. 8-3. Other properties of the Fourier transformation are readily derived in a manner similar to that used in connection with the Laplace transformation. Table 10-6.1 summarizes the most important of the properties.

TABLE 10-6.1 Properties of the Fourier Transformation

No.	$f(t)$	$F(j\omega)$	Description
1	$f_1(t) \pm f_2(t)$	$F_1(j\omega) \pm F_2(j\omega)$	superposition
2	$Kf(t)$	$KF(j\omega)$	homogeneity
3	$\dfrac{d^n f}{dt^n}$	$(j\omega)^n F(j\omega)$	differentiation
4	$\displaystyle\int_{-\infty}^{t} f(\tau)\,d\tau$	$\dfrac{1}{j\omega} F(j\omega)$	integration
5	$f(t - t_0)$	$F(j\omega)e^{-j\omega t_0}$	shifting of time axis
6	$f(t)e^{j\omega_0 t}$	$F[j(\omega - \omega_0)]$	shifting in frequency domain
7	$f(at)$	$\dfrac{1}{|a|}F\left(j\dfrac{\omega}{a}\right)$	time scaling
8	$f_1(t)*f_2(t)$	$F_1(j\omega)F_2(j\omega)$	convolution in time domain

In addition to discussing the properties of the Fourier transformation, it is interesting to obtain the transforms of some of the more common time-domain functions. First let us find the transform of a unit impulse $\delta(t)$. Applying the Fourier transformation, we obtain

$$\mathcal{F}[\delta(t)] = \int_{-\infty}^{\infty} \delta(t)e^{-j\omega t}\,dt = e^{-j\omega t}\big|_{t=0} = 1 \tag{10-123}$$

where the evaluation of the integral follows the method outlined in Sec. 8-3. Thus, we see that the Fourier transformation of the unit impulse is unity. Note that this is the same result as was obtained for the Laplace transform of the unit impulse. Now let us consider the transformation of the unit step function $u(t)$. If we evaluate

$$\int_{-\infty}^{\infty} |u(t)|\,dt \tag{10-124}$$

we note that the integral is not finite; therefore, this function does not satisfy the criterion normally applied for Fourier transformability. Thus, we may conclude that the unit step does not have a Fourier transform. Note that this result is quite different than the result achieved for the Laplace transformation in which the integrating factor $e^{-\sigma t}$ permitted a transform to be found for the unit step. However, the application of a more involved branch of mathematics, called the theory of distributions, can be made to show that a Fourier transform for the unit step function may be defined as

$$\mathcal{F}[u(t)] = \pi\delta(\omega) + \frac{1}{j\omega} \tag{10-125}$$

The first term represents an impulse (in the frequency domain) of strength π occurring at ω equals zero. The second term is the same as the Laplace transformation of a unit step function in which s has been replaced by $j\omega$. There are many other functions which, like the unit step function, do not have a bounded integral and thus are not ordinarily considered to be transformable.

In considering general time-domain functions, it is important to note that the Fourier transformation is defined over the entire time spectrum and not just for positive values of time as the Laplace transformation is. As an example of this, we must distinguish between the functions $e^{-at}(a > 0)$ and $e^{-at}u(t)(a > 0)$. Only the latter function has a Fourier transformation. A list of some typical time-domain functions and their Fourier transforms is given in Table 10-6.2. The entries in this table represent transform pairs. Thus, they may be used as a means of finding the inverse transform of frequency domain functions, just as was done for the Laplace transformation.

Now let us apply Fourier transformation methods to some actual network problems. Suppose that such a problem is specified for a given network in terms of an excitation variable $g(t)$ and it is desired to find some response variable $f(t)$. From our study of time-domain networks in Chaps. 5 and 6 we know that these variables must be related by a differential equation having the form

$$a_0 f(t) + a_1 \frac{df}{dt} + a_2 \frac{d^2 f}{dt^2} + \cdots = b_0 g(t) + b_1 \frac{dg}{dt} + b_2 \frac{d^2 g}{dt^2} + \cdots \qquad (10\text{-}126)$$

We may take the Fourier transformation of both members of this equation by first

TABLE 10-6.2 Basic Fourier Transform Pairs

No.	$f(t)$	$F(j\omega)$
1	$\delta(t)$	1
2	$u(t)$	$\pi\delta(\omega) + \dfrac{1}{j\omega}$
3	1	$2\pi\delta(\omega)$
4	$|t|$	$-\dfrac{2}{\omega^2}$
5	$e^{-at}u(t)$	$\dfrac{1}{a + j\omega}$
6	$e^{-a|t|}$	$\dfrac{2a}{a^2 + \omega^2}$
7	$te^{-at}u(t)$	$\dfrac{1}{(a + j\omega)^2}$
8	$\sin \omega_0 t\, u(t)$	$\dfrac{\pi}{2j}[\delta(\omega - \omega_0) + \delta(\omega + \omega_0)] + \dfrac{\omega_0}{\omega_0^2 - \omega^2}$
9	$\cos \omega_0 t\, u(t)$	$\dfrac{\pi}{2}[\delta(\omega - \omega_0) - \delta(\omega + \omega_0)] + \dfrac{j\omega}{\omega_0^2 - \omega^2}$
10	$\sin \omega_0 t$	$j\pi[\delta(\omega + \omega_0) - \delta(\omega - \omega_0)]$
11	$\cos \omega_0 t$	$\pi[\delta(\omega - \omega_0) + \delta(\omega + \omega_0)]$
12	$e^{-at} \sin \omega_0 t\, u(t)$	$\dfrac{\omega_0}{(a + j\omega)^2 + \omega_0^2}$

defining the quantities

$$F(j\omega) = \mathfrak{F}[f(t)]$$
$$G(j\omega) = \mathfrak{F}[g(t)]$$
(10-127)

Applying the relations of Table 10-6.1, we obtain

$$F(j\omega)[a_0 + a_1(j\omega) + a_2(j\omega)^2 + \cdots] = G(j\omega)[b_0 + b_1(j\omega) + b_2(j\omega)^2 + \cdots]$$
(10-128)

This may be written in the form

$$\frac{F(j\omega)}{G(j\omega)} = N(j\omega) = \frac{b_0 + b_1(j\omega) + b_2(j\omega)^2 + \cdots}{a_0 + a_1(j\omega) + a_2(j\omega)^2 + \cdots}$$
(10-129)

Thus, we see that the relation between the transformed response variable $F(j\omega)$ and the transformed excitation variable $G(j\omega)$ is given by the quantity $N(j\omega)$ defined in Eq. (10-129). This is exactly the same result as was obtained in the application of phasor methods to the solution of differential equations in Sec. 7-3. In addition, the expression in the right member of Eq. (10-129) defining $N(j\omega)$ is identical with the quantity relating input and output phasors given in Summary 7-3.2. The reason for this is quite simple; namely, it is due to the fact that differentiation of a time-domain function is represented by multiplication by the factor $j\omega$ in both the phasor method and the Fourier transformation method. Since this is so, and since integration is similarly related to division by the factor $j\omega$, we may conclude that all the network functions defined in Chap. 7 may be directly applied to the Fourier transformation method of network analysis. The really great fact about this conclusion is that we can apply all the methods based on the use of impedances and admittances developed in Chap. 7 to the problem of finding the quantity $N(j\omega)$ which relates the Fourier transforms of response and excitation variables. The resulting function can then be used to determine the response due to arbitrary nonperiodic inputs by applying the Fourier transformation as well as being used in sinusoidal steady-state analysis. An example follows.

Example 10-6.1

As an example of the application of Fourier transformation methods to network analysis consider the determination of the output voltage $v(t)$ when a pulse of current $i(t)$ with the waveform shown in Fig. 10-6.1 is applied to the parallel RC circuit shown in Fig. 10-6.2. For this problem we may define the quantities $V(j\omega)$, $I(j\omega)$, and $Z(j\omega)$ as follows:

Figure 10-6.1 Input waveform for Example 10-6.1.

Figure 10-6.2 Network for Example 10-6.1.

$$V(j\omega) = \mathfrak{F}[v(t)] \qquad I(j\omega) = \mathfrak{F}[i(t)] \qquad Z(j\omega) = \frac{V(j\omega)}{I(j\omega)}$$

The Fourier transform $I(j\omega)$ is found by applying the methods given in Sec. 10-5. The network function $Z(j\omega)$ is determined by applying the methods developed in Chap. 7. Thus, to find the Fourier transform $V(j\omega)$, we proceed as follows:

$$V(j\omega) = Z(j\omega)I(j\omega) = \frac{R}{1 + j\omega RC} \frac{I_0(1 - e^{-j\omega a})}{j\omega}$$

The time-domain function $v(t)$ is found by taking the inverse transformation of the expression for $V(j\omega)$. This may be done by applying the integral relation for the inverse transformation given in Eq. (10-108). Thus, we may write

$$v(t) = \frac{1}{2\pi} \int_{-\infty}^{\infty} \frac{I_0 R(1 - e^{-j\omega a})}{j\omega(1 + j\omega RC)} e^{j\omega t} \, d\omega$$

For simplicity, let us consider this expression as the sum of two integrals $v_1(t)$ and $v_2(t)$ as follows:

$$v(t) = v_1(t) + v_2(t) = \frac{I_0}{2\pi} \int_{-\infty}^{\infty} \frac{R e^{j\omega t}}{j\omega(1 + j\omega RC)} \, d\omega - \frac{I_0}{2\pi} \int_{-\infty}^{\infty} \frac{R e^{j\omega(t-a)}}{j\omega(1 + j\omega RC)} \, d\omega$$

Applying Euler's relation from (10-85) and rearranging terms in the integrand of the first integral, we may write

$$v_1(t) = \frac{I_0}{2\pi} \int_{-\infty}^{\infty} \frac{R(-j - \omega RC)}{\omega(1 - \omega^2 R^2 C^2)} (\cos \omega t + j \sin \omega t) \, d\omega$$

The imaginary component may be eliminated since the resulting integral must be real. Thus, we may write

$$v_1(t) = \frac{I_0}{2\pi} \int_{-\infty}^{\infty} \frac{-R^2 C}{1 - \omega^2 R^2 C^2} \cos \omega t \, d\omega + \frac{I_0}{2\pi} \int_{-\infty}^{\infty} \frac{R}{\omega(1 - \omega^2 R^2 C^2)} \sin \omega t \, d\omega$$

Integrating and applying the limits, we obtain

$$v_1(t) = I_0 R[u(t)(1 - e^{-t/RC}) - \tfrac{1}{2}]$$

where $u(t)$ is the unit step function which appears because of cancellation of positive and negative sign terms for $t < 0$. The second integral is evaluated by a similar process in which we replace the variable t by the quantity $t - a$. Thus, we obtain

$$v_2(t) = -I_0 R[u(t - a)(1 - e^{-(t-a)/RC}) - \tfrac{1}{2}]$$

Combining the above results, we obtain the following expression for $v(t)$:

$$v(t) = I_0 R[u(t)(1 - e^{-t/RC}) - u(t - a)(1 - e^{-(t-a)/RC})]$$

A plot of this waveform is shown in Fig. 10-6.3.

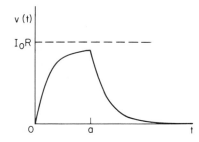

Figure 10-6.3 Output waveform for Example 10-6.1.

In the above example, we have seen that a network function which relates the output and input phasors of a given network, and which is determined by the use of impedance and admittance concepts, is directly usable *without changes of any type* as the network function relating the corresponding output and input Fourier transformed variables. Nor need we stop here. As pointed out in Summary 8-5.2, the same network function (with s substituted for $j\omega$) gives the relation between the Laplace transformed output and input variables (in the absence of initial conditions). Thus, the network function is an invariant element common to sinusoidal steady-state methods, Fourier transformation methods, and Laplace transformation methods. Vive la network function!

SUMMARY 10-6.1

Fourier Transform Analysis of Networks. If a non-periodic function $g(t)$ with Fourier transform $G(j\omega)$ is applied as an excitation to a network, the Fourier transform $F(j\omega)$ of the response $f(t)$ to this excitation is given as

$$F(j\omega) = N(j\omega)G(j\omega)$$

where $N(j\omega)$ is the same function as defined in Sec. 7-7 for use in phasor methods. If s is substituted for the argument $j\omega$, then $N(s)$ is the network function giving the ratio of response to excitation variables defined in Sec. 8-5 for use in Laplace transformation methods. ∎

The results given above allow us to make another interesting observation about the network function. Assume that $N(j\omega)$ is such a function and that we apply a unit impulse $\delta(t)$ as an excitation. Since the Fourier transform of a unit impulse is 1, the resulting response is simply the inverse transform of $N(j\omega)$. If we define this unit impulse response in the time domain as $h(t)$, we may write

$$h(t) = \frac{1}{2\pi} \int_{-\infty}^{\infty} N(j\omega)e^{j\omega t} \, d\omega \qquad (10\text{-}130)$$

Thus, we conclude that the network function is the Fourier transform of the unit impulse response of the network, or, conversely, taking the inverse Fourier transformation of the network function yields the unit impulse response for the network.

One of the most important practical problems in the analysis of networks and linear systems is the correlation between frequency-domain behavior and time-domain behavior. The reason for this is that the frequency-domain data are easier to obtain experimentally than almost any other type. As an example, consider the experiment shown in Fig. 10-6.4 in which a sinusoidal signal generator with a variable output frequency is used as an excitation source for a network and in which some suitable measuring equipment such as an oscilloscope or a set of magnitude and phase measuring meters are used to monitor the output. By varying the frequency of the signal generator over a wide range and recording the data characterizing the resulting output signal from the network, we may readily plot curves which show the magnitude and phase spectra of the network function. Such data, however, are not directly usable in

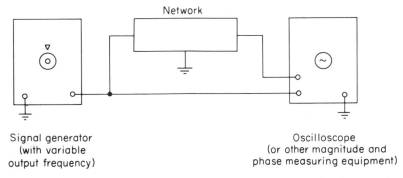

Network

Signal generator
(with variable
output frequency)

Oscilloscope
(or other magnitude and
phase measuring equipment)

Figure 10-6.4 Determining the magnitude and phase characteristic of a network.

applications which require that the network possess some specified *time-domain* behavior, such as a certain output rise time in response to a step input. In the following paragraphs, we shall show how the Fourier transformation method may be applied to establish a correlation between a network's performance in the frequency domain and in the time domain.

Let us begin our discussion by defining the concept of an *ideal transmission characteristic* for a network. Such a characteristic is assumed to reproduce the exact waveform of some input signal with two possible exceptions. Namely, the magnitude of the input signal waveform may be scaled, and the entire waveform may be delayed, i.e., displaced in t. Thus, if an input signal $i(t)$ is applied to a network which has such an ideal transmission characteristic, the output signal $o(t)$ may be expressed as

$$o(t) = Ki(t - a) \tag{10-131}$$

where K is a scaling factor and a is a delay factor. This result is illustrated in Fig. 10-6.5. Depending on whether K is greater than or less than unity, the network may

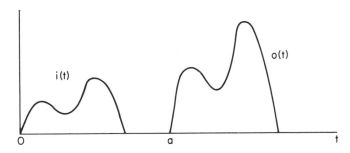

Figure 10-6.5 Ideal transmission of a signal.

be said to have "gain" or "loss," respectively. Applying the Fourier transformation to Eq. (10-131) and using the shifting property presented in Eq. (10-121), we obtain

$$O(j\omega) = KI(j\omega)e^{-j\omega a} \tag{10-132}$$

where $O(j\omega)$ and $I(j\omega)$ are the Fourier transforms of $o(t)$ and $i(t)$, respectively. Thus, the network function $N(j\omega)$ for such a system is defined as

$$N(j\omega) = \frac{O(j\omega)}{I(j\omega)} = Ke^{-j\omega a} \qquad (10\text{-}133)$$

The magnitude and phase spectra for such an ideal transmission characteristic are shown in Fig. 10-6.6. These characteristics are physically unrealizable since they require that the phase be linear over an infinite bandwidth. Such a set of spectra can, however, be approximated over a limited frequency range. Ideally, such a *band-limited* characteristic will appear as shown in Fig. 10-6.7. Such a set of spectra is said to define an *ideal low-pass* characteristic.

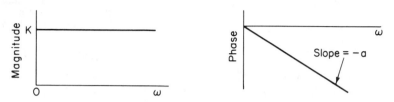

Figure 10-6.6 Ideal transmission characteristic.

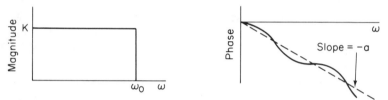

Figure 10-6.7 Band-limited approximation to an ideal transmission characteristic.

Now let us see what happens if we apply a simple pulse having the form shown in Fig. 10-5.4 as an input to such an ideal low-pass network. The amplitude spectrum of the pulse is shown in Fig. 10-5.5. The first lobe of the amplitude spectrum, which is also the part of the spectrum with the largest values, occurs around $\omega = \pm 2\pi/a$. If we let ω_0 be the cutoff frequency of the ideal low-pass filter, then the filter will pass most of the energy contained in the signal if $\omega_0 = 2\pi/a$. In this case, we see that

$$\omega_0 a = 2\pi \qquad (10\text{-}134)$$

where a is the duration of the pulse and ω_0 is the cutoff frequency required in an ideal low-pass filter that transmits most of the energy contained in the pulse. Extending this result, it may be shown that, in general, for an abitrarily shaped input pulse the following relation holds:

$$\begin{bmatrix} \text{Bandwidth required in an ideal} \\ \text{low-pass filter to transmit the} \\ \text{major portion of the energy} \\ \text{contained in an input pulse} \end{bmatrix} \times \begin{bmatrix} \text{Duration of} \\ \text{the input pulse} \end{bmatrix} = [\text{A constant}] \qquad (10\text{-}135)$$

As another example of the use of the Fourier transformation to relate the frequency response characteristics of a network to its time-domain behavior, it may be shown that a relation similar to that given in (10-134) exists between the bandwidth of an ideal low-pass filter and the rise time of the output signal in response to a step input. Thus, we may write

$$\begin{bmatrix} \text{Bandwidth required in} \\ \text{an ideal low-pass filter} \end{bmatrix} \times \begin{bmatrix} \text{Rise time of output} \\ \text{waveform when a step} \\ \text{input is applied} \end{bmatrix} = [\text{A constant}] \qquad (10\text{-}136)$$

In this latter expression, if the rise time is specified as the difference between the times that the output reaches 90% and 10% of its final value, and if there is little overshoot (less than 5%), the value of the constant is 2.2.

Other examples of the use of the Fourier transformation to relate the frequency-domain characteristics and time-domain performance of a network may be found in the references given in the Bibliography.

It is interesting to note some of the other similarities and differences between the Laplace and Fourier transformations. First, let us consider some of the advantages of the Laplace transformation over the Fourier transformation. The Laplace transformation has the capability for introducing initial conditions explicitly in the network equations, while the Fourier transformation does not. This capability is readily verified by considering the expressions for the transformation of derivative functions given in Table 8-3.2 for the Laplace transformation and in Table 10-6.1 for the Fourier transformation. Another advantage of the Laplace transform method is that it is easier to apply to the analysis of networks than the Fourier transform method, since the algebraic manipulations necessary to find a transfer function are more readily made in terms of the variable s than in terms of the imaginary quantity $j\omega$. Finally, in the Laplace transform method, the time-domain behavior of network functions may be predicted from the pole and zero locations of the complex frequency function. In the Fourier transform method this, of course, is not possible. The Fourier transform method also has certain advantages over the Laplace transform one in that it provides more direct information on the frequency-domain behavior of a network. Thus, it is more readily applied to correlate experimentally obtained frequency-domain measurements than is the Laplace transform method.

10-7 CONCLUSION

In this chapter we have presented our final treatment concerning the frequency-domain analysis of networks by introducing the Fourier series and the Fourier transformation. The basic importance of the Fourier series is that it extends all the concepts of sinusoidal steady-state analysis to cover the situation where a non-sinusoidal (but still periodic) excitation is applied to a lumped, linear, time-invariant network. Just as was the case for the sinusoidal steady-state, the use of a Fourier series in analyzing a network only provides information as to the behavior of the network variables after any transient terms have disappeared. That is, it is only concerned with steady-state behavior.

The second technique discussed in this chapter is the Fourier transformation. This transformation is directly derivable from the Laplace transformation (or vice versa). It provides a means of determining the frequency spectrum, i.e., the frequency content of a time-domain waveform. Thus, it provides a readily applied link between the frequency-domain properties of a network and the effect that these properties will have on the energy contained in a specific time-domain waveform. The treatment given in this section can only be considered as an introduction to the subject, since many complete books have been devoted solely to the Fourier transformation.

PROBLEMS

Problem 10-1 (Sec. 10-1)

Decompose each of the periodic functions shown in Fig. P10-1 into the sum of an even and an odd function.

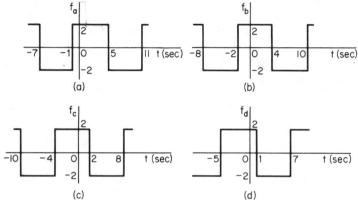

Figure P10-1

Problem 10-2 (Sec. 10-1)

Decompose each of the periodic functions shown in Fig. P10-2 into the sum of an even function and an odd one.

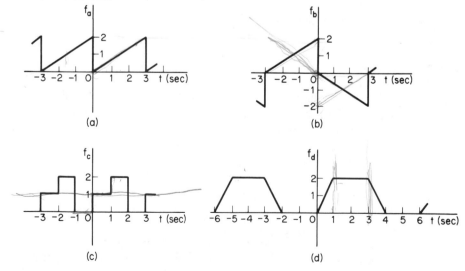

Figure P10-2

The Fourier Series and the Fourier Integral   Chap. 10

Problem 10-3 (Sec. 10-1)

Decompose each of the periodic functions defined below into the sum of an even function and an odd one:

(a) $f_a(t) = 3t, 0 \le t \le 3$, with a period of 3 s.

(b) $f_b(t) = \begin{Bmatrix} 3t, 0 \le t < 2 \\ 0, 2 < t \le 3 \end{Bmatrix}$ with a period of 3 s.

(c) $f_c(t) = \begin{Bmatrix} 3t, 0 \le t < 1 \\ 0, 1 < t \le 3 \end{Bmatrix}$ with a period of 3 s.

Problem 10-4 (Sec. 10-1)

Repeat the three parts of Problem 10-3 using $3t^2$ in place of $3t$.

Problem 10-5 (Sec. 10-1)

The Fourier series for the waveshapes $f_a(t)$ and $f_b(t)$ shown in Fig. P10-5 are given below. Find the Fourier series for the waveshape $f_c(t)$.

$$f_a(t) = \frac{4}{\pi} \sin \pi t + \frac{4}{3\pi} \sin 3\pi t + \frac{4}{5\pi} \sin 5\pi t$$

$$f_b(t) = \frac{8}{\pi^2} \sin \pi t - \frac{8}{9\pi^2} \sin 3\pi t + \frac{8}{25\pi^2} \sin 5\pi t$$

  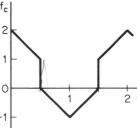

Figure P10-5

Problem 10-6 (Sec. 10-2)

Determine the coefficients a_i and b_i of the first three harmonic terms of the Fourier series for the functions shown in Fig. P10-1 by directly applying the integral formulas for these coefficients to the waveshapes shown in the figure.

Problem 10-7 (Sec. 10-2)

Repeat Problem 10-6 for the functions shown in Fig. P10-2.

Problem 10-8 (Sec. 10-2)

Determine the coefficients a_i and b_i of the first three harmonic terms of the Fourier series for the functions shown in Fig. P10-1. Do this by using the results found in Problem 10-1 to find the coefficients of the odd and even portions of the function separately, then add the results.

Problem 10-9 (Sec. 10-2)

Repeat Problem 10-8 for the functions shown in Fig. P10-2.

Problem 10-10 (Sec. 10-2)

Sketch the magnitude and phase spectra for the first harmonics of each of the functions shown in Fig. P10-1.

Problem 10-11 (Sec. 10-2)

Determine the coefficients a_i and b_i of the first three harmonic terms of the functions shown in Fig. P10-11. Determine the magnitude and phase coefficients for these same terms.

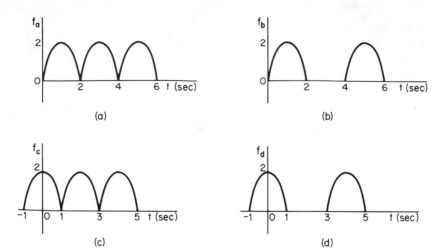

(a)

(b)

(c)

(d)

Figure P10-11

Problem 10-12 (Sec. 10-2)

A 1-Ω resistor has a square wave of current of amplitude ± 1 A with a period of 1 s flowing through it. Find the average power dissipated in the resistor by:

(a) Finding the rms value of the square wave.

(b) Expressing the square wave as a Fourier series, and finding the power dissipated by each of the terms in the series through the fifth harmonic.

Compare the answers obtained by the two methods.

Problem 10-13 (Sec. 10-2)

(a) The waveform shown in Fig. P10-13(a) has the following Fourier series. Rewrite the series using the magnitude and phase form of the series with terms $c_i \cos(i\omega_0 t + \phi_i)$ where $c_i \geq 0$.

$$f(t) = \frac{8}{\pi^2}\left(\sin \pi t - \frac{1}{9}\sin 3\pi t + \frac{1}{25}\sin 5\pi t + \cdots\right)$$

(b) Find the Fourier series of the function shown in Fig. P10-13(b) through the fifth harmonic. Use the series form $a_i \cos i\omega_0 t + b_i \sin i\omega_0 t$.

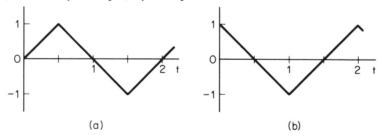

(a)

(b)

Figure P10-13

Problem 10-14 (Sec. 10-3)

Predict upper bounds on the manner in which the magnitude coefficients of the Fourier series for the periodic functions defined below will diminish for higher-frequency harmonic terms, and determine enough coefficients for the series to verify your conclusion.

(a) $f_a(t) = (t + 1)(t - 1)$, $-1 < t < 1$, with a period of 2 s.

(b) $f_b(t) = (t + 1)^2(t - 1)^2$, $-1 < t < 1$, with a period of 2 s.

Problem 10-15 (Sec. 10-3)

Use as many of the following letters as are appropriate to describe the symmetry of each of the waveforms shown in Fig. P10-15.

a—even function   d—only odd harmonics
b—odd function   e—only even harmonics
c—half-wave symmetry

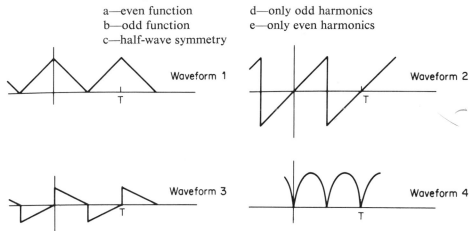

Figure P10-15

Problem 10-16 (Sec. 10-3)

For each of the cases listed below, sketch a non-trivial periodic function (which has *not* been discussed in the text) which meets the criteria.
(a) A function which has only even harmonics.
(b) A function which has only odd harmonics.
(c) A function which is an even function.
(d) A function which is an odd function.
(e) A function that has half-wave symmetry.

Problem 10-17 (Sec. 10-3)

Make a table similar to that shown in Fig. P10-17 indicating which coefficients are present in

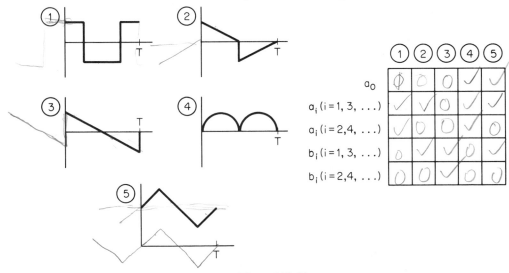

Figure P10-17

the Fourier series for each of the five waveforms. In the table put a zero in each block for which the corresponding series terms are all zero, and the letters NZ (*Not Zero*) in each block for which the series terms are nonzero. The period of each waveform is T. The general form of each series is $a_0 + \sum a_i \cos i\omega_0 t + \sum b_i \sin i\omega_0 t$.

Problem 10-18 (Sec. 10-4)

(a) The Fourier series for one of the waveforms shown in Fig. P10-18(a) and (b) is

$$f(t) = \sin t + \tfrac{1}{3} \sin 3t + \tfrac{1}{5} \sin 5t + \cdots$$

Find the Fourier series of the other waveform.

(b) If the Fourier series given above in part (a) is applied as the input waveform $v_1(t)$ to the network shown in Fig. P10-18(c), find the first three terms of the Fourier series of the output waveform $v_2(t)$.

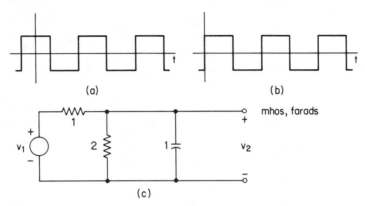

(a)   (b)

$\dfrac{1}{sc}$

(c)

Figure P10-18

Problem 10-19 (Sec. 10-4)

For the circuit shown in Fig. P10-19 do the following.

(a) Find the magnitude and phase of the first three nonzero Fourier coefficients in the output waveform $v_2(t)$ when the indicated periodic input function is applied.

(b) Find the average input power to the network in the first three harmonic frequencies which are present.

(c) Find the average output power in the resistor which is connected in parallel with the capacitor.

Figure P10-19

Problem 10-20 (Sec. 10-4)

Repeat Problem 10-19 using the input excitation shown in Fig. P10-20.

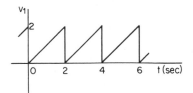

Figure P10-20

Problem 10-21 (Sec. 10-4)

The input waveform to the circuit shown in Fig. P10-21 has large unwanted harmonic components at ω equals 3 and 7 rad/s. Find values for the shunt inductance and the series capacitance so as to eliminate these frequency components from the output waveform $v_2(t)$.

Figure P10-21

Problem 10-22 (Sec. 10-4)

Find the values of G (in mhos) and C (in farads) for the network shown in Fig. P10-22 such that the third harmonic component of $v_1(t)$ is reduced by 0.3 and the fifth harmonic component is reduced by 0.1. Assume $\omega_0 = 1$ rad/s.

Figure P10-22

Problem 10-23 (Sec. 10-4)

A square wave of voltage has the (approximate) Fourier series

$$v_1(t) = \frac{4}{\pi} \sin \pi t + \frac{4}{3\pi} \sin 3\pi t + \frac{4}{5\pi} \sin 5\pi t$$

If this is applied to the circuit shown in Fig. P10-23, find the resulting Fourier series for $v_2(t)$. Put the series terms in the form $b_i \sin (i\omega_0 t + \phi_i)$.

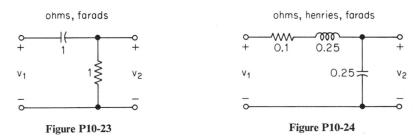

Figure P10-23   Figure P10-24

Problem 10-24 (Sec. 10-4)

The waveform for the voltage $v_1(t)$ in Fig. P10-24 is

$$v_1(t) = \frac{4}{\pi}\left(\cos \pi t - \frac{1}{3}\cos 3\pi t + \frac{1}{5}\cos 5\pi t + \cdots\right)$$

Find the series expression for the voltage $v_2(t)$.

Problem 10-25 (Sec. 10-5)

Sketch the magnitude of the Fourier transform of the function $f(t)$ shown in Fig. P10-25. Indicate the values of frequency at which the amplitude is zero.

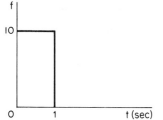

Figure P10-25

Problem 10-26 (Sec. 10-5)

Find the Fourier transform for the exponential pulse $f(t) = F_0 e^{-at}u(t)$ and sketch the magnitude and phase curves for the transform.

Problem 10-27 (Sec. 10-5)

Find the Fourier transform of the triangular pulse $f(t)$ shown in Fig. P10-27 and sketch the magnitude and phase curves for the transform.

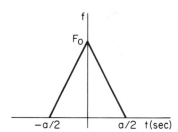

Figure P10-27

Problem 10-28 (Sec. 10-5)

(a) Use the results obtained in Problem 8-15 for $F_1(s)$ to verify the result shown in Eq. (10-136) (and the discussion immediately following).

(b) Repeat for the function $F_3(s)$.

Problem 10-29 (Sec. 10-6)

Use the results of Problem 10-25 to sketch the magnitude of the Fourier transform of the function $f(t)$ shown in Fig. P10-29.

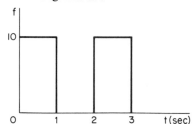

Figure P10-29

The Fourier Series and the Fourier Integral   Chap. 10

Problem 10-30 (Sec. 10-6)

Repeat the problem given in Example 10-6.1 for the case where the input is an exponential pulse similar to the one defined in Problem 10-26.

Problem 10-31 (Sec. 10-6)

Apply the conclusions for amplitude modulation given in Eq. (10-119) to the case where the modulating signal is $f(t) = A_0 \cos \omega_m t$. The result is called a *double-sideband suppressed-carrier* signal. Draw its frequency spectrum.

Problem 10-32 (Sec. 10-6)

In many mathematics texts the Fourier transform is defined using the variable ω rather than $j\omega$. Thus

$$F(\omega) = \int_{-\infty}^{\infty} f(t) \, e^{-j\omega t} \, dt \qquad \text{and} \qquad f(t) = \frac{1}{2\pi} \int_{-\infty}^{\infty} F(\omega) \, e^{j\omega t} \, d\omega$$

In this case the transform pair is symbolized as $f(t) \longleftrightarrow F(\omega)$. Using the defining integrals given above, prove that the following *symmetry relation* holds:

$$2\pi f(-\omega) \longleftrightarrow F(t)$$

Now use the symmetry relation to find the inverse Fourier transforms of the following:

$$(a) e^{-a|\omega|} \ (a > 0); \qquad (b) \ \delta(\omega)$$

Bibliography

INTRODUCTORY CIRCUIT THEORY

The list of books which have been written on the subject of basic circuit theory is a very large one, and the following entries can only be considered as being representative samples. In most cases, however, the titles listed cover approximately the same topics as those covered in this text. Thus, they may be helpful to the student as sources of slightly different viewpoints on the material contained herein. They also include many examples which may be helpful in understanding the basic concepts.

BOBROW, L. S., *Elementary Linear Circuit Analysis*, Holt, Rinehart and Winston, New York, 1981.

BRENNER, E., and M. JAVID, *Analysis of Electric Circuits*, McGraw-Hill Book Company, New York, 1967.

BUDAK, A., *Circuit Theory Fundamentals and Applications*, Prentice-Hall, Inc., Englewood Cliffs, N.J., 1978.

CALAHAN, D. A., A. B. MACNEE, and E. L. McMAHON, *Introduction to Modern Circuit Analysis*, Holt, Rinehart and Winston, New York, 1974.

CHIRLIAN, P. M., *Basic Network Theory*, McGraw-Hill Book Company, New York, 1969.

CLOSE, M., *The Analysis of Linear Circuits*, Harcourt Brace & World, Inc., New York, 1966.

DESOER, C. A., and E. S. KUH, *Basic Circuit Theory*, McGraw-Hill Book Company, New York, 1969.

GUPTA, S. C., J. W. BAYLESS, and B. PEIKARI, *Circuit Analysis with Computer Applications to Problem Solving*, Intext Educational Publishers, Scranton, Pa., 1972.

HAYT, W. H., JR., and J. E. KEMMERLY, *Engineering Circuit Analysis*, Third Edition, McGraw-Hill Book Company, New York, 1978.

HUELSMAN, L. P., *Basic Circuit Theory with Digital Computations*, Prentice-Hall, Inc., Englewood Cliffs, N.J., 1972.

JOHNSON, D. E., J. L. HILBURN, and J. R. JOHNSON, *Basic Electric Circuit Analysis*, Prentice-Hall, Inc., Englewood Cliffs, N.J., 1978.

KARNI, S., *Intermediate Network Analysis*, Allyn and Bacon, Inc., Boston, 1971.

KIM, W. H., and H. E. MEADOWS, JR., *Modern Network Analysis*, John Wiley & Sons, Inc., New York, 1971.

LEON, B. J., and P. A. WINTZ, *Basic Linear Networks for Electrical and Electronics Engineers*, Holt, Rinehart and Winston, New York, 1970.

MURDOCH, J. B., *Network Theory*, McGraw-Hill Book Company, New York, 1970.

PAPOULIS, A., *Circuits and Systems, A Modern Approach*, Holt, Rinehart and Winston, New York, 1980.

SKILLING, H., *Electric Networks*, John Wiley & Sons, Inc., New York, 1974.

VAN VALKENBURG, M. E., *Network Analysis*, Third Edition, Prentice-Hall, Inc., Englewood Cliffs, N.J., 1974.

ACTIVE FILTERS

Due to the many applications for filtering, and the great interest in active-*RC* filters, many books have recently been published in this area. The following list, although not exhaustive, contains many of the most useful works.

BLINCHIKOFF, H. J., and A. I. ZVEREV, *Filtering in the Time and Frequency Domains*, John Wiley & Sons, Inc., New York, 1976.

BOWRON, P., and F. W. STEPHENSON, *Active Filters for Communications and Instrumentation*, McGraw-Hill Book Company (UK) Ltd., London, 1979.

BRUTON, L. T., *RC-Active Circuits, Theory and Design*, Prentice-Hall, Inc., Englewood Cliffs, N.J., 1980.

BUDAK, A., *Passive and Active Network Analysis and Synthesis*, Houghton Mifflin Company, Boston, 1974.

DARYANANI, G., *Principles of Active Network Synthesis and Design*, John Wiley & Sons, Inc., New York, 1976.

HUELSMAN, L. P., and P. E. ALLEN, *Introduction to the Theory and Design of Active Filters*, McGraw-Hill Book Company, New York, 1980.

LAM, H. Y-F., *Analog and Digital Filters, Design and Realization*, Prentice-Hall, Inc., Englewood Cliffs, N.J., 1979.

RHODES, J. D., *Theory of Electrical Filters*, John Wiley & Sons Ltd., London, 1976.

SEDRA, A. S., and P. O. BRACKETT, *Filter Theory and Design: Active and Passive*, Matrix Publishers, Inc., Champaign, Ill., 1978.

TEMES, G. C., and S. K. MITRA, *Modern Filter Theory and Design*, John Wiley & Sons, Inc., New York, 1973.

VAN VALKENBURG, M. E., *Introduction to Modern Network Synthesis*, John Wiley & Sons, Inc., New York, 1960.

VAN VALKENBURG, M. E., *Analog Filter Design*, Holt, Rinehart and Winston, New York, 1982.

APPENDIX *A*

Matrices

and Determinants

~~~~~~

Frequently in network analysis studies we have the need to write sets of simultaneous equations. Such a task can be facilitated by the use of matrices. There are two advantages to such a use:

1. The elements of the set of simultaneous equations are divided so as to emphasize the three different kinds of these elements, namely, the dependent variables, the independent variables, and the coefficients of the equations.
2. A symbolic shorthand notation is provided which may be conveniently used to indicate operations such as substitution of variables and solutions of a set of equations.

In this appendix we shall present a discussion of the general theory of matrices and the rules for manipulating them. The latter subject is frequently referred to as *matrix algebra*.

A-1 DEFINITIONS

A *matrix* is a rectangular array of elements arranged in horizontal and vertical lines. The array is usually enclosed in brackets. The horizontal lines are called *rows* and the vertical lines are called *columns*. The elements of the array may be literal symbols (usually with subscripts), real or complex numbers, or functions of some real or complex variable. If we use a literal symbol to specify a matrix element, it is conventional to use a first subscript to indicate the row in which the element is located and a second subscript to indicate the column. Thus the element a_{ij} is defined as the element located in the ith row and the jth column. Using this notation, we may write the literal form of the general matrix with m rows and n columns as

$$\begin{bmatrix} a_{11} & a_{12} & \cdots & a_{1n} \\ a_{21} & a_{22} & \cdots & a_{2n} \\ \cdot & \cdot & \cdots & \cdot \\ \cdot & \cdot & \cdots & \cdot \\ \cdot & \cdot & \cdots & \cdot \\ a_{m1} & a_{m2} & \cdots & a_{mn} \end{bmatrix} \qquad \text{(A-1)}$$

A first means of characterizing matrices is by the number of rows and columns they contain. To indicate this information we use the term *order*. Thus, a matrix which has m rows and n columns is referred to as a matrix of order (m, n) or, more simply, as an $m \times n$ matrix (note that the number of rows is given first, then the number of columns). Some frequently encountered orders of matrices are usually given special names. For example, a matrix which has only a single column is called a *column matrix* or a *vector*. This notation applies as long as there are two or more rows. Frequently, for such a matrix, only a single subscript is used for the elements of the array. Another type of matrix which is given a special name is one which contains only a single row. This is logically called a *row matrix*. The last special classification given matrices covers the case of a matrix which has the same number of rows and columns. Such a matrix is called a *square matrix*. More specifically, if there are n rows and n columns, we say that such a matrix has order n. Finally, it should be noted that a 1×1 matrix which is comprised of a single element is simply referred to as a *scalar*.

There are several subclasses of square matrices which are frequently encountered. One of these is the *symmetric matrix*. If we use the symbol a_{ij} to refer to the elements of a symmetric matrix, then these elements satisfy the equality

$$a_{ij} = a_{ji} \qquad \text{(A-2)}$$

for all values of i and j. Now let us define the *main diagonal* of a square matrix of order n as the n elements $a_{11}, a_{22}, a_{33}, \ldots, a_{nn}$. From (A-2) we see that a symmetric matrix is one which is symmetrical with respect to the main diagonal. A special case of a symmetric matrix is one in which all the elements except those on the main diagonal are zero. This is called a *diagonal matrix*. If, in addition, the diagonal elements are all unity, the matrix is referred to as an *identity matrix*.

In order to simplify the notation when sets of simultaneous equations are to be considered, it is convenient to use a single symbol to represent a given matrix. In this text we use a boldface capital letter for such a representation. It is usually convenient to make this be the same letter as is used for the individual elements of the array. Thus, if the elements of the array are expressed in literal form as a_{ij}, the matrix is simply referred to as the matrix **A**.

A-2 MATRIX EQUALITY

In order to start our discussion of matrix algebra, we begin by defining a matrix equality. Two matrices are equal if and only if each of the corresponding elements of the matrices is equal. Thus, if **A** and **B** are two matrices with elements a_{ij} and b_{ij}, respectively, $\mathbf{A} = \mathbf{B}$ if and only if $a_{ij} = b_{ij}$ for all values of i and j. As a consequence of this definition, we see that only matrices having the same number of rows and columns can be equated.

A-3 ADDITION AND SUBTRACTION OF MATRICES

The next operation of matrix algebra we shall consider is the summing of two matrices so as to form a third matrix. The sum of two matrices is equal to a third matrix if and only if the sum of the elements in corresponding rows and columns of the two matrices is equal to the value of the element in the same row and column of the third matrix. Thus, if $\mathbf{A}$, $\mathbf{B}$, and $\mathbf{C}$ are matrices with elements a_{ij}, b_{ij}, and c_{ij}, respectively, the relation $\mathbf{A} + \mathbf{B} = \mathbf{C}$ implies that $a_{ij} + b_{ij} = c_{ij}$ for all values of i and j. The operation of addition is clearly defined only for matrices which have the same number of rows and columns. As an example of matrix addition, consider the following.

$$\begin{bmatrix} 1 & 2 & 3 \\ 4 & 5 & 6 \end{bmatrix} + \begin{bmatrix} 2 & 3 & 4 \\ 5 & 6 & 7 \end{bmatrix} = \begin{bmatrix} 3 & 5 & 7 \\ 9 & 11 & 13 \end{bmatrix}$$

Subtraction is similarly defined, i.e., $\mathbf{A} - \mathbf{B} = \mathbf{C}$ implies $a_{ij} - b_{ij} = c_{ij}$ for all values of i and j.

A-4 MULTIPLICATION OF A MATRIX BY A SCALAR

Now let us define the operation of multiplication of an entire matrix by a scalar, i.e., a single number. Such an operation simply multiplies every element of the array by the scalar. Thus, if $\mathbf{A}$ is a matrix with elements a_{ij} and if k is a scalar, the elements of the matrix $k\mathbf{A}$ are ka_{ij}. The term scalar as we use it here may be interpreted as a real number, a complex number, or a function of some variable. As an example, consider the following.

$$\frac{s}{s^2 + 1} \begin{bmatrix} 1 & 2 \\ 3 & 4 \end{bmatrix} = \begin{bmatrix} \dfrac{s}{s^2 + 1} & \dfrac{2s}{s^2 + 1} \\ \dfrac{3s}{s^2 + 1} & \dfrac{4s}{s^2 + 1} \end{bmatrix}$$

It should be noted that this operation is considerably different than the operation of multiplying a determinant by a scalar constant, since the latter operation is equivalent to multiplying only the elements of a single row or a single column by such a constant.

A-5 MULTIPLICATION OF A SQUARE MATRIX BY A COLUMN MATRIX

One of our principal goals in the use of matrices is the representation of sets of simultaneous equations. As an example of this, let us consider the following such set:

$$y_1 = a_{11}x_1 + a_{12}x_2$$
$$y_2 = a_{21}x_1 + a_{22}x_2 \tag{A-3}$$

We will assume that we desire to use the matrix notation

$$\mathbf{Y} = \mathbf{AX} \tag{A-4}$$

to represent such a set of equations, and that the individual matrices $\mathbf{Y}$, $\mathbf{A}$, and $\mathbf{X}$ are defined as

$$\mathbf{Y} = \begin{bmatrix} y_1 \\ y_2 \end{bmatrix} \qquad \mathbf{A} = \begin{bmatrix} a_{11} & a_{12} \\ a_{21} & a_{22} \end{bmatrix} \qquad \mathbf{X} = \begin{bmatrix} x_1 \\ x_2 \end{bmatrix} \tag{A-5}$$

In order to use the notation of (A-4), the multiplication operation indicated as $\mathbf{AX}$ must be defined so as to produce the correct right members for the equations given in (A-3). Thus, we must be able to write

$$\mathbf{Y} = \begin{bmatrix} y_1 \\ y_2 \end{bmatrix} = \begin{bmatrix} a_{11}x_1 + a_{12}x_2 \\ a_{21}x_1 + a_{22}x_2 \end{bmatrix} = \begin{bmatrix} a_{11} & a_{12} \\ a_{21} & a_{22} \end{bmatrix}\begin{bmatrix} x_1 \\ x_2 \end{bmatrix} = \mathbf{AX} \tag{A-6}$$

Now consider the matrix equality relating the two column matrices given as the second and third members of the above equation. For the first element of these two column matrices (the elements in the first row) to be equal requires that

$$y_1 = a_{11}x_1 + a_{12}x_2 \tag{A-7}$$

Note that the right member of this equation may be interpreted as requiring that each element of the first row of the square matrix $\mathbf{A}$ be multiplied by a corresponding element of the column matrix $\mathbf{X}$ and the sum taken of the resulting products. Similarly, for the second element of the two column matrices to be equated requires that

$$y_2 = a_{21}x_1 + a_{22}x_2 \tag{A-8}$$

Thus, in the right member of this expression, we sum the products of the second row of the square matrix $\mathbf{A}$ with the corresponding elements of the column matrix $\mathbf{X}$. This result defines the multiplication of a square matrix of order 2 by a column matrix with two elements. It is easily generalized to the case where $\mathbf{A}$ is an nth-order square matrix and $\mathbf{X}$ and $\mathbf{Y}$ are column matrices with n rows. In such a case, the matrices $\mathbf{Y}$, $\mathbf{A}$, and $\mathbf{X}$ are defined as

$$\mathbf{Y} = \begin{bmatrix} y_1 \\ y_2 \\ \cdot \\ \cdot \\ \cdot \\ y_n \end{bmatrix} \qquad \mathbf{A} = \begin{bmatrix} a_{11} & a_{12} & \cdots & a_{1n} \\ a_{21} & a_{22} & \cdots & a_{2n} \\ \cdot & \cdot & \cdots & \cdot \\ \cdot & \cdot & \cdots & \cdot \\ \cdot & \cdot & \cdots & \cdot \\ a_{n1} & a_{n2} & \cdots & a_{nn} \end{bmatrix} \qquad \mathbf{X} = \begin{bmatrix} x_1 \\ x_2 \\ \cdot \\ \cdot \\ \cdot \\ x_n \end{bmatrix} \tag{A-9}$$

The matrix equation $\mathbf{Y} = \mathbf{AX}$ is now simply defined by the relation

$$y_i = \sum_{k=1}^{n} a_{ik}x_k \qquad i = 1, 2, \ldots, n \tag{A-10}$$

As an example, applying this result to a third-order set of equations, we obtain

$$\mathbf{AX} = \begin{bmatrix} a_{11} & a_{12} & a_{13} \\ a_{21} & a_{22} & a_{23} \\ a_{31} & a_{32} & a_{33} \end{bmatrix}\begin{bmatrix} x_1 \\ x_2 \\ x_3 \end{bmatrix} = \begin{bmatrix} a_{11}x_1 + a_{12}x_2 + a_{13}x_3 \\ a_{21}x_1 + a_{22}x_2 + a_{23}x_3 \\ a_{31}x_1 + a_{32}x_2 + a_{33}x_3 \end{bmatrix} = \begin{bmatrix} y_1 \\ y_2 \\ y_3 \end{bmatrix} = \mathbf{Y} \tag{A-11}$$

A-6 SOME BASIC RULES OF MATRIX ALGEBRA

In this section we present some general basic rules of matrix algebra. In specifying the indicated operations of equality, addition, subtraction, and multiplication, we assume that the order of the component matrices is such that the implied operations are defined as noted in the preceding sections. The proof of these rules is left to the reader as an exercise.

Rule 1   Matrix addition is associative. $\mathbf{A} + (\mathbf{B} + \mathbf{C}) = (\mathbf{A} + \mathbf{B}) + \mathbf{C}$.

Rule 2   Matrix addition is commutative. $\mathbf{A} + \mathbf{B} = \mathbf{B} + \mathbf{A}$.

Rule 3   Matrix multiplication is associative. $\mathbf{AB}(\mathbf{C}) = \mathbf{A}(\mathbf{BC})$.

Rule 4   Matrix multiplication is not, in general, commutative. Except in special cases, $\mathbf{AB} \neq \mathbf{BA}$.

Rule 5   Matrix multiplication is distributive with respect to addition. $\mathbf{A}(\mathbf{B} + \mathbf{C}) = \mathbf{AB} + \mathbf{AC}$.

Rule 6   Any matrix multiplied by an identity matrix equals itself. $\mathbf{1A} = \mathbf{A1} = \mathbf{A}$, where $\mathbf{1}$ is an identity matrix.

A-7 THE INVERSE OF MATRICES

The only major operation found in scalar algebra which has not been defined above for matrix algebra is division. The reason is that *division is not a permitted operation in the algebra of matrices*. Thus, we may *not* use a notation such as $\mathbf{A}/\mathbf{B}$. The operation of matrix algebra which most closely corresponds to the division process of scalar algebra is multiplication by the inverse of a matrix. A treatment of the determination of the inverse of a matrix may be found in Chap. 3.

A-8 DETERMINANTS AND COFACTORS

A *determinant* is a scalar quantity defined in terms of the elements of a square matrix. If the matrix has n rows and n columns, the determinant may be indicated as follows.

$$\det \mathbf{R} = \begin{vmatrix} r_{11} & r_{12} & \cdots & r_{1n} \\ r_{21} & r_{22} & \cdots & r_{2n} \\ \cdot & \cdot & \cdots & \cdot \\ \cdot & \cdot & \cdots & \cdot \\ \cdot & \cdot & \cdots & \cdot \\ r_{n1} & r_{n2} & \cdots & r_{nn} \end{vmatrix} \tag{A-12}$$

This is frequently referred to as a *determinant of order n*. Such a determinant has a definite (scalar) value which is determined by the elements r_{ij}. For low-order determinants, the value of the determinant may be found directly by simple rules. For example, for a second-order determinant, it is well-known that

$$\det \mathbf{R} = \begin{vmatrix} r_{11} & r_{12} \\ r_{21} & r_{22} \end{vmatrix} = r_{11}r_{22} - r_{12}r_{21} \tag{A-13}$$

The equation for evaluating the determinant of the second-order array given in Eq. (A-13) may be remembered by thinking of multiplying the elements on the *main diagonal* (the elements placed along a diagonal line from the upper left corner of the array to the lower right corner), and subtracting the product of the terms on the other diagonal. A similar procedure may be applied to determine the determinant of a third-order array. This is shown below. For convenience, the first and second columns have been repeated at the right of the array.

$$(\text{A-14})$$

$$\det \mathbf{R} = r_{11}r_{22}r_{33} + r_{12}r_{23}r_{31} + r_{13}r_{21}r_{32} - r_{13}r_{22}r_{31} - r_{11}r_{23}r_{32} - r_{12}r_{21}r_{33}$$

An example follows.

Example A-1

The rule given above is easily applied to find the determinant of the matrix for the three-mesh resistance network analyzed in Example 3-3.1. The resistance matrix $\mathbf{R}$ is repeated here for convenience.

$$\mathbf{R} = \begin{bmatrix} 3 & -2 & -1 \\ -2 & 5 & -3 \\ -1 & -3 & 8 \end{bmatrix}$$

Following the rule for the evaluation of the determinant given in Eq. (A-14), we obtain

$$\det \mathbf{R} = (3 \times 5 \times 8) + (-2 \times -3 \times -1) + (-1 \times -2 \times -3)$$
$$-(-1 \times 5 \times -1) - (3 \times -3 \times -3) - (-2 \times -2 \times 8)$$
$$= 44$$

A more general procedure for evaluating the third-order determinant is by *expanding* the determinant. Such an expansion is made in terms of the elements of any row or column and a set of second-order determinants known as cofactors. For example, expanding the determinant along the first row, we obtain

$$\det \mathbf{R} = r_{11}\begin{vmatrix} r_{22} & r_{23} \\ r_{32} & r_{33} \end{vmatrix} - r_{12}\begin{vmatrix} r_{21} & r_{23} \\ r_{31} & r_{33} \end{vmatrix} + r_{13}\begin{vmatrix} r_{21} & r_{22} \\ r_{31} & r_{32} \end{vmatrix} \qquad (\text{A-15})$$

Thus, we have expressed the determinant of a third-order array in terms of the determinants of second-order arrays. A general procedure for expressing the determinant of an nth-order array in terms of the determinants of $(n - 1)$-order arrays is called *Laplace's expansion*. If such an expansion is made along the ith row of an array, it has the form

$$\det \mathbf{R} = \sum_{k=1}^{n} r_{ik}R_{ik} \qquad (\text{A-16})$$

where the r_{ik} are the elements of $\mathbf{R}$ and the R_{ik} are *cofactors*. These cofactors are formed by deleting the ith row and kth column of the array, taking the determinant of the remaining elements (a determinent of order $n - 1$), and prefixing the result by the positive or negative multiplier determined by the expression $(-1)^{i+k}$.[1] For example, for a third-order array with elements, r_{ij}, if we expand along the first row, then we see from Eq. (A-16) that the cofactors R_{11}, R_{12}, and R_{13} are required. These are determined as

$$R_{11} = + \begin{vmatrix} r_{22} & r_{23} \\ r_{32} & r_{33} \end{vmatrix} \qquad R_{12} = - \begin{vmatrix} r_{21} & r_{23} \\ r_{31} & r_{33} \end{vmatrix} \qquad R_{13} = + \begin{vmatrix} r_{21} & r_{22} \\ r_{31} & r_{32} \end{vmatrix}$$

It is readily apparent that these are the cofactors which were used in Eq. (A-15).

By the use of Eq. (A-16) we can express the determinant of an nth order array as a function of determinants of $(n - 1)$-order arrays. These, in turn, may be expressed as a function of the determinants of $(n - 2)$-order arrays, and the process may be continued until the value of the determinant is obtained. Such a process becomes quite tedious (as well as very liable to error) for arrays higher than third degree. In such cases, we are led to seek other methods of solving sets of simultaneous equations which do not require the use of determinants. A method which is readily implemented on the digital computer is the Gauss-Jordan method. This method is used in the GJSEQ subroutine described in Appendix C.

A Laplace expansion may also be made along a *column* of a given array. For example, to evaluate the determinant in terms of the elements of the jth column, we may use the expression

$$\det \mathbf{R} = \sum_{k=1}^{n} r_{kj} R_{kj} \tag{A-17}$$

where the R_{kj} are cofactors as previously defined.

---

[1] The determinant of order $n - 1$ is called a first-order minor.

Complex Numbers

In network studies involving sinusoidal steady-state analysis, we must use numbers and variables which are complex rather than real. The rules for manipulating such complex quantities, however, are quite different than the corresponding rules for manipulating real quantities. In this appendix we will present a summary of such rules. Taken together, these rules define an *algebra for complex numbers.*

B-1 DEFINITIONS

A complex number a may be defined as an ordered pair of two real numbers (a_1, a_2). The two real numbers are referred to as the real part and the imaginary part, respectively, of the complex number a. Thus, we may write

$$a = a_1 + ja_2$$
$$a_1 = \text{Re}\,(a) \qquad\qquad\qquad \text{(B-1)}$$
$$a_2 = \text{Im}\,(a)$$

where Re stands for "the real part of," Im stands for "the imaginary part of," and j has the value $\sqrt{-1}$. A convenient graphical representation for such a complex number is obtained by defining a complex plane in which real quantities are plotted along the abscissa and imaginary quantities are plotted along the ordinate as shown in Fig. B-1.1(a). Such a representation is referred to as the rectangular form of a complex number. Using this rectangular form as a starting point, we may develop a second representation for a complex number. This is obtained by drawing a vector from the origin of the complex plane to the point a as shown in Fig. B-1.1(b) and using this

Figure B-1.1 Representation of a complex number.

(a) (b)

vector to represent the complex number. In terms of such a representation, we may define a complex number a as follows.

$$a = Ae^{j\alpha}$$

$$A = |a| \tag{B-2}$$

$$\alpha = \text{Arg}\,(a)$$

Thus, we may consider a complex number as a vector, and we may use the magnitude and the argument of the vector to define the complex number. Such a representation is referred to as the polar or exponential form of a complex number. Although the angle (or argument) α defined in Eq. (B-2) is given in radians, it is frequently convenient to use angle measure in degrees. In such a case, we may write the polar form of a complex number a as

$$a = Ae^{j\phi^\circ}$$

where φ° is the angle α given in degrees. The relations between the rectangular and polar forms are easily found by comparing the representations given in Fig. B-1.1. Thus, from (B-1) and (B-2) we see that

$$a_1 = A\cos\alpha \qquad A = \sqrt{a_1^2 + a_2^2}$$

$$a_2 = A\sin\alpha \qquad \alpha = \tan^{-1}\left(\frac{a_2}{a_1}\right) \tag{B-3}$$

Inserting the relations for a_1 and a_2 given in (B-1), we obtain the relation

$$e^{j\alpha} = \cos\alpha + j\sin\alpha \tag{B-4a}$$

Similarly, we may show that

$$e^{-j\alpha} = \cos\alpha - j\sin\alpha \tag{B-4b}$$

These are frequently referred to as *Euler's identity*.

A concept that arises frequently in the application of complex numbers is the concept of the *complex conjugate* of such a number. If $a = a_1 + ja_2$ is a complex number, then we may designate a^* as its complex conjugate. We may define a^* by the relation

$$a^* = a_1 - ja_2 \tag{B-5}$$

Thus, we see that the complex conjugate of a given complex number is formed by reversing the sign of its imaginary part. From (B-3) we see that this is equivalent to reversing the sign of the phase angle. Thus, if $a = Ae^{j\alpha}$, then a^* (the complex conjugate of a) is given as

$$a^* = Ae^{-j\alpha} \tag{B-6}$$

B-2 ADDITION AND SUBTRACTION

If a and b are complex numbers defined as $a = a_1 + ja_2$ and $b = b_1 + jb_2$, then the sum and difference of a and b are defined by the relations

$$a + b = (a_1 + b_1) + j(a_2 + b_2)$$
$$a - b = (a_1 - b_1) + j(a_2 - b_2)$$

(B-7)

In other words, complex numbers are added (or subtracted) by separately adding (or subtracting) their real and imaginary parts. In general, it is much more convenient to express complex numbers using the rectangular form if addition is to be performed. As examples of adding and subtracting two complex numbers, consider the following.

$$(1 + j3) + (2 - j4) = 3 - j1$$
$$(1 + j3) - (2 - j4) = -1 + j7$$

If is readily shown that the operations of addition and subtraction are commutative and associative. Thus, if a, b, and c are complex numbers, we may write

$$a + b = b + a$$
$$a + (b + c) = (a + b) + c$$

(B-8)

B-3 MULTIPLICATION AND DIVISION

As a preliminary result which will be useful in studying the multiplication and division of complex numbers, let us consider the case where a complex number $a = a_1 + ja_2$ is multiplied by a real number k. This operation is defined as

$$ka = k(a_1 + ja_2) = ka_1 + jka_2$$

(B-9)

Thus, we see that in such a case the real number separately multiplies the real and the imaginary part of the complex number. Similarly, division of a complex number $a = a_1 + ja_2$ by a real number k may be defined as

$$\frac{a}{k} = \frac{a_1 + ja_2}{k} = \frac{a_1}{k} + j\frac{a_2}{k}$$

(B-10)

It is readily shown that these operations are commutative and associative. Thus, if a is a complex number and k and m are real numbers, we may write

$$(km)a = k(ma) = m(ka)$$

(B-11)

Now let us consider the product of two complex numbers $a = a_1 + ja_2$ and $b = b_1 + jb_2$. If we separately multiply each of the terms of these two complex numbers, and note that $j^2 = -1$, we obtain

$$ab = (a_1 b_1 - a_2 b_2) + j(a_1 b_2 + a_2 b_1)$$

(B-12)

Let us again perform this operation using polar notation. For $a = Ae^{j\alpha}$ and $b = Be^{j\beta}$ we obtain

$$ab = AB\, e^{j(\alpha + \beta)}$$

(B-13)

where the last result is obtained by applying Euler's identity as given in (B-4). Thus, we see that the product of two complex numbers which are expressed in polar form is found by multiplying the magnitudes and adding the arguments of the two numbers. Obviously, it is much more convenient to express complex numbers in polar form if it is desired to multiply these numbers. As an example of the multiplication of two complex numbers, consider the following:

$$(1 + j1)(-2 + j2) = \sqrt{2}\, e^{j45°}2\sqrt{2}\, e^{j135°} = 4e^{j180°} = -4 + j0$$

As an application of the multiplication of two complex numbers, let us see what happens when we multiply a number by its complex conjugate. Let $a = Ae^{j\alpha}$ be the number. Applying the results given above, we obtain

$$aa^* = Ae^{j\alpha}\, Ae^{-j\alpha} = A^2 e^{j0} = A^2 \tag{B-14}$$

We thus conclude that the product of a number and its complex conjugate yields a real number equal to the square of the magnitude of the complex number.

It is readily shown that multiplication of complex numbers is commutative and associative. Thus if a, b, and c are complex numbers, we may write

$$ab = ba$$
$$a(bc) = (ab)c \tag{B-15}$$

Similarly, multiplication of complex numbers is distributive, i.e.,

$$a(b + c) = ab + ac \tag{B-16}$$

Now let us consider the division of two complex numbers $a = a_1 + ja_2$ and $b = b_1 + jb_2$. If the numerator and denominator of the quotient a/b are first multiplied by the complex conjugate of the denominator, the quotient may be expressed as a multiplication of two complex numbers and a division by a real number. Thus, we may write

$$\frac{a}{b} = \frac{ab^*}{bb^*} = \frac{(a_1 + ja_2)(b_1 - jb_2)}{(b_1 + jb_2)(b_1 - jb_2)} = \frac{(a_1 b_1 + a_2 b_2) + j(a_2 b_1 - a_1 b_2)}{b_1^2 + b_2^2} \tag{B-17}$$

This result may also be expressed in polar form as

$$\frac{a}{b} = \frac{Ae^{j\alpha}}{Be^{j\beta}} = \frac{A}{B} e^{j(\alpha - \beta)} \tag{B-18}$$

Thus, we see that the quotient of two complex numbers is simply equal to the quotient of their magnitudes and the difference of their arguments. Obviously, the division of two complex numbers is most conveniently performed if the numbers are first expressed in polar form. As an example of the division of two complex numbers, consider the following.

$$\frac{(1 - j1)}{(2 + j2)} = \frac{\sqrt{2}\, e^{-j45°}}{2\sqrt{2}\, e^{j45°}} = \frac{1}{2} e^{-j90°} = 0 - j\frac{1}{2}$$

B-4 POWERS AND ROOTS

To raise a complex number to a power, we need simply use the polar representation for a complex number and apply the usual rule of exponents. Thus, if $a = Ae^{j\alpha}$ is a complex number, raising it to the nth power ($n = 1, 2, 3, \ldots$) is done as follows:

$$a^n = (Ae^{j\alpha})^n = A^n e^{jn\alpha} \tag{B-19}$$

From the above equation, we see that raising a complex number to a power is done by raising the magnitude to the specified power and multiplying the argument by the value of the power. As an example of this, consider the following:

$$(1 + j1)^3 = (\sqrt{2}\, e^{j45°})^3 = 2\sqrt{2}\, e^{j135°} = -2 + j2$$

Now let us consider taking the root of a complex number. Again we will use a polar representation for the number. We first note that adding or subtracting any multiple of 2π radians to the argument of a complex number does not change its value. Thus, if $a = Ae^{j\alpha}$ is a complex number, we may write

$$a = Ae^{j\alpha} = Ae^{j(\alpha+2\pi)} = Ae^{j(\alpha+4\pi)} = \cdots = Ae^{j(\alpha+2k\pi)} \qquad \text{(B-20)}$$

where, in the right member, k is a positive or negative integer. We may now apply the rule of exponents to find $a^{1/m}$. Using the expression given in the right member of (B-20), we obtain

$$a^{1/m} = A^{1/m}e^{j(\alpha/m + 2k\pi/m)} \qquad \text{(B-21)}$$

If we investigate this result for values of k equal to $0, 1, 2, \ldots, m - 1$, we find that there are exactly m unique values of the argument. Thus, there will be m roots whose locus is a circle of radius $A^{1/m}$ centered at the origin. As an example of this conclusion, consider the determination of the cube root of unity, which in complex form may be written $1 + j0$. Applying (B-20), we obtain three different polar representations for $1 + j0$. These are

$$(1 + j0) = 1e^{j0} = 1e^{j2\pi} = 1e^{j4\pi}$$

Now let $m = 3$ in (B-21). Inserting the values for $1 + j0$ given above, we obtain

$$(1 + j0)^{1/3} = (1e^{j0})^{1/3} = 1e^{j0} = (1 + j0)$$
$$(1 + j0)^{1/3} = (1e^{j2\pi})^{1/3} = 1e^{j2\pi/3} = -0.5 + j0.866$$
$$(1 + j0)^{1/3} = (1e^{j4\pi})^{1/3} = 1e^{j4\pi/3} = -0.5 - j0.866$$

The location of the three distinct cube roots of unity are shown in Fig. B-4.1. It is readily verified that cubing any of these complex numbers produces unity.

Figure B-4.1   Cube roots of unity.

Digital Computer Subroutines for Circuit Analysis

∿∿∿∿∿∿∿∿∿∿∿∿∿∿∿∿∿∿∿∿∿∿∿∿∿∿∿∿∿∿∿∿∿∿∿∿∿

In this appendix, summaries and listings for some FORTRAN subroutines are given. These subroutines are readily applied to extend the range of the circuit analysis techniques developed in this text. All that is required is for the user to add relatively simple main programs.[1] In general, these subroutines have been designed with the minimum level of sophistication necessary to effectively perform their function. The obvious advantage of such an approach is the minimization of computer core requirements as well as the reduction of compilation and execution times. The student or the professor who wishes a higher level of performance will find that these programs form an excellent basis upon which embellishments may be readily added.

C-1 THE SUBROUTINE GJSEQ

This subroutine is used to solve a set of up to ten simultaneous equations. It can readily be applied to the solution of resistance networks on either a mesh or a node basis.

Identifying Statement: SUBROUTINE GJSEQ (V, R, AMPS, N)

Purpose: To solve a set of simultaneous equations having the form

$$\mathbf{v} = \mathbf{Ri}$$

i.e., given the elements of the matrices $\mathbf{v}$ and $\mathbf{R}$, to find the elements of $\mathbf{i}$.

Additional Subprograms Required: None.

[1] Some examples of such main programs as well as additional information concerning the operation of the subroutines may be found in L. P. Huelsman, *Basic Circuit Theory with Digital Computations*, Prentice-Hall, Inc., Englewood Cliffs, N.J., 1972.

Input Arguments:

    V   the one-dimensional array of variables V(I) in which are stored the values of the known quantities v_i, the elements of the matrix **v**

    R   the two-dimensional array of variables R(I, J) in which are stored the values of the coefficients r_{ij} of the square matrix **R**

    N   the order of the matrix **R**, i.e., the number of simultaneous equations to be solved

Output Argument:

    AMPS   the one-dimensional array of variables AMPS(I) in which are stored the values of the unknown variables i_i of the matrix **i**

Notes:

1. The dimensions of the variables used in this subroutine are V(10), R(10, 10), and AMPS(10).
2. The subroutine uses a Gauss-Jordan reduction algorithm. The original **R** matrix is preserved.

```
      SUBROUTINE GJSEQ (V, R, AMPS, N)
C     SUBROUTINE TO SOLVE A SET OF SIMULTANEOUS EQUATIONS V=R*AMPS
C     USING GAUSS-JORDAN REDUCTION. THE R MATRIX IS PRESERVED
C            V - COLUMN MATRIX OF KNOWN VARIABLES
C            R - SQUARE NON-SINGULAR MATRIX
C         AMPS - OUTPUT COLUMN MATRIX
C            N - NUMBER OF EQUATIONS
      DIMENSION V(10), R(10,10), AMPS(10), RA(10,11)
C
C     ENTER R MATRIX IN RA ARRAY AND ENTER V AS
C     N+1TH COLUMN OF RA ARRAY
      DO 23  I = 1, N
      DO 22  J = 1, N
   22 RA(I,J) = R(I,J)
   23 RA(I,N+1) = V(I)
C
C     PORTION OF PROGRAM TO REDUCE AUGMENTED
C     MATRIX RA TO SOLVE A SET OF N SIMULTANEOUS EQUATIONS.
C     NP IS NO. OF COLUMNS IN AUGMENTED MATRIX,
C     ALFA AND BETA ARE TEMPORARY STORAGE
      NP = N + 1
      DO 12  I = 1, N
C
C     SET MAIN DIAGONAL ELEMENTS TO UNITY.
      ALFA = RA(I,I)
      DO  5  J = I, NP
    5 RA(I,J) = RA(I,J) / ALFA
C
C     SET ELEMENTS OF ITH COLUMN TO ZERO
      DO 11  K = 1, N
      IF (K - I) 8, 11, 8
    8 BETA = RA(K,I)
      DO 10  J = I, NP
   10 RA(K,J) = RA(K,J) - BETA * RA(I,J)
   11 CONTINUE
   12 CONTINUE
C
C     SET OUTPUT MATRIX AMPS EQUAL TO LAST COLUMN OF
C     AUGMENTED MATRIX RA
      DO 31  I = 1, N
   31 AMPS(I) = RA(I,NP)
   32 RETURN
      END
```

C-2 THE SUBROUTINE MXRK4

This subroutine is used to provide numerical integration of a set of up to ten simultaneous first-order differential equations. The equations can be non-linear or time-varying as well as linear and time-invariant.

Identifying Statement: SUBROUTINE MXRK4 (TI, AI, NA, TSTOP, ITER, A)

Purpose: To solve the first-order matrix differential equation

$$\mathbf{a}' = \mathbf{g}[t, \mathbf{a}(t)]$$

and thus to find the elements of the column matrix $\mathbf{a}(t)$ at the value of t specified as t_{stop}, starting from t_i, a known value of t, and $\mathbf{a}_i$, where $\mathbf{a}_i = \mathbf{a}(t_i)$.

Input Arguments:

TI   the initial value t_i of the independent variable t

AI   the one-dimensional array of variables AI(I) in which are stored the initial values of the variables of the first-order differential equation, i.e., the elements of the column matrix $\mathbf{a}_i$, where $\mathbf{a}_i = \mathbf{a}(t_i)$

NA   the number of elements in the column matrix, $\mathbf{a}(t)$, i.e., the number of variables

TSTOP   the final value t_{stop} of t for which $\mathbf{a}(t)$ is to be found

ITER   the number of iterations used by the subroutine in going from t_i to t_{stop}

Output Argument:

A   the one-dimensional array of variables A(I) in which are stored the values of the first-order matrix differential equation evaluated at t_{stop}, i.e., the elements of the column matrix $\mathbf{a}(t_{\text{stop}})$

Additional Subprograms Required: This subroutine calls the subroutine identified by the statement

SUBROUTINE GN (T, A, G)

which must be used to define the differential equation, i.e., to specify $\mathbf{g}[t, \mathbf{a}(t)]$. The arguments A and G in the subroutine are both one-dimensional arrays.

Input Arguments:

T   the time variable t

A   the one-dimensional array of variables A(I) representing the variables a_i in the state variable equations

Output Argument:

G   the one-dimensional array of variables G(I) specifying the individual differential equations, e.g., G(1) gives the value of the equation for da_1/dt, G(2) gives the equation for da_2/dt, etc.

Notes:

1. The dimensions of the variables used in these subroutines are AI(10), A(10), and G(10).
2. The subroutine uses a fourth-order Runge-Kutta predictor method.

```
      SUBROUTINE MXRK4 (TI, AINIT, NA, TSTOP, ITER, A)
C     SUBROUTINE FOR SOLVING A FIRST-ORDER MATRIX
C     DIFFERENTIAL EQUATION BY A FOURTH-ORDER
```

```
C       RUNGE-KUTTA METHOD.  A SUBROUTINE GN(T,A,G)
C       MUST BE PROVIDED TO DEFINE THE PROBLEM.
C          TI - INITIAL VALUE OF T
C       AINIT - ARRAY OF INITIAL VALUES OF A(I)
C          NA - NUMBER OF VARIABLES A(I)
C       TSTOP - FINAL VALUE OF T
C        ITER - NUMBER OF ITERATIONS BETWEEN TI
C                 AND TSTOP
C           A - ARRAY OF FINAL VALUES OF A(I)
        DIMENSION AINIT(30), A(30), ATEMP(30),
       1G1(10), G2(10), G3(10), G4(10)
C
C       COMPUTE DT, DT/2, AND DT/6, AND INITIALIZE T
        HITER = ITER
        DT = (TSTOP - TI) / HITER
        DT2 = DT / 2.
        DT6 = DT / 6.
        T = TI
C
C       INITIALIZE THE ARRAY OF VARIABLES A(I)
        DO 7 I = 1, NA
      7 A(I) = AINIT(I)
C
C       ENTER A DO LOOP TO COMPUTE ITER INTERMEDIATE
C       VALUES OF THE VARIABLES A(I)
        DO 23  J = 1, ITER
C
C       COMPUTE THE INTERMEDIATE ARRAYS G1, G2, G3,
C       AND G4, USING THE APPROPRIATE VALUES OF T
        CALL GN(T, A, G1)
        DO 11  I = 1, NA
     11 ATEMP(I) = A(I) + G1(I) * DT2
        T = T + DT2
        CALL GN(T, ATEMP, G2)
        DO 15  I = 1, NA
     15 ATEMP(I) = A(I) + G2(I) * DT2
        CALL GN(T, ATEMP, G3)
        DO 18  I = 1, NA
     18 ATEMP(I) = A(I) + G3(I) * DT
        T = T + DT2
        CALL GN(T, ATEMP, G4)
C
C       COMPUTE THE VALUES OF THE VARIABLES A(I) AT
C       THE JTH INTERMEDIATE POINT
        DO 22  I = 1, NA
     22 A(I) = A(I) + (G1(I)+2.*(G2(I)+G3(I))+G4(I)) * DT6
     23 CONTINUE
        RETURN
        END
```

C-3 THE SUBROUTINE CGJSEQ

This subroutine is used to solve a set of up to ten simultaneous equations with complex coefficients. It can readily be applied to the solution of networks defined under conditions of sinusoidal steady-state (phasors) on either a mesh impedance basis or a node admittance basis.

Identifying Statement: SUBROUTINE CGJSEQ (V, Z, A, N)

Purpose: To solve a set of simultaneous equations having the form

$$\mathbf{V} = \mathbf{ZA}$$

i.e., given the elements of the matrices $\mathbf{V}$ and $\mathbf{Z}$, to find the elements of $\mathbf{A}$. The quantities v_i, the elements of $\mathbf{V}$, z_{ij}, the elements of $\mathbf{Z}$, and a_i, the elements of $\mathbf{A}$ are specified as complex quantities. Thus, this subroutine is ideally suited to application to phasor problems in ac network theory.

Additional Subprograms Required: None.

Input Arguments:

 V the one-dimensional complex array of variables V(I) in which are stored the complex values of the known quantities v_i, the elements of the matrix $\mathbf{V}$; for sinusoidal steady-state phasor problems, these variables represent the sum of the phasor output voltages of the voltage sources located in the *i*th mesh of the network

 Z the two-dimensional complex array of variables Z(I, J) in which are stored the complex values of the coefficients z_{ij} of the square matrix $\mathbf{Z}$; for sinusoidal steady-state phasor problems, these quantities represent the values of the self and mutual complex impedances

 N the order of the matrix $\mathbf{Z}$, i.e., the number of simultaneous equations to be solved

Output Argument:

 A the one-dimensional complex array of variables A(I) in which are stored the complex values of the unknown variables a_i of the matrix $\mathbf{A}$; for sinusoidal steady-state phasor problems, these variables represent the phasor values of the mesh current in the *i*th mesh

Notes:

1. The dimensions of the variables used in this subroutine are V(10), Z(10, 10), and A(10).
2. The subroutine uses a Gauss-Jordan reduction algorithm similar to that used in the subroutine GJSEQ. The original $\mathbf{Z}$ matrix is preserved.
3. If the algorithm fails, i.e., if an attempt is made to divide the elements of any row by a very small number, an error message is printed.

```
      SUBROUTINE CGJSEQ (V, Z, AMPS, N)
C     SUBROUTINE TO SOLVE THE SET OF SIMULTANEOUS EQUATIONS V=Z*AMPS
C     SUBROUTINE USES GAUSS-JORDAN REDUCTION. Z MATRIX IS PRESERVED
C           V - COLUMN MATRIX OF KNOWN COMPLEX VARIABLES
C           Z - SQUARE NON-SINGULAR COMPLEX MATRIX
C        AMPS - OUTPUT COMPLEX COLUMN MATRIX
C           N - NUMBER OF EQUATIONS
      COMPLEX V(10), Z(10,10), AMPS(10), ZA(10,11), ALFA, BETA
C
C     ENTER Z MATRIX IN ZA ARRAY AND ENTER V AS
C     N+1TH COLUMN OF ZA ARRAY
      DO 23  I = 1, N
      DO 22  J = 1, N
   22 ZA(I,J) = Z(I,J)
   23 ZA(I,N+1) = V(I)
C
C     REDUCE MATRIX ZA TO SOLVE SIMULTANEOUS EQUATIONS.
```

```
C      NP IS NUMBER OF COLUMNS IN AUGMENTED MATRIX
       NP = N + 1
       DO 12  I = 1, N
C
C      SET MAIN DIAGONAL ELEMENTS TO UNITY, AFTER
C      TESTING DIAGONAL ELEMENT FOR ZERO VALUE
       ALFA = ZA(I,I)
       IF (CABS(ALFA).LT.0.0001) GO TO 33
       DO  5  J = I, NP
     5 ZA(I,J) = ZA(I,J) / ALFA
C
C      SET ELEMENTS OF ITH COLUMN TO ZERO
       DO 11  K = 1, N
       IF (K - I) 8, 11, 8
     8 BETA = ZA(K,I)
       DO 10  J = I, NP
    10 ZA(K,J) = ZA(K,J) - BETA * ZA(I,J)
    11 CONTINUE
    12 CONTINUE
C
C      SET OUTPUT MATRIX AMPS EQUAL TO LAST COLUMN OF MATRIX ZA
       DO 31  I = 1, N
    31 AMPS(I) = ZA(I,NP)
       RETURN
C
C      PRINT ERROR MESSAGE IF METHOD FAILS
    33 PRINT 34
    34 FORMAT (1H0, 29HMAIN DIAGONAL ELEMENT IS ZERO)
       RETURN
       END
```

C-4 THE SUBROUTINE SSS

This subroutine is used to find the values of the magnitude and phase (or the real part and imaginary part) of a network function $N(j\omega)$ (consisting of a ratio of polynomials with real coefficients) at a specific value of ω. As such, it may be called repeatedly, using a sequence of values of ω, to generate frequency response data. It calls the subroutine CVALPL described in Sec. C-5.

Identifying Statement: SUBROUTINE SSS (M, A, N, B, OMEGA, VREAL, VIMAG, VPHASE, VMAG)

Purpose: To determine the real part, imaginary part, phase, and magnitude of a network function $N(j\omega)$ having the form

$$N(j\omega) = \frac{A(j\omega)}{B(j\omega)} = \frac{a_1 + a_2 j\omega + a_3 (j\omega)^2 + \cdots + a_{m+1}(j\omega)^m}{b_1 + b_2 j\omega + b_3 (j\omega)^2 + \cdots + b_{n+1}(j\omega)^n}$$

which determines the ratio of a response phasor to an excitation phasor under sinusoidal steady-state conditions.

Additional Subprograms Required: This subroutine calls the subroutine CVALPL.

Input Arguments:

M	the degree m of the numerator polynomial $A(j\omega)$
A	the one-dimensional array of variables A(I) in which are stored the coefficients a_i of the numerator polynomial $A(j\omega)$ of the rational function $N(j\omega)$

	N	the degree n of the denominator polynomial $B(j\omega)$
	B	the one-dimensional array of variables B(I) in which are stored the coefficients b_i of the denominator polynomial $B(j\omega)$ of the rational function $N(j\omega)$
	OMEGA	the frequency ω in radians per second at which the polynomial is to be evaluated

Output Arguments:

	VREAL	the value of the real part of $N(j\omega)$
	VIMAG	the value of the imaginary part of $N(j\omega)$
	VPHASE	the value (in degrees) of the phase of $N(j\omega)$
	VMAG	the value of the magnitude of $N(j\omega)$

Notes:

1. If $N(j\omega)$ is infinite, a statement to that effect is printed and any previously computed values of VREAL, VIMAG, VPHASE, and VMAG are given as output.
2. The variables of the subroutine are dimensioned as follows: A(10), B(10).

```
      SUBROUTINE SSS(M, A, N, B, OMEGA, VREAL, VIMAG, VPHASE, VMAG)
C     SUBROUTINE FOR EVALUATING A NETWORK FUNCTION GIVING THE
C     SINUSOIDAL STEADY-STATE RELATION BETWEEN AN INPUT PHASOR
C     AND A RESPONSE PHASOR
C           M - DEGREE OF NUMERATOR POLYNOMIAL A(JW)
C           A - ARRAY OF NUMERATOR POLYNOMIAL COEFFICIENTS
C           N - DEGREE OF DENOMINATOR POLYNOMIAL B(JW)
C           B - ARRAY OF DENOMINATOR POLYNOMIAL COEFFICIENTS
C       OMEGA - VALUE OF W AT WHICH THE NETWORK FUNCTION
C               IS TO BE EVALUATED
C       VREAL - THE REAL PART OF THE NETWORK FUNCTION
C       VIMAG - THE IMAGINARY PART OF THE NETWORK FUNCTION
C      VPHASE - THE PHASE OF THE NETWORK FUNCTION (IN DEGREES)
C        VMAG - THE MAGNITUDE OF THE NETWORK FUNCTION
      DIMENSION A(10), B(10)
      COMPLEX P, VA, VB, V
C
C     DEFINE A COMPLEX VARIABLE WHOSE IMAGINARY PART IS OMEGA
      P = CMPLX(0.0, OMEGA)
C
C     CALL THE SUBROUTINE CVALPL TO EVALUATE THE NUMERATOR
C     AND DENOMINATOR POLYNOMIALS
      CALL CVALPL(M, A, P, VA)
      CALL CVALPL(N, B, P, VB)
C
C     TEST TO MAKE CERTAIN THAT THE DENOMINATOR IS NOT ZERO
      IF (CABS(VB).LE.1.E-08) GO TO 12
C
C     COMPUTE THE VARIOUS OUTPUT QUANTITIES
      V = VA / VB
      VREAL = REAL(V)
      VIMAG = AIMAG(V)
      VMAG = CABS(V)
      IF (VMAG.LT.1.E-08) GO TO 15
      VPHASE = ATAN2(VIMAG, VREAL) * 57.2958
      RETURN
   12 PRINT 13, OMEGA
   13 FORMAT (31HOMAGNITUDE IS INFINITE, OMEGA =,E10.3)
      RETURN
```

```
15 PRINT 16, OMEGA
16 FORMAT (32HOPHASE IS INDETERMINATE, OMEGA =,E10.3/)
   RETURN
   END
```

C-5 THE SUBROUTINE CVALPL

This subroutine is used to find the complex value of a polynomial $A(s)$ (with real coefficients) when it is evaluated at some complex value of the variable s. The subroutine is called by the subroutine SSS described in Sec. C-4.

Identifying Statement: SUBROUTINE CVALPL (M, A, P, VALUE)

Purpose: To determine the complex value of a polynomial $A(s)$ having the form

$$A(s) = a_1 + a_2 s + a_3 s^2 + \cdots + a_m s^{m-1} + a_{m+1} s^m$$

where the argument s is set to some complex value p_0, i.e., to find $A(p_0)$.

Additional Subroutines Required: None.

Input Arguments:

    M   the degree m of the polynomial $A(s)$

    A   the one-dimensional array of variables A(I) in which are stored the values of the coefficients a_i of the polynomial $A(s)$

    P   the complex value p_0 of the variable s at which the polynomial $A(s)$ is to be evaluated

Output Argument:

    VALUE   the value of the polynomial $A(s)$ when s is set equal to the complex value p_0

Note: The array A of the subroutine is dimensioned A(10).

```
      SUBROUTINE CVALPL (M, A, P, VALUE)
C     SUBROUTINE FOR DETERMINING THE VALUE OF A POLYNOMIAL
C     A(1) + A(2)*P + A(3)*P**2 + ... + A(M+1)*P**M
C     FOR A COMPLEX VALUE OF ITS VARIABLE
C         M - DEGREE OF POLYNOMIAL
C         A - ARRAY OF POLYNOMIAL COEFFICIENTS
C               (IN ASCENDING ORDER)
C         P - COMPLEX VALUE OF FREQUENCY VARIABLE
C     VALUE - OUTPUT COMPLEX VALUE OF THE POLYNOMIAL
      DIMENSION A(10)
      COMPLEX B(10), P, VALUE
      MC = M + 1
C
C     STORE COEFFICIENTS AS COMPLEX VARIABLES
      DO 3 I = 1, MC
3 B(I) = CMPLX(A(I), 0.)
C
C     STORE VALUE OF HIGHEST DEGREE COEFFICIENT
      VALUE = B(MC)
      IF (M.LE.0) RETURN
C
C     ACCUMULATE EFFECT OF OTHER COEFFICIENTS
      DO 8 I = 1, M
      K = MC - I
8 VALUE = VALUE * P + B(K)
      RETURN
      END
```

C-6 THE FUNCTION XL

This function is used to provide a logarithmically spaced set of values of a frequency variable. Such a set of values is required when computing the frequency response data for a Bode plot.

Identifying Statement: FUNCTION XL (N, LO, ND)

Purpose: To generate a sequence of logarithmically spaced numbers for use in plots with a logarithmic scale for the frequency variable.

Additional Subprograms Required: None.

Input Arguments:

N   the index n giving the number of times the function has been called; the index is automatically advanced by 1 each time the function is called

LO   the logarithm (to the base 10) of the value of the function when the index $n = 0$

ND   the number of intermediate values that the function should generate for each decade of output values

Output Value:

XL   the logrithmically spaced value generated by the function

Note: For use in connection with the plotting subroutine PLOT5, the argument N should be set to zero before the first time the function XL is called by the main program.

```
      FUNCTION XL(N, LO, ND)
C     FUNCTION FOR GENERATING LOGARITHMICALLY SPACED
C     VALUES OF A FREQUENCY VARIABLE
C           XL - VALUE OF THE VARIABLE
C            N - INDEX INDICATING WHICH VALUE OF THE VARIABLE
C                IS CURRENTLY BEING COMPUTED
C           LO - LOGARITHM OF THE VALUE OF THE VARIABLE WHICH IS
C                COMPUTED FOR N = 0
C           ND - NUMBER OF VALUES OF THE VARIABLE COMPUTED
C                PER DECADE OF FREQUENCY
      AN = N
      ALO = LO
      AND = ND
      AA = AN / AND + ALO
      IF (AA) 6, 6, 8
    6 XL = 1. / 10.**(-AA)
      GO TO 9
    8 XL = 10.**AA
    9 N = N + 1
      RETURN
      END
```

C-7 THE SUBROUTINE PLOT5

This subroutine is used to provide a plot of numerically computed data using the standard computer printer, rather than a specialized plotting device, to generate the plot. Up to five plots may be printed simultaneously.

Identifying Statement: SUBROUTINE PLOT5 (Y, M, N, MAX)

Purpose: To provide a simultaneous plot of several functions $y_i(t)$ for discrete, evenly spaced values of the independent variable t.

Additional Subroutines Required: None.

Input Arguments:

Y   the two-dimensional array of variables Y(I, J) giving the values of the functions $y_i(t)$ at the jth value of the independent variable t

M   the number of functions $y_i(t)$ which are to be plotted

N   the maximum number of values of t to be used in plotting the functions

MAX   the upper value of the ordinate scale used to plot the functions; if the value of this argument is set to 999, the subroutine will automatically scale the plot of each function so that it includes the entire ordinate range

Output: A printer-constructed plot is provided of the functions. The positive direction of the independent variable is taken as downward on the printed page. The letter A is used to indicate the points for y_1, the letter B for y_2, etc. The numerical values of the function y_1 are also printed along the edge of the sheet as a convenience. If the value of a function exceeds the ordinate range of the plot, the symbol $ is printed along the left or right edge of the plot, depending on whether the function was less than or greater than the permitted range.

Notes:

1. The subroutine is dimensioned Y(5, 101). Thus, it permits plotting up to five functions for up to 101 values of the independent variable.

2. An ordinate scale of 100 points is provided. Thus, the lower bound of the scale will be MAX − 100. The scale is printed automatically on the ordinate.

```
      SUBROUTINE PLOT5(Y,M,NF,MAX)
C     SUBROUTINE FOR PLOTTING AND SCALING A
C     5 X 101 INPUT ARRAY (FORTRAN 4)
      DIMENSION Y(5,101), LINE(101),L(11),JL(5)
      DATA (JL(I),I=1,5)/1HA,1HB,1HC,1HD,1HE/,JN,JP,JI,JBLANK,JZ/
     11H-,1H+,1HI,1H ,1HS/
      NS=MAX
      IF (NS.NE.999) GO TO 19
C     RESCALE DATA TO COVER ENTIRE ORDINATE RANGE
      NS=100
      DO 16 J=1,M
      YMAX=-1.E+50
      YMIN=1.E+50
      DO 9 I=1,NF
      IF (Y(J,I).GT.YMAX) YMAX=Y(J,I)
    9 IF (Y(J,I).LT.YMIN) YMIN=Y(J,I)
      RANGE=YMAX-YMIN
      TEMP=100./RANGE
      DO 13 I=1,NF
   13 Y(J,I)=(Y(J,I)-YMIN)*TEMP
      PRINT 15,JL(J),YMIN,YMAX,RANGE
   15 FORMAT (1X,5HPLOT ,A1,5H FROM,E10.3,3H TO,
     1E10.3,9H  RANGE =,E10.3)
   16 CONTINUE
      PRINT 18
   18 FORMAT ( )
   19 DO 20 I=1,101
   20 LINE(I)=JBLANK
      N=1
C     PRINT ORDINATE SCALE
```

```
      DO 23 I=1,11
   23 L(I)=10*I-110+NS
      PRINT 25,(L(I),I=1,11)
   25 FORMAT (3X,10(I4,6X),I4,2X,8HA=VALUES)
      GO TO 28
   27 IF ((N-1)/10-(N-2)/10) 37,37,28
C     CONSTRUCT ORDINATE GRAPH LINE
   28 ND=0
      DO 34 I=1,10
      ND=ND+1
      LINE(ND)=JP
      DO 34 J=1,9
      ND=ND+1
   34 LINE(ND)=JN
      LINE(101)=JP
      GO TO 39
C     CONSTRUCT 1 LINE OF ABSCISSA GRAPH LINES
   37 DO 38 I=1,101,10
   38 LINE(I)=JI
C     CHANGE NUMERICAL DATA TO LETTERS
   39 DO 49 I=1,M
      XNS=NS
      JA=Y(I,N)+101.49999-XNS
      IF (JA-101) 43,48,44
   43 IF (JA) 46,46,48
   44 LINE(101)=JZ
      GO TO 49
   46 LINE(1)=JZ
      GO TO 49
   48 LINE(JA)=JL(I)
   49 CONTINUE
C     PRINT LINE OF DATA
      IF (N.EQ.1) GO TO 51
      IF ((N-1)/10-(N-2)/10) 55,55,51
   51 N1=N-1
      PRINT 53,N1,LINE,Y(1,N)
   53 FORMAT (1X,I4,101A1,1X,  E12.5)
      GO TO 57
   55 PRINT 56,LINE,Y(1,N)
   56 FORMAT (5X,101A1,1X,E12.5)
C     SET LINE VARIABLES TO ZERO
   57 DO 58 I=1,101
   58 LINE(I)=JBLANK
   59 N=N+1
      IF (N-NF) 27,27,61
   61 RETURN
      END
```

C-8 THE SUBROUTINE ROOT

This subroutine can be used to find the real or complex roots of a polynomial of up to ninth degree.

Identifying Statement: SUBROUTINE ROOT (N, B, P)

Purpose: To find the roots of a polynomial $B(s)$ of degree n having the form

$$B(s) = b_1 + b_2 s + b_3 s^2 + \cdots + b_n s^{n-1} + b_{n+1} s^n$$

i.e., to find the quantities p_i where

$$B(s) = K \prod_{i=1}^{n} (s - p_i)$$

Additional Subroutines Required: None.

Input Arguments:

    N   the degree of polynomial $B(s)$

    B   the one-dimensional array of variables B(I) in which are stored the values of the coefficients b_i of the polynomial $B(s)$

Output Argument:

    P   the complex one-dimensional array of variables P(I) in which are stored the values of the roots p_i of the polynomial $B(s)$

Notes:

1. The values of the coefficients of the polynomial $B(s)$ are preserved invariant in the B array.
2. The subroutine uses a Lin-Bairstow method for determining the roots.
3. The variables of this subroutine are dimensioned as follows: B(10), P(10).

```
      SUBROUTINE ROOT (N, B, P)
C     SUBROUTINE TO DETERMINE THE ROOTS OF A POLYNOMIAL OF THE FORM
C     B(1) + B(2)*S + B(3)*S**2 + ... + B(N+1)*S**N   USING THE
C     LIN-BAIRSTOW METHOD. THE POLYNOMIAL B(S) IS PRESERVED
C                N - DEGREE OF THE POLYNOMIAL B(S)
C                B - INPUT ARRAY OF COEFFICIENTS OF B(S)
C                P - COMPLEX OUTPUT ARRAY OF ROOTS OF B(S)
      DIMENSION A(10),B(10),C(10),D(10),E(10)
      COMPLEX P(10)
      M = N
      MP = M + 1
      DO 4   I = 1, MP
    4 A(I) = B(I)
    5 IF (ABS(A(1)).GT.1.E-06) GO TO 13
      DO 7   I = 1, M
    7 A(I) = A(I+1)
      P(M) = (0.,0.)
      M = M - 1
      IF (M.EQ.0) RETURN
      MP = M + 1
      GO TO 5
   13 IF (M - 2) 67, 64, 14
   14 ITER = 0
      Q1 = A(1) / A(3)
      Q2 = A(2) / A(3)
   17 ITER = ITER + 1
      C(MP) = A(MP)
      D(MP) = 0.
      E(MP) = 0.
      C(M) = A(M) - Q2 * C(MP)
      D(M) = -C(MP)
      E(M) = 0.
      IF (M.LT.4) GO TO 33
      MM = M - 3
      DO 32   I = 1, MM
      MI = M - I
      MI1 = MI + 1
      MI2 = MI + 2
      C(MI) = A(MI) - Q2*C(MI1) - Q1*C(MI2)
```

```
      D(MI) = -C(MI1) - Q2*D(MI1) - Q1*D(MI2)
32 E(MI) = -C(MI2) - Q2*E(MI1) - Q1*E(MI2)
33 Q1A = A(1) - Q1 * C(3)
      Q2A = A(2) - Q1 * C(4) - Q2 * C(3)
      G11 = -C(3) - Q1 * D(4) - Q2 * D(3)
      G12 = -C(4) - Q1 * E(4) - Q2 * E(3)
      G21 = -Q1 * D(3)
      G22 = -C(3) - Q1 * E(3)
      DET = G11 * G22 - G12 * G21
      DQ1 = (-G11 * Q1A + G21 * Q2A) / DET
      DQ2 = ( G12 * Q1A - G22 * Q2A) / DET
      Q1 = Q1 + DQ1
      Q2 = Q2 + DQ2
      IF (ABS(DQ2).LT.1.E-6) GO TO 47
45 IF (ITER.GT.30) GO TO 69
      GO TO 17
47 IF (ABS(DQ1).GT.1.E-06) GO TO 45
48 DISC = Q2 * Q2 - 4. * Q1
      IF (DISC.LT.0.) GO TO 55
      DISCR = SQRT(DISC)
      P(M) = CMPLX((-Q2 + DISCR) / 2., 0.)
      P(M-1) = CMPLX((-Q2 - DISCR) / 2., 0.)
      M = M - 2
      GO TO 59
55 DISCR = SQRT(-DISC)
      P(M) = CMPLX(-Q2 / 2., DISCR / 2.)
      P(M-1) = CONJG(P(M))
      M = M - 2
59 IF (M.EQ.0) RETURN
      MP = M + 1
      DO 62   I = 1, MP
62 A(I) = C(I+2)
      IF (M-2) 67, 64, 14
64 Q1 = A(M-1) / A(M+1)
      Q2 = A(M) / A(M+1)
      GO TO 48
67 P(1) = CMPLX(-A(1) / A(2), 0.)
      RETURN
69 PRINT 70
70 FORMAT (//34H ROOT SUBROUTINE DOES NOT CONVERGE//)
      RETURN
      END
```

C-9 THE SUBROUTINE FSER

This subroutine is used to determine the coefficients of the Fourier series for an arbitrary periodic time-domain function $f(t)$. The latter is specified by implementing its mathematical description in a Fortran function with the identifying statement FUNCTION FCN(T). To perform the integration required to determine the coefficients, the subroutine FSER calls the integration subroutine ITRPZ described in Sec. C-10.

Identifying Statement: SUBROUTINE FSER (NF, PERIOD, ITER, A0, A, B, C, PH)

Purpose: To determine the coefficients of the Fourier series representation for a non-sinusoidal periodic function $f(t)$, i.e., to find the coefficients a_i, b_i, c_i, and ϕ_i, where

$$f(t) = a_0 + \sum_i (a_i \cos i\omega_0 t + b_i \sin i\omega_0 t) = a_0 + \sum_i c_i \cos (i\omega_0 t + \phi_i)$$

and where ω_0 is the fundamental harmonic frequency.

Additional Subprograms Required: The subroutine calls the trapezoidal integration subroutine ITRPZ, which in turn calls the function Y. This function must be provided in the form given below. A function FCN(T) must also be provided to define the periodic function $f(t)$ over the range $0 \leq t \leq T$, where the period $T = 2\pi/\omega_0$.

Input Arguments:

NF   the number of the highest harmonic coefficient of the Fourier series that it is desired to compute

PERIOD   the value of time required for one period of the non-sinusoidal periodic function $f(t)$

ITER   the number of iterations to be used by the trapezoidal subroutine ITRPZ in determining each of the coefficients of the Fourier series

Output Arguments:

A0   the value of the coefficient a_0

A   the one-dimensional array of variables A(I) in which are stored the values of the cosine coefficients a_i of the Fourier series

B   the one-dimensional array of variables B(I) in which are stored the values of the sine coefficients b_i of the Fourier series

C   the one-dimensional array of variables C(I) in which are stored the values of the magnitude coefficients c_i of the Fourier series

PH   the one-dimensional array of variables PH(I) in which are stored the values of the phase coefficients ϕ_i (in radians) of the Fourier series

Notes:

1. The variables of the subroutine are dimensioned as follows: A(10), B(10), C(10), PH(10).
2. The function Y(T) which must be used with the subroutine is given below.

```
      FUNCTION Y(T)
C     THIS FUNCTION IS DESIGNED TO BE USED WITH THE
C     SUBROUTINE FSER TO COMPUTE THE INTEGRAND OF THE
C     INTEGRALS WHICH DETERMINE THE FOURIER SERIES
C     COEFFICENTS. IT IS CALLED BY THE SUBROUTINE ITRPZ.
C     A FUNCTION FCN(T) MUST BE PROVIDED TO SPECIFY THE
C     NON-SINUSOIDAL PERIODIC FUNCTION
      COMMON OM, H, K
      IF (K. GT. 1) GO TO 4
      Y = COS (OM*H*T)*FCN(T)
      RETURN
    4 Y = SIN(OM*H*T)*FCN(T)
      RETURN
      END
```

```
      SUBROUTINE FSER(NF, PERIOD, ITER, A0, A, B, C, PH)
C     SUBROUTINE FOR DETERMINING THE COEFFICIENTS OF THE
C     TERMS IN THE FOURIER SERIES FOR A PERIODIC
C     NON-SINUSOIDAL FUNCTION SPECIFIED AS FCN(T).  THIS
C     SUBROUTINE MUST BE USED WITH A SPECIAL FUNCTION Y(T)
C     WHICH IS CALLED BY THE SUBROUTINE ITRPZ
C             NF - FINAL HARMONIC TERM TO BE EVALUATED
```

```
C     PERIOD - TIME FOR ONE COMPLETE CYCLE
C       ITER - NUMBER OF ITERATIONS USED BY SUBROUTINE
C                 ITRPZ FOR EACH COEFFICIENT
C         A0 - OUTPUT ARGUMENT, VALUE OF DC COEFFICIENT
C          A - OUTPUT ARRAY, VALUES OF COSINE COEFFICIENTS
C          B - OUTPUT ARRAY, VALUES OF SINE COEFFICIENTS
C          C - OUTPUT ARRAY, COEFFICIENT MAGNITUDES
C         PH - OUTPUT ARRAY (PHASE, RADIANS)
      DIMENSION A(10), B(10), C(10), PH(10)
      COMMON OM, H, K
      OM = 6.28318 / PERIOD
C
C     COMPUTE THE DC COEFFICIENT A0
      H = 0.
      K = 1
      CALL ITRPZ(0., PERIOD, ITER, A0)
      A0 = A0 / PERIOD
      H = 1.
      LH = 1
    9 IF (LH.GT.NF) RETURN
C
C     COMPUTE THE HTH COEFFICIENT FOR THE COSINE TERMS
      K = 1
      CALL ITRPZ(0., PERIOD, ITER, A(LH))
      A(LH) = A(LH) * 2.0 / PERIOD
      K = 2
C
C     COMPUTE THE HTH COEFFICIENT FOR THE SINE TERMS
      CALL ITRPZ(0., PERIOD, ITER, B(LH))
      B(LH) = B(LH) * 2.0 / PERIOD
      C(LH) = SQRT(A(LH)**2 + B(LH)**2)
      PH(LH) = ATAN2(-B(LH), A(LH))
      H = H + 1.
      LH = LH + 1
      GO TO 9
      END
```

C-10 THE SUBROUTINE ITRPZ

This subroutine is used to integrate a time-domain function whose mathematical description is implemented in a Fortran function with the identifying statement FUNCTION Y(T). The ITRPZ subroutine is called by the FSER subroutine described in Sec. C-9.

Identifying Statement: SUBROUTINE ITRPZ (TA, TB, N, A)

Purpose: To find the definite integral of a specified function $y(t)$, that is, to find A, where

$$A = \int_{t_a}^{t_b} y(t)\, dt$$

by using trapezoidal integration.

Additional Subprograms Required: This subroutine calls the function identified by the statement

FUNCTION Y(T)

This function must be used to define $y(t)$, i.e., the integrand.

Input Arguments:
> TA   the lower limit t_a of the variable of integration t
> TB   the upper limit t_b of the variable of integration t
> N   the number n of trapezoidal segments used to approximate the area, i.e., the number of iterations used in performing the trapezoidal integration

Output Argument:
> A   the value of the integral

```
      SUBROUTINE ITRPZ (TA, TB, N, A)
C     TRAPEZOIDAL INTEGRATION SUBROUTINE
C           TA - INITIAL VALUE OF INDEPENDENT VARIABLE T
C           TB - FINAL VALUE OF INDEPENDENT VARIABLE T
C            N - NUMBER OF TRAPEZOIDAL SECTIONS USED
C            A - VALUE OF INTEGRAL FROM TA TO TB
C     A FUNCTION Y(T) MUST BE USED TO DEFINE THE INTEGRAND.
C     N MUST BE 2 OR GREATER
C
C     INITIALIZE THE VALUE OF TIME T TO THE VALUE TA
C     AND CALCULATE DT, THE INCREMENT IN TIME
      T = TA
      TN = N
      DT = (TB - TA) / TN
C
C     INITIALIZE THE VALUE OF THE INTEGRAL
      A = Y(TA) / 2.
C
C     COMPUTE THE N-1 TERMS IN THE SUMMATION
      DO 8  I = 2, N
      T = T + DT
      A = A + Y(T)
    8 CONTINUE
C
C     FINISH THE COMPUTATION OF THE INTEGRAL
      A = (A + Y(TB) / 2.) * DT
      RETURN
      END
```

C-11 THE SUBROUTINE FPLT

This subroutine is used to provide a plot of one period of a time-domain function specified by the coefficients of a Fourier series. It prepares the data in a form directly usable by the subroutine PLOT5 described in Sec. C-7.

Identifying Statement: SUBROUTINE FPLT (NT, PERIOD, A0, NC, C, PH, NP, P, SCALE)

Purpose: To compute and store the data required for plotting the waveform of one period of a non-sinusoidal periodic function $f(t)$ which is specified in the form

$$f(t) = a_0 + \sum_i c_i \cos (i\omega_0 t + \phi_i)$$

where ω_0 is the fundamental harmonic frequency.

Additional Subprograms Required: None.

Input Arguments:
> NT   the number of points desired to be used for plotting one period of the non-sinusoidal periodic function $f(t)$

PERID   the value of time required for one period of $f(t)$

A0   the value of the Fourier series coefficient a_0

NC   the number of harmonic terms to be included in calculating the points used for plotting $f(t)$

C   the one-dimensional array of variables C(I) in which are stored the values of the magnitude coefficients c_i of the Fourier series

PH   the one-dimensional array of variables PH(I) in which are stored the phase coefficients ϕ_i (in radians) of the Fourier series

NP   the position in the plotting array P(I, J) in which the computed data are to be stored, i.e., the value to be given to the index I

SCALE   the scaling factor which is to be used when data are stored in the plotting array P(I, J); if no scaling is desired, this argument must be set to unity

Output Argument:

P   the two-dimensional array of variables P(I, J) in which the values of the non-sinusoidal periodic function $f(t)$, as computed at different values of t, are stored for plotting; specifically, these values are stored as P(NP, J) where J has a range from 1 to NT

Note: The variables of the subroutine are dimensioned as follows: C(10), PH(10), P(5, 101).

```
      SUBROUTINE FPLT(NT, PERIOD, A0, NC, C, PH, NP, P, SCALE)
C          NT - NUMBER OF TIME VALUES TO BE PLOTTED
C      PERIOD - TIME FOR ONE PERIOD OF FUNCTION
C          A0 - CONSTANT FOURIER COEFFICIENT
C          NC - NUMBER OF HARMONIC TERMS TO BE USED
C           C - ARRAY OF FOURIER COEFFICIENT MAGNITUDES
C          PH - ARRAY OF FOURIER COEFFICIENT PHASES
C          NP - STORAGE LOCATION IN PLOTTING ARRAY
C           P - OUTPUT ARRAY FOR USE IN SUBROUTINE PLOT5
C       SCALE - SCALING FACTOR FOR PLOTTED DATA
      DIMENSION C(10), PH(10), P(5,101)
      RAD = 6.28318 / PERIOD
      TNT = NT - 1
      DT = PERIOD / TNT
      T = 0.0
C
C     MAKE COMPUTATIONS AT NT DIFFERENT VALUES OF TIME
C     AND STORE THE RESULTS FOR PLOTTING
      DO 11 I = 1, NT
      F = A0
C
C     DETERMINE THE TOTAL EFFECT OF NC HARMONIC TERMS
C     AT EACH VALUE OF TIME
      DO 9  J = 1, NC
      X = J
    9 F = F + C(J) * COS(RAD * X * T + PH(J))
      P(NP,I) = F * SCALE
   11 T = T + DT
      RETURN
      END
```

Solutions to Selected Problems

Chapter 1

Problem 1-2. $q(t) = 1 - e^{-4t}$ C. **Problem 1-3(c).** $w(0.003) = 0.8595 \times 10^{-9}$ J.
Problem 1-10. $i_1(t) = 3 + 2t$, $i_4(t) = 3 - \sin t$, $i_5(t) = -6 - 2t + \sin t$ A.
Problem 1-14. $v_2(t) = 1 - 7t + e^{-t}$, $v_3(t) = \sin t - 7 - e^{-t}$ V.
Problem 1-19. $i_2(t) = 1 + 6t - t^2 + 2 \sin t$ A.   **Problem 1-24.** 4 nodes, 6 branches, 7 cutsets.

Chapter 2

Problem 2-1(b). $0.01 \sin^2 (2\pi t)$ W.
Problem 2-4. Power in 5-V source is $\frac{25}{8}$ W, power in 10-V source is $-\frac{50}{8}$ W.
Problem 2-8. $i(t) = 1000\sqrt{2} \sin 377t$ A.   **Problem 2-10.** $\frac{11}{8} \Omega$.
Problem 2-20. $i_e = \frac{5}{2}$ A, $G_e = \frac{3}{2}$ mho.   **Problem 2-23.** $i_o(t) = \frac{9}{5} \cos 2t$ A.

Chapter 3

Problem 3-1. $i_1 = 2$, $i_2 = -1$ A.
Problem 3-6. $r_{11} = 11$, $r_{44} = 14$, $r_{14} = -5$; $v_1 = -1$, $v_4 = 5$ V.
Problem 3-10. $i_1 = 6$, $i_2 = 7$, $i_3 = 8$ A.   **Problem 3-15.** $a = 31$, $b = d = 19$, $c = e = 11$.
Problem 3-17. $v_1 = 2$, $v_2 = -6$ V.   **Problem 3-23.** $v_1 = 2$, $v_2 = -2$, $v_3 = 1$ V.
Problem 3-25. $i_a = \frac{25}{13}$ A.   **Problem 3-35.** $i_1 = 3$, $i_2 = -1$ A.
Problem 3-42. $R_{eq} = \frac{7}{2} \Omega$.   **Problem 3-44(a).** $R_{eq} = \frac{8}{11} \Omega$, $V_{eq} = \frac{10}{11}$ V.
Problem 3-49. $V_{eq} = 5$ V, $R_{eq} = \frac{3}{2} \Omega$.   **Problem 3-59.** Gain $= (G_1/G_2) + 1$.

Chapter 4

Problem 4-7. $i(t) = (6 + 2t) \cos 2t + \sin 2t$ A.   **Problem 4-9(a).** $f_a(t) = -3u(-t + 2)$.
Problem 4-10(a). $f_a'(t) \times 3\delta(-t + 2)$,   **Problem 4-11(b).** $\int f_b(t) \, dt = 4(t + 1)u(t + 1)$.
Problem 4-13(a). $v(t) = (t - 2)u(t - 2)$ V.

Problem 4-19. $q(10) = \frac{5}{2}$ C, $\phi(10) = \frac{1}{2}$ Weber-turn, $v(10) = \frac{1}{20}$ V.

Problem 4-23. $i(0+) = -1$ A, $di(0+)/dt = 2$ A/s.   **Problem 4-27(a).** $C = \frac{26}{33}$ F.

Problem 4-31(a). $L_{eq} = L_1 + L_2 + 2$ MH.

Chapter 5

Problem 5-2. $V_o = 44.5$ V.   **Problem 5-12.** $v(t) = -e^{-t} + \cos 2t + 2 \sin 2t$ V.

Problem 5-21. $i(0) = -\frac{2}{3}$ A.   **Problem 5-27.** $i(t) = 3e^{-t}$ A.

Problem 5-34. $v_C(t) = 2 - e^{-2t}$ V, $i(t) = e^{-2t}/3$ A.

Problem 5-40. $v(t) = 2 - 2e^{-t}$ V $(0 \leq t \leq 2)$; $v(t) = 6 - 31.556\, e^{-t}$ V $(2 \leq t \leq \infty)$.

Problem 5-43. Current source in parallel, $i(t) = 15\delta(t)$ A.

Chapter 6

Problem 6-2. $i_L(t) = -5e^{-2t} + \frac{3}{2}e^{-4t}$ A.   **Problem 6-6(a).** $v(t) = e^{-2t}(\cos t + 2 \sin t)$ V.

Problem 6-14. $v(t) = -e^{-t} \sin t$ V.   **Problem 6-22.** $v(t) = 3e^{-2t} - 3e^{-4t}$ V.

Problem 6-29. $v(t) = -5e^{-t} + 15e^{-3t}$ V.   **Problem 6-37.** $v_{sw}(0+) = 2$, $v_{sw}(\infty) = 20$ V.

Problem 6-52. $s^3 + 3s^2 + 3s + 1 = 0$.

Problem 6-56. $v_2(t) = (4e^{-t}/5) + e^{-t/2}(-\frac{4}{5} \cos t + \frac{7}{5} \sin t)$ V.

Chapter 7

Problem 7-4. $K/\sqrt{3}$ for both waveshapes.   **Problem 7-9.** $v(t) = \sqrt{5} \cos (2t - 63.5°)$ V.

Problem 7-15. $v_2(t) = 10 \sin t$ V.

Problem 7-20. $i_1(t) = (\sqrt{2}/6) \cos (2t - 135°)$, $i_2(t) = \frac{2}{3} \cos (2t - 90°)$ A.

Problem 7-31(a). $Y(j\omega) = (3 - 5\omega^2 + 7j\omega)/(6 + 5j\omega)$ mhos.

Problem 7-49. $\mho = 1/\sqrt{2}\,\underline{/90°}$, $Z = \frac{3}{2} + j\frac{1}{2}$.

Problem 7-55. $\mho_2/\mho_1 = 1/(-j\omega^3 - 5\omega^2 + 6j\omega + 1)$.   **Problem 7-73(a).** $n = 5$.

Problem 7-76. $R = 500\ \Omega$, $L = 0.159$ H, $C = 0.318\ \mu$F.   **Problem 7-95.** $C = 16\ \mu$F.

Problem 7-99. $|\mathscr{J}| = 6.19$ A(rms), $P = 1152$ W.

Chapter 8

Problem 8-2(a). $v_1''(t) = v_2''(t) + \frac{3}{2}v_2'(t) + \frac{1}{4}v_2(t)$.   **Problem 8-5(a).** $F_a(s) = e^{-2s}/s$.

Problem 8-9(d). $f(t) = e^{-t} (\cos t - 2 \sin t)$.

Problem 8-10(c). $f(t) = \delta'(t) - 3\delta(t) + e^{-t}(6 - 4t + t^2)$.

Problem 8-17. $v_2(t) = -e^{-t} + e^{-t}(\cos t + \sin t)$.

Problem 8-35. $18v_2''(t) + 33v_2''(t) + 20v_2'(t) + 4v_2(t) = 3v_1''(t) + 3v_1'(t) + 4v_1(t)$.

Problem 8-41(b). $s = -\frac{1}{4} \pm j\sqrt{7}/4$.   **Problem 8-45.** $A(s) = F(s) = D(s)$, $B(s) = E(s) = H(s)$.

Problem 8-51(c). Reciprocal.   **Problem 8-52(c).** Transmission zeros (simple) at -1, $\pm j1$.

Problem 8-67(a). $V_{Thev.} = 3/(2s^2 + 3s)$.

Chapter 9

Problem 9-1(c). $z_{11}(s) = (6s^2 + s)/(2s^2 + 6s + 1)$.

Problem 9-5(c). $y_{11}(s) = (6s + 1)/(6s^2 + s + 3)$.   **Problem 9-11(b).** $g_{11}(s) = s + 2$.

Problem 9-16(d). $A(s) = (s^2 + 3s + 1)/(s^2 + 2s + 1)$.

Problem 9-19(d). Z and $ABCD$ parameters.

Chapter 10

Problem 10-6(a). $b_1 = 4\sqrt{3}/\pi$, $a_1 = 4/\pi$, $a_3 = 8/3\pi$, others zero.

Problem 10-14(a). Coefficients decrease as $1/n^2$.

Problem 10-15. Waveform 3 has symmetries c, d.

Problem 10-18(a). $f_{Left}(t) = \cos t - \frac{1}{3} \cos 3t + \frac{1}{5} \cos 5t$.   **Problem 10-21.** $L = \frac{1}{9}$ H, $C \times \frac{1}{49}$ F.

Problem 10-25. $|F(j\omega)|$ is zero at 2π, 4π, 6π, $\cdots$, rad/s.

Index

lagging and leading, 354
  rates for, 358
  in three-phase circuits, 362-64
Product, of even and odd functions
  541
Proper rational function, 411
Pulse, formed from unit step
  functions, 119

Q

Q (*see* Quality factor, *see also*
  Reactive power)
Quadrant, of arctangent function,
  273
Quadrature form, of a sinusoid, 272
Quality factor, 322
  relation to bandwidth, 462
  relation to damping factor, 336
  relation to pole positions, 461-62

R

R, 23
R-2R network, 40
Radians, as measure of phase, 274
  relation to degrees, 327
Ramp function, 120
  as integral of step function,
    120-21
  Laplace transform of, 401
Rao, K. R., 416
Rational function (*see also* Network
  function)
Rational function, 409-10
  behavior at infinity, 409-10
    inverse transform of, 410-11
  gain constant of, 411
  pole of, 411
  proper, 411
  total number of poles and zeros,
    411
  zero of, 411
Rationalized MKS system, 2
RC network, natural frequencies of,
  240-41
  location of poles, 453
  second-order, roots of
    characteristic equation, 224
Reactance, of a capacitor, 293
  as a function of frequency, 300
  of an inductor, 293
  as part of impedance, 293
Reactive power, 354
  in three-phase circuits, 362-64
  units of, 354-56
Real part, equality of, 283
  interchange of order of taking
    with differentiation, 283
Real power (*see* Power)
Real translation theorem, for
  Laplace transform, 403

Reciprocal network (*see*
  Reciprocity)
Reciprocity, ABCD-parameter
  relations for, 520
  definition of, 448
  g-parameter relations for, 513
  h-parameter relations for, 514
  y-parameter relations for, 508
  z-parameter relations for, 495
Rectangular form, of a sinusoid,
  272
Rectification, 43
Reference conventions (*see also*
  Reference directions)
Reference conventions, for a current
  source, 27
  for two-terminal elements, 7
  for a voltage source, 26
Reference directions (*see also*
  Reference conventions)
Reference directions, associated, 7
  for charge, 5-6
  for current, 6
  in mesh equations, 50, 55
  for power, 7-8
  for voltage, 6-7
  for voltage and current phasors,
    292
  in two-terminal element, 295
Reference node, 62
Reference polarities (*see* Reference
  directions)
Reluctance, 143
Residue, expression for
    determining, of a non-simple
    pole, 414
  in partial-fraction expansions,
    412
  for simple complex poles, 416
  for simple real poles, 414
Resistance, definition of, 23
  equivalent, 83
  gain constant of an ICVS, 29
  internal, of a current source, 27
    of a voltage source, 26
  mutual, 58
  negative, 98
  of networks containing controlled
    sources, 83
  of paralleled resistors, 33
  as part of impedance, 293
    as a function of frequency,
      299-300
  self, 58
  of series connected resistors, 32
  time-varying, 24
  transfer, of an ICVS, 29
Resistance network, with current
    sources only, 62
  delta-connected, 74
  flow chart for analysis of, 74
  flow chart for mesh equation

formulation, 79
  flow chart for node equation
    formulation, 81
  i-v characteristic for, 32, 86
  Norton equivalent circuit of, 88
  Thevenin equivalent circuit of,
    84-86
  with voltage and current sources,
    74
  with voltage sources only, 48
  wye-connected, 74
Resistive part, of an impedance,
  293
Resistor, circuit symbol for, 24
  definition of, 23
  delta connection of, 74
  as a dissipative element, 25
  equivalent, 83
  impedance of, 293
  of infinite resistance, 27
  i-v characteristic, 24
  linear, 24
  model of, 25
  negative-valued, 25
  non-linear (*see* Non-linear
    resistor)
  parallel connection of, 33
  phase of voltage and current,
    294
  physical, 26
  power in, 25-26
  range of values, 26
  series connection of, 32
  time-varying, 24
  wattage rating, 26
  wye connection of, 74
  of zero resistance, 26
Resonance (*see* Resonant circuit)
Resonant circuit, 319
  center frequency of, 319
  parallel second-order, 325-26
  use of phasor diagrams, 323-24
  resonant frequency of, 319, 322
  series second-order, 319-25
  3-dB frequencies, 319
Resonant frequency, of series
    resonant circuit, 319, 322
Response, boundedness of, 451
  sinusoidal steady-state, as
    particular solution, 271
  zero-input, 164
  zero-state, 164
Response phasor, 311
Rhodes, J. D., 593
Right-hand rule, 137
RL network, location of poles, 454
  natural frequencies of, 243
  second-order, roots of
    characteristic equation, 224
RLC network, location of poles,
  454
  natural frequencies of, 244

RLC network (*cont.*)
second-order, roots of
characteristic equation, 224
rms value, definition of, 277
of a non-sinusoidal periodic
function, 563-64
in phasors, 355
Rohrer, R. A., 250
Root, of a complex number, 605
Root-locus plot, for an active
network, 453
of characteristic polynomial
roots, 220
Root-mean-square (*see* rms)
ROOT subroutine, 616-18
Rotation of phases, 361
Row, of a matrix, 594
Row matrix, 595
Runge-Kutta method, 608

S

S, 114
Sampling property, of an impulse
function, 399
Scalar, 596
Scaling theorem, 407
Schwarz, R. J., 558
Second-order circuit, active-RC,
235-36
coefficients of differential
equation, 212-17
discriminant for RC, 224
discriminant for RL, 224
energy stored in, 218
excited by initial conditions and
sources, solution of, 231
forced response of, 234
frequency of oscillation, 214
initial conditions, number of, 210
use of loops in writing state-
variable equations, 246
oscillation of, 214-19
overdamped, 221
state-variable equations for,
245-47
transient response of, 234
undamped, 221
underdamped, 221
zero-input response of, 234
zero-state response of, 234
Sedra, A. S., 593
Self-admittance, 307
Self-conductance, 67
Self-impedance, 303
Self-inductance, 137
Self-resistance, 58
Separation of variables, 158, 390
Series connection, of capacitors,
133
of current sources, 37

of impedances, in sinusoidal
steady-state analysis, 297
in a transformed network, 431
of inductors, 135
charged, 136-37
of resistors, 32
in a transformed network, 432
of two-port networks, 496
of voltage and current source, 38
of voltage sources, 35
Series network elements, for
producing transmission
zeros, 455-57
ABCD parameters of, 522
Series resonant circuit, second
order, 319-25
shock hazard of, 325
Shifted unit step function, 118
Laplace transform of, 401-2
Shifting, in the time domain, 182
Shifting theorem, for Laplace
transformation, 403-4
Shock hazard, of a charged
capacitor, 127-28
of a series-resonant circuit, 325
Short circuit, 26
Short-circuit admittance parameters
(*see* y parameters)
Short-circuit current transfer
function, 496
Short-circuit natural frequencies,
446
Shunt connection (*see* Parallel
connection)
Shunt network elements, for
producing transmission
zeros, 455-57
ABCD parameters of, 517-19
Shunt-peaked transistor stage, 309
Simultaneous equations, with
complex coefficients, 287-90
digital computer solution of,
290
integro-differential, 113
conversion to algebraic in
sinusoidal steady-state, 293
linear algebraic, 113
in mesh analysis, 51
for resistance networks, 57
digital computer solution of,
62
solution by Cramer's rule, 61-62
in three-phase circuits, example,
367
Sine and cosine form, of Fourier
series, 548-49
Sine function, Laplace transform of,
400
Single-sided Laplace
transformation, 408
Sinor (*see* Phasor)

Sinusoidal frequency, relation to
natural frequency, 219
Sinusoidal function, A-B plane
representation of, 273-74
addition of, 275
amplitude of, 272
angular frequency of, 272
argument of, 272
average value of, 276
constants for household
electricity, 324
magnitude and phase form, 272
peak magnitude of, 272
period of, 272
phase angle of, 272
phasor representation of, 281
power dissipated by, 276
quadrature form, 272
rectangular form, 272
relation between forms of,
272-73
rms value of, 277
Sinusoidal oscillations, in second-
order circuits, 219
Sinusoidal steady-state, average
power, 276, 351-54
complex power in, 354
equivalence of circuits, 300-301
use of exponentials, 278-80
impedance triangle, 351
instantaneous power in, 351-54
maximization of power transfer,
358-59
network functions, 311
Norton equivalent circuit in, 310
as particular solution, 271
phasors with rms values, 355
power factor, 351
reactive power, 354
relation to Laplace transform,
430
solution by phasors, 285
Thevenin equivalent circuit in,
310
Sinusoidal steady-state response,
graphical evaluation of,
457-59
phasor determination of, 285
6dB/ octave slope, 328
Skew-symmetric function, 540
Skilling, H., 593
Small-signal equivalent circuit, 42
Solar panels, 105
Solenoid, 130
Solid-state diode, 24
Solution, of a first-order differential
equation, 164-66
Solution for a first-order network,
flow chart for, 176
Solutions to selected problems,
623-24